SEP 8 1955

TOXICITY OF
INDUSTRIAL ORGANIC SOLVENTS

Revised in consultation with the Toxicology Committee

By ETHEL BROWNING

*Reprinted by permission of the Controller
of Her Britannic Majesty's Stationery Office
British Crown Copyright*

1953
CHEMICAL PUBLISHING CO., INC.
212 Fifth Avenue   New York, N. Y.

Revised American Edition
1953

CHEMICAL PUBLISHING CO., INC.
212 Fifth Avenue   New York 10, N. Y.

Printed in the United States of America

MEDICAL RESEARCH COUNCIL

The Right Hon. the EARL OF LIMERICK, K.C.B., D.S.O. (*Chairman*)
Sir GEOFFREY VICKERS, V.C. (*Treasurer*)
Group Captain C. A. B. WILCOCK, O.B.E., A.F.C., M.P.
Sir PERCIVAL HARTLEY, C.B.E., M.C., D.Sc., F.R.S.
Professor J. MCMICHAEL, M.D., F.R.C.P.
Professor W. E. LE GROS CLARK, M.D., D.Sc., F.R.C.S., F.R.S.
Professor F. G. YOUNG, D.Sc., F.R.S.
Professor G. L. BROWN, C.B.E., M.Sc., M.B., F.R.S.
Professor Sir JAMES LEARMONTH, K.C.V.O., C.B.E., Ch.M., F.R.C.S.
Professor G. R. CAMERON, D.Sc., M.B., F.R.C.P., F.R.S.
Professor AUBREY J. LEWIS, M.D., F.R.C.P.
Sir JAMES SPENCE, M.C., D.Sc., M.D., F.R.C.P.
    Sir HAROLD HIMSWORTH, K.C.B., M.D., F.R.C.P. (*Secretary*)
    A. LANDSBOROUGH THOMSON, C.B., O.B.E., D.Sc. (*Second Secretary*)

TOXICOLOGY COMMITTEE

Professor E. J. KING, D.Sc. (*Chairman*)

A. J. AMOR, C.B.E., M.Sc., M.D.

Professor J. H. GADDUM, Sc.D., M.R.C.S., F.R.S.

R. A. E. GALLEY, Ph.D., F.R.I.C.

DONALD HUNTER, M.D., F.R.C.P.

E. R. A. MEREWETHER, C.B.E., M.D., F.R.C.P.

W. D. M. PATON, M.A., B.M.

Sir RUDOLPH A. PETERS, M.C., M.D., F.R.S.

J. DAVIDSON PRATT, C.B.E., B.Sc.

Professor J. R. SQUIRE, M.D., F.R.C.P.

J. WALKER, D.Sc.

Professor R. T. WILLIAMS, D.Sc.

J. M. BARNES, M.B. (*Secretary*)

PREFACE TO THE SECOND EDITION

In response to a request from the Home Office in 1935, the Council undertook to investigate the possibility that various volatile substances, used industrially as solvents and often in enormous quantities, might injure the health of the workers handling them. A special Committee on the Toxicity of Industrial Solvents, under the chairmanship of Sir Joseph Barcroft, was appointed to advise on this matter.

At the time when the Committee was formed, there was, to quote the Preface to the First Edition of this Report, " a clear need for research into the possible effect of substances that seem open to suspicion, either through having a chemical constitution closely allied to that of compounds already proved to be dangerous or through complaints of ill-health among those who constantly use them. As a first step, however, it was plainly desirable to take stock of existing knowledge of the subject."

The Committee therefore asked Dr. Ethel Browning to compile a summary of the available information on such solvents as were then in general use. Issued in 1937 as No. 80 in the Council's Series of Industrial Health Research Board Reports, it was almost the only publication of its kind in existence and, although not exhaustive, was for practical purposes sufficiently detailed to show the extent of the problem and to indicate the most suitable points of attack. Requests for copies came from all parts of the world, and the demand continued for some years after the edition had gone out of print. Consequently, in 1946 it was decided that the Report should be re-issued, having first been revised and also enlarged to include those solvents which had come into use since the first edition was prepared. Fortunately, Dr. Browning was willing to undertake the formidable task of revision, and the Council would like to take the opportunity of expressing their thanks to her for the care with which she has completed it.

The original Committee had not met during the War, and the responsibility for supervising the new edition, on which Dr. Browning was already working, fell to the Toxicology Committee, appointed in 1947 under the chairmanship of Professor G. R. Cameron. In discussing what additions or changes might be made, the Committee decided that the new edition, like the old, should be limited to substances used primarily as solvents, and that the Report should retain its original form as a compendium of information rather than attempt a critical review of the published literature.

In recent years a number of new books dealing with industrial toxicology have been published. Most of them cover a wider field than this Report, but do not deal so comprehensively with the industrial solvents. This new edition, with its extensive bibliography covering the years up to 1948, should be a valuable source of references for medical officers and others responsible for the health of workers concerned with the manufacture, transport and use of organic solvents.

MEDICAL RESEARCH COUNCIL,
    38 Old Queen Street,
        London, S.W.1.

CONTENTS

INTRODUCTION

CHAPTER I: Hydrocarbons

	PAGE		PAGE
1. Benzene	3	10. Petroleum spirit	74
2. Toluene	46	11. Benzine	79
3. Xylene	55	12. White spirit	93
4. Ethylbenzene	63	13. *cyclo*Hexane	94
5. Cumene	64	14. Methyl*cyclo*hexane	98
6. Tetrahydronaphthalene	65	15. Turpentine	100
7. Decahydronaphthalene	68	16. Dipentene	106
8. Methylated naphthalenes	69	17. *cyclo*Pentadiene	106
9. Coal tar solvent naphtha	70	18. Di*cyclo*pentadiene	107

CHAPTER II: Chlorinated Hydrocarbons

1. Methylene dichloride	122	8. Trichloroethylene	169
2. Chloroform	124	9. Perchloroethylene	182
3. Carbon tetrachloride	128	10. Propylene dichloride	185
4. *sym.*-Dichloroethane	149	11. Amyl chloride	186
5. Tetrachloroethane	154	12. Amylene dichloride	187
6. Pentachloroethane	163	13. Monochlorobenzene	187
7. Dichloroethylene	165	14. *o*-Dichlorobenzene	189

CHAPTER III: Alcohols

1. Methyl alcohol	202	11. Methyl*iso*butylcarbinol	236
2. Wood spirit	216	12. *cyclo*Hexanol	236
3. Ethyl alcohol	217	13. Methyl*cyclo*hexanol	239
4. *n*-Propyl alcohol	222	14. Allyl alcohol	241
5. *iso*Propyl alcohol	224	15. Benzyl alcohol	242
6. *n*-Butyl alcohol	227	16. Diacetone alcohol	244
7. *sec.*-Butyl alcohol	230	17. Ethylene chlorohydrin	246
8. *iso*Butyl alcohol	231	18. Monochlorohydrin	250
9. *tert.*-Butyl alcohol	231	19. Dichlorohydrin	251
10. Amyl alcohol	232		

CHAPTER IV: Ethers

1. Ethyl ether	261	5. Dioxan	268
2. $\beta\beta'$-Dichloroethyl ether	266	6. Methylal	274
3. *iso*Propyl ether	267	7. Acetal	274
4. Propylene oxide	268	8. Paraldehyde	276

CHAPTER V: Esters

1. Methyl formate	281	7. Ethyl acetate	289
2. Ethyl formate	281	8. *n*-Propyl acetate	293
3. *n*-Butyl formate	284	9. *iso*Propyl acetate	294
4. Amyl formate	285	10. *n*-Butyl acetate	295
5. Benzyl formate	286	11. *sec.*-Butyl acetate	298
6. Methyl acetate	286	12. *iso*Butyl acetate	298

CHAPTER V: Esters—continued

- 13. Amyl acetate 299
- 14. sec.-Hexyl acetate .. 306
- 15. cycloHexyl acetate .. 306
- 16. Methylcyclohexyl acetate 308
- 17. Butoxyl 308
- 18. Benzyl acetate 309
- 19. n-Butyl propionate .. 311
- 20. Amyl propionate .. 311
- 21. n-Butyl butyrate .. 312
- 22. Methyl benzoate .. 312
- 23. Ethyl benzoate 313
- 24. Ethyl lactate 313
- 25. Butyl lactate 314
- 26. Amyl lactate 314
- 27. Ethyl hydroxyisobutyrate 315
- 28. Diethyl carbonate .. 315
- 29. Dialkyl carbonates .. 316
- 30. Diethyl oxalate 316

CHAPTER VI: Ketones

- 1. Acetone 320
- 2. Methyl acetone .. 327
- 3. Acetone oils 328
- 4. Methyl ethyl ketone .. 328
- 5. Methyl isobutyl ketone .. 331
- 6. Mesityl oxide 331
- 7. cycloHexanone 333
- 8. Methylcyclohexanone .. 334
- 9. Isophorone 335

CHAPTER VII: Glycols and Their Derivatives

- 1. Ethylene glycol 340
- 2. Ethylene glycol monomethyl ether 345
- 3. Ethylene glycol monoethyl ether 348
- 4. Ethylene glycol monoethyl ether monoacetate .. 351
- 5. Ethylene glycol diethyl ether 352
- 6. Ethylene glycol mono-n-butyl ether 353
- 7. Ethylene glycol monoacetate 355
- 8. Ethylene glycol diacetate 356
- 9. Diethylene glycol .. 356
- 10. Diethylene glycol monoethyl ether 358
- 11. Diethylene glycol mono-n-butyl ether 360
- 12. Diethylene glycol monoacetate 360
- 13. Dipropylene glycol .. 361

CHAPTER VIII: Amines and Coal Tar Bases

- 1. cycloHexylamine .. 365
- 2. Dicyclohexylamine .. 365
- 3. Ethanolamines .. 366
- 4. Pyridine 367
- 5. Picoline 371

CHAPTER IX: Nitro-compounds

- 1. Nitromethane 373
- 2. Nitropropanes 373
- 3. Nitrobutanes 375
- 4. Nitrobenzene 376

CHAPTER X: Miscellaneous Compounds

- 1. Carbon disulphide .. 380
- 2. Acetic acid 392
- 3. Acetic anhydride .. 392
- 4. Cresols 393
- 5. Dimethyl sulphate .. 395
- 6. Silicones and silane intermediates 397

TOXICITY OF INDUSTRIAL ORGANIC SOLVENTS

INTRODUCTION

THE following Report summarizes the existing available information about the effects on animals and man of the various solvents used in industry. The literature from which the Report has been compiled is widely scattered in various books and scientific periodicals, and although much trouble has gone towards its composition, completeness of the references cannot be vouched for. Moreover, new work is being published at frequent intervals, and the bibliography must rapidly cease to be up to date.

The main object in the collection and publication of these data is the avoidance of poisoning from the use of these solvents, some of which are used on a vast scale in industry. To suggest the ease with which intoxication could occur, and to emphasize the importance of finding appropriate measures for its prevention, the following figures are quoted from statistical records. The Annual Abstracts of Statistics shows that in the United Kingdom in 1950 the consumption of ethyl alcohol as 68 O.P. spirit was 36·8 million gallons; that of methyl alcohol was 72,000 tons. The Ministry of Fuel and Power's Statistical Digest records that, in 1951, the total production of different fractions of benzol, excluding motor and aviation spirits, was 15·2 million gallons, of which 9·2 million gallons were the 'pure' solvent; the total production of toluol was 8·7 million gallons, of which 4·9 million gallons were 'pure'. Similar figures for other solvents are not readily available, but these few are surely impressive enough to indicate the need for investigation.

It is often difficult to determine, should illness occur in a worker or group of workers, whether or not this is due to the toxic action of chemical agents, and, if so, which particular agent is responsible. This difficulty is often increased by the slow development of symptoms which may occur after prolonged or intermittent exposure to low concentrations. The direct method of determining the possible and relative toxicity of a substance to man is by experiments with that substance on man himself, under conditions as nearly as possible parallel to those under which a worker is exposed. Experiments on man, however, are not justified, and the indirect method using experiments on animals in the laboratory must therefore be adopted.

Owing to the expense and difficulties of housing large animals, preliminary investigations are nearly always made on large numbers of small animals, although no general rules can be laid down which relate the sensitivities to poisons of small rodents and of man. There are many poisons which produce completely different effects in different species of animals, but it is very exceptional for a substance which is toxic to various species of rodents to be innocuous to man, and vice versa. It is to be anticipated that there will exist qualitative and quantitative differences in response to any poison between man and the lower animals, just as they are known to exist between different species of lower animals themselves. Nevertheless, there can be no doubt that experiments on small animals are of value in giving an indication of the type of toxic action as well as the relative toxicity of the substance under investigation. Thus the fact that butyl cellosolve (ethylene glycol mono-*n*-butyl ether) is five times as toxic to mice as ethylene glycol at least gives a warning that the former compound may have greater potential dangers for man than the latter.

In attempting to relate experimentally determined effects on small animals to probable or suspected effects on man, certain factors have to be taken into consideration. The first of these is the size of the animal. As regards the systemic actions of poisons, depending as they do mainly upon the concentration of the poison which occurs in the blood and tissues, it is generally true that a given amount of poison will produce greater effects in a small than in a large animal. No entirely satisfactory method is known for correlating accurately dosage and body weight, but a useful approximation is obtained by the usual method of stating the dosage as per kilogramme of body weight; however, it will be readily understood that if the minimum lethal dose of a solvent for a rat weighing 0·2 kg. is 1 ml., it does not follow that the corresponding dose for a man weighing 70 kg. will be 350 ml. Even such an approximate relation does not hold good for the local effects of poisons, before they are absorbed. For example, a given concentration of a volatile solvent in air may produce local effects on the lungs, the intensity of which bears no relation to the weight of the animal.

It is now known that individual variation is found in response to all drugs and in all populations of animals. Measurements of toxic effects on animals usually express the average or median toxic dose or concentration, i.e., the quantity that produces the observed effect in half the population studied. Unfortunately the range over which individual variation is scattered differs widely in the case of different drugs. Some individuals will be less, some more, sensitive to a toxic action than the majority. This is of especial importance in human beings when it is necessary to avoid producing toxic effects on any particular individual. There is also the important question of exceptional sensitivity or idiosyncrasy to poisons. It has been found that certain drugs, which have been widely used for many years without apparent danger, can nevertheless produce very serious toxic effects in certain exceptional individuals. It is possible that cases of personal idiosyncrasy would occur with all drugs though they may be only recognized with those that are widely used. Similar cases of idiosyncrasy will no doubt arise from the use of substances in industry, and slight signs of intoxication occurring with any new solvent should consequently be regarded seriously, even if other individuals have sustained greater exposure without any apparent injury.

Consideration of the available literature on the actions of solvents has shown that many experiments have been made with little direct relation to the conditions that may possibly occur in industry. For example, the effect of the subcutaneous injection of a large dose of substance may be entirely different from, and have little bearing upon, the effects which may be produced by prolonged inhalation of small quantities. The Committee had some hesitation in including an account of such experiments, but for some substances they were the only ones available. The results have therefore been included partly for completeness and partly for such evidence as they afford.

One other difficulty should be mentioned. The number of chemical compounds used as solvents is large, but unfortunately the number of proprietary names is much larger, and a toxic substance may be concealed under such a name. It has not been possible to give a glossary of proprietary names, and although many are mentioned in the text and appear in the index, this report will, in general, only be of service in judging the safety of a solvent in those cases where the chemical composition of the solvent is known.

CHAPTER I

HYDROCARBONS

1. Benzene

(Benzol)

C_6H_6, i.e. ⬡

PROPERTIES

BENZENE, or benzol, a hydrocarbon of the aromatic series, and a coal tar product, is to be carefully distinguished from benzine, a mixture of varying and uncertain proportions of hydrocarbons, chiefly hexane and heptane, which is a distillate of petroleum.

A certain amount of confusion has arisen from time to time in reports of cases of toxic injury due to benzine, e.g. the case of fatal aplastic anaemia in a cellulose sprayer in 1934 (Special Article, *Lancet*), in which the toxic agent was referred to as benzine, when benzol was meant. Simonin (1934) points out that the distinction between benzene and benzine is not merely one of theoretical importance. He quotes as an instance the occurrence of a severe collective intoxication in a boot factory in 1932, involving forty-four workers, eight of whom died, caused by the misinterpretation of an order for benzine, benzene being delivered instead (Merklen and Isräel, 1934).

The effects of chronic poisoning by these two substances, especially with regard to the blood picture, are clearly distinguishable, while the acute narcotic effect of benzine on animals has been shown to be less than half that of benzol (Bamesreiter, 1932).

It has been suggested by some French workers, notably Duvoir (1922), that benzine should be called *petrol essence* and that the word *benzinism* should be replaced by *petrolism*.

Benzol is graded commercially as crude or refined, according to the percentage which distils below 100° C. *Pure benzol* should be the pure substance strictly designated as *benzene* (C_6H_6), B.P. 80° C., Sp. Gr. 0·884 at 15° C., M.P. 5° C., soluble in water to the extent of 0·082 g. per 100 ml. at 22° C. It is an excellent solvent for rubber, gums, resins, and fats of all kinds. *Commercial benzol* is practically never pure; it contains traces of xylene, toluene, phenol, thiophene, carbon disulphide, acetonitrile, and, according to Ellis and Meigs (1921), probably pyridine, and many other substances. (For method of analysis see Gooderham, 1935). A process for removing thiophene by the action of acidified hypochlorite solution with only a small attendant loss of benzene has been described by Ardagh and Bowman (1935). There are three usual commercial types.

(*a*) *Commercial crystallizable*. 100 per cent benzene, B.R. 80–81° C., Sp. Gr. 0·879 at 20° C. This variety appears to produce very severe toxic effects, many cases of poisoning having been reported from its use. A typical example is the series recorded by Heim de Balsac and Agasse-Lafont (1933), of which eight cases were fatal, and by Merklen and Isräel (1934) who describe their cases as "haemorrhagic aleukaemia" from the use of crystallizable benzene. These investigators urge that the use of crystallizable benzene should be prohibited.

(*b*) *Commercial* 90 *per cent benzol*. 90 per cent by volume distils below 100° C. It contains 13–15 per cent toluene, 2·3 per cent xylene, and sometimes, traces of olefins, paraffins, sulphuretted hydrogen, and other bodies.

(*c*) 50 *per cent benzol*. 50 per cent of constituents distil below 100° C. and 90 per cent below 120° C.—a highly mixed product.

The usual *commercial* 90 *per cent benzol* is a colourless, mobile, refractive liquid, with a not unpleasant odour. It burns with a luminous but smoky flame; it is extremely volatile, especially if slightly heated. Its relative time of volatilization compared with that of ethyl ether is 3 : 1 (I. G. Farbenindustrie, 1930). It is a good solvent for a number of cellulose esters and ethers, especially ethyl cellulose, for most oils, ester gums, benzyl abietate, copal ester, coumarone, benzyl resin, and many other resins. It does not dissolve cellulose acetate or nitrate, copal, or shellac.

MANUFACTURE

Benzol is given off during the distillation of coal in a closed vessel, part remaining in the tar and part occurring in the gas. The chief means of recovery is by "stripping" the benzol from the coke oven gas, a method which has of recent years largely superseded distillation of the tar from gas works and coke ovens. During 1934, of the total production of crude benzol at coke ovens, gas works, low temperature carbonization works and tar distilleries, about 80 per cent was obtained from the gas, and 20 per cent from distillation of tar (Secretary for Mines, 1934).

A full description of the distillation method was given by Lehmann (1910) and of the "stripping" process in the Final Report on Benzol of the National Safety Council (1926).

USES AND APPLICATIONS

The use of benzol received a great impetus during the 1914–18 war, and since that time its extension has been wide and rapid. While in 1922, 68 million gallons were used in America, by 1928 the figure had reached 115 million gallons (McCord, 1929). During the 1939–45 war, benzol was again increasingly used owing to the scarcity of other solvents.

Benzol may be regarded as having two more or less distinct fields of application in industrial processes:

(*a*) Where it is handled in large quantities in closed mechanical systems.

(1) The distillation of coal and coal tar in the production of benzol.

(2) The blending of motor fuels.

(3) The chemical industries, including oil extraction, the manufacture of dyes and dye intermediates, paints, varnishes and stains, and paint and varnish removers.

(*b*) Where it is used as a solvent or diluent.

(1) The rubber industry, in solutions for rubber cement in the manufacture of straw hats, cardboard boxes, waterproof goods, shoes, cameras, and the sealing of cans; in the manufacture of rubber tyres, the metal case on which the tyre is built being usually coated with rubber cement in which considerable quantities of benzol are used.

(2) The manufacture of artificial leather; textiles are spread with a coating of a viscous liquid consisting essentially of a solution of nitrocellulose, benzol in amounts up to 60 per cent being added as a diluent.

(3) In the dyeing and cleaning industry for degreasing, mordanting and removing grime from clothing and other articles; also generally for cleaning

purposes in workshops. In many dyeing and cleaning establishments, however, other solvents have largely replaced benzol.

(4) In the paint and varnish industry as a diluent for lacquers, a constituent of quick-drying paints, in bronzing and gilding pottery, in varnishing reservoir vats, ships, motor cars, etc., and in floor and woodwork stains and floor waxes.

(5) In the aviation industry as a constituent of the dope solution, to the extent of about 20 per cent.

(6) In the linoleum and celluloid industries.

(7) In artificial manure and glue manufacture.

(8) In electrical fitting and accumulator works.

(9) In chemical laboratories as a solvent.

(10) In the alkaloid industry for the extraction of atropine, hyoscine, codeine, etc.

(11) In the photogravure printing process.

BENZOL VAPOUR

Concentration in Air of Factories

The actual concentrations of benzol in air to which workers have been exposed have been found to vary greatly. The earliest figures, given by Lehmann (1910), ranged from 25 to 106 p.p.m. of the air in benzol washing and distillation plants. Legge (1919–20) found the quantities in the atmosphere of a balloon fabric spreading room to range from 210 to 1,050 p.p.m. in different parts of the room, and in a pneumatic tyre manufacturing room from 800 p.p.m. with an exhaust fan in operation to 2,800 p.p.m. with the windows open. In a Milan raincoat factory, where three fatal cases occurred, Pugliese (1922) found 1,000 p.p.m.; in the printing works referred to by Saita and Dompé (1947) they found the concentration of benzol varied from 188 to 908 p.p.m. In the National Safety Council Report of 1926, examination of the air of 18 different workrooms, representative of the various processes in which benzol is used in rubber, artificial leather, wire insulating, dry cleaning, and sanitary can manufacture, revealed a very wide variation of its benzol content, 0 to 4,140 p.p.m., the latter in a dry cleaning process during the summer when no other solvent than benzol was used. It should be noted that the method used for the estimation of benzol in the survey failed to distinguish between benzol and other solvent vapours such as alcohol, methyl acetone, etc., but the figures for solvent vapours were all computed in terms of benzol. These studies brought out the importance of local exhaust ventilation in reducing the atmospheric content of benzol. The Report concludes that "rooms in which benzol is evaporated into the air without local exhaust ventilation will in most cases show high concentration of the fumes in the air of the rooms. With ideal local exhaust ventilation on the other hand even large quantities of benzol can be used without heavy atmospheric contamination". After the occurrence of two fatal cases in Edinburgh in 1918, it was found that without ventilation the concentration of benzol in the room reached 16,800 p.p.m., and with 30 changes of air per hour was reduced to 550 p.p.m.

Risk of Explosions

The vapour of benzol, which has a flash point (Fl.P.) of $10°$ F. ($-12°$ C.), forms an explosive mixture with air in proportions from 5 to 8 per cent. Escape of the vapour or liquid may occur through faulty design of the plant or faulty ventilation, and ignition may be produced by the presence of naked lights, the production of sparks by the use of tools or appliances, such as electric fans and motors, and by iron-shod boots on stone floors, trolley wheels, etc. Precautions to avoid these dangers are described in the Chemical Works Regulations Reg. 4 b (1922), also in *Safety Circulars* Nos. 29, 43, and 51 of the Association of British Chemical Manufacturers (1929, 1931) and in *Quarterly Safety Summary* (1930, 1931, 1932).

Concentration in relation to Toxic Effects

The maximum allowable concentration above which toxic symptoms may be expected to occur was estimated by the National Safety Council (1926) as 100 p.p.m., and this figure was again taken by the American Standards Association in 1941. It has been suggested by Greenburg and Moskowitz (1945) with regard to the war-time use of benzene as a solvent for synthetic rubber that 50 p.p.m. is a safer desirable maximum, since, they state, "death may occur with exposures to only 25 p.p.m., and at least one such case has been observed". The concentration which produces fatal results in a very short time appears to be 19,000 to 20,000 p.p.m. In a report made by McNair to the Home Office* on a fatal case of benzol poisoning occurring in 1930, he states that to form a mixture which would be fatal would require 2 volumes in 100 of air, representing 7 oz. of benzol per 100 cubic feet. He also states that 1 volume in 100 of air can be recognized by smell, while *Safety Circular* No. 51 states that more than 1 mg. per l. (0·029 per cent) is necessary before the smell is distinctly perceived.

The concentrations of benzol and their effects and time of exposure (Table 1) are taken from the table drawn up by Sayers and Dallavalle (1935).

TABLE 1
Effects from different concentrations of benzol vapour

Concentration in air (p.p.m.)	Effects
19,000	Fatal in very short time
3,000	Dangerous with exposures of $\frac{1}{2}$ to 1 hour
3,130–4,700	Maximum concentration for exposures of $\frac{1}{2}$ to 1 hour
1,570–3,130	Slight symptoms with exposures of several hours
100	Maximum concentration allowable

More recently (1949) the standard limit suggested has been lowered to 35 p.p.m. (Congress of American Hygienists).

Estimation in Air

Numerous methods for the determination of the amount of benzol in air have been described. The majority of the earlier ones are reviewed by Tausz (1924) and include the nitration method, introduced by Harbeck and Lunge (1898), used by Lehmann (1910) and adapted by Smyth (1931) for concentrations as low as 30 p.p.m., and the method of activated charcoal considered superior to all others by Tausz himself, and used by the National Safety Council in 1926. More recent methods include those of Ficklen and Cook (1933) and of

---

* References to cases reported to the Home Office are taken from confidential records, to which the author had access, of the Factory Department, Home Office, later transferred to the Ministry of Labour and National Service.

Cook and Ficklen (1935); the former, the pernitric method, has since been found not applicable because of the interference of toluene; the latter, the oxidation method with peroxide in presence of iron salts, is described below. A simple test, involving the absorption of the benzene vapour in formaldehyde–sulphuric acid, has been devised by the Department of Scientific and Industrial Research (1939).

(1) *The nitration method.* The method described by Smyth (1929) consisted in catching the benzene vapours in a mixture of sulphuric and nitric acids, neutralizing, separating the dinitrobenzene formed in the acids by steam distillation from a buffered solution when toluene was present in the air, and then titrating the nitro compound with titanous chloride. This method was shown to be 93·1 per cent correct for pure benzene vapour at 11,000 p.p.m., and 99·2 per cent correct at 200 p.p.m. In order to adapt the method for concentrations as low as 30 p.p.m., Smyth (1931) took larger samples of air for analysis than formerly, so as to obtain enough dinitrobenzene for titration. He found the method 99–100 per cent correct with pure benzene vapours at concentrations of from 30 to 585 p.p.m., even in the presence of various aliphatic alcohols and acetates. In the presence of not more than three times as much toluene as benzene, the accuracy was over 85 per cent down to a concentration of 30 p.p.m. of benzene.

It is pointed out by Cook and Ficklen (1935) that the method requires transportation of air samples to a central laboratory for determination of the amounts of benzene present. A modification of it is, however, described by Schrenk, Pearce and Yant (1935) by which small samples of air can be used and as little as 0·001 mg. of benzene can be determined, and further modifications by Dolin (1943), and by Duvoir and Fabre (1946). The latter claimed that a variation in the technique serves to distinguish benzene from toluene with an error of less than 10 per cent. The titrating reagent used is methyl ethyl ketone in a 40 per cent solution which gives a violet colour not given by the higher homologues of benzene. Dolin's technique has been still further elaborated by Milton (1945), making the reaction still more specific for benzene in the presence of xylene and toluene.

(2) *The activated charcoal method* (National Safety Council 1926). Special precautions to render the sampled air moisture-free before passing it into the charcoal absorption tube consisted in the provision of a first tube filled with soda-lime for the absorption of acid vapours and a second with calcium chloride for absorption of water vapour. The charcoal tube itself contained approximately 7 g. of 8–14 mesh activated charcoal. The charcoal was also dried overnight in a constant temperature oven at 105° C. The charcoal tubes were equilibrated before use by the passage of compressed air, flowing at the rate of approximately 6 l. per minute, through a chain composed of the following elements: cotton-wool tube, soda-lime tube, calcium chloride tube, glass tubing manifold and six charcoal tubes. A sample of the air of the workroom, usually 20 l. in amount, was aspirated through the prepared charcoal, and the increase in weight of the charcoal was taken to represent the approximate amount of solvent vapours in the atmosphere sampled. Where benzol is the only solvent used, this increase in weight represents that due to benzol vapours only, but since the charcoal absorbs other solvent vapours also the method is not specific for benzol.

(3) *The oxidation method with hydrogen peroxide in presence of iron salts* (Cook and Ficklen, 1935). In this method the air to be sampled is streamed at the rate of 2 l. per minute on to the surface of glass beads in a U-tube packed round with solid carbon dioxide; the tube is later removed and immersed in hot water when the collected benzene is volatilized, passed through a bubbler unit containing 25 ml. of water for the removal of the more water-soluble vapours, and thence into a trap immersed in a solid carbon dioxide and acetone cooling bath, the benzene and a small portion of less water-miscible solvents being frozen out in the trap. For the estimation of the benzene, 5 ml. of 0·5 per cent ferrous sulphate followed by 2 ml. of 1 per cent hydrogen peroxide is directed down the inlet tube of the trap, the contents of which are then shaken and transferred to a test-tube. If benzene is present in an amount of 0·005 ml., a characteristic brown coloration is produced in 2 to 5 minutes; if in amounts of 0·010 ml. to 0·050 ml., a black amorphous precipitate also appears. Upon the addition of 1 ml. of $2N$ nitric acid the black amorphous precipitate will dissolve and the solution may then be diluted with water and compared with standards in a colorimeter.

A number of other solvents, having boiling points relatively near that of benzene (toluene, xylene, methanol, ethyl acetate, trichloroethylene, carbon tetrachloride, cellosolve, acetone) were tested, and did not give the reaction when no benzene was present. Carbon disulphide gave a greenish amorphous sulphur precipitate, which, if benzene is being simultaneously estimated, must be filtered from the solution after the addition of nitric acid, the benzene being then estimated from the coloration of the filtrate.

Ficklen and Cook (1933) observe that, owing to the complicated nature of the reaction, the method is only approximately quantitative, but sufficiently so to permit estimation of the concentration of benzene in air within the limits required in an industrial hygiene investigation.

(4) *The formaldehyde-sulphuric acid test* (Department of Scientific and Industrial Research, 1939). A sample of the air to be tested is drawn through a tube containing the reagent —concentrated sulphuric acid containing a trace of formaldehyde—by means of a hand pump of definite capacity. From the number of strokes of the pump required to produce the standard orange-brown colour, the concentration of benzene vapour present is obtained by reference to a given table. The colour with the formaldehyde-sulphuric reagent is also given by the vapours of toluene and of coal-tar naphthas. Crude benzols, which may contain compounds such as thiophene and unsaturated hydrocarbons, might give a slightly different, yellow or red shade of colour, but the quantities of such substances usually present in commercial grades of benzenes should not interfere with the comparision of the depth of colour of the reagent and standard.

(5) *The interferometer.* Data on the use and value of this instrument are supplied by the United States Bureau of Standards, and by the catalogue of the Arthur H. Thomas Co. It is stated (Answer to a letter, *J. Amer. med. Ass.*, 1935) that the instrument is of restricted value under many industrial circumstances, owing to the fact that benzene, benzine, naphtha, toluene, and xylene are rarely used as single entities, but as mixtures with various esters, acetates alcohols and glycols. This fact precludes any ready standardization of the apparatus to meet the diversity of vapour conditions likely to be encountered.

ESTIMATION IN BLOOD AND IN TISSUES

The chief methods of estimating the amount of benzol in the blood and tissues of animals or human beings subjected to exposure to benzol vapour have been described by Peronnet (1934).

(1) *Method of Lazarew, Brussilowskaja and Lawrow* (1931a, b). The benzol is extracted by a current of air flowing over a sample of blood, and the amount present in the air measured by the method described by Matveev, Pronin and Frost (1930), which consists in combustion of the air in an electric oven and titration of the CO_2 formed by conductivity.

(2) *Method of Gadaskin* (1928). The oxalated blood is nitrated by means of sulphuric and nitric acid and the *meta*dinitrobenzene formed is reduced to *meta*phenylenediamine, which by oxidation with dimethyl*para*phenylenediamine gives a violet colour.

(3) *Method of Peronnet* (1934). The benzol extracted from a slightly heated sample of blood is nitrated to *meta*dinitrobenzene, which is then diluted with distilled water, neutralized and rendered alkaline. Alcohol is added, followed by a solution of laevulose in 70 per cent alcohol and soda. The violet colour formed is compared in a colorimeter with that produced by a benzene solution of known concentration. This method is not applicable in the presence of xylene or toluene.

A modification of this method has been used by Koppenhöfer (1935) in estimating the benzol content of the blood and tissues in a fatal human case of benzol poisoning, and by Duvoir and Fabre (1946) in estimating separately benzene and toluene in the blood.

TOXICITY

As an acute poison benzene is a narcotic, producing severe, or even fatal, depression of the central nervous system. Its chronic toxic effect is that of injury to the bone-marrow, producing great variety in the blood picture, with the salient feature of aplastic anaemia.

It is clearly apparent, both from the study of cases of benzol poisoning in human beings, and from experimental work on animals, that acute and chronic benzol poisoning are two entirely distinct phenomena.

"In acute benzol poisoning the symptoms are due to a severely irritant and destructive effect upon the central nervous system, whereas in chronic benzol poisoning the relatively slight effects on the central nervous system are not at all comparable with the severity of the degenerative changes occurring in the haematopoietic system. In acute poisoning benzol thus acts as a convulsive neurotoxin and later as an asphyxiant narcotic." (Batchelor, 1927.)

Chronic poisoning presents a wide variety of subjective symptoms, both in number and in degree of severity, and although the classical picture of anaemia of aplastic type is still the most widely recognized example of benzene poisoning, extensive investigations of recent years have tended to bring out the salient fact that this is by no means the invariable one. This concept of expected variation tends to alter the value of the criteria of benzene poisoning, or benzene absorption, such as those laid down in the Report of the National Safety Council on Benzol, 1926.

Toxicity in Industrial Processes

In the earlier literature, cases were reported with some frequency of acute, sometimes fatal, poisoning from swallowing benzol. Such were the two cases recorded in St. George's Hospital Reports for 1877–78, and those of Averill (1889), Simonin (1903) (in this case the patient survived, but developed swelling and oedema of the skin), Hetzer (1922), in which the poisoning was followed by intense toxic gastritis and later pyloric stenosis, and Nick (1922), in which recovery followed the injection of 5 ml. of a 10 per cent lecithin emulsion.

Although it has been stated by Hamilton and Johnstone (1945) that "acute benzol poisoning is of little importance under modern industrial management", owing to unforeseen imperfections of plant, or neglect of the precautions provided, severe and even fatal cases do occur. During the last seven years thirty-one cases, nine of them fatal, have been reported to the Factory Department. Most of the total number of fatalities recorded in the literature have been caused by the entry of workmen into tanks containing benzol, or having contained it at some time.

With regard to the possibility of reducing the incidence of acute benzol poisoning from this cause, Marian-Wolfen (1925) suggests certain measures for the cleaning of benzol containers. He states that removal of benzol residues from tanks is not possible by means of water alone, that in fact such a method increases the danger, since the benzol, being of comparatively low specific gravity and scarcely soluble in water, forms a layer on the walls of the vessel which is later vaporized. The passage of steam through the container, however, with the provision of several openings for the escape of the steam and volatilized benzol, and of an exit at the lowest level of the container for the escape of the benzol-containing water of condensation will, he claims, constitute an entirely successful procedure for cleaning out vessels containing benzol and removing the residue; this method should, if carried out universally, much reduce the danger of severe and fatal cases of benzol poisoning.

It may be mentioned, however, that though the number of acute fatalities has certainly decreased since the need for adequate protection has been realized and acted upon, one of the nine cases reported to the Factory Department occurred in 1945 during the steam-cleaning of a condenser, the procedure advocated by Marian-Wolfen.

According to Pfeil (1932), cases of poisoning reported from the exhaust gases of machines in which benzol is used as a motor fuel are not actually due to the benzol but to carbon monoxide, since exhaust gases contain no benzol.

In the rubber industry four cases of poisoning, two of which were fatal, from the use of shoe cements containing a large proportion of benzol were reported by Hunter and Hanflig in 1927. With regard to the sealing of cans, Hamilton

(1931) remarks that since the introduction in 1926 of rubber latex as a substitute for rubber dissolved in benzol, many works have substituted latex for the benzol rubber sealing mixture, but that no less than one third of all food cans are still sealed with benzol. In tyre-building, five cases (three fatal) of poisoning with aplastic anaemia and haemorrhages occurred (Harrington, 1917). Six deaths and twelve severe and several slighter cases of poisoning occurred in a rubber factory in Vienna in 1932, although all the apparatus used was of the closed type. The trouble arose apparently from the transportation of goods moist with benzol from one room to another (Fischer, 1932). An Austrian Government Order of May 1, 1934, prohibited workers in processes using benzol, toluol, xylol, or trichloroethylene from working more than 4 hours a day, and, when large quantities are used, not more than 2 hours a day. The processes covered by the more stringent restrictions include the cold vulcanization of rubber and the production of rubber substitutes.

Two of six cases of benzol poisoning reported by Hunter and Hanflig (1927) and a fatal case mentioned by Hamilton (1928a) occurred in the manufacture of patent leather.

In connexion with quick-drying paints, it is emphasized by Flury and Zernik (1931) that the use of benzol is attended with special danger from the fact that it volatilizes quickly from the painted surface, especially when the surface is warm, as in metal parts of machines, or when the paint is sprayed on. Details of the conditions and risks of this process are given by Smyth and Smyth (1928) summarizing the findings of the Pennsylvania Department of Labor and Industry (1926) and of the National Safety Council Committee on Spray Coating (1927). It was finally recommended by the Committee that no lacquers be used for spraying which contain over 0·5 per cent of benzol.

An investigation by Kranenberg and Peeters (1928) of workers in an aeroplane factory showed great and characteristic changes of the blood picture in a large number. In one case, after six years' exposure, there was a relative lymphocytosis of 62 per cent and a reduction of haemoglobin to 65 per cent.

Attention has been drawn by Bloomfield (1928) to the possibility of poisoning in laboratories where tests are conducted with rubber, paint, varnish and oil products, and involve centrifuging the material to be tested with benzol. He found a considerable amount of benzol in the air of such laboratories, 28 to 223 p.p.m., and examination of workers revealed in three cases a change in the ratio of polymorphonuclear leucocytes to lymphocytes.

Nine cases of chronic benzol poisoning in workers in the alkaloid industry were reported by Mitnik and Genkin (1931).

In England benzol is rarely used in high concentration in printing inks, but an investigation made by Saita and Dompé (1947) in Italy revealed that the percentage of benzol in the ink solvents and diluents varied from 66 to 93 per cent. Of the fifty-five operatives examined, 67·3 per cent showed some abnormality of the blood picture, and at one of the factories there had been two serious cases of benzene poisoning, one of which was fatal. In England also a series of cases of severe intoxication has occurred during recent years (p. 42).

TOXIC EFFECTS IN ANIMALS
Absorption and Excretion

Absorption

(a) *Through the lungs.* Rabbits, according to the investigations of Lehmann (1910), have a smaller capacity for absorption of benzol through the lungs than human beings. Whereas men absorbed 80 per cent of the benzol vapour inhaled, rabbits absorbed only 63 per cent. No definite ratio between the relative amount absorbed and the total amount present in the air was found, but Robinson and Climenko (1941) have since observed that rabbits exposed to about 1,000 p.p.m. for 2 hours show a blood concentration of 25 to 30 mg. per 100 ml.

(b) *Through the skin*. The investigations of Lazarew, Brussilowskaja, Lawrow and Lifschitz (1931c) have shown that benzol in the form of vapour, as well as in liquid form, can be absorbed through the skin of animals. In liquid form, benzol produced great irritation. Immersion of the ears of rabbits for $2\frac{1}{2}$ to 3 hours was followed by inflammation and oedema, then pus formation, and finally by mummification of the ear, while immersion of the whole body of a mouse was followed by death after about 3 hours.

When animals were kept in a chamber filled with benzol vapour, their heads being passed outward through an opening so that inhalation of the vapour was impossible, no general symptoms of intoxication were observed. Evidence of absorption through the skin, however, was obtained in three ways:

(1) The hydrocarbon content of the air expired by tracheotomized rabbits with one foot immersed in benzol was estimated. The estimation was carried out by the method of Matveev, Pronin and Frost (1930). The benzol appeared in the expired air after 3 or 4 minutes in amounts of 1·2 to 1·5 mg. per l., the total amount expired in the course of 115 minutes' immersion of the foot being 138 mg. The speed of absorption of benzol through 1 sq. cm. of skin was therefore calculated as more than 0·016 mg. per minute (0·013 to 0·76 mg. in a later investigation by Lazarew *et al.*, 1931c).

(2) Phenol, an oxidation product of benzol in the urine, was estimated. The urine of animals kept as described above in a chamber filled with benzol vapour showed a strong phenol reaction (Millon's reagent).

(3) The amount of benzol in the blood of rabbits after immersion of the ear in benzol was estimated by Lazarew *et al.* (1931a, b), and the average amount during 30 minutes' immersion plotted as a curve, which rose to between 50 and 100 mg. per kg. of blood taken from the external jugular vein.

Lazarew and co-workers correlate the absorption of benzol through the skin with its water solubility as well as its fat solubility. They estimate its water solubility as 0·033 to 0·040 g. per l. and explain its greater speed of absorption as compared with that of benzine by the lower water solubility of the latter (0·007 to 0·015 g. per l.).

(c) *Distribution in the body tissues and fluids after absorption*. The experiments of Yant, Schrenk, Sayers, Horvath and Reinhard (1936), and Schrenk, Yant, Pearce, Patty and Sayers (1941) have shed much light on the biochemical action of benzene. Of special importance is the partial explanation they afford of the injurious effect of benzene on the haematopoietic system—the predilection of benzene for fat, of which the bone-marrow essentially consists. The concentration of benzene in the fat, bone-marrow and urine of the exposed animals was approximately twenty times that in the blood.

Other conclusions reached were:

(1) The initial rate of absorption of benzene by the blood was extremely rapid, a matter of minutes.

(2) The final equilibrium value was attained very gradually, after several hours of exposure.

(3) The final elimination took place very slowly, the time increasing as the length of exposure increased. In some cases complete elimination had not

taken place until about 13 hours after termination of exposure. The fat, acting as a reservoir, is saturated slowly, and also loses its benzene slowly.

(4) The red blood cells contained approximately twice the concentration of benzene found in the plasma, and the muscles and vital organs about one to three times that in the blood.

(5) Benzene absorbed into the body by inhalation was apparently excreted into the stomach, and the high benzene concentration in the urine indicated that the benzene is concentrated by the kidneys.

Excretion

While the greatest amount of benzene is excreted unchanged by the lungs, a certain amount is oxidized in the body and excreted in the urine in the form of phenol, catechol, hydroquinone, and muconic acid (Jost, 1932). In animals the excretion of benzene in the urine has usually been estimated by the amount of phenol excreted. The phenol excretion of rabbits after 2 hours' daily inhalation of benzene in a concentration of 13·6 mg. per l. has been estimated by Sartorius and Sudhues (1933). The average phenol excretion rose almost immediately from a normal level of 2 mg. per day to 50 mg., and later to nearly 100 mg. When inhalation was stopped the phenol excretion sank to the normal level only after 4 or 5 days. This finding appears to indicate, according to Sartorius and Sudhues, that many cell systems must constitute special storage centres for benzene, or its ethereal sulphates.

Change in sulphates in urine. Some investigations by Yant *et al.* (1936) show that a change in the normal ratio of inorganic to organic sulphates, i.e. a decrease in inorganic sulphates, in the urine is a constant and early sign of benzene poisoning in animals. The analysis of urine specimens from 79 dogs exposed to concentrations of from 100 to 800 p.p.m. of benzene vapour for periods of 100 to as long as 1,000 days in one instance showed that a rapid and marked decrease occurred in the percentage of inorganic sulphates in the total sulphates in the urine. Single examples of their figures will suffice to show the quantitative aspect of this decrease (Table 2).

TABLE 2

Decrease in urinary inorganic sulphates after exposure to benzene

Concentration (p.p.m.)	Daily exposure (hours)	Exposure period (days)	Percentage of inorganic sulphates	
			Before exposure (average)	After exposure (average)
100	8	42	89·4	57·0
500	8	317	94·7	28·8
800	1	17	86·7	62·5
800	8	192	96·5	6·3

A distinct decrease in the percentage of inorganic to total sulphates occurred even with conditions of exposure which did not produce anaemia or leucopenia. With conditions which produced benzene poisoning a great decrease occurred weeks and months in advance of anaemia, leucopenia, and the common signs and symptoms of benzene poisoning. The mechanism of the response is believed to be due to oxidation of the benzene to phenol or phenolic derivatives, which in turn are conjugated in the liver with sulphate ions to form ethereal sulphates, thereby causing a shift to the right in the system: inorganic sulphates ⇌ conjugated sulphates. The shift or decrease in inorganic sulphates is related quantitatively to the severity of the exposure to the point of complete elimination of the inorganic sulphates (p. 26).

Acute Poisoning

It has long been known that benzene in large doses produces in animals a severe narcosis, often accompanied by symptoms of irritation of the central nervous system, tremor, muscular twitching, etc., and, above the minimum lethal dose, followed by death. It was believed that the cause of death was invariably respiratory paralysis, but some investigations by Nahum and Hoff (1934) suggest that, although death may occur from respiratory failure during the stage of narcosis, it may also occur suddenly from myocardial failure produced by the action of benzol, both by causing the liberation of adrenaline and at the same time sensitizing the myocardium to its toxic action.

The acute effects of benzene upon animals have been studied from results of intraperitoneal injections and of inhalation. The symptoms arising from irritation of the central nervous system are essentially similar in both modes of administration, but the intense effect is most apparent when benzene is administered intraperitoneally, owing to its rapid absorption.

Effect of Intraperitoneal Injection

Lethal dose. For guinea-pigs, 0·73 ml. per kg. body weight (Chassevant and Garnier, 1903); for rats, 1·5 to 1·75 ml. per kg. (Batchelor, 1927).

Symptoms. Profound tremor and muscular convulsive twitchings were produced by doses as low as 0·25 ml. per kg. body weight, and in addition, a definite narcotic effect, with drowsiness, instability of gait, impaired equilibrium, weakness, paresis, decreased susceptibility to pain stimuli, and cessation of voluntary movement; as the spinal cord was affected, loss of reflexes resulted.

Post-mortem findings. Batchelor observed intense acute congestion of the peritoneum and abdominal viscera, with much fibrin deposit on these surfaces, considerable sero-sanguinous intraperitoneal exudate, injection of the gastric mucosa with small haemorrhages and ulceration, and acute congestion of the lungs. No appreciable effect was noted on the blood cell count.

Effect of Inhalation

In large doses, the effect of inhalation of benzol vapour upon animals is that of a nerve poison, with a characteristic neuro-irritant effect, as evidenced by muscle twitching, convulsions, general tremor (compared by Bénech (1897) to that of paralysis agitans), and a state of hypertonicity of the body and musculature (Batchelor, 1927), shown in many cases by a peculiar rigid S-shaped formation of the tail (Lazarew, 1929a).

A toxic action upon the heart is also postulated by Nahum and Hoff (1934), i.e. progressive anoxaemia of the myocardium, and a sensitization of the myocardium to the effects of adrenaline, so that death may occur from ventricular fibrillation.

Different species of animals vary in their susceptibility to acute benzol poisoning, the rabbit being more resistant than the mouse (Lehmann, 1912), while the cat appears to have a special sensitivity (Lederer, 1932; Engelhardt, 1931). A sinus arrhythmia, with ventricular asystoles, due to the action of benzene, either on the myocardium itself or on the vegetative nervous system has been observed by Caccuri (1940).

According to Lehmann and others, animals of the same species, especially cats, also show a great variation in individual susceptibility, death occurring sometimes with relatively small dosage and short exposure, while on the other hand, feeble animals receiving large doses may show only slight weakness and others may recover from severe narcosis. It appears from the work of Nahum and Hoff (1934), and also that of Caccuri (1940) that some of these individual susceptibilities may be explained by variations in adrenaline response; periods of hyperexcitability being accompanied by increased liberation of adrenaline, to the action of which the myocardium is already sensitized by the action of benzol.

Lethal and narcotic concentrations
These are given in Table 3.

TABLE 3
Lethal and narcotic concentrations of benzene by inhalation

Animals	Lethal concentration (mg./l.)	(p.p.m.)	Narcotic concentration (mg./l.)	(p.p.m.)	Author
Mice	45	14,000	—	—	Lazarew (1929a)
,,	—	—	38	12,000	Führer (1921b)
Cats	170	53,000	60	19,000	Lehmann (1912)
,,	—	—	30	9,500	Bamesreiter (1932)
Dogs	146	46,000	—	—	Luig (1913)
Rabbits	46 (after 3 hours)	14,500	—	—	Lehmann (1912)
Guinea-pigs	—	—	40	12,600	Peronnet (1934)

Symptoms

The onset of acute symptoms is apparent at concentrations ranging from 15 mg. per l. (4,700 p.p.m.) for mice to 22 mg. per l. (7,000 p.p.m.) for cats (Bamesreiter, 1932).

Table 4 taken from Lehmann (1912) shows the progress of effects with different concentrations in rabbits.

TABLE 4
Effect of inhalation of benzene on rabbits

Concentration (mg./l.)	(p.p.m.)	Time of exposure	Time before apparent effects				Further progress
			Convulsions	Lateral position	Light narcosis	Deep narcosis	
37	12,000	5 hr.	2¼ hr.	35 min.	—	5 hr.	Slow recovery.
46	14,500	3 hr.	80 min.	—	—	2 hr.	Death after 3 hr.
92	29,000	70 min.	9 min.	9 min.	40 min.	50 min.	Slow recovery.

(It should be noted here that the results obtained by Bamesreiter (1932) with regard to the narcotic symptoms produced by inhalation of benzol vapour by cats give a toxicity greater than that found by Lehmann. Lehmann, in a note appended to Bamesreiter's article, observes that the discrepancy is due to the fact that Bamesreiter has used a modification of the original *Würzburger* apparatus, the new *Ludwigshafen* apparatus, described by Gross and Kuss (1931).)

Bamesreiter's results are summarized as follows:

Lateral posture	..	above 22 mg. per l. after 6 hours' exposure
Light narcosis	..	above 28 mg. per l. after 6 hours' exposure.
Deep narcosis	..	above 30 mg. per l. after 6 hours' exposure.

Post-mortem findings. The organs of animals dying from inhalation of lethal concentrations of benzol vapour have not shown any specially characteristic changes; haemorrhage of the lungs has been observed in a few cases. The most outstanding feature was the fact that the blood remained fluid for a long time after death (Beinhauer, 1896; Lehmann, 1912; Heffter, 1915). No microscopic changes in the blood were found in Lehmann's animals; an odour of benzene could be detected in the lung cavity if opened promptly after death.

Local irritant effect

Mucous membranes. In rats, irritation was caused by concentrations of 1,000 to 2,400 p.p.m. or even less (Batchelor, 1927).

Cornea. In rabbits, irritation with turbidity and a blister-like appearance after repeated exposure to 38·6 mg. per l.

Lungs. Histamine may be liberated (Garan, 1938).

Effect on the blood

The effect of a single exposure to 7,500 to 12,000 p.p.m. for 2 hours was found by Climenko and Macleod (1942) and Robinson and Climenko (1941) to be an immediate temporary leucopenia followed within 22 hours by a leucocytosis. There was no evidence of neutropenia, but on the contrary, varying degrees of neutrophilia, with a shift to the right. There was also a fall in the number of circulating erythrocytes. These investigators assume from the reaction of the white cells that the bone-marrow under benzol exposure is refractory to the normal endogenous stimuli which maintain the equilibrium of juvenile to senile cells; it proved, however, responsive to the more potent exogenous stimulus of sodium nucleinate.

Effect on the Body Temperature

According to Gaede (1944), the effect of administration of benzene to animals, orally, rectally, or intravenously, in doses of 5 to 20 drops, is an immediate rise of temperature. This, it is suggested, is due to stimulation of the temperature-regulating centre of the brain. With narcotic doses, the temperature falls, this being due to paralysis of the centre. Inhalation of 100 p.p.m. causes a rise of temperature during the first half hour, followed by a rapid fall.

Effect of Mixtures of Benzene and other Solvents

An investigation by Svirbely, Dunn and von Oettingen (1943), on the acute toxic effect of certain mixtures of solvents containing benzene, toluene and xylene, has shown that so far as the lethal effect is concerned, benzene is less toxic than either toluene alone, or the blends of benzene, toluene and xylene. Pulmonary irritation was more predominant with benzene alone, but pathological changes were not prominent owing to the short exposure and resulting early death of the animals.

Chronic Poisoning

Both by subcutaneous injection and by inhalation experiments on animals, the toxic effects of benzol have been investigated and described by a large number of workers beginning with Santesson (1897), Langlois and Desbouis

(1907), Lehmann (1910) and Selling (1916). It was chiefly Selling's extensive experiments, carried out by subcutaneous injection, that led to the clear conception of benzol as a leucotoxin, destroying not only the circulating leucocytes in the blood, but also the parenchymatous cells of the bone-marrow, spleen and lymph glands. The destructive effect, he observed, was more pronounced on the myelocytic than on the lymphatic elements. In addition, Selling's experiments brought out the following points: a less obvious effect on erythrocytes, due to injury of the erythroblastic tissue in the bone-marrow, an initial stimulating effect on the bone-marrow with a resulting initial leucocytosis, and regeneration of the injured tissues when removed from the influence of benzol.

In essentials these results have been confirmed by later workers, but in some cases the details have not been reproduced. Thus Ferguson, Harvey and Hamilton (1933) were not able to produce in rats and rabbits a well-marked neutropenia with less effect on the erythrocyte system. In fact anaemia was even more evident than leucopenia, and an erythropoietic disturbance was even more manifest in the peripheral blood picture than a leucopoietic. According to a statement made by Hamilton (1934) these findings of Ferguson *et al.* are characteristic rather of the usual findings in human beings suffering from benzol poisoning than of those observed by most workers with experimental animals. She states that animal experiments do not show the destruction of red blood corpuscles which in human beings is often more conspicuous than the destruction of white corpuscles.

The discrepancy may be explained by the opinion expressed by Ferguson and others, and their agreement with a statement made by Selling which has been less emphasized than the main results of his investigation. Selling observed that "in general the response of the bone-marrow is not specific. It does not respond to a loss of white blood cells solely by production of granulocytes, nor to the loss of red blood cells solely by the production of erythroblasts." The conclusion of Ferguson and co-workers appears to be a confirmation of this observation. They state: "Our opinion indeed is crystallizing in the direction that the action of benzene or its homologues is not as singular in incidence upon the blood cells as it is generally made out to be, and that its effect is a varied one, not necessarily neutropenia, anaemia, nor thrombopenia, a stimulant as much as a destroying agent, and, generally speaking, more or less nonspecific."

This also appears to be the opinion of McCord, Cox and O'Boyle, as the result of the extensive survey embodied in their report to the Industrial Health Conservancy Laboratories in 1932. Their general conclusions as to the toxic effects of benzol on animals may be summarized as follows:—leucopenia is not a consistent feature of benzol poisoning in animals and is not pathognomonic of it; no pathognomonic features absolutely characteristic of benzol poisoning are to be found in life or post mortem; concentrations of 5 parts of benzol per 1,000 of air have set up signs of intoxication in rabbits equal to or more severe than 10 parts per 1,000; all grades of benzol, whether or not containing impurities, such as thiophene, carbon disulphide or acetonitrile, exert a toxic action on rabbits.

Some of the intoxication produced in animals, both by subcutaneous injection and by inhalation, cannot strictly be labelled chronic since the doses used led immediately to more or less acute effects.

In Batchelor's (1927) experiments, for example, doses of 1 ml. per kg. body weight were injected on successive days, and in some cases untoward general symptoms appeared early and resulted finally in death. Similarly the inhalation experiments of Ferguson and others can scarcely be regarded as illustrative of chronic poisoning, for the effect was acute and the exposure had to be short in order to avoid the death of the animal. These investigators themselves suggest that in this fact lies part of the explanation that the blood picture which followed, at least in the majority of cases, could not be called a neutropenia, or even an agranulocytic anaemia. Batchelor's inhalation experiments, however, included the inhalation of lower as well as higher concentrations of benzol, from 2,440 p.p.m. down to 460 p.p.m., and while no symptoms, except weakness, anorexia, and loss of weight, resulted from concentrations of 800 and 460 p.p.m., definite injury to the haemopoietic and excretory organs was produced by these lower concentrations.

The inhalation experiments of Lederer (1932) are of special interest in this connection, since apparently dogs are considerably less sensitive than other species to benzol poisoning, and Lederer was able to carry on the investigations over a period of $2\frac{1}{2}$ years, with inhalations of 6 hours daily of concentrations of 7 to 10 mg. per l. (2,200 to 3,100 p.p.m.).

Symptoms

The symptoms of poisoning by relatively small amounts of benzene in animals are slight, e.g. weakness, anorexia and loss of weight. When, as in the case of Lederer's dogs, symptoms of nervous instability, tremor, muscular hypertonicity, restlessness, appear at the beginning of the experiment, with the disappearance of gross pathological manifestations tolerance appears to be fairly rapidly established.

Changes in the Bone-marrow and Internal Organs

Bone-marrow. Both in the more acute and in the strictly chronic intoxications, the changes in the bone-marrow are essentially those leading to aplasia affecting all cell types, but the appearances differ in individual animals according to whether preliminary stimulation or regeneration has taken place. Thus, in some of the animals investigated by Engelhardt (1931) with inhalations of 10 to 25 mg. per l. (approx. 3,000 to 8,000 p.p.m.), the bone-marrow showed very little change; in some there was degeneration and destruction of leucocytic elements, and in others irritative phenomena—new red cell formation, immature leucocytes, etc.

The regenerative phase has perhaps been best described by Selling (1916) in his original series of animals.

The regenerative process in the bone-marrow begins with the formation of small, circumscribed cell groups and islands, composed of large lymphocytes, granulocytes or erythroblasts. In any given island, after cellular differentiation has once begun, the type of cell, whether erythroblast or granulocyte, remains constant. The islands increase in size and number, fuse with each other and lead to a complete filling of the reticulum with parenchymal elements. In the earliest stages the younger cell types (myeloblasts and megaloblasts) are relatively abundant. In the later stages these cells are still present, but in comparison with more highly differentiated cells they are relatively scarce.

These reparative changes have been further emphasized by Heitzmann (1931), who describes a preliminary atrophy of the bone-marrow followed by a hyperplasia, especially of the leucocytic elements, in animals subjected to subcutaneous injection. Heitzmann, however, lays particular stress on the destruction of leucocytic elements in the bone-marrow as the characteristic effect of chronic poisoning by inhalation of benzol while the greater sensitivity of myeloid tissue has also been stressed by Larionow (1932).

In Schillowa's (1933) experiments with rabbits receiving subcutaneous injections of benzol, the characteristic lesion of the bone-marrow was hyperplasia, chiefly of pseudo-eosinophils, megakaryocytes and erythroblasts.

Spleen. The characteristic injury to the spleen appears to be aplasia (atrophy of Malpighian corpuscles and stroma) with pigmentation and increase in connective tissue, but myeloid metaplasia and hyperplasia have also been observed.

Selling (1916) and Brandino (1922) recorded the occurrence of myeloid metaplasia as a regenerative phenomenon. In the more acute cases of intoxication in animals examined by Ferguson *et al.* (1933) no indication of myeloid metaplasia and no evidence of haemosiderin deposition were found, but in the animals subjected to chronic inhalation by Engelhardt there was much deposition of blood pigment in the spleen, and Heitzmann, who examined them, states that this indicates a toxic action of benzol on the red blood corpuscles. Lignac (1932), who claims to have produced acute leukaemia in mice by injection of small doses of benzol, states that while the myeloid tissue in the spleen undergoes destruction, the lymphoid tissue is increased (hyperplasia follicularis lienis). Schillowa (1933) on the other hand emphasizes a marked hyperplasia of the reticulo-endothelial elements, especially the endothelial sinuses, as the characteristic change in rabbits injected with benzol, and describes the formation of rhombic crystals, analogous to Charcot-Leyden crystals, as the result of phagocytosis of eosinophils.

Liver. Among the earlier workers, Pappenheim (1913) reported the presence of a parenchymatous hepatitis, but this was not confirmed by Neumann (1915). In Batchelor's experiments the liver showed evidence of injury and focal necrosis with inhalation of concentrations as low as 815 p.p.m., but in Lederer's dogs the liver injury was comparatively slight, affecting chiefly the reticulo-endothelial cells, and in Engelhardt's cats and rabbits the chief changes were congestion and deposition of haemosiderin. Silberberg reported negative findings (1928).

Kidneys. Among the earlier workers only Pappenheim observed injury of any severity in the kidneys, while according to Batchelor's (1927) investigation, injury is slight with inhalations of low concentration; he observed cloudy swelling and tubular casts only with 440 and 815 p.p.m. McCord, Cox and O'Boyle (1932), however, lay great emphasis on nephritis as an outstanding characteristic. They state that the special point of attack is the convoluted tubule.

Bladder. McCord, Cox and O'Boyle also state that inflammatory conditions of the bladder, in which large numbers of pus cells may be present in the urine, are of fairly frequent occurrence in animals poisoned with benzol.

Gastro-intestinal tract. Inflammatory reactions of the stomach and other parts of the intestinal tract are also recorded by McCord and co-workers. They state that in the stomach these reactions are preceded by purpura of the mucous membrane and may be followed by the formation of ulcers.

Thymus. According to Lewin (1928), benzol produces atrophy of the thymus in rabbits, the histological changes being those of accidental involution. An increase of interlobular connective tissue is accompanied by the new formation of a network of collagenous bundles within the lobules. Regeneration is possible only if complete inversion of the gland has not taken place.

Lymph glands. Only slight hyperplasia with pigmentation and phagocytosis was found with fairly high concentrations (Batchelor, 1927), though Brandino (1922) recorded a notable destruction of lymphocytes. Schillowa (1933) noted a diminution of lymphoid tissue and phagocytosis of pseudo-eosinophils but no nuclear pyknosis or necrosis of the lymphoid nodules. Muto (1931), however, reported complete atrophy of the lymph follicles.

Lungs. Lesions of the lungs—catarrhal bronchitis, emphysematous areas, decrease in the lymphatic tissue and thrombosis of vessels—have been described by Schmidtmann (1930), pneumonic processes, bronchial catarrh of a desquamative nature and commencing cirrhosis of the lung by Mgebrow (1930), and fatty emboli and emphysema by Schillowa (1933).

Brain. Degenerative processes in the vessels of the brain cortex and in the ganglion cells have been reported by Amenossow and Blinkow (1929).

Genital organs. According to Barzilai (1933), the characteristic change in the genital organs of female mice was dilatation of the blood vessels of the endometrium.

Effect on the Circulatory System

Paralysis of the vasomotor system, with resultant rapid lowering of the blood pressure, is recorded as an effect of benzol poisoning in animals by Dautrebande (1933, 1935b) and Dautrebande and Waucomont (1933), after inhalation of benzol vapour through the trachea. These workers believe that the site of action of benzol is upon the muscle fibres of the peripheral vessels, not on the nerve endings. Their opinion is supported by experiments with fragments of isolated organs, heart, intestine, ureter, which, when placed in Ringer's solution containing benzol, immediately cease contraction (Dautrebande and Waucomont), and by the negative results obtained on applying known peripheral vasomotor stimulants, adrenaline, ephedrine, pituitrin, after benzol (Dautrebande, 1935a).

According to Litzner (1932), benzol produces on the heart progressive diminution of contraction, especially of the ventricles, while Dautrebande (1935b) states that benzol inhalation produces bradycardia, which appears even during the occlusion of the two common carotids. This bradycardia is always increased by injection of adrenaline or hordenine, and after adrenaline, benzol intoxication may produce a fatal syncope by ventricular fibrillation. This conclusion can be well correlated with the results of Nahum and Hoff (1934) on the acute effects of benzol poisoning in animals (p. 13).

Effect on the Blood

The results of innumerable experiments on animals, mostly by subcutaneous injection of benzol, have, on the whole, confirmed Selling's original observation that the striking and characteristic effect of benzol poisoning is a reduction of leucocytes. The essential fact brought out by later researches, especially those of Wallbach (1929, 1931), Weiskotten and co-workers (Weiskotten, Schwartz and Steensland, 1915, 1916; Weiskotten and Steensland, 1919; Weiskotten, Gibbs, Boggs and Templeton, 1920; Weiskotten, 1930) and others, is that the toxic action affects all forms of leucocytes rather than being specifically toxic to granulocytes. The striking effect on granulocytes is not so much their depression as the appearance of immature forms (Hunt and Weiskotten, 1930; Weiskotten, 1930). In the later stages the red blood cells and platelets are also involved, and it is from these two marked differences between the condition of the blood in benzol poisoning and in true agranulocytosis that a conception of the mechanism of benzol poisoning has been evolved. This subject is discussed by McCord (1929), Dameshek (1929), Kracke (1932), and Kracke and Parker

(1935); the latter claimed to have produced true agranulocytosis in animals by subcutaneous injections of benzol, but the present position appears to be that summarized in the opinion of Reznikoff and Fullarton (1933):

> This "shift to the left" may be due to an attempt on the part of the bone-marrow to compensate for leucocytic destruction or to depression of maturation. The fact that bone-marrow studies do not show any hyperplasia and do show some depression for the most part indicates that the depressing action of benzol is on the formative and maturative function of the marrow in addition to any possible leucolytic effect. . . . In addition to eventual involvement of the red blood-cells and platelets which is present in true granulocytopenia, the striking predominance of immature polymorphonuclears during the period of progressive leucopenia when benzol is administered suggests a different mechanism from that taking place during granulocytopenia. It may be that granulocytopenia affects primarily the formative functions of the granulocytogenic elements of the marrow and perhaps the delivery of granulocytes from the marrow into the circulation; whereas benzol injures the maturative as well as the formative function.

In the earlier investigations constant results were only obtained by subcutaneous injections. Thus, while Langlois and Desbouis (1907) recorded a definite diminution of leucocytes and increase in erythrocytes after inhalation of benzol vapour, Luig (1913) obtained practically negative results with chronic inhalation of 5 to 10 mg. per l. Later investigations, however, by Weiskotten and co-workers (1920), Engelhardt (1931) and Batchelor (1927), have shown that the typical leucopenia observed after subcutaneous injection can be reproduced after inhalation, though, according to Engelhardt, to a smaller degree.

The results obtained by Ferguson, Harvey and Hamilton (1933), either by injection or inhalation, do not present the picture recorded by many workers, of a well-marked neutropenia, with less effect on the erythrocyte system. These workers tentatively explain the variability of their results as "one of phase", owing to the fact that the toxic action of benzol is not a specific action but operates according to circumstances on the stem cells of the myeloid leucocyte or the erythrocyte. The somewhat similar results obtained by Beyer (1933) in which the picture was that rather of a terminal anaemia preceded by a leucocytosis than of a definite leucopenia are attributed by them to the fact that the doses given (0·01 g. per kg. by subcutaneous injection) were 100 times less than those used by the earlier workers.

Susceptibility of animals of different species and of the same species. Many of the discrepancies in the findings of different workers in animal experiments are apparently to be ascribed to both individual and species susceptibility. Thus Wallbach (1931) was unable to produce in guinea-pigs and white mice the characteristic leucopenia obtained under similar conditions in rabbits, and states (1933) that only in man, dogs and rabbits is leucopenia an invariable result of benzol poisoning; other animals are refractory to the action. He also points out that underlying constitutional conditions, especially infection, e.g. infection at the site of injection of benzol solutions, may have a considerable effect on the leucocytic response.

Schillowa (1933) lays special emphasis on the variation of susceptibility in animals of the same species, as judged by the blood picture and the degree of bone-marrow injury. She states that in the benzol-susceptible animal leucopenia and thrombopenia occur, with an aplasic condition of the bone-marrow and necrobiotic changes in the granulocytes. In less susceptible animals the blood picture shows little change, while the bone-marrow may show considerable hyperplasia of myelogenous tissue.

It has already been observed that cats have been found to show a special sensitivity to acute poisoning by benzol. According to Sartorius and Sudhues (1933) they are not suitable for investigation of the results of chronic benzol poisoning because they are liable, even with comparatively small doses by inhalation, to develop symptoms of irritation of the respiratory tract, leading to secondary infection and purulent broncho-pneumonia.

White cells

Total count. The total white cell count, according to the great majority of observers, is reduced after both injection and inhalation of benzol, sometimes to a remarkable extent. In Batchelor's experiments with inhalations of low concentrations (460 to 815 p.p.m.) the reduction was as much as from 73 to 92 per cent. In one of Engelhardt's animals, a rabbit, after 13 injections of 0·5 ml. per kg., the count reached the low level of 400 per c. mm. Weiskotten and co-workers (1920), using concentrations of benzol in air which they describe as similar to those found in industrial conditions (average amount vaporized per hour 16·6 ml. in 3 l. air), found that the lowest level reached (1,260 per c. mm.) was never as low as that reached after subcutaneous injections.

An initial leucocytosis followed by the characteristic leucopenia was described by the earlier workers (Selling, 1916; Langlois and Desbouis, 1907) but was not confirmed by Neumann (1915). Later researches, however, appear to show that an initial leucocytosis may occur, but is not constant, and is to be attributed to an initial irritative effect on the haemopoietic system (Engelhardt, 1931). Schillowa (1933) found in all her animals a large increase of leucocytes during the first 3 months.

Diphasic leucopenia, i.e. a secondary fall in the leucocyte curve following a primary rise after benzol injection is discontinued, has been observed by Weiskotten and Steensland (1919) and by Batchelor (1927) in the case of subcutaneous injection. This phenomenon in the case of inhalation was not observed by Weiskotten and co-workers (1920). A permanent lower level of leucocytes (average 6,166, as compared with 11,302 before exposure) was reached when the inhalations were discontinued and no further "deuterophase" was observed.

Lymphocytes. There have been differences of opinion as to whether a relative lymphocytosis is a characteristic feature. Batchelor, dividing the white cells into polynuclear (including polynuclear neutrophils, eosinophils and basophils) and mononuclear (including lymphocytes proper and large mononuclear forms) showed that there was a relative, as well as an absolute, decrease in the polynuclear forms, and a relative increase in the mononuclear forms. This relative lymphocytosis has also been recorded by Selling, Fontana (1921) and others, and has been interpreted as an indication that the lymphocytes were more resistant to the toxic action of benzol than the polymorphonuclear leucocytes. Among those who have found no relative lymphocytosis are Neumann (1915), Muto (1931), and Nicolajew and Schparo (1929), the latter stating that the lymphocytes are even more reduced than the polymorphonuclear leucocytes. Engelhardt's findings were variable, but on the whole a relative lymphocytosis was present in the majority of the animals. Sklawunos (1925), however, was able to produce a very marked leucopenia in animals without a corresponding diminution of lymphocytes, so that finally the white cell formation consisted almost entirely of lymphocytes. Lignac (1932) claims to have produced leukaemia and allied diseases in white mice by subcutaneous injections of benzol.

Eosinophils. An increase of eosinophils, or rather "pseudo-eosinophils", has been described in animals, especially by Schmidtmann (1930) and Schillowa (1933). Beyer (1933) states that these constitute one of the differences in the normal blood picture of the human being and the rabbit; they are described as in the category of granulocytes, taking on Romanowsky-Giemsa stain with large light red granules.

Degenerative changes. The degenerative changes in the white cells occurring in human benzol poisoning have also been observed in animals, but with similar differences of opinion in the hands of different workers. Woronow (1929) recorded destruction of leucocytes but no basophil granulation due to benzol itself, this phenomenon only occurring with its homologues, a finding which is contradicted by Engelhardt (1931).

Most workers, however, are agreed that the "shift to the left", indicating an immature polymorphonuclear response of the bone-marrow, is a characteristic feature in the blood of animals at some stage (Hunt and Weiskotten, 1930; Wallbach, 1931; Engelhardt, 1931; Reznikoff and Fullarton, 1933, etc.). The findings of the latter, in a cat injected with benzol, are given in Table 5 as typical of this response when it does occur.

TABLE 5

Effect on the blood of injection of benzol into cats

Day of expt.	Benzol subcut. (ml.)	Leucocytes (thousands per c. mm.)	Immature polymorphs. (per cent)	Mature polymorphs. (per cent)	Lymphocytes (per cent)	Monocytes (per cent)	Eosinophils (per cent)	Basophils (per cent)
1	2	39.0	30	21	45	2	2	0
2	2	44.4	73	18	6	3	0	0
3	2	20.2	60	18	12	8	2	0
4	2	25.2	50	13	16	7	4	0
5	cat dead							

Changes designated by Gloor (1929) as "nuclear pyknosis"—disappearance of normal segmentation, concentration of the chromatin, clumping of the individual parts of the nucleus, have been recorded in rabbits by Beyer (1933). He states that it is uncertain whether these changes are identical with the toxic granulation described by Naegeli (1931), but that they indicate a toxic affection of the bone-marrow elements.

Red cells

In inhalation experiments Engelhardt (1931) found the red cell count variable, sometimes remaining constant, at other times diminished or increased. In Batchelor's (1927) injection experiments the reduction in the red cell count (which ranged from 0 to 52 per cent) occurred considerably later than that of the white cells; the red cells usually began to fall off and reached a minimum several days after the discontinuance of the injections, and remained low for some time afterwards. Batchelor, and Lévy (1935), attribute the final anaemia partly to destruction of peripheral red blood cells, and partly to aplasia of the bone-marrow, and state that when the haematopoietic system is severely affected there may be a complete absence of immature forms. Silberberg (1928) and Orzechowski (1929) consider the variation in numbers and formation of red cells of much less importance than some of the white cell changes.

An initial increase, or hyperglobulia, in some of the experimental animals, was noted by the earlier workers, as well as by Batchelor and by Orzechowski (1929) and others. This has been attributed to a preliminary stimulation of the bone-marrow. A secondary increase, 4 or more weeks after the cessation of injections, was also observed by Batchelor. Anisocytosis and polychromatophilia were found in some cases, but normoblasts very rarely. Engelhardt (1931) observed normoblasts in only two cases. The appearance of reticulocytes, however, was a feature of the findings of Ferguson, Harvey and Hamilton (1933). They seem to have made their appearance irregularly, in "showers", and these workers suggest that they may indicate the state of bone-marrow or haematopoietic activity in this form of anaemia as in many others. The same phenomenon observed by Paul, Friedlander and McCord (1927) is attributed by them to an irritant action of benzol on the haematopoietic tissues. Robinson and Climenko (1941) observed with inhalations of 1,000 p.p.m. for 2 hours an immediate fall in the number of red cells, followed by a sharp, brief rise and a return to normal in about 21 days. They also recorded a reticulocyte rise, up to 15 per cent, but do not attribute this to marrow hyperplasia.

Haemoglobin level and colour index. A reduction in the haemoglobin content, corresponding more or less to the reduction of red cells, appears to occur in animals. Fontana (1921) noted reduction of haemoglobin to the extent of about one tenth, with a reduction of red cells to 3,000,000, as also did Secchi (1914) and Mauro (1925). Engelhardt (1931), with subcutaneous injections, found a slow decrease of erythrocytes, accompanied in most cases by no essential decrease in haemoglobin, and in one case by an increase, so that the colour index remained fairly high. In inhalation experiments the two factors varied more or less simultaneously.

Resistance to haemolysis. In injection experiments Engelhardt (1931) found the resistance somewhat increased, but with inhalation, the findings were not characteristic.

Sedimentation rate. A slight increase in the sedimentation rate, but not until late in the course of the toxic action, was observed by Engelhardt.

Blood platelets

A marked fall in the number of platelets, preceded in some cases by a rise, has been noted by some observers, but it does not appear to be so characteristic in animals as in human beings. Duke (1913) found that large doses (5 ml.) of benzol acted first as a stimulant, then as a poison, to the platelet-forming organs; the count first rose to 1,780,000 then fell rapidly to about 61,000. In small doses (2 ml.) it acted only as a stimulant, causing a gradual rise in platelet count, which later fell, but not below the normal. Orzechowski (1929) found no change in the thrombocyte count, but Hultgren (1926) and Hurwitz and Drinker (1915) confirmed the findings of Duke. The reduction of platelets observed by the latter workers approximated to that recorded by Duke, though wide variations occurred, the average number following benzol injections in rabbits (average 2 ml. per kg. body weight) being 233,800 (normal average 683,400). They were unable to reduce the count below 31,000 and the animals showed no haemorrhagic symptoms such as occur after a more extreme lowering of the count. Where the platelets and leucocytes were much diminished in number the marrow usually showed well-marked aplasia, but destructive changes were rarely noted without accompanying signs of regeneration.

Hurwitz and Drinker (1915) were impressed by the fact that the blood platelets may remain at a high level at a time when the white cells have almost disappeared from the circulation. They suggest that this fact may be explained either by a very rapid regeneration of the megakaryocytes of the bone-marrow, forerunners of the blood platelets, or a higher resistance of these elements to the toxicity of benzol than of the forerunners of the polymorphonuclear leucocytes and erythrocytes. Schillowa (1933) also found that the number of blood platelets corresponded with the number of megakaryocytes in the bone-marrow. Apitz and Hühn (1942) have observed thrombopenia in rats poisoned by benzol.

Factors of coagulation

Although it has been observed by Duke (1913) that the blood of animals poisoned by benzol may clot at the normal rate even when the platelets are reduced to 10 per cent of the normal, it has also been shown that when the thrombocytes are at a very much reduced level the blood tends to show delayed coagulation, and the findings of the early observers, especially Santesson (1897), of symptoms of purpura haemorrhagica in clinical benzol intoxication gave special point to investigation of this aspect of benzol poisoning in animals.

The coagulation factors, prothrombin, antithrombin and fibrinogen, were investigated by Hurwitz and Drinker (1915). It was found that injections of benzol produced a diminution of circulating prothrombin, but that the antithrombin and fibrinogen were little changed from the normal. The reduction in prothrombin is regarded as being due partially to the reduction of blood platelets, which contain prothrombin, but only partially. There was no parallelism between the extent of bone-marrow injury, the number of blood platelets, and the relative amount of prothrombin in the blood; therefore it appears that either some other tissue or organ in addition to the bone-marrow is concerned in prothrombin formation or that a minimum amount of myeloid tissue suffices to keep the quantity of prothrombin above a dangerous level.

Immunity Reactions

Many workers, including Rusk (1914), Schiff (1914), Simonds and Jones (1915), Hektoen (1916a), Wallbach (1933), Rich and McKee (1934) and Schnitzer and Goddard (1943), have described the inhibiting influence exerted by benzol on the natural immunity reactions of the body.

Breakdown of Resistance to Infection

Lowered resistance to pneumonia in rabbits has been observed by Winternitz and Hirschfelder (1913) and Kline and Winternitz (1913), to tuberculosis in rabbits by White and Gammon (1914), to the common bacteria of inflammation, to the chemical irritants by Camp and Baumgartner (1915) and to streptococcal infection by Schnitzer and Goddard (1943).

Most of the early investigators attributed this breakdown of natural resistance either to the reduction of phagocytic activity associated with the leucopenia produced by benzol poisoning, or to the inhibition of antibody formation. To these factors must now be added another—a profound modification of the inflammatory response in animals poisoned with benzol, postulated by Schnitzer and Goddard (1943).

Phagocytic activity. Hektoen (1916a) and Wallbach (1933) found that in rabbits with benzene leucopenia the cytophagic index of the leucocytes for staphylococcus under the opsonic influence of normal rabbit serum fell to levels as low as 0·5 to 0·3, with leucocyte counts of about 1,800 per c.mm.

Antibody formation. Rusk (1914) recorded that the injection of benzol, either at the same time as or before the antigen, greatly reduced the formation of lysin for sheep corpuscles and of precipitin for horse serum. Simonds and Jones (1915) also observed a reduction of lysin for dog corpuscles, and of agglutinin and opsonin for typhoid bacilli. These findings were confirmed by Hektoen (1916a), but he, nevertheless, found that the toxic action of benzene on leucopoiesis did not interfere with a fully developed immunity.

Modification of inflammatory response. Schnitzer and Goddard (1943) observed a profound modification in the inflammatory response of tissues of benzol-poisoned animals. The normal conditions of influx and outflow in the infected areas were reversed, so that the leucocytes, even if available, could not reach the site of infection. This alteration in the inflammatory mechanism occurs only in animals which have not been immunized. In both actively and passively immunized animals, local antibody formation appears to be sufficiently unimpeded to prevent the fatal outcome which occurs in non-immunized animals.

Effect on Metabolism

A deleterious effect of benzol on the general metabolism of animals seems to be indicated by the experiments of Underhill and Harris (1923). They found a sharp rise in the urinary elimination of both creatine and total nitrogen following subcutaneous injection of benzol into rabbits. Assuming that creatine is an index of endogenous metabolism (Mendel and Rose, 1911), it would appear that benzol acts not only on the blood elements, but exerts a catabolic influence on body tissues as a whole.

Alkali reserve

According to Cavagliano (1932) no change takes place in the alkali reserve, estimated by the method of van Slyke, of animals in which the characteristic blood changes have been produced by chronic inhalation.

Vitamin C metabolism

It appears from the investigations of Meyer (1937) that benzol intoxication in animals is associated with an increased consumption of vitamin C, though the evidence for this hypothesis was obtained only indirectly from the fact that the administration of large doses of vitamin C had a favourable influence on the symptoms induced by subcutaneous injection of benzene. The relation between benzene poisoning and the favourable effect of vitamin C administration is suggested by the researches of Frada (1937), Fischbach and Terbrüggen (1938), and others in the possible detoxicating effect of vitamin C in its stimulating action on liver function (see also p. 34).

Phosphatase activity

According to Ambrosio (1942), there is a perceptible diminution of phosphatase activity of the liver and kidneys in benzol poisoning in animals. He suggests that this may possibly be related to lesions of the liver and kidneys, and to endocrine and metabolic disturbances caused by benzol poisoning.

Effect of Mixtures of Benzene and other Solvents

Some experiments carried out by Svirbely, Dunn and von Oettingen (1944) appear to indicate that the chronic effect of benzene when mixed with its higher homologues, xylene and toluene, is more injurious to animals than that of benzene alone. Rats, dogs and monkeys were exposed for 7 hours daily for 5 days a week for 28 weeks to concentrations of 1,000 p.p.m. of mixtures containing varied concentrations of benzene, xylene, toluene, and other hydrocarbons. No significant differences were found in the blood picture following exposure either to the benzene or to the mixtures, but the pathological changes in rats, deposition of haemosiderin in the kidneys and toxic appearances in the spleen, were more severe with the mixed solvents.

TOXIC EFFECTS IN MAN
Absorption and Excretion

Absorption

Absorption of benzene takes place chiefly by inhalation. Whether there is any absorption through the skin in human beings as in animals has been questioned, though Litzner (1932) and others believed it to be possible. An experiment was carried out in 1947 by the Research Association of British Rubber Manufacturers, in which benzene liquid and vapour were applied to the thoracic skin of subjects prevented from inhaling the vapour. No detectable benzene was present in the expired air and the ratio of inorganic to total urinary sulphate was not changed. This confirms the observation of Cesaro (1946), that application of benzene to the skin was not followed by the disturbance of the ordinary organic/inorganic sulphate ratio. In any case, absorption through the skin is less important than through the respiratory tract. According to Lehmann and Flury (1943), when benzene is inhaled by human beings 80 to 85 per cent is absorbed within the first half-hour of exposure.

Excretion

Excretion of benzene takes place chiefly through the lungs (according to Feil (1933), nine tenths through the lungs, and scarcely one tenth in the urine). Benzene is not excreted as such in the urine, but is previously oxidized in the body to phenol, catechol, or hydroquinone, and is excreted in combination with sulphuric or glycuronic acid (Nencki and Giacosa, 1880; Schmiedeberg, 1881; Jost, 1932). By oxidation of the benzene ring, a further portion is converted into muconic acid (Jaffe, 1909), but according to Jost (1932), this substance is not of great importance in chronic benzol poisoning since it is only formed by rapid oxidation, and then only to a very small extent (Forssman and Frykholm, 1947). Jost has examined these constituents in the urine of workers in the printing industry, exposed to benzol, xylol and toluol, and in the case of benzol, he found a distinct increase in sulphuric and glycuronic acids, as well as in phenol. His actual figures, and the normal figures given by Tollens (1909) respectively, are given in Table 6.

These results may be correlated with the findings of Yant, Schrenk, Sayers, Horvath and Reinhard (1936), and which have already been described in detail (p. 12). These investigators state that normally the total sulphates in the urine consist of 85 to 95 per cent of inorganic sulphates, and 5 to 15 per cent conjugated. After exposure to benzol it is believed that phenolic products of oxidation or metabolism of the benzol become available, and that the conjugation reaction is increased with a shift of the system inorganic sulphates⇌conjugated sulphates to the right.

TABLE 6

End-products in the urine after exposure to benzene

	Normal daily average (g.)	Benzol poisoning daily average (g.)
Sulphuric acid	0·18	0·407
Glycuronic acid	0·62	1·08
Phenol:		
(1) Volatile phenol (phenol and cresol)	0·03–0·07	0·17
(2) Non-volatile (dihydroxybenzene)	0·015–0·05	0·07

It has been suggested by Yant and co-workers (1936), and by Kammer, Isenberg and Berg (1935) that the nature and reliability of the change in the sulphate ratio in the urine in animals is such that it could be taken as a measure in the exposure in workers before injury has occurred, and that estimation of the ratio would form a satisfactory method of medical supervision in a possible benzene hazard.

Jephcott and Bulmer (1939) point out, however, that while the urinary sulphate test has some value in estimating the relative degree of exposure to benzene, it does not indicate the effect of exposure on the worker, and there is no invariable correlation between the haematological and urinary findings. This last statement was confirmed by Hamilton-Paterson and Browning (1944) during an investigation of women using rubber solutions containing benzene. Of seventeen women showing neutropenia, which these authors consider the earliest and most reliable sign of benzene absorption, the partition of organic and inorganic sulphates was in all cases within normal limits.

Whether excretion of benzene takes place through the faeces is not certain, but an interesting point arising during the above-mentioned investigation by Hamilton-Paterson and Browning was the complaint by a number of the women examined that their faeces smelt strongly of benzene. Peacock (personal communication), on the basis of his work with fat-soluble substances, observes that these are to some extent excreted in the bile, and that some of the benzene may similarly find its way into the intestine and so be excreted with the faeces.

Urinary sulphur analysis and laevorotation as a measure of benzene exposure. Owing to the discrepant values found by different authorities for the relative percentage of organic and inorganic sulphates in the urine, other criteria have been suggested by Gueffroy and Luce (1937). These percentages are obtained from estimations of the neutral sulphur in the urine and the increase of urinary laevorotation.

An increase in the neutral sulphur is associated with the increased excretion of mercapturic acid characteristic of the metabolic disturbance caused by benzol poisoning; an increase of urinary laevorotation is said to indicate excessive excretion of phenylglycuronic acid.

Gueffroy and Luce found the urinary neutral sulphur to be up to 30 per cent in some workers exposed to benzol, as compared with the highest normal value of 12 per cent, and these high values were in some cases associated with a lowered leucocyte total.

The polarization test showed no close agreement with the blood findings but is regarded as a sensitive indicator of the disturbed metabolism due to benzol absorption. A combination of both tests is suggested as being more reliable than the organic-inorganic urinary sulphate ratio.

Blood phenol as a criterion of benzene absorption. Since estimation of the urinary sulphates does not take into account that part of the phenol which may be combined with glycuronic acid during the detoxication by the liver of the toxic metabolites of benzol, Seghini (1941) suggests that estimation of the phenol compounds in the blood may be considered a more accurate index of benzol absorption. His method is based on the original technique of Theis and Benedict (1924) using a photometer with a scale of values corresponding to known quantities of phenol and giving the dosage in mg. per cent. The values given for twenty controls not exposed to benzol, and for thirty operatives employed in waterproof manufacture, are shown in Table 7.

TABLE 7
Concentration of phenol in the blood of persons exposed to benzene and a control group

	Controls (per cent)	Exposed persons (per cent)
Free phenol	0·86–1·24	0·96–1·76
Conjugated phenol	0·02–0·18	0·36–1·12
Total phenol	0·94–1·32	1·86–2·62

An increase in polyphenols and in urochrome A in the urine of workers exposed to benzene has also been noted by Forssman and Frykholm (1947), but these investigators consider that until there is more known about the factors which influence excretion of these substances, their analysis cannot be relied upon as a criterion of benzene exposure.

Acute Poisoning

Lethal and Toxic Concentrations

Flury (1928) gives the following figures:

20,000 p.p.m. (64 mg./l.) may be fatal if inhaled for 5 to 10 minutes;
7,500 p.p.m. (24 mg./l.) may be fatal if inhaled for 30 minutes to 1 hour;
3,000 p.p.m. (10 mg./l.) may be tolerated if inhaled for 30 minutes to 1 hour.

Predisposing Factors

The histories of the fatal cases, as described in the reports of the Benzol Poisoning Committee of the National Safety Council (1922), of McCord, Cox and O'Boyle (1932) and of Hamilton (1925), show great variation in individual susceptibility. Sometimes a man exposed for a short period died, while another, exposed for a longer period and more intensely, survived.

Physical exertion associated with toxic conditions often ended fatally. In many instances members of a rescue party were fatally poisoned while the original victims, after a period of unconsciousness, survived. Such were the cases reported by Lewin (1907) and by Hamilton (1925), and they led to the belief that muscular exertion tends to increase the susceptibility to poisoning and to decrease correspondingly the prospects for recovery in acute cases.

Emotional reaction has also been shown to be a factor in increasing the severity of the intoxication. In a case quoted by Feil (1933), the victim rushed suddenly out of the space where he had been working, cried out that he was burning and dropped dead, while a very interesting account of a "*Massenvergiftung*" by benzol in a Russian factory was described by Dworetsky (1914) where the medical authorities considered that of the 230 persons involved, the greater number were affected by hysteria only, and that a few suffered from benzol poisoning together with hysteria.

The interesting results of the experiments carried out by Nahum and Hoff (1934) on the mechanism of sudden death in animals intoxicated with benzol appear to shed much light on this question of individual susceptibility in human beings. They point out that in muscular exercise there is a reflex cardio-acceleration with a liberation of sympathin, while in excitement there is also a well-marked liberation of adrenaline into the blood stream. Their experiments on animals showed that benzol effects the liberation of adrenaline and sensitizes the myocardium to its action. The adrenaline and sympathin contribute to the production of ventricular extra-systoles, and the amount liberated determines the severity of the attack. Thus the danger of death from ventricular fibrillation is increased by either muscular exertion or emotional excitement or both, and Nahum and Hoff consider that many of the individual variations in susceptibility reported in the literature may be due largely to variations in the adrenaline response.

Symptoms

The gravity of the symptoms of acute benzol poisoning is generally in accord with the amount absorbed and the length of exposure. In the cases reported to the Factory Department since 1935, the fatal cases had been exposed, for periods varying from 45 minutes to 3 hours, to concentrations which must have been very high, since practically all of them took place in closed spaces containing the vapour evaporating from residues of benzol. Feil (1933) describes several forms of acute benzol poisoning.

(*a*) *Mild form*. A state of euphoria is followed, if the subject is not removed from exposure, by giddiness, headache, nausea, vomiting, staggering gait, sensation of tightness in the chest, and inability to escape from the site of the poisoning. Of the twenty non-fatal cases reported to the Home Office between 1921 and 1939, twelve might be regarded as in this class, eight of the patients were rendered partially unconscious, and four complained only of dizziness, headache or nausea.

(*b*) *Severe form*. After inhalation of large amounts of the vapour, the principal symptoms, convulsive movements, paralysis, unconsciousness with dilated and non-reacting pupils, are related to the central nervous system, and in the most severe cases are followed by death. Of the twenty-eight cases that from 1921 to 1939 were reported to the Home Office, eight were fatal, and eight other patients were rendered completely unconscious. According to Fischer (1932) this form of intoxication is not unlike that produced by other fat-soluble narcotics; he notes especially that there is no blue-grey colour of the face as in poisoning by nitrobenzene, though Hamilton observed cyanosis of mucous membranes and finger tips in some cases.

(*c*) *Atypical forms*. Intoxication is preceded by either coma or a state of violent excitement and delirium. Such is the case described by Feil (1933), and also one reported by Harrington (1917) of a man who is described as having acted "as if he had gone crazy", becoming wildly excited and later unconscious.

After-effects in Non-fatal Cases

In the majority of patients who recover from acute intoxication there are no immediate after-effects, except temporary symptoms such as pain in the head and chest, shortness of breath, giddiness, loss of appetite, sometimes nausea, and even vomiting; late manifestations and sequelae have, however, been recorded. The most severe case is probably that recorded by the Chief Inspector of Factories and Workshops in 1918 of an ex-patient who, 2 nights after his return to work, relapsed into unconsciousness from which he never recovered. The chief sequelae of acute poisoning recorded are:

(a) Dizziness and uncertain gait lasting about 12 days, as in the cases reported by Genhard (1910) and Wyss (1910).

(b) Respiratory catarrh and pleurisy (Genhard, 1910; Schaefer, 1909; Kobert, 1906).

(c) Nervous disorders, depression, insomnia, bad dreams (Wyss, 1910), nervous exhaustion (Lewin, 1907).

(d) Skin changes; a residual yellow pallor was observed in one of Lewin's cases (1907); Genhard (1910) reported an exanthem over the back.

(e) Cardiac distress; spells of cardiac distress lasting 4 weeks after recovery were reported by Cronin (1924) and by Lewin (1907), who also observed a blowing murmur.

Post-mortem Findings

The earliest record of an autopsy on a case of acute benzol poisoning is that of Sury-Bienz (1888), in which the chief findings were bright red spots on the body and a dark red, fluid condition of the blood, which remained fluid for a long time after death.

The chief results of post-mortem examinations since that time, including the very complete and detailed examination of a case made by Koppenhöfer (1935), may be summarized as follows:

(1) *Petechial haemorrhages.* The findings of Sury-Bienz in this respect were confirmed by Beinhauer (1896). Minute haemorrhages in the pleural, gastric and intestinal mucosa, and in the pancreas, were also noted by these two workers, by Buchmann (1911) in the brain and pericardium, in the lungs by Heffter (1915), Binder (1921) and Ziel (1925), in the subcutaneous fatty tissue and serous membranes by Floret (1926), in the brain, pleura, gastro-intestinal mucous membrane, kidney, pelvis, ureter and bladder by Koppenhöfer.

(2) *Fluidity of the blood.* The observations of a dark or cherry red fluid condition of the blood by the earlier workers was confirmed by Binder (1921), Heffter (1915) and Floret (1926). Heffter remarked that in spite of the blood remaining fluid for a long time after death, there was no evidence of haemolysis. The result of a test for methaemoglobin in the blood made in a case reported by Martland (quoted by Hamilton, 1931) was negative, but Carter (1928) stated that the spectroscopic appearances of the fluid blood in a fatal case were those of oxyhaemoglobin and that reduction was easily effected by ammonium sulphide. Koppenhöfer (1935) also demonstrated the presence of oxy- but not of methaemoglobin.

(3) *Congestion or cyanosis of the internal organs.* Floret (1926) describes the phenomenon as "all the organs being full of blood", as also does Ziel (1925). Cyanosis of the liver, spleen and kidneys in one case, and of the brain in another, was observed by Martland; in another case, cited by Hamilton (1931), there was fluid blood in the right side of the chest with marked distension. Koppenhöfer (1935) observed hyperaemia of all the internal organs and an albuminous fluid exudate in the liver and lungs.

(4) *Changes in the respiratory organs.* Small areas of interstitial emphysema in the lungs, reddened and irritated bronchi (Martland), oedema of the lungs (Binder, 1921; Koppenhöfer, 1935), and "bloody mucus" in the air-passages (Heffter, 1915; Sury-Bienz, 1888) are among the findings reported, while in a case reported to the Home Office in 1925, the lungs showed

recent adhesions and cyanosis. In this case, traces of pyridine (0·5 per cent) were present in the vapour to which the man, a worker in a dye factory, was exposed.

(5) *Benzene in the internal organs.* There has been some difference of opinion as to whether benzene can be detected, either by the odour, or by chemical analysis, in the organs after death. Martland noted a distinct odour on section of the lungs, while in the Home Office case already mentioned, the trachea, larynx and liver smelt of benzene. In another case examined by Martland (quoted by Hamilton, 1931), there was no distinctive odour in the lung cavity, or in the brain, although the autopsy was performed while the body was still warm. In Koppenhöfer's case (1935), the post-mortem examination was made 24 hours after death, and he states that a strong typical aromatic odour issued not only from the muscles as they were cut, and from the body cavity, but even more decidedly from the brain and spinal cord.

With regard to the actual estimation of benzene in the tissues, and in spite of the statements of Heffter (1915), Beinhauer (1896), and Buchmann (1911), that benzene could not be detected chemically in the bodies of subjects dying from acute poisoning, Koppenhöfer has been able to estimate its presence, not only qualitatively but quantitatively. He used a modification of the method described by Peronnet (1934), obtaining needle-shaped crystals of dinitrobenzene, and producing the intense violet coloration by adding a 1 per cent solution of laevulose to an alkaline alcoholic solution of these. In contrast to the opinion of Heffter (1915) and of Joachimoglu (1915), that the greatest amount of benzene should be present in the brain, and that this should denote a special affinity of benzene for the lipoid tissue of the brain and spinal cord, Koppenhöfer found the highest percentage of benzene to be in the blood (Table 8). He calculated that the whole blood of the body of this man weighing 72 kg. would contain approximately 1 ml. of benzene. He believes that in the most severe cases of benzene poisoning, at any rate, the essential lesion is a colloid-chemical disturbance of the blood rather than a disturbance of the lipoidal tissue of the central nervous system.

TABLE 8

Percentage of benzene in different organs (Koppenhöfer, 1935)

Organ	Benzene (mg./100 g.)
Blood	14·0
Spinal cord	12·6
R. kidney	10·3
Spleen	9·4
Brain	7·5
R. lung	5·5
Liver	5·3

(6) *Changes in the urine.* No benzene was detected in the urine taken from the bladder after death by Heffter or by Martland, but both, as well as Beisele (1912) and Simonin (1903), found an increased amount of "phenol bodies" (phenylsulphuric acid in Heffter's case). Heffter states that in cases which are rapidly fatal, phenol bodies are not usually found, since the change to conjugated acids is fairly slow. Urobilin, diminished urea and diminished chlorides were observed by Simonin (in a case of poisoning by accidental ingestion of benzol). No blood, albumin or haematoporphyrin were detected by Beisele.

Treatment

The usual measures are artificial respiration, administration of oxygen, etc., but intravenous injections of lecithin have been claimed by Nick (1922) to have saved life in one case.

Chronic Poisoning

While animal experiments have undoubtedly elucidated some of the problems of benzol poisoning in human beings, their results cannot be said to be applicable in all respects to human intoxication. The variation of individual and species susceptibility, and the liability to infection of animals injected with benzol are only two of the factors which make the complete picture of benzol intoxication unlike that found in human beings. Among the facts which have emerged from the more recent researches on the subject, perhaps the most outstanding is the variability in the clinical syndrome and blood picture hitherto regarded as typical. Undoubtedly the incidence of symptoms is a less reliable indication of poisoning than the blood picture, since there is often a striking variation between the two. Hamilton (1925) quotes a case where, with the typical symptoms of bleeding from the gums and nose, retinal haemorrhage and purpura haemorrhagica, the only abnormal feature of the blood picture was the absence of blood platelets. Much more frequent, in Browning's experience, is the comparative absence of symptoms when the blood picture reveals a very definite evidence of benzol poisoning. The blood picture of a woman, whose case is described on p. 39 as one of delayed benzol poisoning 2 years before the fatal outcome, revealed severe leucopenia and anaemia, but at the time she did not complain of any disturbance of health. In rare cases, such as that described by Friemann (1936), deterioration may even be so rapid that serious, or fatal, illness may set in very shortly, even after a blood examination has revealed no abnormality.

Related Factors

Sex, pregnancy and age. It has been rather generally believed that young subjects, particularly women, are specially liable to chronic benzol intoxication (Feil, 1933; Pulford, 1931; Meda, 1922; and others). Whether women are actually more susceptible than men appears to be uncertain, since groups of opposite sexes working under exactly similar conditions do not seem to have been investigated. In Dimmel's investigation (1933) both men and women were affected, and though the conditions of work in this rubber factory were such that it was impossible to say definitely that both sexes were equally exposed, Dimmel found no grounds for believing that women showed more susceptibility than men. According to Danysz (1942), women invariably show a slight deterioration of the blood picture after 8 to 12 weeks' exposure, especially with regard to a fall in the total red cells, which men often escape.

That women may suffer more from subjective symptoms than men having an equal degree of leucopenia is suggested by records secured by Smith (1928) in a camera-manufacturing factory, where among a group of thirty-five women only six made no complaint of symptoms such as headache, dizziness, nervousness, weakness and great fatigue, and where eleven positive cases and one suspected case of benzol poisoning were found. Of fifteen men exposed under the same conditions, whose white cell counts were below 5,625, only one had symptoms associated with the work.

An increased susceptibility in pregnant women has been definitely observed (Meda, 1922; Hamilton, 1925; Smith, 1928). The latter author quotes a case in which no symptoms appeared until the woman became pregnant; she then suffered from severe nausea, vomiting, bleeding from the nose, gums and rectum, and into the skin. After the birth of a premature child she had a severe uterine haemorrhage and died. According to Feil (1933) pregnancy should constitute an absolute contra-indication to the use of benzol.

Youth, especially in women, was stated in the Report of the National Safety Council (1926) to be a predisposing factor, but this was not found to be the case by the New York State Department of Labor, 1927. The percentage of positive cases was lowest in the youngest group and highest in the oldest. Smith remarks that the opinion of earlier workers in this respect may have been due to the extreme youth of the women whom they studied. In the factory from which

Selling's cases came for instance, all the fourteen girls exposed were between 14 and 16 years of age, whereas in the factories studied in the Department of Labor Report the average age was 28.

Length of exposure. The findings of the Department of Labor seem to indicate that where susceptibility to poisoning exists it tends to develop during the first year, the percentage of positive cases being lowest in the group with an exposure of 3 months or less, but showing little difference in groups with an exposure of 3 to 12 months, 1 to 4 years, and 4 years and over, respectively. According to Lévy (1935) a latent period may extend from 3 weeks to 6 months or even up to 3, 6 or 12 years. Among the cases reported to the Home Office since 1926, the time of exposure seems to have had little relation to the severity of the intoxication. Three cases exposed for $1\frac{1}{2}$, 2 and $4\frac{1}{2}$ years, respectively, were comparatively little affected, while two cases of 4 months and 31 years exposure, respectively, both proved fatal. Feil (1933) emphasizes the fact that long-continued exposure may result in a progressive hypersensitivity, while Danysz (1942) states that the critical period of evidence of deterioration of the blood picture arises 2 to 3 years after the first exposure.

Intensity of exposure. The actual amounts of benzol in the air of factories in which cases of benzol poisoning have been reported have already been discussed. In the fatal cases reported by Legge in 1918 the average amount was 550 p.p.m. It is interesting to observe that in the factory in which one of the deaths occurred, that of a man employed for 6 months in coating the metal rims of tyres with a benzol solution of rubber, the ventilation had been diminished during the last 2 months by the carrying out of structural alterations. The Report of the National Safety Council (1927) states that benzol poisoning occurs with greater frequency in cold weather when natural ventilation is usually reduced to a minimum by closed windows and doors so that the concentration in the atmosphere reaches a maximum. Atmospheric conditions of temperature and humidity also play an important part. At times of heat and high humidity, other things being equal, spontaneous and sporadic outbreaks are most apt to appear. In an investigation of photogravure operatives in Milan, Saita and Dompé (1947) considered that the abnormalities of the blood picture found in thirty-seven of the fifty-five workers examined were more or less closely correlated with the period of exposure, and also with the kind of work performed, the atmospheric concentrations varying in different parts of the room from 188 to 908 p.p.m.

According to Lévy (1935) and Merklen and Israël (1934), the intensity of exposure determines to some degree the type of blood picture. The latter states that the "classical aleukaemia" is only produced with long exposure and lack of individual resistance, while short and moderate exposure produces the *formes frustes.*

Individual predisposition. According to Lévy this is the most important factor in determining whether a given time and intensity of exposure will produce symptoms of poisoning among a group of individuals. Weil (1935) has gone so far as to call it the "sol hématique"—an innate constitution of the haemopoietic tissues predisposing to their injury.

An apparent family susceptibility has been noted by Reifschneider (1922). He found three cases in one family; two of these proved fatal, the third patient with secondary anaemia was removed before other symptoms developed.

Conditions of ill-health. General lowering of vitality, respiratory diseases, especially tuberculosis, alcoholism by virtue of its effect on the liver and brain, heart disease, nervous disorders, nephritis by reducing the elimination of toxins, and obesity, are among the predisposing factors listed by the National Safety Council and by Feil (1933).

Vitamin C deficiency. On the basis of animal experiments already described on p. 25, and the observations of Friemann (1936) and Meyer (1937) that workers exposed to benzene show a decreased urinary excretion and an increased consumption of vitamin C, it has been suggested that toxic avitaminosis may be a factor in benzene poisoning. Many French observers believe that lack of vitamin C is an important predisposing factor in benzene poisoning. Roubinet (1939), for example, states that vitamin C deficiency appears to damage the liver and suprarenal glands; he urges, as a prophylactic measure, that vitamin C saturation should be maintained in all workers exposed to benzene. Seyfried (1942) recommends vitamin C as a therapeutic measure, and states that there is a "slight haemorrhagic diathesis" among benzene workers which varies with the seasonal vitamin C intake, and that the therapeutic effect of vitamin C in their condition is rapid and complete.

Blood group. According to Chevallier and Désoille (1947), statistics on the incidence of benzol poisoning in persons with various blood groupings suggest that individuals of Group O are the most susceptible and should therefore not be exposed.

Symptoms

Most common form

The development of a syndrome pointing to damage of the blood-forming organs is preceded by slight preliminary symptoms—headache, giddiness, drowsiness, lassitude, loss of appetite, nausea, even vomiting. Sometimes more definite symptoms appear at this stage, pallor, anaemic condition, metrorrhagia or menorrhagia in women, ecchymoses. Lévy (1935) states that at this stage an examination of the blood would show leucopenia and neutropenia. Generally it is the occurrence of haemorrhages which reveal the condition as one of benzol poisoning—epistaxis, bleeding from the gums, persistent and heavy metrorrhagia, subconjunctival and retinal haemorrhages, ecchymoses and purpuric eruption. Haemoptysis, or even haematemesis, may occur; in a case reported to the Home Office in 1929 other symptoms (lassitude and oppression in the chest) were comparatively slight, and the changes in the blood picture were not of great severity (leucocyte count 4,800, with neutrophils 50 per cent and lymphocytes 39 per cent), but were preceded by haematemesis. Ulcerations of the buccal and tonsillar mucosa, scorbutic in appearance, are not uncommon, though according to Pulford (1931) they never progress to the gangrenous and sloughing condition found sometimes in agranulocytosis. The patient appears characteristically pale, dyspnoeic and anxious, has a rapid pulse and often a raised temperature.

Neurological symptoms. Smith (1928) considers the symptoms related to the nervous system found in a number of women examined (Table 9) as evidence of the neurotoxic action of benzol. Similar evidence was provided in a case, also a woman, reported by Fauré-Beaulieu and Lévy-Bruhl (1922). The nervous symptoms took the form of increased tendon reflexes, bilateral clonus, positive Babinski reflex, impairment of deep sensitivity, pseudo-tabetic lesions with paraesthesia, ataxia and paraplegia and motor impairment, signs indicating lesions of the posterior columns and pyramidal tracts. Such lesions, according to the National Safety Council Report, were probably the combined effects of the prolonged and persistent anaemia, and of the possible direct action of the benzol on the central nervous structure, inasmuch as it has been found that benzol is excessively held by these tissues.

TABLE 9

Frequency of symptoms of benzene poisoning among women

Complaint	No.	Percentage
Headache	18	60·0
Excessive fatigue	14	46·7
Dizziness	13	43·3
Nausea	10	33·3
Anorexia	9	30·0
Weakness	8	26·7
Nervousness	6	20·0
Numbness and tingling	5	16·6
Frequent urination	4	13·3
Nose bleeding	4	13·3
Disturbed sleep	4	13·3
Indigestion	4	13·3
Frequent menstruation	3	10·0
Shortness of breath	2	6·7
Skin eruptions	2	6·7
Vomiting	2	6·7
Pain in abdomen	2	6·7

Neuritis of the median nerve and retrobulbar neuritis with dimness of vision and headache in a worker in spray lacquering have been observed by Landé and Kalinowsky (1928) and by Goldmann (1930).

Epilepsy has also been related directly to benzol poisoning by Korvin (1933). She quotes the case of a female worker exposed to benzol who developed typical epileptic attacks after some months. No other clinical symptoms of benzol poisoning, except lassitude and loss of appetite were present, but she states that the diagnosis was strongly supported by the presence of a typical blood picture, a leucopenia of 2,000 and a relative lymphocytosis of 50 per cent. She also mentions a rather similar case reported by Albrecht in 1932, which was at first diagnosed as a brain tumour, but in which the diagnosis of benzol poisoning was made more probable by the favourable progress after removal from exposure.

Localized myelitis in a benzol worker has also been observed by Saenger (1914). Spastic paresis, nystagmus and a positive Babinski reflex were present and recovery took place on removal from exposure.

Frequency of urination was mentioned by Starr (1922) as a common symptom among the milliners whom he examined, and was attributed by him to the possible excretion of phenols in the urine. This symptom was also observed by Smith (1928) as a transient phenomenon.

More severe form

Many cases prove fatal in spite of removal from exposure to benzol. The picture is that of a progressive severe anaemia, ushered in by purpuric manifestations; death occurs from a few days to one or two weeks after the appearance of the purpuric symptoms. This was the type of case which first drew attention to the existence of benzol poisoning in industry, as in the famous cases of Santesson (1897) and in those reported by Le Noir and Claude (1897), Selling (1916), Hogan and Shrader (1923), Legge (1919–20, the first cases reported in Great

Britain), Harrington (1917), Flandin and Roberti (1922), and many others. It was the typical syndrome in a series of cases reported by Heim de Balsac and Agasse Lafont (1933) in workers using an adhesive where benzol was the solvent. Forty operatives were affected, eight of whom died; thirty-six showed purpuric manifestations, eruptions, haemorrhage from the gums and nose, ecchymoses, and metrorrhagia. A feature of this outbreak was the delay in the onset of the purpuric symptoms, which appeared some days after the initial slight symptoms of headache, giddiness, digestive troubles, anorexia and asthenia.

Cause of the purpuric manifestations. Although variations in the condition of the blood itself, such as thrombopenia or anaemia, are regarded as having some influence on the tendency to haemorrhage in benzol intoxication, they are apparently not the whole cause (Litzner, 1932). Mitnik and Genkin (1931) have in fact observed no haemorrhage in the presence of severe thrombocytopenia, but conversely, haemorrhage in the presence of a normal blood platelet count; they also found no diminution in the fibrinogen and thrombin content of the blood. Injury to the capillary vessels has been put forward as an important factor in the cause of the haemorrhage by Santesson (1897), Landé and Kalinowsky (1928) and Mitnik and Genkin (1931). Santesson related this injury to fatty degeneration of the capillary endothelium, and Mitnik and Genkin postulate a preliminary functional disturbance, in the sense of a strong vasomotor spasm with long continued stasis. Litzner (1932) also states that the Rumpel-Leeds sign, (production of capillary haemorrhage by means of stasis) is an early symptom of benzol poisoning, and that this phenomenon, together with a lengthening of the time of bleeding in the presence of a normal coagulation time, constitutes further evidence of capillary injury.

Terminal infection. This is often the cause of death in cases of great severity; it is believed to be due to a decrease in the leucocytic defences, as shown in the case of animals. Gangrenous periostitis and osteomyelitis were observed in a fatal case by Löwy (1926), and severe haemorrhagic ulcerative gingivitis by Laignel-Lavastine, Lévy and Désoille (1928). This feature was also shown in two fatal cases (1934) reported to the Home Office. In the first subject, a leather sprayer, the anaemia was so severe as to arouse suspicions of pernicious anaemia, and at autopsy there was found a terminal sepsis of the tonsil and lower end of the oesophagus. In the second, a female worker in leather cloth and lacquers, death was due to aplastic anaemia with septicaemia, the latter arising from a septic endometritis.

Slighter forms

Cases of less severity with slight symptoms and no very great disturbance of the blood picture may recover under treatment and with removal from exposure. Such were the cases reported by Teleky and Weiner (1924) occurring in the manufacture of rubber goods. The symptoms were variable, headaches, nausea, eructation, vomiting, tendency to bleed from membranes, irritation of conjunctivae, menstrual irregularities and anaemia with reduced platelets and inversion of the leucocyte-lymphocyte formula. Substitution of benzine for benzol resulted in definite improvement both of the symptoms and of the blood picture.

Latent form

Falconer (1931) quotes a case where after 15 years' exposure the only demonstrable sign of benzol poisoning was a reduction of blood platelets to 100,000 per c.mm. The patient was removed from exposure and 2 years later developed an acute respiratory infection, followed by the full picture of benzol poisoning, which proved fatal. Falconer states that an infection such as influenza may precipitate frank benzol poisoning with leucopenia, anaemia, purpura, and haemorrhages several months after exposure. Cases of progress of the disease without further exposure have also been reported by Santesson, four of whose cases developed symptoms only after they had left the factory, and by Rohner, Baldridge and Hansmann (1926). In one of the cases reported by the latter authors, eye irritation was the only symptom during exposure, but haemorrhage and a fatal aplastic anaemia developed a month later.

Duration of Symptoms after Removal from Exposure

From the observations of Gounelle and Dumas (1935) it appears that benzol intoxication produces a lasting effect on the organism, manifested both by symptoms existing for long periods after removal from exposure and by the persistence of an abnormal blood picture. Examination of four women, 16 to 18 months after removal from exposure because of purpuric manifestation and typical blood changes (Dumas, 1934; Isräel, 1934), showed that they were still suffering from asthenia, fatigue, pallor, slight attacks of purpura, gingivitis with loss of teeth, and in two instances metrorrhagia and menorrhagia. Gounelle and Dumas (1935) remark that the metrorrhagia prolongs the injurious effect of the benzol and tends to produce a chronic hypochromic anaemia, which they have found resistant to iron therapy, except in so far as it produces a purely palliative result.

Effect on the Skin

Although a chronic toxic effect of benzol on the skin, of an eczematous nature, has been described by Landé and Kalinowsky (1928) and Oppenheim (1930), these are regarded by most workers when they occur as probably allergic in nature, or because of innate or acquired susceptibility, according to Engelhardt (1931). Landé and Oppenheim have described them as characterized by folliculitis, cornification of the seborrhoeic glands, hyperkeratosis and hyperpigmentation, and have regarded them as arising from the fat-solvent action of benzol on the superficial layers of the skin, leading to irritation and inflammation of the deeper layers. No such changes were observed by Dimmel (1933) in any of the sixty-six cases of benzol poisoning recorded by him.

Effect on the Blood Picture

While the most characteristic blood picture in benzol poisoning remains that described by the earlier writers, a leucopenia with neutropenia, thrombocytopenia and some anaemia, it becomes increasingly evident that this is by no means the invariable picture. Wide variations, such as hyperchromic anaemia of pernicious type, myeloid leukaemia with splenomegaly, and typical leukaemia have been described by Weil (1933; 1935), Deloré and Borgomano (1928), Loeper, Fabre and Borreau (1946), Lechelle, Coste, Thieffrey and Cuadrano (1940) and by others. Marked deviations in the red cell picture have also been observed; for example, Lévy (1935) states that anaemia is a much less constant, and much later feature, than disturbance of the white cell count, while Schwartz and Teleky (1941) state that "the maintenance of a normal count of red cells in the presence of other signs of damage in the blood is better explained by a stimulated compensating regeneration than by an uninjured erythropoiesis".

Side by side with the undoubted destructive effects of benzol on the bone-marrow, there evolves a regenerative activity, and the resulting picture in any individual case is a reflection of the balance of these two opposing processes. It has been pointed out by Schwartz and Teleky (1941) that "there is no greater stimulus for regeneration than the products of decaying cells of the same kind. . . . Either the stocks of new-formed cells are time and again destroyed, in which case the result will be a hypo- or aplastic bone marrow, or regeneration surpasses destruction and the result is a gradual transformation of aplastic into hyperplastic bone marrow." This process has been specially emphasized by Jackson, Parker and Lemon (1940), Mallory, Gall and Brickley (1939) and by Bowers (1947). Although the histological picture was not that of a true leukaemia, the bone-marrow in Bowers' fatal case was grossly hyperplastic and there were extensive areas of extra-medullary haemopoiesis. Similar appearances observed by Jackson have been stated by him to be characteristic of the condition which he calls "agnogenic myeloid metaplasia". If, for instance, the regenerative, or stimulative process is uppermost, a leucocytosis or hyperglobulia may be present, which, according to the accepted criteria of "white cells not less than 5,000", and "red cells decrease of 25 per cent", would mean that any worker showing these high values for white and red cells should not be considered in any danger from further exposure. It may be, however, that these high values are evidence of excessive regeneration under the continuous stimulus of benzene, which, as in many other cases of biological over-stimulation, may be followed later by exhaustion and atrophy. In such a case further exposure would constitute a serious danger. Schwartz and Teleky (1941) suggest that another possible reason for the increase during recent years of atypical cases of benzol poisoning may be the greatly increased use of mixed solvents instead of benzene alone. It is possible that the effect on the blood of two or more substances may produce a different clinical picture from that of one acting alone, even though other constituents of the benzene mixture may be less destructive to haemopoietic activity than benzene itself.

White cells

Total count. The typical effect of benzol poisoning is a reduction of white cells, sometimes to a very low level. The lowest recorded appears to be 104 per c.mm. (Hogan and Shrader, 1923). In one of Selling's cases the total leucocyte count was 480, in Harrington's 500, and in Laignel-Lavastine's 600, while values from 1,000 to 2,000 are very frequently reported. It is stated by some authors, e.g. Lévy (1935), that this reduction in white cells may go on to a complete agranulocytosis, indistinguishable from true agranulocytic anaemia except by the history of exposure to benzol. In this connexion the work of Kracke and Parker (1934) on the effect of drugs containing the benzene ring is of interest. They state that such drugs, especially amidopyrin, may produce true agranulocytic anaemia, also caused by amytal, barbital, etc. Merklen and Israël (1934) point out that in man, as in animals (see Reznikoff and Fullarton, 1933), while hypogranulocytosis may result from the destruction of leucocytes by benzol, a liberation of immature granulocytic forms, myeloblasts, myelocytes, may be simultaneously provoked.

So great was the leucopenia in the cases of benzol poisoning recorded by the earlier workers that in 1912 Koranyi suggested taking advantage of this leucopenic action of benzol by using it as a therapeutic agent in leukaemia. Favourable results from this treatment were reported by several workers, including **Kiralyfi**

(1913), Billings (1913), Barker and Gibbes (1913) and others, but its dangers were pointed out by Pappenheim (1913) and Klemperer and Hirschfeld (1913); greatly varying results were reported in the hands of other workers such as Myers and Jenkins (1913). In later years, according to the Report of the National Safety Council (1926) "benzol therapy has more or less fallen into disrepute, due to lack of uniformity of action and the more or less transient nature of the results obtained." In view of the leukaemic effect of benzol reported by Weil (1932) (see below), the variation in the results of its administration is not surprising. In an investigation carried out by Hamilton-Paterson and Browning (1944), neutropenia was shown to be the commonest and earliest sign of benzene absorption, the leucopenia and a low polymorphonuclear count being concurrent.

Atypical total counts. Leucocytosis has been reported by a few workers both as an initial phenomenon and as a consequence of intercurrent infection. Rabe and Hirschland (1920) observed an initial increase of leucocytes after administration of minute doses to men and women while Lowy (1926) recorded an increase up to 16,000 in slight, fairly acute cases of poisoning. The presence of suppuration may also produce an increase of polymorphonuclear leucocytes (Paul, Friedlander and McCord, 1927; Meda, 1922).

Leukaemia on the other hand has been reported by Deloré and Borgomano (1928) and Weil (1933), and Loeper, Fabre and Borreau (1946). Weil describes two forms: (*a*) acute—as in Deloré and Borgomano's case where a man, an employee in a chemical works, died within 3 weeks of the appearance of multiple enlarged glands, slightly enlarged spleen, anaemia with haemorrhages into the skin and mucous membranes, and gangrene of the palate; the white cells numbered 542,000, and myelocytes were present; (*b*) chronic—as in a case reported by Weil (1932), where a woman worker in a rubber factory died after 2 years of myelogenous leukaemia, with typical post-mortem findings. The white cell count in this case was 68,000. In the case recorded by Loeper, Fabre and Borreau, the leukaemia occurred 15 months after exposure to benzol had ceased. There was a profound anaemia and the white cell count was up to 58,000 per c.mm. Appreciable amounts of benzene were detected in the blood. A somewhat similar case was reported in 1947 to the Chief Inspector of Factories in which the leukaemia formed the terminal phase of an aplastic anaemia which had existed when the woman, employed in cementing rubber rings with a benzol rubber solution, was examined 2 years previously, but during these 2 years she had not been exposed. Saita (1945), describing a fatal case of aleukaemic leukaemia following exposure to benzol, considers that this syndrome should be regarded as a disease deserving compensation whether the terminal leukaemia follows directly or indirectly upon a primary benzol aplastic anaemia.

Differential count. A relative lymphocytosis is found in most cases of benzol poisoning, the neutrophils being decreased in comparison with the lymphocytes. Instead of the normal ratio of 65 to 70 per cent polymorphonuclears to 20 to 30 per cent lymphocytes, values of 50 to 60 per cent polymorphonuclears and 40 to 50 per cent lymphocytes have frequently been recorded.

Atypical differential counts. (a) Polynucleosis, a rise in the polymorphonuclear leucocytes (75 to 82 per cent) accompanying slight leucopenia, is postulated by Danysz (1942) and by Duvoir and Dérobert (1946) as a sign of benzene absorption in a certain number of cases.

(b) Lymphocytosis and monocytosis, not always accompanied by neutropenia, have been observed by Bernard (1942), Garnier and Cordier (1942), Laignel-Lavastine *et al.* (1928) and Brindeau (1931a). In a case described by Bernard, a female worker handling benzene, the total white cells were 15,300 with a lymphocytosis of 79 per cent (12,000 total). In this and other similar cases described by him, and also by Garnier and Cordier, the anomaly disappeared rapidly on removal of the patient from exposure.

(c) Monocytosis, a rarer phenomenon, is regarded by Mazel, Picard and Bourret (1944) as not so much a direct consequence of benzol intoxication as an exaggeration of a disturbance of bone-marrow function found during recent years among workers exposed to other toxins, or even not exposed to any, but possibly suffering from alimentary disequilibrium. These authors insist, however, that mononucleosis occurring in any benzol worker should be regarded as an indication for prohibition of exposure.

(d) Immature white cells in the peripheral blood were noted by Mitnik and Genkin (1931), Jackson *et al.* (1940) and by Bowers (1947), who reports a fatal case showing hyperplasia of the bone-marrow with metamyelocytes 3 per cent, myelocytes 1 per cent and myeloblasts 2 per cent.

(e) An increase in eosinophils has been recorded by some authorities, especially in France (Heim de Balsac and Agasse-Lafont, 1933; Dimmel, 1932, 1933; Duvoir and Dérobert, 1942). Duvoir and Dérobert, taking the lower limit of normal as 5 per cent, observed eosinophilia in 21·8 per cent in a series of 555 cases, the incidence being slightly higher in women than in men, and showing some correlation with time of exposure. No convincing explanation of this phenomenon is forthcoming, though some authorities appear to regard it as a sign of special resistance of the eosinophil cell to the toxic agent. This, Duvoir and Dérobert remark, is somewhat paradoxical in view of the fact that the eosinophil granules are believed to consist of a protein-lipoid compound in which the lipoid constituent predominates. Dimmel (1932) regards eosinophilia as a favourable prognostic sign, but Duvoir and Dérobert look upon it as a late manifestation. Other observers, including Teleky and Weiner (1924), and Browning have failed to observe any significant eosinophilia in chronic benzene poisoning.

(f) An increase in basophil leucocytes has been recorded by Smith (1928).

Red cells

As a rule a reduction of red cells occurs during benzol poisoning, leading to a moderate degree of anaemia, but in certain cases anaemia of great severity is the outstanding symptom. Weil (1933) describes two forms of benzol anaemia: (*a*) slight, as described by Delarue (1919) and Chambovet (1921), with pallor of mucous membranes, fatigue, anorexia, dyspnoea on effort, and in women menorrhagia and metrorrhagia; (*b*) severe hyperchromic of the pernicious type, with a red cell count down to 1,860,000, and a colour index of 1·6. Similar cases have been described by Brindeau (1931a), Rivet and Guédé (1928) and some fatal cases have been reported to the Home Office (1934).

Total count. One of the lowest red cell counts recorded is that of a case of Brocher (1929)—630,000—while other very low values are 880,000 (Hogan and Shrader, 1923), 900,000 (Hayhurst and Neiswander, 1931) and 1,000,000 (Hamilton, 1925; Landé and Kalinowsky, 1928). The majority of cases, however, appear to range between 2 and 4 million per c.mm.

Abnormalities. Degenerative changes, including anisocytosis, poikilocytosis (Rohner and co-workers, 1926; Schneider, 1930; Hayhurst and Neiswander, etc.), stippling, punctate basophilia, and even the appearance of nucleated red cells have been reported with some frequency (Hunter and Hanflig, 1927; Ronchetti, 1922; von Oettingen, 1919; Brindeau, 1931a; Bowers, 1947).

HYDROCARBONS

Resistance. According to Schneider (1930) the resistance of the red cells to haemolysis in benzol poisoning is never decreased and is sometimes slightly increased.

Nucleated red cells. In a fatal case of benzene exposure, Bowers (1947) noted a progressive rise in nucleated red cells in the circulated blood.

Reticulated cells. A slight reticulocytosis is recorded by Mitnik and Genkin (1931), also by Pulford (1931).

Haemoglobin level and colour index. In most cases the haemoglobin level sinks concurrently with the red cell count so that the colour index is below 1, but in the severe cases, such as those described by Weil, etc., the haemoglobin remains relatively high and a colour index of above 1 results. In 2 of the fatal cases reported to the Home Office in 1934, the colour indices were 1 and 1·06 respectively. Dimmel (1932) found a high colour index to be a feature of the severe cases recorded in his investigation, lying between 1·1 and 1·3, and falling below 1 with improvement in the condition. Hamilton (1934) states that out of 75 cases the colour index was found to be low in 17, high in 22 and normal in 36 and that as a rule the lower the red cell count the higher the colour index. A case recorded by Selling (1916) was an exception to this rule, having very profound anaemia (640,000 red cells per c.mm.) with a low colour index (0·6).

Blood platelets

In nearly all cases where a platelet count has been made, thrombocytopenia has been recorded, the platelets sometimes falling to levels less than $\frac{1}{60}$ of the normal, e.g. 4,000 per c.mm. in a case of Mitnik and Genkin (1931), 600 in one of Brocher (1929). These low values usually occur in cases with severe purpuric manifestations but not always. Hamilton (1931), for example, observed no haemorrhage in a case with marked loss of platelets, and Mitnik and Genkin reported haemorrhage with a normal platelet count. Nikulina and Titowa (1934) agree that reduction of the number of blood platelets does not always result in haemorrhagic symptoms. They record one case with a thrombocyte count of 18,790 per c.mm. with no signs of bleeding, but state that with a high degree of thrombocytopenia the tendency to haemorrhage increases, and that this tendency is dependent also upon the length of exposure, i.e. it is present chiefly in persons exposed to benzol for a short period in spite of comparatively high thrombocyte counts, while amongst those exposed for a long period, a tendency to haemorrhage is only present with a high degree of thrombocytopenia. These workers found that when exposure to benzol is prolonged the thrombocyte count tends to increase, but they are unable to explain this finding.

They have also attempted to correlate the degree of thrombocytopenia with the leucopenia, anaemia and relative lymphocytosis of typical chronic benzol poisoning, but have come to no more definite conclusions than that while no strict parallelism between these conditions can be postulated, thrombocytopenia is an earlier and more constant symptom than anaemia, that the number of thrombocytes tends to increase with a rising lymphocyte count, and that thrombocytopenia develops *pari passu* with leucopenia.

According to Helmer (1944), thrombocytopenia is the most persistent feature of chronic benzene poisoning. In a series of serious cases in a factory manufacturing rubber raincoats in Sweden he found that after more than a year of work many of the workers showed a level not exceeding 200,000, though red cells, haemoglobin and leucocytes had long returned to normal.

Blood coagulation

According to Simonin (1934) the coagulation time may be normal or delayed. Hayhurst and Neiswander (1931) found that the coagulation time was 4 minutes in a case with bleeding from the skin and mucous membranes. Brindeau (1931a) found it delayed in 2 cases with severe symptoms resembling pernicious anaemia, and Rohner and co-workers (1926) estimated it at 9 minutes in a fatal case, but Dimmel (1933) observed no delay in a series of 66 cases of varying severity.

Coagulation itself, according to Simonin (1934), may be atypical, showing irretractability of the clot.

Bleeding time is also variable, but is apparently more often normal than prolonged. It was prolonged (25 minutes) in the case recorded by Hayhurst and Neiswander, and those of Brindeau (1931b) and of Rohner and co-workers. Dimmel states that with extreme thrombocytopenia it is prolonged, otherwise normal or doubtful.

Changes in the bone-marrow

The changes in the bone-marrow described in the earlier literature of cases which came to autopsy were usually those of profound aplasia, affecting both the erythroblastic and leucoblastic elements. Selling (1911, 1916) described a bone-marrow practically depleted of leucocytes, and in cases recorded in the Home Office Reports for 1918 and 1931 respectively, the bone-marrow was described as showing an aplastic anaemia identical with that produced by trinitrotoluol poisoning and as "practically all replaced by yellow fat". However, later cases have shown that hyperplasia is by no means an infrequent sequela of chronic benzene poisoning. This was clearly shown in the series of cases described by Mallory *et al.* (1939) whose observations included the histological examination of tissues from 14 autopsies and 45 biopsies. Of these examinations only 6 showed hypoplasia of the bone-marrow while 9 showed hyperplasia. Extramedullary haemopoiesis was also found in the liver, spleen, and lymph nodes. Bowers (1947) describes a fatal case in which the bone-marrow hyperplasia shown by sternal puncture during life was confirmed post mortem, and extramedullary haemopoiesis, both myeloid and erythroid, was also present.

Persistence of benzene effect

Studies by Goldwater and Tewksbury (1941) indicate that damage to the blood picture caused by benzene may persist for at least 2 years following cessation of exposure. Of 108 men re-examined 24 months after removal from exposure, 4 still had distinctly abnormal blood pictures. In a number of cases where re-examination was made after 2 months' interval of non-exposure, the finding that the blood picture showed further deterioration, even in spite of treatment with liver and vitamin B complex, suggested that the deleterious effect of benzene may continue to operate for some time after exposure has ceased. A similar observation was made by Browning in a number of photogravure process workers examined at intervals of 3 months after their first examination had been followed either by complete removal from exposure or continuance of work in conditions where exposure had been greatly decreased by improvement in exhaust ventilation and by the substitution of a solvent containing not more than 5 per cent of benzene (Table 10).

TABLE 10

Blood picture of two workers severely exposed to benzene

Examination	Exposure	R.B.C. × 10³ per c.mm	W.B.C. per c.mm.	Hgb. per cent	Colour index	Polymorpho-nuclears per cent	Lymphocytes per cent	Eosinophils per cent	Basophils per cent	Large monocytes per cent
No. 1.										
1	4 years	4,472	2,250	86	0·9	43	39	8	0·5	9·5
2	2 months after cessation	4,332	2,000	87	1·01	43	44·5	5·5	0·5	6·5
3	8 months after cessation	4,164	3,150	98	1·07	54·5	36	3	0	6·5
4	10 months after cessation	4,120	2,900	82	1	55	38·5	3·5	0	3
No. 2.										
1	8 years	3,990	3,425	70	0·9	39	49·5	3	1	7·5
2	2 months after conditions improved	3,676	3,400	72	1	49·5	43·5	1·5	0	5·5
3	6 months later; conditions further improved	3,648	3,350	66	0·9	32·5	62	3·5	0·5	1·5
4	8 months later	4,360	3,400	76	0·9	42·5	51	3·5	0	3

The "lag phenomenon", or appearance of symptoms long after exposure has ceased and in cases where no signs of intoxication were present during the period of exposure, has been reported by several observers. Usually the appearance of symptoms has coincided with some infectious disease, or with pregnancy, as in a fatal case reported to the Factory Department in 1945 and also one recorded by Perrault, Dérobert and Tiret (1944). In a case described by Hunter (1939), symptoms occurred one year after cessation of exposure, following "heat-stroke" and a septic infection of the leg; in a case reported by Smith (1943) the interval was 4 years, and in one reported by Meyer and Ginsberg (1942) the interval was 10 years. It may be assumed that the injury inflicted by exposure to benzol may not be apparent until extra demands upon the haematopoietic system are made by infection and disease. In the case recorded by Perrault *et al.* (1944), the presence of benzene was actually demonstrated in the bone-marrow at autopsy in amounts of 20 γ per 100 g., 14 months after cessation of exposure, thus leading them to conclude that benzol, like many other chemical substances, may preserve its potential toxicity for long periods of time. The fact that in another fatal case investigated by Rachet, Jumière and Dérobert (1944), no benzene was found in the bone-marrow does not, it is considered, invalidate the conclusion. The oxidation-reduction mechanism of the organism differs in individual cases, and these workers believe that in this case the local oxidation was so intense that the benzol had completely disappeared after causing irreparable damage to the bone-marrow.

Table 11 shows the features of the blood picture found in the chief investigations published.

TABLE 11

Blood picture after benzene poisoning

Author	Haemoglobin (per cent)	R.B.C. (millions per c.mm.)	Thrombocytes (per c.mm.)	W.B.C. (per c.mm.)	Differential count	Other features
Santesson (1897)	80	3·7	—	Leucopenia	—	—
	20	0·6				
Selling (1916)	8	0·64	Thrombopenia	480	—	—
McClure (1916)	25	1·46	Thrombopenia	1,100	—	—
Harrington (1917)	60	2·8	—	500	—	—
Home Office Report (1918)	35	2·8	—	2,000	—	—
Newton (1920)	80	5·7	—	1,200	—	—
	—	3·6		1,250		
	—	4·0		1,700		
Pugliese (1922)	29	1·7	Thrombopenia	1,700	—	—
Hogan and Schrader (1923)	39	1·24	—	600	—	—
	12	0·9		104	Lymphocytes 56%	
	16	0·88		950	Lymphocytes 52%	
Hamilton (1931)	20	1·0	—	2,000	—	—
Brücken (1923)	32	1·51	24,240	2,460	Lymphocytes 42%	Monocytes 10%
	55	—		3,870	Lymphocytes 32%	
Teleky and Weiner (1924)	60	3·58	—	3,500	Lymphocytes 42%	—
Rohner et al. (1926)	20	0·85	70,000	1,400	Lymphocytes 48%	Aniso- and poikilocytosis
Winslow (1927)	Under 30	0·8–5·4	—	1,450–6,140	Usually lymphocytosis	—
Smith (1928)	Reduced	Slightly reduced	—	Leucopenia	Lymphocytosis	Myelocytes
Landé and Kalinowsky (1928)	35	1·0	34,000	2,000	Lymphocytes 76%	Anisocytosis
	70	4·2	—	3,200	Lymphocytes 35%	Poikilocytosis, basophilia
Laignel-Lavastine et al. (1928)	20	1·24	42,000	600	—	Monocytes 51%
Brocher (1929)	11	0·63	6,000	2,000	Lymphocytes 39%	—
Schneider (1930)	47	2·2	114,000	4,640	Lymphocytes 44%	Nucleated R.B.C.
	56	2·34	116,000	1,320	Lymphocytes 44%	Poikilocytosis, basophilia
Sorrentini (1930)	80	2·6	—	3,000	—	—
Fischer (1932)	—	2·0	—	1,000	—	—
Adler-Herzmark (1930)	—	3·0	—	2,000–3,000	—	—
Hayhurst et al. (1931)	10	0·9	100,000	850	—	Anisocytosis. Bleeding time 25 min. Coagulation time 4 min.
Brindeau (1931a)	50	1·638	—	1,770	—	Monocytosis
	22	0·9	—	1,300	—	Aniso- and poikilocytosis, nucleated R.B.C.
Löwy (1926)	30	1·3	—	1,800	Lymphocytes 38%	—
	26	1·6	4,000	1,600	Lymphocytes 82%	Punctate basophilia
Mitnik and Genkin (1931)	33	2·4	—	3,800	Lymphocytes 24%	—
	63	3·9	128,000	4,000	Lymphocytes 29%	Reticulocytes increased
	69	4·1	—	4,300	Lymphocytes 29%	
	73	4·85	194,000	4,500	Lymphocytes 40%	
Espeut and Salinger (1930)	Normal	Normal	—	Slight leucopenia	Lymphocytes 32% and 50%	—
Kranenberg and Peeters (1928)	65	—	—	More than 50% reduction	Lymphocytes > 39%	—
Nikulina and Titowa (1934)	70–79	3·5–4·0	19,283	4,737	Lymphocytosis	—
	70–79	3·5–4·0	25,497	6,300	Lymphocytosis	
Merklen and Israël (1934)	60–85	2·5–4·5	3,800–230,000	2,700–8,600	Lymphocytes 20–55%	Monocytes 3–22%. Myeloblasts, myelocytes
Pulford (1931)	21	1·2	94,000	1,940	Lymphocytosis	Slight reticulocytosis
Newton (1920)	80	4·7	—	1,200	—	Monocytes 39%
Dimmel (1932)	18–110	0·78–5·3	0–360,000	170–9,600	Lymphocytes 8–60%	Monocytes up to 18%. Eosinophils up to 53%. Degenerated and young leucocytes; megalocytes

Lesions of the Internal Organs

In such lesions as have been observed in autopsies of fatal cases it is difficult, as pointed out by Litzner (1932), to decide whether they are due to direct injury by benzol or to secondary injury from severe blood changes.

Heart. Myocardial infarcts (Le Noir and Claude, 1897) and fatty degeneration of the muscles (Selling, 1916) have been recorded, while in the two (1918) cases in the Home Office Reports haemorrhages were found under the endothelium of the heart.

Liver. Fatty degeneration (Selling, 1916, Loeper, 1941) and areas of necrosis (Rohner and co-workers, 1926) are recorded and also enlargement with excessive iron deposition (fatal case, Home Office Report, 1931).

Spleen. The reports of post-mortem examinations of the spleen appear to show less change than in experimental animals. Landé and Kalinowsky (1928) found it small and anaemic with no demonstrable histological lesions. In the 1931 Home Office fatal case the spleen showed excessive iron deposition. A spleen removed therapeutically by Hegler (1933) was very little enlarged; the sinuses were filled with erythrocytes and reticulo-endothelial cells.

Adrenals. Petri (1930) states that the adrenals in benzol poisoning are deficient in fat and show areas of necrosis.

Gastro-intestinal tract. In both the 1918 Home Office cases submucous haemorrhages were found throughout the intestinal tract, in the 1931 case melanosis of the small intestine and ulcers of the large intestine, and in the 1934 case submucous haemorrhages of the stomach.

Gastritis, demonstrable by endoscopy, in the form of a more or less intense irritation of the mucous membrane, is stated by Chevallier and Moutier (1947) to appear usually within the first few weeks of exposure. It shows no parallelism with blood changes, and when tolerance is developed, even without removal from exposure, the gastritis disappears.

Treatment

Therapeutic measures of various kinds have been tried, but apparently with little success if the poisoning has become severe. Dimmel (1932) states that in cases of severe injury there are no means by which it is possible to save life. The following measures have been tried:

(1) Blood transfusion has been found to have no permanent value in severe cases of poisoning.

(2) Irradiation of the bone-marrow and spleen had no success in Dimmel's cases.

(3) Splenectomy was followed by recovery in a case reported by Hegler (1933).

(4) Liver therapy and a vitamin-rich diet have been recommended in cases with a tendency to hyperchromic anaemia (Schneider, 1930; Dimmel, 1932, etc.).

(5) Protein shock and adrenaline therapy with a view to stimulating the bone-marrow was also tried by Dimmel but with no definite results.

DIFFERENTIAL DIAGNOSIS OF POISONING

In diagnosing benzol poisoning from the three chief conditions in which leucopenia and anaemia are found together, it appears that the only indubitable distinguishing feature is a history of exposure to benzol. The criteria of the National Safety Council (1926) for benzol poisoning were as follows:—"A history of exposure to benzol and a white blood count below 5,600 is accepted as reasonable evidence of poisoning."

The three chief conditions from which benzol poisoning may have to be distinguished are: (1) agranulocytosis; (2) aplastic anaemia (idiopathic); (3) thrombocytopenic purpura. Dimmel (1932) adds also septic aleukaemia, from which, he states, benzol poisoning in its final stages is practically indistinguishable, if accompanied by secondary infection.

In agranulocytosis, according to Pulford (1931), the spongy oozing gums may ulcerate and become gangrenous and slough, which does not happen with benzol poisoning. Other differential diagnostic points are that in agranulocytosis, fever is an early and persistent sign while in benzol poisoning it is usually terminal; the coagulation time, bleeding time, platelet count and retractility of clot are usually unaffected in agranulocytosis.

In aplastic anaemia, the symptoms and blood picture are very similar to those of benzol poisoning, but, according to Pulford, the duration of the disease is usually less.

In thrombocytopenic purpura, purpura is the more outstanding feature, while oozing is the more characteristic feature of benzol poisoning. In thrombocytopenic purpura also the spleen may be palpable, and the white cell count is likely to be higher in the later stages than in those of benzol poisoning. These diagnostic features are also discussed by Sweeney (1928), Askey (1928), McCord (1929), Dameshek (1929) and Kracke (1932).

The various diagnostic points have been summarized by Pulford (1931) as in Table 12.

TABLE 12

Benzol poisoning and comparable diseases with anaemia and leucopenia

	Benzol poisoning	Agranulocytosis	Thrombocytopenic purpura	Aplastic anaemia
R.B.C. reduced	+++	+	+	++
W.B.C. reduced	+++	++++	+	+++
Lymphocytes	+++	++++	-----	Normal
Platelet count reduced	++	-----	+++++	+
Coagulation time increased	-----	-----	Normal	-----
Bleeding time increased	+	-----	+++	-----
Contractility of clot prolonged	+	-----	+++	-----
Tourniquet test	++	-----	++++	-----
Spleen enlarged	-----	-----	++	-----
Reticulation of R.B.C.	+	-----	Normal	-----
Normoblasts	+	-----	-----	-----
Fragility of R.B.C.	-----	-----	-----	-----

R.B.C. = Red blood corpuscles. W.B.C. = White blood corpuscles.

2. Toluene

(*Toluol, Methylbenzene*)

$C_6H_5 \cdot CH_3$, i.e., [benzene ring with CH$_3$ substituent]

PROPERTIES

A colourless fluid with an odour similar to that of benzene, but sharper, with properties very similar to those of benzene, for which, according to Keyes (1925) it is the most frequent substitute. B.R. of commercial product, 109–111° C. (Durrans, 1933, 1935). Sp.Gr. 0·86–0·871. Fl.P. 7–13° C. Evaporation time 6·1 (compared with ethyl ether, 1). Saturated air contains 100 mg. per l. at 20° C. Water dissolves 0·047g. per 100 ml. at 20° C.

Good solvent for ethyl cellulose, dibenzyl cellulose, rubber, ester gum, mastic, dammar and elemi, fats, oils. Partial solvent for soft copal but not for kauri, hard copal or sandarac, or for cellulose esters, but solutions of these latter in most of the solvents will usually tolerate the addition of a greater proportion of toluol than of any other hydrocarbon diluent.

MANUFACTURE

Toluol is produced, like benzol, from the distillation of coal tar. The relation of the amount of benzol and toluol produced during the process varies with the volatile content of the coal, the type of oven, and the heat employed; low heat increases the production of toluol. The manufacture of toluol in Great Britain from coke ovens increased from 1,182,000 gallons in 1930 to 1,258,000 gallons in 1933, not including that from gas works, which in 1930 was 262,000 gallons.

USES

(1) Toluol is among the "medium boilers" of the fluid components of lacquers, and is used especially as a diluent, its vapour pressure and power of dissolving resins making it very suitable for the purpose.

(2) In photogravure colour-printing processes where commercial toluol is used to dilute the ink.

(3) In the manufacture of explosives.

(4) In the rubber industry, although toluol is used to a much smaller extent than benzol. During the 1939–45 war, toluol was used in combination with petroleum naphtha as a solvent for synthetic rubber (neoprene).

(5) In linoleum manufacture, toluol is sometimes a constituent (up to 35 per cent) of the cellulose solution and (up to 20 per cent) of the cleansing solvent.

(6) In the dyeing and cleaning industry.

(7) In the impregnation of cartridge paper.

TOXICITY

It appears to be agreed that while toluene, in concentrations of 200 p.p.m. or over, may produce some disturbance of nervous control, it does not cause any definite injury to the haematopoietic system such as that exerted by benzene. Its action on the liver is less severe than that of carbon tetrachloride.

The differences of opinion of various writers on the toxicity of toluene, and especially on its toxicity relative to that of benzene, have evidently arisen to some extent from their having used animal experiments as a criterion. From animal experiments it appears that toluene has an acute narcotic effect like that of benzene, though without the convulsive neurotoxic effect of the latter; in high concentrations it is dangerous to life, even more so than benzene; slight narcosis appears earlier with toluene than with benzene (Engelhardt and Estler, 1935). In the less acute and in the chronic forms of poisoning, however, toluene appears to be considerably less toxic than benzene. Thus the position was summarized by Henderson and Haggard (1927) as follows:—"In the literature the relative toxicity of benzene and toluene is uncertain. Probably there are no great differences in (acute) toxicity. . . . Toluene is less active than benzene in causing chronic poisoning."

Würm (1931) believes that the smaller injuries caused by the homologues of benzene are due to differences in volatility, and that if the action of toluene is sufficiently prolonged it can be as toxic as benzene, while Engelhardt and Estler

(1935) remark that the oxidation products of toluene (chiefly benzoic acid) are relatively non-toxic, and that its more ready oxidation in the organism also provides for its removal. Actual investigations of the health of workers with toluene certainly appear to show that the effects of toluene, especially on the blood-forming organs, are less characteristic and less severe than those of benzene, though it causes symptoms of clinical disturbance, chiefly related to the nervous system, and may cause enlargement of the liver.

Toxicity in Industrial Processes

An investigation by Adler-Herzmark (1933) of workers using a toluol-containing lacquer revealed some symptoms and blood findings which appeared to point to bone-marrow injury.

Toxic effects of a psychotic nature have been reported by Panse and Bender (1934) in colour-printers, and blood changes similar to those of benzol by Meyer (1928, 1931) and Stocké (1929), who gives a special warning of the danger of toluol in this industry. An examination by H.M. Factory Inspectors of two colour-printers in 1933 in whom injury was suspected showed no abnormality either clinically or in the blood picture. Stocké points out that in many printing works, the constitution of the solvent used is uncertain; in some works a solvent called xylol has been found to be pure toluene, while in others mixtures of xylol and toluol are employed, or additions of benzol and paraffin hydrocarbons are used.

During the 1914–18 war the demand for toluol for the manufacture of explosives was enormously increased, but cases of poisoning from its use do not appear to have been reported. One workman employed in the manufacture of cartridge cases was examined by H.M. Inspector of Factories in 1931, but the only blood changes found were slight anaemia and poikilocytosis of the red cells.

Reports of toxic effects in rubber workers are very few in comparison with those of benzol on account of its more limited use. Three cases of slight injury to the health of workmen making crêpe rubber soles were reported to the Home Office in 1929, accompanied by a blood picture similar to that of benzol poisoning, but it was suggested in these cases that the effects might in reality be due to benzol, since the boiling point of the toluol supplied was lower than that of pure toluene, varying from 88 to 105° C. instead of 110·5° C.

An examination by the Home Office in 1931 of four men in the linoleum industry using solutions containing toluol revealed no evidence of intoxication either clinically or from the blood picture.

A survey by Litzner and Edlich in 1934 of workers in a dye factory where pure toluene was used as a solvent under pressure revealed severe subjective symptoms and slight blood changes similar to those produced by benzol.

TOXIC EFFECTS IN ANIMALS

Absorption and Excretion

The absorption of toluene is at first rapid but slow later, and its elimination is also rapid at first, excretion being chiefly by the lungs (Jost, 1932), and only small quantities being present in the blood after 2 hours (Greenburg, Mayers, Heimann and Moskowitz, 1942). The blood concentration increases with the air concentration, and nervous impairment may occur when the blood concentration reaches about 0·5 mg. per 100 ml. (von Oettingen, Neal and Donahue, 1942a;

von Oettingen, Neal, Donahue, Svirbely, Baernstein, Monaco, Valaer and Mitchell, 1942b). Part of the toluene inhaled is oxidized in the body to benzoic acid which reacts with glycine to form hippuric acid. Von Oettingen and coworkers found that exposure to 50 to 800 p.p.m. was followed by increased excretion of hippuric acid, roughly parallel to the intensity of exposure. Dogs exposed to 200, 400 and 600 p.p.m. showed an increase in hippuric acid excretion from the normal value of 250 mg. per 8 hours to 465, 571 and 670 mg. This was markedly increased by the administration of 2 g. of glycine and the toxic effects were moderately alleviated by 50 mg. of ascorbic acid.

Acute Poisoning

Lethal and Narcotic Concentrations

Lethal dose. For white rats, 1,600 p.p.m. (6 mg. per l.) of pure toluene vapour (Batchelor, 1927); for mice, 8,000 to 8,100 p.p.m. (30 mg. per l.) (Lazarew, 1929a) and 5,300 p.p.m. (Svirbely, Dunn and von Oettingen, 1943) compared with 10,400 p.p.m. of benzene. (Values refer to single 7-hour exposures.) A comparison of lethal doses is shown in Table 13.

TABLE 13
Comparison of lethal doses of benzene, toluene and xylene

Author	Solvent	Lethal dose (p.p.m.)
Batchelor	Benzene	2,440 (neurotoxic)
	Toluene	1,600
	Xylene	1,600
Lazarew	Benzene	14,000
	Toluene	8,000–8,100
	Xylene (para)	3,450–8,000
	,, (meta)	11,500

Batchelor's result of apparent lower toxicity of benzene applies only to lethal concentrations; below this point much more definite injury resulted from inhalation of benzene than of toluene.

Thus the order of increasing lethal toxicity, according to both these workers and to Lehmann (1912) is approximately benzene→xylene→toluene. This was not the order found by Chassevant and Garnier (1903) by intraperitoneal injection; these authors found that the lethal doses were as follows: toluene, 0·50 ml. per kg. body weight; benzene, 0·73 ml. per kg.; xylene (*meta*) 1·65 ml. per kg. Chassevant and Garnier related the varying toxicity to the molecular weight, the number of substitutions, and the position of substitution of the benzene ring. They found that for molar equivalents the mono-substituted derivatives such as toluene, are more toxic than benzene, and always more toxic than the bi-substituted, such as xylene. Flury and Zernik (1931) point out, however, that although theoretically the toxicity increases with the molecular weight, so that xylene should be the most toxic and benzene the least, the solubility in the blood decreases with the molecular weight, with the result that the toxic concentration of the vapours of the three substances remains practically the same.

Narcotic dose. Most workers from Lehmann onwards have found toluene (and xylene) more toxic than benzene—i.e. producing narcosis in smaller concentrations and earlier. The concentrations of toluol vapour for commencing narcosis in mice, according to Batchelor (1927) and Smyth and Smyth (1928), is about 1,250 to 1,600 p.p.m. (5 to 6 mg. per l.) while, according to Lazarew (1929a), light narcosis (side position) is reached at a concentration of 2,700 to 3,200 p.p.m. (10 to 12 mg. per l.). Estler (1935) also observed narcosis generally in cats and rabbits at about 10 mg. per l.; with mice, death occurred frequently some time after inhalation of these concentrations had ceased. Toluene and xylene are certainly more toxic in this respect than benzene, and toluene more so than xylene, as can be seen from the following figures given by Lazarew (1929a) for commencing narcosis: benzene, 15 mg. per l.; toluene, 10 to 12 mg. per l.; *meta*-xylene, 10 to 15 mg. per l.

Estler (1935), however, found the toxicity to be in ascending order—benzene→toluene→xylene, for fairly low concentrations (7·5 mg. per l. being the lowest concentration which produced narcotic symptoms). In low concentrations (1,000 p.p.m., 3 to 4 mg. per l., or less) no narcotic effects were observed by Batchelor, and none by Estler at 5 mg. per l.

It should be mentioned here that Mgebrow (1930) lays special emphasis on the difference, as regards narcotic effect, between the subcutaneous and inhalation administration of toluene. He states that no narcotic effects were observed from a sub-lethal subcutaneous dose (he gives the lethal dose as 2·05 ml. per kg. body weight) of toluene, the only effect being loss of weight.

Symptoms

Intraperitoneal injections of toluol in Batchelor's experiments produced much less severe symptoms than benzol given in the same amount (0·25 ml. to 1 ml. per kg. body weight); inactivity and apathy occurred but no loss of consciousness. With doses high enough to produce narcosis there was but slight evidence of the neuro-irritation characteristic of acute benzol poisoning. There was some loss of weight, but not so severe as with benzol. Estler (1935) noted coldness of the bodies of the animals, and Wojciechowski (1910) a scarlet coloration of the mucous membranes of cats subjected to inhalation of 21·7 g. per l. Von Oettingen *et al.* (1942a,b) found no change in the circulation, respiratory rate or spinal pressure in animals exposed to concentrations of 200 to 600 p.p.m., but with concentrations of 850 p.p.m. they observed an increase in the respiratory rate and a decrease of the respiratory volume.

Lesions of the Internal Organs

Kidney and liver. With subcutaneous injections and with concentrations of 1,600 p.p.m. there occurred some lesions in the kidney (a mild diffuse nephritis) and some focal necrosis of the liver, but the lesions were much less severe than with benzol (Batchelor, 1927). Smyth and Smyth (1928) also observed signs of early toxic degeneration of the liver and kidneys with inhalations of high concentration, and some pulmonary inflammation. Enlargement of the liver due, not to hyperaemia but to a change in the liver cells of animals exposed to concentrations of 600 to 5,000 p.p.m., was observed by von Oettingen and co-workers.

Bone-marrow. In Batchelor's experiments there was also some hyperplasia of the bone-marrow, a moderate hyperplasia of the Malpighian corpuscles in the spleen and lymph node follicles, and pigmentation of the spleen, indicating blood cell destruction. The apparent stimulation of the bone-marrow by toluene has been specially emphasized by Mgebrow (1930) and by Ferguson, Harvey and Hamilton (1933). The latter workers observed that the marrow was red and cellular, more cellular than that of animals subjected to benzene inhalation, and showed a polymorphonuclear type of reaction with an abundance of eosinophil myelocytes and normoblasts, and megakaryocytes with abnormal nuclei.

Spleen. Ferguson and co-workers observed congestion, but little haemosiderosis and no myeloid metaplasia; the reticulum cells were prominent in contrast to the lymphocytes.

Lungs. Engelhardt (1935) observed a fairly high incidence of pneumonia, but not so high as with benzol.

Effect on the Blood

The effect of toluol given by injection (Batchelor, 1927; Engelhardt, 1935), or by inhalation of comparatively large amounts (Ferguson and co-workers, 1933) appears to be rather that of stimulation than destruction of the haemopoietic organs. Batchelor found but little, if any, effect on the blood count, though the pigmentation in the spleen indicated some red cell destruction. Ferguson, Harvey and Hamilton concluded that toluene is less toxic in this respect than benzene, but has an injurious action on the young bone-marrow cells, and consequently on the peripheral blood cells. Engelhardt found some anaemia with polychromasia and normoblasts, a leucocytosis in some cases but no leucopenia, and sometimes a relative lymphocytosis and some toxic granulation of the leucocytes in his animals injected with 0·5 ml. toluene per kg. body weight. He related the toxic granulations to the local irritation produced by the injection.

Chronic Poisoning

With concentrations of 620 to 1,000 p.p.m., inhalation of toluol produces practically no symptoms in animals (Batchelor, Smyth and Smyth, etc.).

Effect on the Blood

Observations of earlier workers on the changes produced in the blood of animals were inconsistent except in one respect, viz., that these changes were less severe and less characteristic than with benzol. Batchelor (1927), Engelhardt (1935), and Ferguson and co-workers (1933) all noted changes similar to those produced by benzene, except that leucopenia was not a constant feature of toluene injury, while some German workers, especially Mgebrow (1930), laid emphasis on the anaemia produced by toluene, regarding it as a haemolytic effect. The results obtained by Ferguson, Harvey and Hamilton (1933) by daily subcutaneous injections of 1 ml. per kg. were also variable, though they interpreted them as showing that "toluene is a poison of the same type and action as benzene". The only later workers who claim to have produced significant blood changes in animals are Jakobsen (1939) and Seghini (1941). Seghini recorded anaemia, leucopenia with lymphocytosis followed by leucocytosis. Jakobsen (1939) emphasized the effect on the granulocyte count, that of an initial fall, followed by a rise, and accompanied by a rise in immature forms, in animals exposed to lacquer solutions "more than half" of which were stated to consist of toluene.

Cameron, Paterson, de Saram and Thomas (1938) produced no significant change in the blood of rats after fourteen 8-hour exposures to inhalation. Von Oettingen and co-workers (1942a, b) observed no evidence of injury to the blood-forming organs in dogs and rats.

Red cells and haemoglobin

A slight preliminary increase of red cells in some animals, but no definite subsequent decrease, was observed by Batchelor in his inhalation experiments of concentrations of 620 to 1,000 p.p.m. (2·4 to 4 mg. per l.). Engelhardt, using slightly higher concentrations (10 mg., and in some cases, 25 mg. per l.) on cats and dogs, found a reduction of both red cells and haemoglobin, especially with the higher concentrations, but sometimes the haemoglobin level was increased. No such changes, however, were found by von Oettingen *et al.* when using still higher concentrations (2,500 to 5,000 p.p.m.) in rats.

White cells

In contrast to the leucopenia recorded by Batchelor (1927) who reported a few cases with a reduction of 28 to 56 per cent, von Oettingen *et al.* (1942a, b) found only a slight decrease of the total white count, with a moderate increase of segmented cells in rats inhaling 2,500 to 5,000 p.p.m. Engelhardt (1935) found practically always a leucocytosis, which might in some cases have been associated with the lung inflammation which was a fairly constant feature.

Differential count

Changes in the lymphocyte proportion have been observed but not with any great consistency. Engelhardt observed sometimes an initial lymphocytosis, sometimes a relative lymphopenia, while von Oettingen *et al.* found a temporary lymphocytosis in dogs at the end of exposure, and a decrease in lymphocytes in rats inhaling 2,500 to 5,000 p.p.m.

It is to be concluded that chronic exposure to toluene causes little, if any, injury to the haemopoietic system of animals.

Immunity Reactions

The effect of toluene on antibody production in animals is, according to Hektoen (1916b), very much less injurious than that of benzene. He found that repeated injections of about 1 ml. per kg. body weight lessened antibody output in the earlier stages of antibody production, but under certain conditions caused prolonged persistence of antibodies in the blood. There was no change in the phagocytic activity of the white cells.

TOXIC EFFECTS IN MAN

Until comparatively recently there has been a more or less general belief that toluene has both acute and toxic effects similar to those of benzene, though less severe. That is to say, the chief feature of its acute toxicity was that of a narcotic, and that of its chronic toxicity that of a haematopoietic depressant. It would now appear that much of this belief with regard to its chronic effect was based on reports of cases in which the exposure was that of commercial toluol containing up to 20 per cent benzol, and that the blood changes observed were therefore a benzene effect rather than one due to toluene itself.

From the results of a careful investigation of both men and animals by the U.S. Public Health Service (1942) it appears that the chief danger to be apprehended from repeated exposure to toluol is that due to its action on the nervous system rather than to any marked haematopoietic effect, and that though in high concentrations it causes enlargement of the liver, in continued exposure to concentrations even as high as 5,000 p.p.m., it is not liable to cause definite injury to the blood-forming organs, to the liver, or to any other organs.

Acute Poisoning

Non-fatal cases. Stocké (1929) has recorded cases of non-fatal acute toluol poisoning in which the solvent used was shown, by chemical analysis, to have been pure toluene. This is in contrast to other cases, such as those of Leschke (1932), where it was uncertain whether toluol or xylol had been used. The symptoms reported by Stocké are headache, acute intoxication ("drunkenness"), nausea, vomiting, loss of consciousness, disturbance of equilibrium, and paraesthesia. He also found a positive Romberg sign. Recovery took place on removal from exposure.

Since 1940, eight cases of acute intoxication by toluol have been reported to H.M. Chief Inspector of Factories. In only two patients was there loss of consciousness, and in one of these the unconsciousness may have been due to concussion from a fall rather than from the effect of toluol itself. The chief symptoms in all cases were headache, nausea, with later vomiting and giddiness, and faintness. Somewhat similar symptoms are recorded by Sack (1941) in a worker distilling toluene, and in whom vertigo was present a year later. In the cases of unconsciousness reported to H.M. Chief Inspector of Factories blood examinations were made within 2 days, 4 days, and 1 month respectively after the accident. Both patients showed slight leucopenia with a relative lymphocytosis, which, in one case, was still present at the end of the month.

Chronic Poisoning

Symptoms

Symptoms related to the nervous system have always been more prominent in the records of chronic toluol poisoning than the blood changes. The only fatal case of agranulocytic anaemia possibly due to exposure to toluol is that recorded by Ferguson, Harvey and Hamilton (1933).

Nervous symptoms. Headache, giddiness and lack of sleep were the commonest symptoms in Stocké's series of workers in a printing industry, and in those examined by Adler-Herzmark (1933) and Litzner and Edlich (1934). In the latter authors' investigation, all the workers showed more or less severe subjective symptoms after $2\frac{1}{2}$ years. Nervous irritability and incapacity to work were also prominent features. Intolerance to alcohol is described both by Stocké and by Litzner and Edlich as a specially characteristic feature of toluol poisoning, in contradistinction to benzol; Litzner and Edlich regard it as a valuable diagnostic sign. They also observed cardiac pain in three cases. A systematic record of the nervous symptoms arising with different concentrations of toluol has been made by von Oettingen *et al.* (1942a, b). This showed that with the lowest concentrations, 50 to 100 p.p.m., the first effects were fatigue and mild headache; as the concentrations rose above 200 p.p.m. a prominent feature was muscular inco-ordination. Higher concentrations, 600 to 800 p.p.m., produced much more severe symptoms, including mental confusion, staggering gait, nausea and severe headache.

Gastro-intestinal symptoms. Loss of appetite, nausea and vomiting were prominent among Stocké's and Litzner and Edlich's workers, and also those of von Oettingen *et al.*

Haemorrhage. In none of the German cases have haemorrhages such as occur in benzol poisoning been recorded, but in a case reported by Ferguson, Harvey and Hamilton (1933), where the solvent used in an india-rubber factory contained 45 per cent of toluol, there were epistaxis, bleeding from the gums, haemorrhagic spots on the tongue and subcutaneous haemorrhages on the hands and face. The patient, who had a history of an accident to the face, later developed gastro-enteritis and pneumonia with fatal outcome, but Ferguson and co-workers admit the possibility that his condition was not due to industrial poisoning, though in view of the blood findings (see below) they conclude that such an origin was highly probable. In one of the three cases reported to the Home Office in 1929 there was slight cyanosis with recent epistaxis.

Local irritation. Irritation of the throat (pharyngitis) was observed in two of Litzner and Edlich's cases and irritation of the arms in one. Throat irritation and lung catarrh were also among the symptoms of a case reported by Fühner and Pietrusky (1934), but it must be mentioned that the solvent used for nitro-cellulose lacquer contained as much as 52 per cent of ester mixtures, chiefly butyl acetate.

Enlargement of the Liver

In a series of spray and brush painters examined by Greenburg *et al.* (1942) palpable enlargement of the liver was present in 30·2 per cent as compared with 7 per cent in a control group. The enlargement was not accompanied by any clinical or laboratory evidence of disease, and it was suggested that it was a compensatory enlargement only. It was confirmed by the examination of animals exposed to 600 to 5,000 p.p.m. by von Oettingen *et al.*, but they found no evidence that it was due to hyperaemia; in their opinion it was associated with a change in the density of the liver cells, perhaps due to endosmosis.

Effect on the Blood

The only fatal, and indeed severe, case of injury to the blood-forming organs is the one recorded by Ferguson, Harvey and Hamilton (1933) in which the authors themselves admitted the possibility of idiopathic agranulocytic anaemia. The solvent used contained 45 per cent toluene, but no benzene, and examination of co-workers of the patient revealed no blood changes which can be regarded as other than normal variations. No serious case of involvement of the haemato-poietic system from chronic toluol poisoning has been reported to the Factory Department.

White cells

The most constant finding appears to be a lymphocytosis, either relative, and sometimes associated with neutropenia, or absolute, as in the series of cases investigated by Greenburg *et al.* (1942). Jakobsen (1939) interpreted the relative lymphocytosis found in his examination of lacquer workers exposed to cellulose solutions, of which the chief constituent was toluene, as a compensatory reaction for the presumed destructive effect of toluene on the granulocytes. A similar diminution of neutrophils has been postulated by Adler-Herzmark (1933), Meyer (1928), Litzner and Edlich (1934). Langelez, Peremans and Bastenier (1940) are among the few authorities who consider that toluene has the same toxic action on the white blood cells as benzene. They base this opinion on the results of the examination of two groups of workers using a solution of 75 per cent alcohol and 25 per cent toluene, and a toluol varnish respectively. Certainly no such effect was found by von Oettingen *et al.* (1942) in their carefully controlled experimental observations of human beings inhaling graded concentrations of toluene containing not more than 1·01 per cent benzene. These three normal volunteers were exposed for 8 hours, with ½ hour's interval, to concentrations of 50, 100, 200, 300, 400, 600 and 800 p.p.m., maintained in a two-compartment exposure chamber of 29,220 litres capacity by evaporating toluene at a certain rate and mixing it with air circulating through. Total white cell and differential counts made at the beginning and end of each exposure showed no significant variation.

Blood examinations made by Browning on a number of factory operatives exposed to toluene, even though in some cases slight contamination with benzene could not be excluded, confirm the view that toluene has not the severe deleterious effect on the white cells that is found so frequently with benzene. Of twenty-three cases examined in 1946–7, only one showed slight leucopenia, and two others relative lymphocytosis. In all these cases the toluene solution used in the composition of a spray paint was known to contain 5 to 8 per cent benzol.

Red cells and haemoglobin

Slight anaemia in workers using a toluol spray lacquer was observed by Adler-Herzmark (1933), but none was reported by Stocké (1929), or by Meyer (1928, 1931). A mild macrocytic anaemia, observed by Greenburg *et al.* (1942) was considered by them to be a diagnostic sign of chronic toluene poisoning. The same condition was found in four of the twenty-three patients examined by Browning in 1946–7.

3. Xylene

(*Xylol, Dimethylbenzene*)

$CH_3 \cdot C_6H_4 \cdot CH_3$

ortho-Xylene   *meta*-Xylene 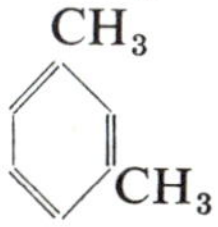   *para*-Xylene

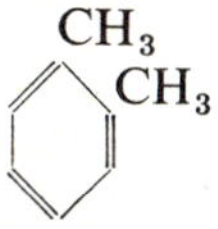

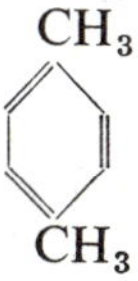

PROPERTIES

Commercial xylol is usually a mixture of three isomeric xylenes—*ortho*-xylene, *meta*-xylene and *para*-xylene, of which the meta-derivative generally preponderates to the extent of 75–80 per cent. It may also contain small quantities of toluene, ethylbenzene and trimethylbenzenes, especially pseudocumene.

The boiling points of the three derivatives are all close together: *ortho*, 142·3° C., *meta*, 139° C. and *para*, 138° C. The boiling range of commercial xylene is between 135° and 145° C. Sp.Gr. 0·862–0·865. Fl.P. 24° C. The presence of toluene lowers the flash point, while the trimethylbenzenes lower the rate of evaporation. Purified xylene of high flash point has a boiling range of 138–140° C. Fl.P. 30° C.

Xylol is a solvent for ester gum, copal ester, benzyl abietate, rubber, castor and linseed oils, and dibenzyl cellulose. It is not a solvent for cellulose esters, but in the presence of anhydrous alcohol it dissolves some forms of the nitrate.

USES

(1) In the photogravure colour-printing industry, where it is used generally as a constituent of solvents which may contain also toluol, benzol, etc.

(2) Mixed with methylated spirit, xylol is sometimes used as a degreaser in place of trichloroethylene.

(3) A solvent containing xylol up to 30 per cent and containing also butyl acetate, ethyl acetate, etc. is used in spreading cellulose solution in leather manufacture.

(4) Resin dissolved in xylol is used in silk-finishing.

(5) Xylol is added to graphite, asbestos, mica, rubber and litharge to make a plastic mass in the manufacture of brake linings.

(6) Xylol is used as a constituent of quick-drying paints, lacquers and cellulose sprays in varying proportions. Sometimes the proportion of xylol reaches 50 to 60 per cent, as in the cellulose spraying factory where workers were examined by the Home Office in 1930, in others the exact proportion is unknown, as in the paint called "Inertol" from which numerous cases of poisoning have been reported in Germany and which is said to consist of up to 35 per cent of "benzol and its homologues, chiefly xylol".

(7) In the rubber industry either as a substitute for benzol, or as a mixture with toluol, benzol, etc.

(8) In the manufacture of quartz crystal oscillators.

XYLOL VAPOUR

Concentration in Air in Industrial Processes

The actual concentration of xylol in the air of rooms where it is being used in different processes of colour-printing was investigated by Brezina and Schmidt (1934). They found that the greatest concentration (13 to 50 mg. per l.; 3,000 to 11,500 p.p.m.) occurred immediately over the colour trough, and it was still high (2·4 mg. per l., 550 p.p.m.) at head level near the machine. On both sides of the machine at head level, the chief site of work, the concentration was also generally higher than desirable (1·5 to 4 mg. per l., 350 to 920 p.p.m.); the middle of the room showed a concentration of about 1·5 mg. per l., while in more distant parts the amount of xylol vapour was negligible. It was found possible to lower these concentrations significantly by suction ventilation producing different velocities of air currents, and when the apparatus was stopped, the concentration rose considerably. Near the machine, for example, a concentration of 2·4 mg. per l. (550 p.p.m.) was reduced to 0·1 to 1·6 mg. per l. (23 to 360 p.p.m.) by a current of air moving at 14 cm. per second. In the quartz crystal industry, concentrations up to 440 p.p.m. were observed by Schulte (1945).

Maximum Permissible Concentration

The maximum allowable concentration suggested by Greenburg and Moskowitz (1945) is 200 p.p.m.

Estimation in Air

A spectrophotometric method of estimating the amount of xylene, even when mixed with toluene, in the air has been recently described by Luszczak (1935).

TOXICITY

As in the case of toluene the exact degree of toxicity of xylene, both in animals and human beings, is still uncertain. Most workers until recently have believed it to be more toxic in producing acute narcosis in animals than benzene and less toxic than toluene, but Estler (1935) regards it as more toxic than either. Judged by its chronic effects in human beings xylene appears to be certainly less toxic than benzene, its effect on the blood especially being decidedly less, but whether it is more or less toxic than toluene is uncertain.

It is probable that some of the cases in the colour-printing industry reported as xylol poisoning are not due to the effect of xylene alone, e.g. in the cases reported by Stocké (1929), where only chemical analysis which revealed the solvent used to be pure toluene, prevented these cases from being reported as "xylol" poisoning. Hirsch (1932) states that in Germany 2 million kilogrammes of xylol-containing solvents are used yearly in the colour-printing industry. Hirsch gives the chemical analysis of solvents used by the workers in whom he reported toxic effects as follows: colour solvent, 16 per cent xylol, diluting fluid, 64 per cent xylol in one sample and 87 per cent in the other. The remainder consisted of benzol and toluol. Hirsch considered the injurious effects to be due to the xylol. Where the solvents used appear to have contained a low or uncertain proportion of xylol they are described under the heading *Mixtures of Xylol Toluol and Benzol*, p. 61.

Two cases in the leather industry with nervous symptoms predominating were reported to the Home Office in 1934.

TOXIC EFFECTS IN ANIMALS

Excretion

According to Jost (1932), earlier researches showed that mixtures of the three isomers of xylene were oxidized to *ortho*-, *meta*-, and *para*-toluic acid respectively. Apparently only one methyl group of the xylene molecule is oxidized, and the resultant product is excreted partly as toluric, partly as free toluic acid. Jost states that it is uncertain whether xylene can be converted to cresol or not, but it is significant that he found no increase in phenol, sulphuric acid or glycuronic acid in the urine of colour-printers, as in the case of those exposed to benzol.

Lethal and Narcotic Doses

Lethal dose. The three xylenes differ in their toxicity, the increasing order according to Chassevant and Garnier (1903) being *ortho*-xylene→*meta*-xylene→ *para*-xylene. The lethal doses were found to be 2·22, 1·65 and 1·36 ml. per kg. body weight respectively. The lethal concentration in inhalation of xylene vapour, according to Lazarew (1929a), gives a different order, increasing from *para*-xylene (11,500 p.p.m. or 50 mg. per l.) to *ortho*-xylene (10,350 p.p.m. or 45 mg. per l.) to *meta*-xylene (7,000 p.p.m. or 30 mg. per l.). Batchelor (1927) found the lethal concentration about the same as that of toluene (1,600 p.p.m.) and less than that of benzene (2,440 p.p.m.).

Narcotic dose. According to Lazarew (1929a) and Estler (1935), narcosis commences with inhalations of from 10 to 15 mg. per l. (15 to 20 mg. for *ortho*-xylene, according to Lazarew). Batchelor (1927) was unable to produce narcosis with concentrations of less than 5 mg. per l.

Acute Poisoning

Symptoms

According to Batchelor (1927), the picture presented by the animals subjected to intraperitoneal injection of xylol in dosage equal to that which in the case of benzol produced severe neuro-irritant symptoms was one of narcosis only. The narcotic effect was more complete and prolonged than that seen with benzol or toluol, though appearing somewhat later. Estler (1935), however, who considers xylol more acutely toxic than toluol or benzol, describes ataxia as a prominent symptom, developing later into paralysis of the hind limbs. He also mentions a characteristic chattering of the teeth and lower jaw, coldness of the animal, and a red coloration of the mucous membranes, observed earlier by Wojciechowski (1910).

Lesions of the Internal Organs

Neither with subcutaneous injections, nor with inhalation, did Batchelor's animals show any characteristic lesions of the internal organs, with the exception of slight injury to the kidneys and liver, and hyperplasia of the bone-marrow. Smyth and Smyth (1928) found evidence of inflammation of the lungs.

Effect on the Blood

A slight reduction in the red cell count, from 9 to 15 per cent of the normal value, was the only effect noted by Batchelor in his injection experiments with xylol. Engelhardt (1935) found the changes in the blood of dogs and cats injected with 0·5 ml. of xylene per kg. body weight similar to those observed with toluene, viz. some reduction of red cells and haemoglobin, with polychromasia and normoblasts, no leucopenia but a leucocytosis, usually a relative lymphocytosis and some toxic granulation of the white cells. The displacement to the left was however greater with xylene than with toluene. Leucocytosis has been stated also by Woronow (1929) to be a constant result of xylene injection in animals.

Chronic Poisoning

Effect on the Blood

No changes appeared to be produced in Batchelor's animals by inhalation of concentrations of xylol below the narcotic dose, but Farber (1933) claims to have produced in rabbits some variations in the granulocytes and red cells, and slight hyperplasia of the bone-marrow.

Red cells. Practically no change in the red cell count was observed by Batchelor with inhalations of 620 p.p.m. of xylol (2·7 to 7 mg. per l.), the effect being less than that of toluol. Engelhardt found the changes similar to those produced by toluol—slight reduction of red cells and haemoglobin.

White cells. Batchelor's experiments showed slighter evidence of injury to white cells than with toluol—a slight leucopenia in some cases. Engelhardt found a leucocytosis in nearly all cases, sometimes an initial relative lymphocytosis, but more generally a lymphopenia, and very rarely degenerative changes in the white cells.

It is to be concluded that the effect of chronic xylol poisoning on the haemopoietic system of animals is much less than that of benzol, and probably less than that of toluol. This conclusion had already been reached by the Medical Research Committee in 1919, on the basis of injection experiments carried out by Durham. The examination of a xylene compound by Dale, with a view to substituting it for benzene, led him to state that "from a practical point of view it seems that xylol compound could be used as a rubber solvent, with at any rate much less danger of causing aplastic anaemia than benzol."

TOXIC EFFECTS IN MAN

Acute Poisoning

No fatal cases from inhalation of vapour consisting wholly or chiefly of xylene have been recorded. The fatal case reported by Koelsch, quoted by Stocké (1931), was due to the use of "Inertol" described under the heading *Mixtures of Toluol, Xylol and Benzol*, p. 61. A fatal case of xylol poisoning in a copper etcher "through neglect of adequate precautions" is also mentioned by Schilling

(1935). He gives no details of the case, but it is classed as an acute disorder of the myeloid system. Non-fatal cases show symptoms characteristic of a nerve poison—giddiness, "intoxication", even loss of consciousness. Later symptoms described as "neurasthenic" were first reported by Rosenblatt (1902), in a man employed in the rubber industry. They included fatigue, giddiness, palpitation, burning in the head, "drunkenness", numbness of the hands and feet, shivering, dyspnoea and anxiety. More severe poisoning may lead to complete unconsciousness, as in two cases reported to the Home Office in 1934. In both these cases recovery from the acute poisoning was followed by excitement, gastric pain and headache, loss of sleep and nervousness, and in one case tremor and "pins and needles of the hands and feet". It should be remarked that the solvent used by these men, workers in leather manufacture, contained 30 per cent of xylol as well as butyl acetate, ethyl acetate, and diamyl phthalate. Nausea and vomiting were also present in Rosenblatt's case, and in the case of a lacquer worker reported by Brezina (1921).

Chronic Poisoning

Symptoms

The symptoms of chronic xylol poisoning in man appear to be very similar to those of toluol, but two additional effects have been described by German workers, viz. cardiac and circulatory injury (Hirsch, 1932) and kidney injury (Rosenthal-Deussen, 1931). Since the latter effect was due to "Inertol", however, this is described under *Mixtures of Xylol, Toluol and Benzol*, p. 61, though Rosenthal-Deussen and Stocké (1929) appear to correlate the particular effect with xylol rather than toluol. The cases of severe "aplastic anaemia", reported by Glibert (1935) and by Verhoogen (1934), the latter's case a fatal one, cannot safely be ascribed to pure xylene poisoning, since the composition of the xylol is not given.

Nervous symptoms. Headache, fatigue, loss of sleep, stupor, giddiness, nervous irritability, and paraesthesia were present in about one third of the 399 colour-printers examined by Nelken (1931), while headache, dizziness and lassitude were also complained of in colour-printers examined by the Home Office in 1929, and giddiness and sleepiness in a silk finisher in 1935. In one colour-printing works where enquiries were made by the Home Office in 1934, it was stated that "a new worker was likely to be affected by the fumes for the first two or three days and might have to take a day off work, complaining of distaste for food, but acclimatization soon occurs."

In a series of cases investigated by Browning during 1945-7, the principal nervous symptoms found were sleepiness, headache and lassitude. Only 3 cases compained of actual giddiness.

Gastro-intestinal symptoms. Anorexia, nausea and flatulence were present in the Home Office cases, and were also prominent in Nelken's cases. In Browning's series of cases slight indigestion was a fairly frequent complaint, but actual vomiting was rare.

Local irritation. Irritation of the eyes (one case reported to the Home Office (1929) and 25 cases in Nelken's series), irritative lesions of the skin, mild desquamatory dermatitis in one case (a silk finisher) reported to the Home Office in 1935, and pustules, furuncles and abscesses in 13 of Nelken's cases, and dryness of the nose and mouth are the chief irritative symptoms recorded. Nasal catarrh was observed in 3 cases in Browning's series.

Circulatory disorders. Abnormalities of the heart and blood vessels accompanied by low blood pressure have been described by Hirsch (1932) in a large number of colour-printers, and he makes the tentative suggestion that these may be directly related to xylene injury. Two cases occurred in this factory: one man died with symptoms of aplastic anaemia, including epistaxis and bleeding from the gums, and the other from sepsis. Both cases were found post mortem to have valvular endocarditis and thickening of the aorta. An examination of other workers in the same factory revealed, radiographically, abnormalities in the size and form of the heart, and dilatation of the aorta, particularly striking in relation to their age; many had low blood pressure, and some an increase of urobilin and urobilinogen in the urine. Hirsch discusses the possibility that xylol poisoning may constitute a favourable condition for the production of inflammatory or degenerative lesions of the heart and blood vessels, and that it may lead to functional injury of the vessel walls with loss of their elastic elements. He mentions that in his own experience xylene produces a characteristic feeling of heat, reddening of the face and dryness of the throat, which he believes to be due to dilatation of blood vessels.

Effect on the Blood

The majority of observers agree that xylene has considerably less effect on the haematopoietic system than benzene. Only a few German workers, notably Adler-Herzmark and Selinger (1931), Adler-Herzmark (1933), and Hirsch (1932) dissent from this view and suggest that "all three compounds, benzene, xylene, and toluene, are to be considered equal as far as industrial occupational hazards are concerned."

No severe anaemia was noted in Browning's series, only 7 patients showing a count below 4·5 millions, and none below 4 millions, and only 3 a haemoglobin level below 85 per cent.

Table 14 shows the comparison between this series of xylene workers and a larger series of photogravure workers exposed to benzene, where the injury to the haemopoietic system was characteristic.

TABLE 14

Comparative effect of benzene and xylene on the blood

Solvent	Total no. of cases	R.B.C.				Hgb.		W.B.C.						Poly-morphs	
		4–4·5 million		Below 4 million		Below 85%		Above 10,000		4,000–5,000		Below 4,000		Below 50%	
		No.	%	No.	%	No.	%	No.	%	No.	%	No.	%	No.	%
Benzene	70	22	31·5	7	10	19	27	4	5·5	14	20	10	14·5	30	43
Xylene	44	7	16	—	—	3	7	—	—	3	7	—	—	7	16

It will be seen that over 80 per cent of the xylene workers showed a blood picture within normal limits, while abnormalities were shown by very few compared with the benzene workers.

It should be remarked that there is no definite evidence that the cellulose lacquerers, on the results of whose examination the statement of the German workers quoted above is based, were exposed to xylene alone, uncontaminated by toluene, or which is more important, by benzene. The same applies to the very few cases of severe or fatal aplastic anaemia following exposure to xylene recorded in the literature by Glibert (1935) and Verhoogen (1934). In

Browning's experience, the characteristic severe haematopoietic depressive effect of benzene has been completely absent in all the workers examined, when the xylene used was known to be uncontaminated with benzene.

White cells. Leucocytosis is more often recorded than leucopenia, which is rarely reported. Nelken found leucopenia 2·55 times as often in colour-printers as in other printers not using xylol, and Brocher (1929) found leucopenia in severe cases of poisoning only, but Adler-Herzmark found the white cell count less disturbed in xylol than in toluol poisoning, and Hirsch found an absolute reduction of white cells in none of his cases, and an increase of over 8,000 in 7 out of 34.

In Browning's series of 44 cases, none showed a fall below 4,000 c.mm., and only 3 fell below 5,000.

A relative lymphocytosis has been recorded by most workers, including Brocher, Hirsch, Adler-Herzmark and Selinger (1931), but not of a high degree —the highest percentage in Hirsch's series was 58 per cent. Nelken (1931) on the other hand found an almost equal relative lymphocytosis in the workers using xylol and in those that did not. In Browning's series, 7 of the 44 showed relative lymphocytosis, but in only 2 of these was it associated with neutropenia.

Red cells and haemoglobin. No serious anaemia has been recorded except in the cases described by Glibert (1935) and Verhoogen (1934). Six cases of "aplastic anaemia" in colour-printers were recorded by the former, with symptoms of vertigo, haemorrhage from nose and gums, abnormal pallor, weakness, reduced haemoglobin and red cells, aniso- and poikilo-cytosis and chromatophilia, a relative lymphocytosis but no leucopenia. The analysis of the "xylol" is not given, Glibert merely remarking that xylol used in colour-printing shows "variability of composition" and that xylol called yellow or light is pure benzene, while "green" or "heavy" is pure xylene, and that "pseudo-xylol" contains toluene with 15 per cent benzene. The fatal case recorded by Verhoogen showed an intense anaemia and leucopenia (red cells 800,000 and white cells 600) as in a severe case of benzol poisoning. Verhoogen states that the intoxication was due to xylol used in a printing works but does not give the analysis of the xylol. Adler-Herzmark (1933) states that the red cell picture is more disturbed (i.e. by the occurrence of abnormal forms—granulated and nucleated cells) than in toluol poisoning. Hirsch (1932) found the red cell count below 5 million in 15 out of his 34 cases. In only a few of Nelken's (1931) cases was the number of red cells estimated, and these showed no values less than 4·3 millions. The haemoglobin values were, however, reduced below 80 per cent in about 16 per cent of Nelken's cases and in over half of those of Hirsch. Hirsch suggests that the enlargement of the heart observed in many of his cases may possibly be correlated with the anaemia produced by xylol, but this anaemia appears to have been slight and not constant, while other workers, including Brocher (1929), have failed to find any anaemia at all.

EFFECT OF MIXTURES OF XYLOL, TOLUOL AND BENZOL

While many of the observations classified above under either toluene or xylene are probably not to be regarded as purely effects of one or other substance, there are at least three investigations resulting in the description of symptoms which have not been attributed to either of the separate substances, in which the solvent used could not with any certainty be called either xylene or toluene. They will therefore be given as examples of the effects of mixed solvents containing xylene, toluene, and perhaps other substances.

Intaglio-printing Solvent

The symptoms found by Brachmann (1937) in an investigation of 105 intaglio printers using a mixture of 50 per cent xylene and 50 per cent toluene are shown in Table 15.

TABLE 15

Symptoms of poisoning by a mixture of toluene and xylene

Symptoms	Percentage of cases
Headache	30
Fatigue	50
Insomnia	9
Irritability	23
Vertigo	36
Oscillation and burning of eyes	18
Dry nose	9
Bleeding from gums	9
Dry throat	40
Lack of appetite	36
Thirst	50
Intolerance of alcohol	36
Vomiting	27
Pressure in gastric region	45
Excessive sweating	27
Sensation of cold	18

Colour-printing Solvent

Symptoms of endogenous psychosis were the predominant feature in a colour-printer examined by Panse and Bender (1934). The solvent used contained 12 per cent of xylol and toluol, 6 per cent of esters, some glycol derivatives, and some unsaturated hydrocarbons. In addition to the symptoms of nervous and gastro-intestinal disorder already described by other authors, this worker showed considerable psychic disturbance—restlessness, disorientation, loss of memory, auditory hallucinations, extreme irritability. He recovered on changing his work, and since there was no family history of psychosis or of any previous attack, Panse and Bender attribute the symptoms to xylol and toluol poisoning. The blood picture was that of a leucocytosis with relative lymphocytosis.

Paint Solvents containing some Benzene and, possibly, some Toluene

In a series of 37 cases examined by Browning (1945, unpublished) where the operatives were employed in the manufacture of paints and inks containing xylol of uncertain composition, the level of disturbance of the blood picture was higher than that of the series of workers exposed to xylene alone, but considerably lower than that of the 70 workers exposed to benzene. Among these 37 operatives slight leucopenia (in no case less than 4,000 per c.mm.) was present in 14 (37·5 per cent), neutropenia in 7 (19 per cent), anaemia of relatively slight degree (4 million r.b.c. per c.mm.) in 6 (16 per cent). There was one case of leucocytosis (12,500 per c.mm.).

"*Inertol*"

"Inertol" is widely used as a weatherproof paint in Germany and many cases of injury from its use, including one fatal case (though, as will be seen below, death was in this case not directly attributable to exposure to "Inertol" alone) have been recorded. It is stated by Rosenthal-Deussen (1931) to contain 35 per cent of benzol and its homologues, chiefly xylol, and in the works in which the cases described by her occurred, the air contained 30 mg. per l. of the vapour. The symptoms were in general similar to those described for toluol and xylol —giddiness, "intoxication", malaise, weakness, loss of appetite, vomiting, and eye irritation—but in addition there were symptoms which pointed to injury of the kidneys. Out of 20 workmen questioned by Rosenthal-Deussen, all stated that they had "coffee-brown" urine, and 2 reported "blood-red". One patient died after an operation for ileus, and post mortem, swelling of the parenchymatous tissue of the kidney was found. A further investigation of the effects of "Inertol" is described by Stocké (1931) who lays further emphasis on the signs of kidney injury, not hitherto described in cases of xylol or toluol poisoning.

Pain in the kidney region, coffee-brown or blood-red urine, and difficulty in micturition are mentioned as the chief signs of kidney lesion, while in one case, red blood cells, albumin and sugar were found in the urine. Stocké considers that such kidney injury may have arisen from the unusually high concentration of xylol and toluol vapour inhaled. Polyneuritis is suggested as a possible consequence of exposure to "Inertol" (Lüthy, 1940), but the evidence is inconclusive. The blood picture in this case, a man who complained of weakness of the legs after painting a cellar under conditions of poor ventilation, showed a leucocytosis of 10,000 per c.mm.

4. Ethylbenzene

(*Phenyl Ethane*)

$C_6H_5 \cdot C_2H_5$

PROPERTIES

A colourless liquid. B.P. 136·5°C. Sp.Gr. 0·87. Fl.P. 59°F. (15°C.) Insoluble in water, but soluble in alcohol or ether.

USES

(1) As a diluent in lacquers.
(2) As an "anti-knock" agent (Patty, 1949).

TOXICITY

From the point of view of acute toxicity as judged from the results of animal experiments, ethylbenzene appears to be slightly more toxic than benzene. No investigations into its chronic toxicity have been made, and no report of injury from its industrial use has been recorded. It has, however, a marked irritative effect on the skin and eyes.

TOXIC EFFECTS IN ANIMALS

Acute Poisoning

For mice, the lethal dose by inhalation is 45 mg. per l. (1·04 vols. per cent) as compared with 45 mg. per l. (1·36 vols. per cent) for benzene. Death results primarily from injury to the central nervous system (Yant, Schrenk, Waite and Patty, 1930). Prostration occurred at 15 mg. per l. (0·35 vols. per cent) compared with 15 mg. per l. (0·95 vols. per cent) for benzene (Lehmann and Flury, 1943).

Chronic Poisoning

No other animal experiments have been recorded.

Metabolism

Absorption is chiefly by inhalation. Ethylbenzene, like toluene, appears to be converted into hippuric acid in the body (Nencki and Giacosa, 1880).

TOXIC EFFECTS IN MAN

No systemic injury has been reported, but according to Oettel (1936) ethylbenzene is the most severe skin irritant of the benzene series. Its irritant effect on the eyes, occurring at concentrations of about 200 p.p.m., gives warning of dangerous concentrations; at about 1,000 p.p.m. the irritation is severe but transient; at 2,000 p.p.m. profuse lachrymation and moderate nasal irritation occur.

5. Cumene

(iso*Propylbenzene*)

$C_6H_5 \cdot CH(CH_3)_2$

PROPERTIES

Cumene is one of the higher homologues of benzene. The pure product, prepared synthetically, has a B.R. of 151·8–152·5° C. Sp.Gr. at 25° C. 0·837. Insoluble in water, but soluble in ethyl alcohol or ether. The technical product, obtained by fractionating petroleum, is usually 95 per cent or more pure, and has a B.R. of 151–156·2° C. (Ward and Kurtz, 1938).

USES

(1) As a constituent of solvents for cellulose lacquer, though not extensively.
(2) In organic synthesis.

TOXICITY

Cumene is a potent narcotic, with slow induction and long duration, but since it is of relatively low volatility, its acute industrial hazard should be easily controllable. Its slow elimination suggests the possibility of a cumulative effect, but so far none has been reported in human beings.

TOXIC EFFECTS IN ANIMALS

Acute Poisoning

The narcosis induced in animals by inhalation of cumene is entirely depressant in nature; there is no initial central stimulating effect as with benzene, and narcosis develops more slowly and lasts longer. The investigations by Werner, Dunn and von Oettingen (1944) indicate a higher acute toxicity for cumene than that postulated by Lazarew (1929a) who gives the dose for prostration of mice as 20 mg. per l., and the dose for loss of reflexes as 25 mg. per l. According to Werner *et al.*, the M.L.D. is 10 mg. per l. (2,000 p.p.m.). On this basis cumene appears to be more acutely toxic than either benzene (19·9 mg. per l.) or toluene (33·1 mg. per l.), but comparing them on the basis of their relative saturation concentrations, the "hazard index" (saturation concentration divided by M.L.D.) is as follows:

	Cumene	4
	Toluene	10
	Benzene	12

Subacute Poisoning

Animals which died after 7 hours' exposure to cumene showed fatty changes in the liver and kidneys, and phagocytosis of the follicles of the spleen (Werner, Dunn and von Oettingen, 1944). At lower concentrations than benzene or some benzene-toluene-xylene mixtures, some animals showed more liver damage, but at similar concentrations of toluene, the damage was about the same. It is suggested that the slow elimination may be associated with a cumulative effect.

Metabolism

Feeding experiments carried out by Nencki and Giacosa in 1880 indicate that only a slight portion of cumene undergoes oxidative degradation, and that there is some increase in conjugated urinary sulphates.

TOXIC EFFECTS IN MAN

No ill-effects have been reported.

6. Tetrahydronaphthalene
(*Tetralin*)

$C_{10}H_{12}$, i.e.

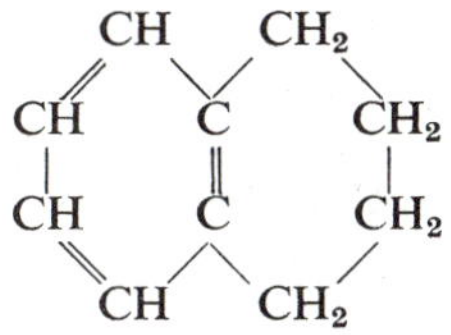

[Structural formula of tetralin]

PROPERTIES

A colourless liquid, with a characteristic smell, somewhat like that of naphthalene, formed by the hydrogenation of naphthalene in presence of catalysts. Neither corrosive nor highly inflammable, its flash point being relatively high. Oxidizes on exposure to air, leaving explosive resinous residues. Only slightly volatile; its rate of evaporation is much slower than that of turpentine (1 : 6). Sp. Gr. 0·973–0·980. B.R. 205°–215° C. Fl.P. 78° C. It is a powerful solvent for oils, resins, waxes, rubber and fats.

If tetralin is treated with more hydrogen without carrying the reaction to completion, "Tetralin Extra" is obtained, which is a mixture of tetralin and dekalin (decahydronaphthalene) only. "Tetralin Extra" is a more powerful solvent than either tetralin or dekalin alone. Sp. Gr. 0·90. B.R. 185–205° C. Fl.P. 60° C.

USES

(1) Owing to its powerful solvent action and slow rate of evaporation, tetralin is specially suitable for the preparation of wax polishes, floor and boot polishes, etc., and of paint removers.

(2) As a solvent for rubber.
(3) When mixed with dekalin or white spirit, as a paint thinner.
(4) As a degreaser and as a substitute for turpentine and petroleum spirit in Germany.
(5) As a fumigant for the clothes moth, since it is cheap, non-injurious to fabrics and can be sprayed (Colman, 1934).

TOXICITY

The only toxic effects of tetralin observed in human beings have been a few cases of dermatitis, some of vague nervous disturbance, irritation of mucous membranes and the production of a green coloration of the urine, which is not apparently associated with any injury to the kidneys. These effects have been recorded chiefly by German investigators, Koelsch (1926) and Rockemann (1922), while English observers, Heaton (1923) and Coleman and Bilham (1922) regard tetralin as relatively innocuous. Until about 1942, experiments on animals appeared to support the English view, but the experiments in Italy of Cardani (1942a), and those in France of Badinand, Paufique and Rodier, (1947) have led these investigators to compare tetralin, and also dekalin, (p. 68) with the halogenated hydrocarbons (chloroform, carbon tetrachloride, tetrachloroethane and dichloroethane) with respect to their toxic action on both the liver and the kidneys.

TOXIC EFFECTS IN ANIMALS

The results of early investigators indicated that tetralin, whether taken by mouth or given by gastric administration, had an injurious effect upon the kidneys and produced some metabolic disturbance evidenced by the excretion of green urine. They also tended to show that by inhalation little toxic effect was to be expected. Later investigations, however, indicate that though the green urine is more definitely noticed after oral administration, the toxic effect of tetralin is similar by whichever route it is administered, and that injury to the liver and the kidneys can follow inhalation.

Method of Administration

By ingestion or gastric administration. The passing of dark greenish urine has been observed by nearly all observers, except Pohl and Rawicz (1919). Rockemann (1922) found that even quantities (2 g. per day for 14 days) which produced no symptoms did produce diminution of the quantity of urine with the characteristic green coloration, though with no histological change in the kidneys; oliguria, albuminuria, and casts followed a dosage of 5 g. per day. In the experiments of Cardani (1942a), animals, ingesting 0·25 g. daily mixed with their food, lost weight, passed dark greenish urine containing red blood corpuscles, urobilin, indican and cylindrical cells, and died after 35 days. They were at first apathetic, then restless, and had intense diarrhoea.

By cutaneous application. When tetralin was applied to a depilated area for 5 minutes every day Cardani found that his animals died after 16 days; the area of skin became eczematous, and post-mortem examination showed congestion of the liver and zonal necrosis of the kidneys.

By inhalation. In contrast to the reports of earlier workers, Geppert (1926), and the Ministry of Health in Berlin (1920) cited by Heaton (1923), found that tetralin vapour, except for being slightly narcotic, is relatively non-toxic.

Cardani (1942a) and Badinand et al. (1947) found the effects of inhalation very similar to those of ingestion; inhalation of 1·48 mg. per l. for 8 hours daily led finally to death. The chief difference, in Cardani's investigation, was the non-appearance of the green colour in the urine following inhalation; he suggests that this is due to transformation products formed during digestive absorption.

Symptoms

According to Cardani, the symptoms of chronic tetralin poisoning are essentially the same whatever the method of administration. After a week the animals show loss of weight, roughening of the fur, restlessness or apathy, anorexia, sometimes torpor, sometimes tremors. When tetralin is taken by the mouth they also have intense diarrhoea. When it is applied to the skin they develop eczema of the crust-forming type. The urine is diminished in amount, is greenish after ingestion, dark-coloured after inhalation, and contains albumin, red blood cells and cylindrical cells.

Eye injuries. Cardani (1942a) observed in rabbits corneal opacity and diffuse clouding of the retina after administration of 0·2 g. daily, while Badinand and co-workers (1947) found in guinea-pigs definite signs of cataract after 6 days of administration.

Effect on the Blood

Some anaemia, neutropenia, and after long exposure, a tendency to leucopenia was observed by Cardani (1942a) and Cesaro (1941). Cesaro suggests that the anaemia is associated with a diminution of the resistance of the red cells owing to penetration of their lipoid surface layer by tetralin.

Metabolic Effects

Tetralin appears to exert a metabolic effect, in the sense of a reduction of the oxidative processes of the body as shown by changes in the urinary quotients $C : N$, $O : N$, and $O : C$ (Kanitz, Lohmeyer and Scholz, 1935).

These workers also state that with long-continued administration, tetralin attacks the intermediary metabolism and leads to increased protein excretion. The green colour appearing in the urine after ingestion is believed to be an expression of transformation products such as tetralol which, according to Lewin (1920), Rockemann (1922), Pohl and Rawicz (1919), is excreted as a conjugated glycuronic acid. This conjugated tetralol glycuronic acid was found by Geppert (1926) in the urine of one animal which had inhaled tetralin from an anaesthetic mask that had been warmed to increase the concentration.

Lesions of the Internal Organs

Cardani observed very definite signs of injury to the liver and kidneys following both ingestion and inhalation, thus confirming and amplifying the findings of Kanitz and co-workers (1935) of "slight swelling of the liver and kidneys, with yellow pigmentation of the peripheral layers." The liver showed acute toxic atrophy in all stages, from hyperaemia to fatty degeneration, and atrophy of the hepatic cells, especially in the lobules, but with no signs of regeneration. The kidneys showed necrotic nephrosis, especially in the tubules, and all stages of inflammation from hyperaemia to necrosis. Marked epithelial nephritis was also observed by Badinand and co-workers (1947).

Ulceration of the gastric mucosa was found in animals ingesting tetralin, and focal bronchopneumonia in those subjected to inhalation.

Symptoms

According to the personal experience of Heaton (1923), working all day in a closed space with paint containing tetralin produced no discomfort or after-effects, while the Report of the Ministry of Health in Berlin (1920) states that inhalation of the concentrated vapour for half an hour produced only headache and sickness which passed off on removal to fresh air. Other workers, however, while not observing very serious symptoms from exposure to the fumes of tetralin, particularly from the use of floor wax, have observed the green coloration of the urine, and Rockemann especially suggests caution about the use of such waxes in the presence of individuals who may be suffering from disorder of the kidneys.

This warning receives confirmatory emphasis from the findings of severe renal lesions in animals as described above.

According to Koelsch (1926) the vapours of tetralin produce in man slight stupor, headache, nausea and vomiting, and conjunctival and throat irritation with sneezing and cough. These symptoms were not, however, observed in the cases reported by Rockemann (1922) and Geppert (1926) as examples of the effect of tetralin-containing floor wax. Geppert records that of a number of children sleeping in a room freshly waxed with a polish containing tetralin, those in the lowest beds had green urine, but no other symptoms. In Rockemann's series of similar cases, the children placed in the freshly waxed rooms (he states that 3 kg. of wax corresponding to 1·5 kg. of tetralin was used) not only had green urine but were very restless, though he admits that the restlessness may have been a nervous reaction to the strong smell, rather than a direct effect of the tetralin vapour. Skin lesions, eczematous in nature, have been observed by Galensky (1922, cited by Cardani) in varnishers who had prolonged contact with tetralin.

Excretion

In man the green pigment in the urine has apparently a different origin from the tetralol glucuronic acid formed in animals. This pigment with oxidizing reagents gives rise eventually to dihydronaphthalene and naphthalene (Pohl and Rawicz, 1919).

7. Decahydronaphthalene
(*Decalin*)

$$C_{10}H_{18}, \text{ i.e.} \quad \begin{array}{c} CH_2 \\ / \quad \backslash \\ CH_2 \quad CH \quad CH_2 \\ | \quad | \quad | \\ CH_2 \quad CH \quad CH_2 \\ \backslash \quad / \quad \backslash \quad / \\ CH_2 \quad CH_2 \end{array}$$

PROPERTIES

Decahydronaphthalene is a solvent very similar to tetrahydronaphthalene, but with a lower boiling point and flash point, a stronger odour of different character and a less strong solvent action.

B.R. 183–192° C. Sp.Gr. 0·887–0·890. Fl.P. 57° C. Evaporation time, slower than turpentine (1 : 3) but more rapid than tetralin. Good solvent for gums and resins, especially dammar, manilla and mastic, but not for hard gums like kauri and copal.

USES

In the paint industry dekalin is used as a thinner for all types of paint and enamel (Heaton 1923).

TOXICITY

Although no toxic effects in human beings have been reported other than some cases of eczema and one case showing an indication of kidney injury, the results of animal experiments would appear to indicate that dekalin has an action on the kidneys and on the liver somewhat similar to that of tetralin, though without producing the characteristic green coloration of the urine that tetralin does.

TOXIC EFFECTS IN ANIMALS

Lethal Concentration

The lethal toxicity of dekalin is about the same as that of tetralin, but death occurs slightly more rapidly with dekalin (Cardani 1942b).

By oral administration. According to Cardani (1942b) administration of 0·2 g. daily caused death in 20 to 22 days.

By inhalation. Inhalation of 1·8 mg. per l. for 8 hours daily caused death after 8 to 23 days, preceded by clonic convulsions.

By cutaneous application. The area of application showed much hardening and crusting. The urine was dark-coloured, and contained albumin, urobilin, red blood cells and cylindrical cells.

Lesions of the Internal Organs

The liver showed toxic changes, varying from congestion and commencing atrophy following oral and cutaneous administration to acute yellow atrophy in one animal following inhalation. The kidneys showed necrosis by all three routes.

TOXIC EFFECTS IN MAN

Eczema was observed by Koelsch (1926) and by Cardani (1942b). The case described by Cardani was that of a man employed in cleaning paving stones with dekalin and other detergents. He showed intense pruritis and vesicular eczema of the areas most in contact with dekalin (the flexor surfaces of the forearms and the sacral region) and skin sensitivity tests gave a reaction to dekalin, but not to other detergents. Kidney involvement was suggested by the fact that his urine contained minimal quantities of albumin and urobilin and the sediment a few leucocytes.

8. Methylated Naphthalenes

$C_{11}H_{10}$, i.e. [1-methylnaphthalene structure with CH_3] and [2-methylnaphthalene structure with CH_3]

Liquids varying in colour from light straw to dark brown, according to the degree of refinement. "Methyl naphthalene", which consists of 90 per cent 1-methylnaphthalene and 10 per cent 2-methylnaphthalene is a dark brown liquid with a B.P. of 243° C.

TOXICITY

Methylnaphthalenes have both an irritant and a photo-sensitizing action on the skin. Animal experiments have not shown them to have any carcinogenic effect, but according to Svirbely (1946) inhalation of vapours with a particle size of 2 to 5 microns causes systemic effects as well as irritation of the skin.

TOXIC EFFECTS IN ANIMALS

According to I. G. Elberfeld Toxicology Index (1931) concentrated application to the skin of animals for one hour produced marked redness. No carcinogenic effects were observed after twenty-five applications in oil for $9\frac{1}{2}$ hours three times weekly.

TOXIC EFFECTS IN MAN

Irritation

Dunn and Brockett (1948) observed marked irritation of the skin which increased directly with the degree of refinement. I. G. Elberfeld Toxicology Index (1931) contains the account of two persons working with methylated naphthalenes who were affected with severe skin lesions.

Photo-sensitization

Unlike the capacity for producing skin irritation, the photo-sensitizing property decreases with the degree of refinement. Dunn and Brockett (1948) consider this to be so because the tarry residue of the cruder methylnaphthalenes contains a photo-sensitizing constituent. Photo-sensitization was demonstrated by means of patch tests. An unpleasant pricking sensation occurred within a few minutes, followed by erythematous reaction, and sometimes also by oedema and a "pig-skin" appearance. These signs were most distinct within 3 to 4 hours after exposure and were only slightly faded by the following day. The photo-sensitizing substance was found to have an absorption spectrum with the peak in the region of 3,300–3,400 Å and comparable to that found in coal tar and petroleum.

9. Coal Tar Solvent Naphtha

CONSTITUTION

Coal tar solvent naphtha is to be sharply distinguished from the petroleum distillate naphtha, though its composition shows wide variations. It consists chiefly of the higher homologues of benzene, viz. xylene and cumenes, but generally also contains small quantities of hydrocarbons of the paraffin series. According to Hamilton (1934) the so-called "Hiflash naphtha" is not naphtha at all, but a mixture of toluene, xylene, and the still heavier cumenes.

PROPERTIES

Solvent naphtha is a colourless fluid with a characteristic odour. Durrans divides it into two grades, with differing boiling ranges and specific gravities and flash points:

(*a*) *Light grade*. Consists chiefly of toluene, the three xylenes, ethylbenzene and propylbenzene, B.R. 110–160° C. Sp.Gr. 0·865–0·875. Fl.P. 21° C. Solvent for ester gum, copal ester, rubber, bitumen, oils, benzyl abietate, cellulose ether but not for cellulose esters.

(b) *Heavy grade* (*Solvene*). Consists chiefly of pseudocumene, mesitylene, ethyl *p*-xylene and coumarone. B.R. 160–190° C. Sp.Gr. 0·88–0·91. Fl.P. about 38° C. Solvent for ester gum, benzyl abietate, pitch bitumen and polymerized benzyl resin, but not for cellulose esters.

USES

Solvent naphtha, owing to its degreasing quality and good solvent action on rubber, is widely used over a range of industries similar to those in which benzol is used. The output of solvent naphtha in the United States in 1933 was 1,524,000 gallons, while that of crude naphtha was 3,062,000 gallons.

(1) In the rubber industry for the manufacture of waterproof coats, waterproof rope and sail cloth, golf and tennis balls, adhesive solution for shoes, coating of steel sheets, motor car upholstery and the manufacture of rubber tyres.

(2) Naphtha is not much used in the general paint industry, but it has properties which make it suitable in the manufacture of certain types of varnishes and nitrocellulose products, and particularly in bituminous paints. A quick-drying naphtha paint is used in the process of silvering mirrors.

TOXICITY

When man is severely exposed to solvent naphtha, symptoms of acute poisoning very similar to those of benzene are produced, but in animals both its narcotic and neuro-irritant effect are less. Its chronic effects, both in men and in animals, are considerably less severe than those of benzene, toluene or xylene.

TOXIC EFFECTS IN ANIMALS
Acute Poisoning

Lethal and Narcotic Doses

Lethal dose. The effects of inhalation of amounts of coal tar solvent naphtha sufficiently large to produce death or even definite narcosis have not been observed, since a concentration of 567 p.p.m. was about the highest that could be obtained with complete volatilization of all its constituent portions (Batchelor, 1927). The lethal dose by intraperitoneal injection was found by Batchelor to be about 2·5 ml. per kg. body weight (as compared with 1·5 to 1·75 ml. for benzene). The naphtha used was known as "Hiflash naphtha", of boiling point about 156° C.

Narcotic dose. True narcosis was obtained only by intraperitoneal injection of doses greater than 2 ml. per kg. body weight. Doses of 0·25 ml. produced instability, inco-ordination, weakness and paresis, and, as the dose increased, paralysis and extreme lethargy resulted, but at no time did the neuro-irritant effect typical of benzene narcosis appear, and there was no loss of consciousness or of reflexes until near the lethal point. The narcosis was later in onset than with either benzene or toluene and was possibly of longer duration.

Rambousek (1913) considered naphtha to be practically non-toxic, because, on testing the cumenes, which formed a large proportion of the solvent naphtha used by him, he could produce no symptoms after an hour's exposure to concentrations of 60 to 70 p.p.m.

Symptoms

Some investigations on the effect on guinea-pigs and rabbits of narcotic inhalations, subcutaneous injections and skin application of bitume-mastic,* which in human cases of poisoning is classed in the Home Office reports as a naphtha product, were carried out by des Essarts (1932). Inhalation of large doses in a closed space proved fatal to rabbits within an hour, after a period of dyspnoea, followed by convulsions and paralysis of the hind quarters. Blood examinations showed the presence of a slight leucocytosis with an eosinophilia of 4 per cent; the red cells were unchanged. With subcutaneous injection local ulceration was accompanied by loss of weight and cachexia; the blood showed a diminution of the red cells, with slight leucocytosis and eosinophilia. The internal organs, both with inhalation and subcutaneous injection, showed congestion with some fatty degeneration of the liver and kidneys.

Chronic Poisoning

The investigations of Gardner (1925) and of Batchelor (1927) appear to show that with chronic exposure solvent naphtha is relatively harmless, especially with regard to its effect on the blood system. Gardner, using a high-boiling (134–172° C.) coal tar distillate, found the only effects of its inhalation to be those of slight irritation of the mucous membranes of the cornea and respiratory tract, and a mild degree of anaemia; the white cell count was practically unaffected.

Batchelor examined its effects both by subcutaneous injection and by inhalation and in both cases found it to be the least toxic of the benzene series (benzene, xylene, toluene, "Hiflash naphtha"). With subcutaneous injection of 1 ml. per kg. body weight the local irritation produced was greater than that with xylene or toluene but less than that with benzene. Local induration with frequent sloughing occurred, which in time presented a tarry appearance and consistency, probably due to the formation of oxidation products. He states that from the amounts present in the tissues, together with the odour of the naphtha and the higher boiling point, it is probable that it is more slowly absorbed than the other solvents.

The loss of weight of the animal following repeated injections was less than with any of the other solvents except xylene—an average of 4 per cent as compared with 26 per cent for benzene, 6 per cent for toluene, and 3 per cent for xylene.

The effect on the blood picture was slight—a reduction in red cells of 5 per cent and no leucopenia.

The bone-marrow showed hyperplasia, "caused probably by the stimulating effect of blood destruction and blood loss through repeated small bleedings, but also in part due to the possible stimulating effect of the naphtha itself." The liver showed periportal focal necrosis in contradistinction to the more central focal necrosis produced by the other solvents; the kidney lesions varied from a diffuse cloudy swelling to a mild diffuse nephritis; and the spleen showed some congestion and hyperplasia of the Malpighian corpuscles, but much less pigmentation than with the other solvents.

After inhalation of concentrations of 567 p.p.m. the effects were practically negligible, with the exception of slight apathy and weakness, loss of weight, and mucous membrane irritation. The blood cell count was unaffected.

TOXIC EFFECTS IN MAN
Acute Poisoning

Symptoms

Few cases which can definitely be said to be examples of the acute toxic effect of pure solvent naphtha are reported in the literature, but several cases of unconsciousness in men using paint and varnish in which naphtha was the chief solvent have been reported to the Home Office since 1921; apparently the toxic effects of bitume-mastic* paint, of which naphtha is a frequent constituent, are well known. An account of bitume-mastic,* with records of a number of cases of poisoning from its use, is given by des Essarts (1932), but he does not relate these effects specifically to solvent naphtha: he states that bitume-mastic* shows on analysis no fixed formula, but contains ammonia, pyridine, aromatic hydrocarbons and phenols. The symptoms described, however, are very similar to those present in the cases due to handling naphtha-containing paint reported to the Home Office. These may be classified as:

(1) *Mild form.* Slight giddiness, a feeling of intoxication, occasionally slight gastro-intestinal disturbance and bronchial irritation. The latter symptoms were present in two cases reported to the Home Office (1921, 1924) due to the use of a naphtha blow lamp, and one (1934) due to the use of a naphtha paint, while des Essarts described particularly symptoms very similar to drunkenness, with loquacity and incoherence.

(2) *More severe form.* Unconsciousness, followed by fairly rapid recovery, sometimes with sequelae of laryngeal and bronchial irritation, is the commonest form of acute poisoning by solvent naphtha.

Of the twenty-one cases reported to the Home Office between 1921 and 1934, thirteen were of this type, while des Essarts describes five due to bitume-mastic.* In three of these latter a blood examination was made, and des Essarts states that the eosinophilia observed (4 to 11 per cent) is a characteristic feature of these cases. He quotes Pirot as having also observed eosinophilia in similar cases.

(3) *Very severe form.* Unconsciousness deepening into coma with cyanosis and stertorous breathing may occur if exposure is severe and prolonged. Des Essarts records such a case, where the coma was apparently uraemic, since the blood urea was high and the urine contained much albumin. Only one fatal case appears to have been reported to the Home Office (1925) and in this case death was actually due to the fact that one third of the body area was severely burnt, the man having been immersed in a tank of naphtha. The naphtha in this case was of the heavy grade, distilling at 170° C. with Sp.Gr. of 0·740. Another fatal case reported by Saverin in 1940 was that of a man found dead in a tank which he had been painting with a paint containing solvent naphtha.

Chronic Poisoning

Hamilton (1925) describes the results of an inquiry in the United States into the health of rubber workers using solvent naphtha, and some slight symptoms were observed during an examination made by the Chief Inspector of Factories in 1921 of twenty-one girls employed in the manufacture of tennis balls. There are no other cases of chronic poisoning from solvent naphtha reported in the literature, and des Essarts states that none has been recorded from the use of bitume-mastic* paint.

Symptoms

Hamilton records the incidence of both nervous and digestive symptoms among the workers in the rubber factory. The nervous symptoms included headache, dizziness, excitement, poor sleep, twitching of the muscles, coldness and numbness of the fingers and toes. Among the digestive symptoms were loss of appetite, indigestion, nausea, colic, and loss of weight. She also quoted the case of a workman who had been employed for 3 years dipping articles into a rubber-naphtha solution and who had frequent attacks of dizziness, was nervous and araemic and had some loss of power in the right arm.

* From the original French of des Essarts. It is understood that "Bitumastic" is a proprietary name and registered Trade Mark throughout the world and should therefore not be confused with bitume-mastic nor used in any sense as a generic term.

Among the twenty-one girls examined by the Home Office in 1921, a few complained of slight symptoms—nervousness, headache, lassitude, a "sickly feeling", dizziness, etc., while eight showed some pallor, which might have been due to other causes, It appeared that headache, lassitude and giddiness were apt to occur at first but passed off in a few days. A blood examination was made in five of these cases, but the results were not conclusive—two only showed a low red cell count with slight leucopenia and a disturbance of the leucocyte-lymphocyte ratio.

In an investigation of some operatives in the waterproof industry in 1942, Browning (unpublished) found no disturbance of health. The only abnormality of the blood picture was slight anaemia which occurred in one factory where the solvent naphtha was known to contain a small percentage of benzene.

10. Petroleum Spirit
(*Benzine, Benzoline, Petrol, Gasoline*)

CONSTITUTION AND PROPERTIES

There is much confusion in the literature as to the physiological effects of the use as solvents of substances listed under any of the above names, and it is unfortunately increased by the fact that many observers fail to give the exact constitution of the substance under investigation. In many cases they are apparently regarded as one and the same substance; in others a specific distinction is made between benzine and petroleum spirit. Durrans (1933, 1935), for example, states that petroleum spirit or ligroin is known also as lythene, benzoline, benzine and gasoline, and that it consists mainly of hexanes and heptanes, while Lazarew (1929b) classifies benzine itself in several groups, only one of which (Pennsylvanian or Caucasian benzine) consists chiefly of pentanes, hexanes and heptanes. The French workers, especially Duvoir (1928) and Simonin (1934), include both benzine and petroleum spirit in the term *essences de pétrole* and Simonin further distinguishes amongst these the *essences de pétrole* called *essences minérales*, which include petroleum ether, ligroin, automobile spirit, cleaning spirit, varnish spirit, gasoline, rubber solvent petrol, etc.

Properties. B.R. 60–120° C. Sp.Gr. 0·67–0·70. Consists chiefly of aliphatic hydrocarbons, hexane, heptane, octane, etc., but the proportion of unsaturated aliphatic and aromatic hydrocarbons varies with the type of petroleum, the distillation and "cracking" methods, and the admixtures of other materials for the purpose of increasing the solvent property.

USES

The wide range of industries in which petroleum spirit is used as a solvent includes:

(1) Rubber industry.
(2) Motor and engineering works.
(3) Electrical works.
(4) Paint and varnish industry.
(5) Glue works for the extraction of fat from bones.
(6) Shoe industry as a solvent for the rubber adhesive.
(7) Leather industry for degreasing.
(8) Printing works for cleaning machines and as a solvent for colours used in poster-writing.
(9) Manufacture of waterproof plaster for chiropodists, etc.

HYDROCARBONS

TOXIC CONCENTRATION IN ATMOSPHERE

According to Drinker, Yaglou and Warren (1943), mild symptoms of giddiness may appear when the concentration of petrol vapour (gasoline) reaches 0·1 per cent, irritation of the eyes with 0·3 per cent, and coughing, unsteadiness of gait, and marked irritation of the eyes with 0·7 per cent. Intoxication, identical with the effect of alcoholic ether, was produced in 4 to 7 minutes by 1 per cent. According to Bowditch, Drinker, Drinker, Haggard and Hamilton (1940), the maximum allowable concentration is 1,000 p.p.m. (0·1 per cent) as compared with toluene and xylene, 200 p.p.m., and benzene, 100 p.p.m. This limit, however, greatly depends on the content of aromatic hydrocarbons, which is usually 5 to 10 per cent but which may be as high as 30 per cent. Even with 0·5 per cent benzene it has been shown by Smyth and Smyth (1928) that the permissible limit of 100 p.p.m. for benzene could readily be exceeded. In Sterner's (1941) investigation at a spraying workshop where gasoline was used as a paint solvent, the actual concentration of total aromatic compounds ranged from 500 to 800 p.p.m.

TOXICITY

In heavy concentration petrol is an acute narcotic, producing severe, sometimes fatal, depression of the central nervous system. Less heavy, prolonged exposure has been said to produce various manifestations of nervous and digestive disorder, but these are neither very definite nor characteristic. Blood changes are also indefinite, and when present probably depend on the aromatic hydrocarbon content of the petrol.

So far as the acute toxic effects in man are concerned there appears to be little difference between other varieties of petroleum spirit and benzine, and where variations in the symptoms of chronic poisoning are recorded, these seem to arise from the difference in constitution, especially with regard to admixtures of benzene and its homologues. It is interesting to note that among the Home Office records of acute poisoning from petrol, a fatal case occurring in 1928 was caused by a mixture of petrol (77·5 per cent) and benzol (22·5 per cent); in two other cases (1922 and 1925) where unconsciousness resulted, a petrol-benzol mixture was used, while in another fatal case in a glue factory (1927) benzine had been used in the extractor.

Since the great majority of observations of both human and animal poisoning, especially during recent years, are specifically described under the heading of BENZINE these observations have been collected in a special section (p. 79).

TOXIC EFFECTS IN ANIMALS

Acute Poisoning

Quantitative experiments on inhalation by animals of petrol vapour itself do not appear to have been carried out, but some early investigations by Poincaré (1885) are interesting and valuable as a guide to the effects on human beings exposed to the vapour, since they were made under conditions to some extent comparable with those prevailing in industrial use.

Lethal and Narcotic Doses

By gastric administration
Lethal dose. For rabbits, about 20 ml. per kg. body weight (Legludic and Turlais, 1914).

Narcotic dose. For rabbits, on the average about 15 to 25 g. but this varied according to the kind of petrol used (Lewin, 1888).

Tolerance acquired by successive doses of petrol was a prominent feature of Legludic and Turlais' experiments.

By intravenous injection

Lethal dose. For rabbits, death occurred after 4 to 13 days after the injection of 0·2 ml. per kg. body weight (Matsushita, 1935). This result shows a higher toxicity than that given by Legludic and Turlais who found an average of 0·4 ml. per kg. body weight to be lethal.

Symptoms

The symptoms produced by gastric or intravenous administration of petrol are essentially those of an acute narcotic poison, accompanied, according to Matsushita (1935), by signs of irritation, twitching, toxic or clonic convulsions, with restlessness, tachycardia, dyspnoea, and in some cases by exophthalmos. Matsushita lays special emphasis on the nervous symptoms, and from histological examination of the brain and spinal cord of rabbits he concludes that petroleum spirit has a destructive action on the nerve cells of the brain cortex, the medulla oblongata and the spinal cord, producing also dilatation of the blood vessels of the brain, with haemorrhages especially in the neighbourhood of the ventricles. Inhalation of petrol directly from a mask produced, in Poincaré's experiments, agitation, then somnolence, loss of appetite, vomiting, and sometimes intestinal haemorrhage, but such inhalations were not fatal. It is interesting to note that Lewin (1888) observed a distinct difference between the effects of inhalation of petroleum and benzine on animals. Rabbits appeared to tolerate petrol vapour, in amounts of the same order as those ordinarily inspired by human beings working with petrol, with no grave discomfort, while exposure to the vapour of 10 g. of benzine produced agitation followed by somnolence, and 30 to 60 g. proved fatal.

Chronic Poisoning

In Poincaré's experiments, dogs, rabbits and guinea-pigs were allowed to breathe a mixture of air and spontaneously-evaporated petrol in proportions similar to those found in industry; the inhalations were repeated at intervals. The dogs showed only some slight modifications of the circulatory and respiratory rhythm; the rabbits showed some inco-ordination of movement and a tendency to somnolence; the guinea-pigs, though showing only slight symptoms of malaise, loss of appetite, and weakness, all died within a period of 1 to 2 years.

TOXIC EFFECTS IN MAN

Inhalation of the fumes of petrol arising from its use as an industrial solvent has been shown in a number of cases to produce, with severe exposure, acute toxic effects of the nature of drunkenness, narcosis and depression of the central nervous system, similar to those reported by many investigators from the accidental or suicidal ingestion of petrol. Chronic effects have also been reported with varied and not very characteristic manifestations; changes in the blood picture are apparently very rare; Mabille (1896) mentions anaemia but gives no definite data of total or differential counts, while some Russian findings, quoted by Flury and Zernik (1931), are also described without specific details. In Sterner's (1941) cases there was a small decrease in haemoglobin, erythrocyte and cell volume values as compared with a group of controls, with an equally small (though considered by Sterner to be statistically significant) increase in mean corpuscular haemoglobin, mean corpuscular volume, and reticulocyte count. It should be noted, however, that analysis of the specimens of gasoline showed it to have an aromatic hydrocarbon content (xylene and toluene) of 1 to 30 per cent (average 5 to 10 per cent), and that the atmospheric content of hydrocarbons was 300 to 800 p.p.m.

Acute Poisoning

Most of the early observations on the effect of inhalation of petrol fumes were made on workers in petrol mines, or on ships transporting petrol (Wiecyk, 1888; Sharp, 1889; Mabille, 1896; Lewin, 1888, etc.). All these observations, however, emphasize the intoxicant and narcotic effect of petrol, as noted by later workers from the use of petrol in paints, varnishes, and in the cleaning of tanks used for storing the petrol to be used for these purposes (Elliott, 1913; Plummer, 1913; Petrie, 1908; Merle, 1928, etc., and a number of cases reported to the Home Office between 1921 and 1935).

Symptoms

Very acute or fatal form. The majority of fatal cases reported have been caused by accidental (e.g. the cases in children reported by Price, 1933) or suicidal ingestion of petrol, and a large series of cases due to drinking gasoline and kerosene was recorded by Nunn and Martin (1934). In all these cases, in addition to the usual symptoms of intoxication and narcosis, pneumonia has been a special feature (Waring, 1933).

Many very severe and fatal cases from inhalation of petrol fumes in industry have, however, occurred nearly always in men who have entered tanks or wagons containing petrol. Out of twenty-two cases of acute poisoning reported to the Factory Department between 1938 and 1949, four died and eight suffered loss of consciousness. It is noteworthy that Nunn and Martin remark that in their series of cases those who inhaled, as well as swallowed, the kerosene and gasoline showed much more severe symptoms, since the toxic agent reaches the central nervous system so much more rapidly after inhalation. Unconsciousness, followed in the non-fatal cases by delirium, with cyanosis, shallow or stertorous respiration and a thready pulse are the prominent features in this group of cases. Vomiting, inward strabismus, contracted pupils and loss of reflexes were observed by Plummer.

Less severe form. The case reported by Elliott (1913) may be taken as representative of this group of cases—that of a man working in a closed space with paint in which petroleum spirit was used instead of turpentine. The symptoms were very similar to those produced by benzine, viz. a mild intoxication of the central nervous system, leading to restlessness and mental excitement, with rapid, irrational and incoherent speech, a flushed face, rather rapid pulse, some amnesia but no inco-ordination or staggering. According to Flury and Zernik (1931), however, the inhalation of moderate quantities of petrol vapour over a fairly long period of time may produce a more depressive syndrome—lassitude, somnolence, heaviness in the head, noises in the ears, deep sleep, with later loss of consciousness and amnesia. Sterner (1941) found evidence of toxic effects in spray-painters even when wearing cartridge-type respirators. The chief symptoms were headache, nausea, weakness, mental depression, anorexia, and inability for sustained attention and activity. In one case the symptoms of mental depression included an episode of acute depression and attempted suicide after taking alcohol. This patient showed tremor and weakness of the arms and legs, with multiple fibrillary twitchings when fatigued.

Sequelae

Recovery from acute poisoning may be followed by irritability and violent headache (Petrie, 1908; Sterner, 1941), disturbances of speech, cyanosis, gastro-intestinal disturbance (Home Office case, 1934), loss of sensitivity, neuritis (Flury and Zernik, 1931), inflammatory conditions, such as conjunctivitis, bronchitis, pneumonia, haemoptysis in tubercular subjects (Flury and Zernik), pain in the chest and dyspnoea (Home Office case, 1927), and syncopal attacks (Mabille, 1896).

Chronic Poisoning

Symptoms

The reports in the literature of chronic poisoning from petrol fumes are few and vague; most of the authorities who mention them quote from the opinions of earlier workers rather than from personal experience. Duvoir (1928), for example, states that nephritis, conjunctivitis and respiratory disturbances, including congestion, are characteristic of chronic petrol poisoning, but gives no authority for his statement nor does he designate the variety of *essence* used. Flury and Zernik (1931) describe in the same manner symptoms of chronic poisoning as headache, giddiness, neuralgia, lack of concentration, respiratory disorders, disturbances of sensibility, loss of corneal reflex, tremor of eyelids, paraesthesia of hands, increased knee jerks and chronic polyneuritis.

Headache is the chief symptom mentioned by the earlier workers (Sharp, 1889; Legludic and Turlais, 1914, etc.), but Lewin (1888), who investigated the health of petrol workers in the United States, stated that chronic injury from the inhalation of petrol vapour was quite exceptional. Spencer (1922) examined twenty-two workers exposed to petrol vapour. The chief symptoms recorded were irritation of the eyes, headache, drowsiness, heaviness in the head and fatigue. These symptoms, with lack of appetite, nausea, dizziness and irritation of the throat, were also observed by Drinker *et al.* (1943) when the atmospheric concentration reached 0·1 per cent. With regard to the eye irritation which was a prominent symptom in Spencer's cases, Drinker *et al.* state that petrol vapour contains an eye-irritant component which at very low temperatures distils in very small amounts. Caccuri (1946) reports that in nine workers employed in the petroleum industry for 3 to 4 years he found symptoms, which he described as "latent", including asthma, anorexia, moderate dyspnoea, hypotension, haemorrhage from the gums, and menstrual disorders.

Effect on the Blood

Anaemia is mentioned as a result of chronic intoxication by petrol by Mabille (1896) and Porrini (quoted by Legludic and Turlais, 1914). The latter stated specifically that the petrol was used in place of benzol as a solvent for varnish. A detailed account of the blood changes has been given by Abramowskaja and Ter (1924), quoted by Flury and Zernik (1931), who state that these changes consist in a reduction of the red and white cells, and increased haemoglobin and colour index. The findings of Sterner (1941) have already been described. Caccuri (1946) also records anaemia, neutropenia, and increased coagulation and bleeding time. In an examination by Browning (1945, unpublished) of twenty-seven operatives using petrol mixtures of uncertain aromatic hydrocarbon content probably not exceeding 5 per cent, as rubber solvents, fourteen showed anaemia of slight degree and only one a slight leucopenia.

11. Benzine

COMPOSITION

The term "benzine", as used by various writers on its application to industrial processes, apparently includes a group of substances covering a wide range of chemical composition and properties, and the somewhat discrepant effects of benzine described are no doubt due to these differences.

Its composition depends to some extent upon that of the petroleum from which it is distilled, which in its turn varies greatly in the different countries from which it is produced. North American (Pennsylvanian) petroleum, for example, according to Zernik (1933), consists essentially of paraffin hydrocarbons; Russian petroleum consists of naphthenes (cycloparaffins) and unsaturated hydrocarbons; Galician and Rumanian have a composition somewhere between these two extremes. The petroleum distillates produced by "crack" distillation are rich in olefines, those of the Far East in aromatic hydrocarbons, while American oils are specially rich in sulphur compounds. The "casing head gasoline" obtained in America by condensation of natural gas contains butane.

Commercial benzine. The properties of commercial benzine vary, especially according to whether it contains benzol, xylol and toluol (as in American and German benzine) or naphthenes and paraffin hydrocarbons (as in Russian benzine). In the former case the presence of benzene and its homologues, chiefly xylene and toluene, may amount to as much as 20 per cent. In Germany a distinction is made between "light benzine" (B.R. 50–110° C., usually 70–100° C.), "middle benzine" (80–130° C.), and "heavy benzine" (100–140° C.).

The chief varieties of benzine have been classified by Lazarew (1929b) according to their content as follows:

(*a*) Pennsylvanian and Caucasian benzine, containing paraffins—pentane, hexane, heptane, etc.

(*b*) Benzine from Baku, Texas, California, Mexico, Japan and Galicia, containing cycloparaffins—cyclopentane, cyclohexane, etc.

(*c*) Rumanian (up to 22–24 per cent), some Galician benzine, and that from Borneo (up to 40 per cent), containing aromatic hydrocarbons—benzene and its homologues.

In addition to these three groups of constituents, some varieties of benzine, especially "crack" benzine, contain 20 per cent aromatic hydrocarbons in addition to unsaturated hydrocarbons of the fatty series—olefines, acetylenes, terpenes.

PROPERTIES

In general benzine is a colourless liquid, soluble in alcohol and insoluble in water with a vapour 3 times as heavy as air and a time of volatilization of 3·5 as compared with ethyl ether, 1.

The special characteristics of some varieties of Russian benzine are given by Lazarew (1929b) in Table 16.

TABLE 16

Properties of Russian benzine

Variety of benzine	B.R.	Sp. Gr.	Distillation
Baku " Galosch " purified	80–120° C.	0·740–0·746	95 per cent up to 110° C.
Baku Benzine Type II	80–175° C.	0·749	97 per cent up to 175° C.
Grosny (Russian)	80–200° C.	0·740	97 per cent up to 200° C.
Krasnodar (aviation)	45–120° C.	0·690–0·702	95 per cent up to 120° C.

The "rectified benzine" used by Petrini (1941) (see below and p. 87) distils at 55–85° C., Sp.Gr. of 0·70–0·71, and contains no benzene or aromatic hydrocarbons.

USES

(1) As a motor fuel.
(2) As a solvent:
    (*a*) for fat extraction;
    (*b*) in colour-printing;
    (*c*) in the rubber industry;
    (*d*) as a substitute for turpentine, and especially as a diluent for cellulose nitrate solutions (Jakobsen, 1939);
    (*e*) as a cleaning material and for the dry-cleaning of clothes.
(3) As a constituent of parasiticide sprays.

TOXICITY

In large dosage, benzine is a narcotic with an additional excitatory effect. Chronic exposure often produces a special effect of functional neurosis, the "hystero-anaesthesia" observed by many investigators. The blood changes noted, unlike those due to benzene, are not characterized by depression of white cell formation, but most investigators have shown the presence of anaemia in workers who are continuously exposed.

The results of animal experimentation are extremely confusing, owing to the fact that different investigators have used different varieties of benzine. An attempt was made by Lazarew (1929b) to clarify some of the obscurities by testing the effect on mice of some of the constituents known to occur in benzines of different origin, and comparing the effects of the pure substances with those of mixtures of these substances and of benzines containing them.

He found that the toxicity of different varieties of benzine depends upon:

(*a*) *Chemical composition.* The cycloparaffin content is chiefly responsible for acute toxic effects. When the content of benzol and its homologues is low, the benzine is more toxic the more cycloparaffin and less paraffin it contains.

(*b*) *Boiling point.* As the boiling point increases, the "monophasic" toxicity (Lehmann's nomenclature, indicating the concentration of the pure vapour) increases also, while the "diphasic" toxicity (product of the monophasic toxicity and volatility) decreases.

(*c*) *Specific gravity.* The specific gravity serves as an indication of the relative toxicity of a known grade of benzine (i.e. the toxicity increases with the specific gravity) especially when the aromatic hydrocarbon content is low.

Comparison with Benzol

While most workers agree that the lethal toxicity of benzine is less than that of benzol, discrepant results in the actual proportion are again probably due to the benzol content of the benzine used. Petrini (1941), however, using rectified benzine, found the proportion to be 2·4 : 2·8 on the basis of varying concentrations (71 to 91 mg. per l. for benzine, and 25 to 35 mg. per l. for benzol), and 5 : 8·8 on the basis of time of action for identical concentrations.

Benzine is, according to Bamesreiter's (1932) figures, considerably less toxic than benzol from the point of view of narcotic effect (Table 17).

TABLE 17

Relative toxicities of benzine and benzol for cats after inhalation for 6 hours

Solvent	Inactivity (mg./l.)	Light narcosis (mg./l.)	Deep narcosis (mg./l.)
Benzine	40	60	75
Benzol	22	28	30

The time of onset of narcosis was also more rapid for benzol than for benzine. These results are confirmed by Engelhardt (1931) for German benzine, containing chiefly hexane, with traces of pentane, heptane and cycloparaffins. He observed complete narcosis in 12 minutes with 50 mg. per l. for benzol, in 15 minutes for benzine.

Penni and de Steffanis (1940) give 81·8 mg. per l. for severe intoxication with asphyxia. Petrini (1941) gives the relative narcotic concentration for benzine and benzol as 1 : 4·4, the concentration for deep narcosis being 133 mg. per l. for benzol, but he found the time of onset about the same. Thus, while the narcotic action of benzine is less marked than that of benzol the primary irritative action is slightly more marked.

Influence of Temperature

According to Lifschitz (1935), the same concentrations of benzine produce more toxic effects on animals as the temperature of the air rises within a certain range. As the temperature rose 3 to 4° above 17° C., with concentrations of 50, 55 and 60 mg. per l., the mortality of mice was doubled, and with 3° further increase, trebled. A limit was then reached above which a further rise of temperature produced no increase in mortality.

TOXIC EFFECTS IN ANIMALS

Lethal and Narcotic Concentrations

Lethal dose

The lethal concentration varies widely with different varieties of benzine. In Lazarew's investigation the concentrations of Baku benzines which caused death in mice lay between 40 and 80 mg. per l. Considerable differences in the lethal dose for mice were also observed by Lestchinskaja (1933) in different samples of benzine, all sold under the name of "Kalosche". The specific gravity of these samples ranged from 0·736 to 0·745, their boiling point from 65 to 82° C., and their content of aromatic hydrocarbons from 0·73 to 1·5 per cent. Above 74 mg. per l. all the mice died and below 30 mg. per l. all survived, but between these concentrations the different benzines showed wide variations. For example, at 56 mg. per l. one sample of benzine proved to be $2\frac{1}{4}$ times as toxic as another, while at 67 mg. per l. it was only $1\frac{1}{2}$ times as toxic. Lestchinskaja concludes that it is impossible to estimate the "dangerous" concentrations of commercial benzine unless its chemical constitution is invariable. For dogs, Babsky and Leites (1931) found the lethal concentration of Russian benzine 50 to 80 mg. per l. In the experiments of Petrini (1941) the lethal dose of rectified benzine for guinea-pigs was 71 to 91 mg. per l.

Narcotic dose

Light benzine. The figures given by Bamesreiter (1932) appear to show a greater toxicity for light benzine than the earlier results of Lehmann (1912) (Table 18).

TABLE 18
Toxicity of light benzine for cats after inhalation

Author	Side position (mg./l.)	Light narcosis (mg./l.)	Deep narcosis (mg./l.)
Lehmann (1 hour)	80	250	—
Bamesreiter (6 hours)	40	60	75

Heavy benzine. From Lehmann's figures, heavy benzine would appear to be more toxic than light—50 mg. per l. for inactivity and 200 for light narcosis.

Russian benzine. In rats and rabbits, according to Lewin (1932), inhalation of 50 to 60 mg. per l. for 5 to 8 hours daily over a period of 4 days produces emaciation and often, especially in rabbits, paralysis of the hind legs. In dogs 50 to 80 mg. per l. produces restlessness, followed by twitching of the whole body, ataxia, and convulsions (Babsky and Leites, 1931). Cats were found more resistant, tolerating 30 mg. per l. for 7 hours with only slight symptoms and recovery after 15 minutes.

Absorption

Some experiments by Lazarew, Brussilowskaja, Lawrow and Lifschitz (1931c) have shown that while benzine is absorbed chiefly by the lungs, it can also be absorbed by the skin of animals. Skin absorption of benzine was found to be less both in degree and rapidity than that of benzol. Mice and dogs were investigated in a special apparatus containing benzine vapour in concentrations up to 300 mg. per l., their heads being left free. No effects of general poisoning or narcosis were observed. When parts of the bodies of the animals were immersed in benzine, however, the mice died, probably owing to skin irritation, while the dogs developed pustular oedematous inflammation of the ears.

In order to ascertain whether benzine so applied to the skin enters the circulation, the air expired by dogs with one leg immersed in benzine was passed through tubes of silica gel; the weight of the silica gel was then found to be $2\frac{1}{2}$ to 3 times greater than in the controls, the increase in weight representing the amount of hydrocarbon expired. It was found that the concentration in the expired air rapidly increased and reached a maximum in about half an hour (0·8 to 1 mg. per l.); it remained at this level during the whole time the leg was immersed. Altogether not less than 64 mg. was excreted in the course of 115 minutes as compared with 138 mg. for benzol. The speed of absorption was therefore estimated as not less than 0·01 mg. per minute per sq. cm. of skin.

Further evidence that less benzine than benzol is absorbed was provided by an estimation of the amount of these substances in the blood of rabbits after immersion of the ear. Benzine was found to be present in very small quantities in the blood taken from the external jugular vein (10 to 12 mg. per kg. of blood as compared with 50 to 100 mg. for benzol).

Acute and Sub-acute Poisoning

Symptoms

In general the effects of benzine in amounts large enough to produce acute or sub-acute effects on animals are those of a narcotic acting on the brain and producing a preliminary irritative effect (Lehmann, 1912; Fühner, 1921a; Lazarew, 1929b). According to Lazarew, it differs from other "indifferent" narcotics by the fact that its action on the spinal cord is much delayed, so that it is difficult to produce complete narcosis with loss of reflexes. This peculiarity and that of the occurrence of sudden death of some of the animals with tetanic convulsions (noted as early as 1865 by Eulenberg) is common to pure paraffins and cycloparaffins (Lazarew, 1929b). The emaciation of animals subjected to acute poisoning noted by Schustrow and Letawet (1927) and related by them to a direct "fat solvent" action of the benzine has not been confirmed by other workers, e.g. Lazarew, Brullowa, Kremnewa, Larionow, Lubimova and Stalskaja (1931d). In Engelhardt's (1931) experiments the symptoms produced by injection of American benzine in olive oil were chiefly loss of appetite and decrease in weight; death was often due to lung emboli.

Briganti and Ambrosio (1941) also produced narcosis, preceded by a stage of excitability, followed in some cases by generalized convulsions, and accompanied by a loss of weight of about 20 per cent.

According to Gladychevskaja (1936) benzine has a stimulating action like that of benzene on the respiratory centre of frogs, but this stimulation is followed by paralysis.

Increased permeability of cerebrospinal membrane. Some interesting experiments carried out by Schachnowskaja in 1935 appear to indicate that benzine exerts a toxic effect on the cerebrospinal membrane, increasing its permeability, possibly by causing dilatation of the blood-vessels through action on the vasomotor centre. The experiments were carried out under conditions corresponding both to acute and chronic poisoning—the animals being exposed to a single inhalation of 50 to 70 mg. per l. and to repeated inhalations of 4 hours daily over periods of 10 to 20 days and 3 to 4 months respectively. The indicators of permeability used were sodium ferrocyanide (5 per cent solution) and trypan blue (2 per cent solution) injected intravenously in amounts of 10 ml. and 5 ml. per kg. body weight respectively. Normally these substances under these conditions should be held back by the cerebrospinal membrane and should not appear in the cerebrospinal fluid, but Schachnowskaja found that in both acute and chronic poisoning the membrane was rendered permeable to sodium ferrocyanide in practically all cases and to trypan blue in about half. She regards the appearance of these substances in the cerebrospinal fluid after intravenous injection into animals exposed to benzine as a very early criterion of chronic poisoning.

Russian Benzine

The changes in the blood of animals subjected to acute intoxication with inhalations of high concentrations of benzine have been carried out by a group of Russian workers on Russian benzine consisting chiefly of cycloparaffins.

Effect on the blood

Red cells. In the experiments of Schustrow and Salistowskaja (1926) as well as in those of Brullowa, Brussilowskaja, Lazarew, Lubimowa and Stalskaja (1930) anaemia was a pronounced feature. The latter workers found the decrease in red corpuscles occurred both with single acute exposures, beginning after 2 hours, and with repeated exposures. A reduction in haemoglobin (preceded in the case of the single acute injection by a slight increase) accompanied the reduction of red cells in all cases. Structural changes such as polychromasia, punctate basophilia in the red cells and the occurrence of normoblasts and reticulated cells were also observed in repeated acute poisoning.

White cells. An initial leucocytosis followed by a leucopenia was the typical result of acute poisoning. Both the leucocytosis and the leucopenia affected the neutrophils; the lymphocytes were little affected, but in the cases with marked leucopenia not only a relative but an absolute lymphopenia developed. Degenerative changes in the white cells were noted, but less than in benzol poisoning. It was stated by Schustrow and Salistowskaja that these changes might be explained as a destruction of the blood cells by a "washing out" of their lipoid constituents by the solvent action of the benzine, but with this hypothesis Brullowa and co-workers and Lazarew and co-workers (1931d) disagree.

American Benzine

The investigations of Engelhardt (1931), using injections of benzine in oil, as well as inhalations, may also be described as acute poisoning. The benzine used in his experiments was American, consisting chiefly of paraffins. That used by Petrini (1941) was "rectified benzine" in which contamination with benzol appears to have been excluded.

Effect on the blood

The blood picture, according to Engelhardt (1931), was chiefly characterized by a leucocytosis with degenerative changes in the leucocytes and reduction of red cells, with only slight reduction of haemoglobin, so that the colour index was often increased. There was also polychromasia of the red cells and the appearance of erythroblasts.

Haemolysis

Some workers have assumed that acute benzine poisoning causes haemolysis of the red cells. Böhme and Köster (1917) based their assumption on the finding that strong haemolysis occurs *in vitro* when benzine is dissolved in blood or added to pig serum. Dorendorf (1901) found that inhalation of benzine by guinea-pigs was followed by the occurrence of dark pigment in the plasma, erythrocytes and a few leucocytes; the origin of this pigment he regarded as a toxic necrosis and fragmentation of the red cells. Engelhardt, however, found the resistance of the red cells increased for total haemolysis, while for commencing haemolysis it was unchanged. Deposition of pigment occurs in the liver and spleen (Dorendorf and Engelhardt) and erythrophagic cells (Engelhardt) have been observed. Ambrosio (1941) believes that the solvent action of benzine on the lipoid-protein constituents of the erythrocytes decreases their resistance, producing a tendency to haemolysis.

Lesions of the Internal Organs

Acute hyperaemia of the organs, most acute in the lungs and spleen, has been recorded by Lewin (1932) and Engelhardt (1931). The lesions are regarded by Lewin as an indication of a specific reaction of the reticulo-endothelial system, since he observed hypertrophy of these cells in the spleen, bone-marrow and adrenals of rats and rabbits subjected to inhalations of 50 to 60 mg. per l.

In the spleen there was proliferation of the sinus endothelial cells, and myeloid metaplastic groups under the capsule. Heitzmann (1931) observed atrophic changes in the spleen in some cases. Similar appearances were observed in the adrenals and bone-marrow; in the latter there was an increase of all elements, especially of myelocytes and leucocytes.

The liver in severe cases showed fatty infiltration. The kidneys also showed slight fatty infiltration in Lewin's acutely poisoned animals; in Engelhardt's investigation degenerative changes in the kidney were only observed with high concentrations (135 mg. per l.).

The heart muscle in mice with acute, short-lasting poisoning shows severe degenerative changes (Kolesnikow, 1932). The serous fluids sometimes contain red blood corpuscles, generally unlaked. Schwartz (1931) relates this phenomenon to a toxic action of benzine on the vessel walls (see also Schachnowskaja, p. 86).

A selective effect of benzine on the vascular supply (arterioles, venules, capillaries) of the kidneys, liver, pancreas, heart, and adrenals is also postulated by Briganti and Ambrosio (1941). They believe that this causes congestion with focal haemorrhages, and a direct toxic effect on the parenchymal tissue. They found no fatty degeneration or inflammatory sclerotic change, but hyperplasia of the lymphoid follicles of the spleen, stomach and intestine, and of the reticulo-endothelial system of the spleen, lungs and liver.

Lowering of Resistance to Infection

According to Lazarew and co-workers (1931d), acute benzine poisoning produces a lowering of resistance in guinea-pigs to tuberculosis and in white mice to rat typhoid (Danysz virus) infection, the duration of life in animals infected with cultures of the infecting bacilli being shortest in the most severe poisoning.

Acquired Tolerance

It has been stated by Schustrow and Salistowskaja (1926), and confirmed by Lazarew and co-workers (1931d), that animals subjected to repeated inhalation of benzine in amounts sufficient to produce severe symptoms at first acquire a certain tolerance to the poison when the inhalations are repeated. This tolerance shows itself with regard to: (*a*) epileptiform convulsions: the animals reach a point where they can tolerate doses which at first caused convulsions; (*b*) general condition: the initial loss of weight ceases and is sometimes followed by gain in weight, while the initial malaise disappears; (*c*) blood changes: the number and structure of the red cells and the haemoglobin percentage return to normal; (*d*) lack of resistance to infection: the initially lowered resistance of guinea-pigs to tuberculosis and of mice to typhoid rises again.

"Conditioned Reflex"

According to Babsky and Leites (1931), repeated slight attacks of benzine poisoning may produce a "conditioned reflex" since animals which had been exposed regularly for short periods to concentrations of 60 to 80 mg. per l. showed nervous symptoms, such as convulsions and rigidity, when introduced into the inhalation chamber without further actual exposure to benzine.

Chronic Poisoning

As in the case of acute poisoning, the picture of chronic benzine poisoning in animals depends to a great extent upon the variety of benzine. This applies especially to the blood picture, where the variations appear to be related very closely to the benzol content of the benzine used.

Russian Benzine containing Cycloparaffins and Paraffins

Effect on the blood

The majority of recent workers find that Russian benzine, in small doses over a long period, produces no profound effects on animals, either with regard to symptoms or the blood picture.

The results of Schustrow and Salistowskaja (1926) were not in agreement with the latter opinion; they found that the benzine used by them, both "pure" and that containing 0·8 per cent of benzol, produced changes very similar to those of benzol, but with an anaemia which they considered to be hyperchromic and of a haemolytic nature, owing to the "defatting" action of benzine on the erythrocytes. Brullowa and co-workers (1930), Lewin (1932) and Lazarew and co-workers (1931d) were, however, unable to confirm these findings; they found no great variations in either the red or white cells. Schachnowskaja (1935) observed only slight anaemia, but there were structural changes in the red cells (increase of basophilic and vital staining cells and the appearance of some normoblasts) which she considers symptomatic of irritation of the bone-marrow.

Changes in the internal organs

Deposition of pigment in the heart muscle and liver and degenerative changes in the liver of the nature of commencing cirrhosis, and in the kidneys of an arteriosclerotic nature, have been described by Schwartz (1931) and Penni and de Steffanis (1940). In some cases Schwartz observed also eosinophilia in the spleen and lymph glands; Lewin (1932) observed some hyperaemia and slight hyperplasia of the spleen, but Larionow and Lazarew (1931) found such changes only with subcutaneous injection.

Lowering of resistance to infection

Chronic poisoning, like acute poisoning, was found by Brullowa and Lubimowa (1928) to lower the resistance of animals to infection by tubercle bacilli. Lewin observed catarrhal pneumonia in some of his animals which he suggests may be evidence of this lowered resistance. Larionow and Lazarew, however, found no lowered resistance in their animals with chronic poisoning, (see *Acquired Tolerance* p. 85).

It must be emphasized again that the brands of benzine used by the Russian workers contained varying amounts of aromatic hydrocarbons so that their results cannot be considered as an effect of pure benzine.

American Benzine containing Paraffins

Engelhardt (1931) observed no serious symptoms following inhalation of 24 to 100 mg. per l. of petroleum benzine containing chiefly hexane (69 per cent) with traces of pentane, heptane, *cyclo*pentane and *cyclo*hexane.

Effect on the blood

The blood picture showed one essential difference from that of benzol poisoning—a leucocytosis rather than a leucopenia. There was no change in the red cells other than a slight tendency to reduction, the haemoglobin level running parallel with the red cell count. The white cell differential count showed a more or less marked displacement to the left and a relative lymphopenia. A few toxic granular leucocytes were observed in a few animals.

Changes in the internal organs

In contrast to benzol, this variety of benzine appeared to cause proliferation rather than destruction of the leucopoietic system. In the bone-marrow there was more destruction of red cells than of white. The kidneys showed injury (necrotic epithelium of glomeruli) only with high concentrations.

Benzine containing Benzol

Several varieties of motor fuel benzine ("Dapolin", "Shell", "Strax","Olexin" —benzine-benzol mixtures, "Esso"—benzine-alcohol mixture) were tested by Schmidtmann (1930) on rabbits, guinea-pigs, rats and mice. The symptoms were essentially those of catarrhal bronchitis, progressing in some cases to a fatal bronchopneumonia.

Effect on the blood

The blood picture was that of an initial leucocytosis, followed by a severe leucopenia, and also by some anaemia.

Changes in the internal organs

The lungs showed the most severe and constant change—emphysema, bronchitis, epithelial metaplasia. The spleen showed myeloid changes with erythrophagocytosis in the early stages, atrophy of the pulp and increased connective tissue in the later stages, with hyaline degeneration and fibrosis of the lymphoid follicles. The liver and kidneys showed variable changes, sometimes periportal fibrosis of the liver, sometimes necrosis of liver cells, sometimes slight deposition of pigment in the kidney. According to Larionow and Lazarew these results, differing greatly from their own, are to be ascribed to the benzol content of the benzines tested by Schmidtmann.

Tendency to thrombosis

According to Kuntzen (1932), exhaust gases from benzine-benzol mixtures raise the tendency to thrombosis and embolism in rats subjected to their inhalation in small quantities. He attempts to correlate this finding with the fact that cases of thrombosis and embolism in human beings have increased during the last decade and that the increase may be due to the exhaust gases of automobiles using benzine-benzol mixtures.

Rectified Benzine

Contamination with benzene appears to have been excluded by Petrini (1941) who used benzine rectified by distillation at 55–85° C., having a Sp. Gr. of 0·70–0·71 and containing no benzene or aromatic hydrocarbons.

Effect on the blood

Inhalation of 39 mg. per l. for 6 hours daily in these experiments showed effects on the blood in some respects comparable to those of benzene, though less severe, i.e. a decrease in haemoglobin and red cells, and leucopenia, but the leucopenia was transitory, and was followed some hours after the end of the experiment by a definite leucocytosis, returning gradually to normal. Petrini believes that the transient leucopenia is due, not to actual destruction of leucocytes, as in the case of benzene, but to their migration into various organs and tissues under the immediate toxic action of the benzine. This phenomenon, he thinks, accounts for the discrepant results of other workers, especially the Russians, who had in fact noted a leucocytosis with a peak at 6 to 7 hours, returning to normal in 20 to 24 hours after the end of the experiment.

Changes in the internal organs

Petrini observed more or less congestion and fatty degeneration of the liver, hyperaemia and atrophy of the Malpighian follicles of the spleen, and epithelial degeneration and haemorrhagic extravasation of the lungs and kidneys after inhalations of 39 and 12 mg. per l. of benzine for 6 hours daily. The bone-marrow changes were not uniform, but certainly of a less depressive nature than with benzene.

Effect on Fat Metabolism

The results of various groups of investigators are somewhat conflicting, but it appears from the work of Nikulin and Hetmann (1933) that the variations observed are to be regarded as a protective reaction of the organism against the toxic effect of benzine, rather than as suggested by Schustrow and Letawet (1927) an actual "defatting" by the lipoid-solvent action of benzine. A final decrease, preceded by an initial increase, of the total blood fat in animals poisoned by benzine, was the result obtained by Schustrow and Letawet. Nikulin and Hetmann, while observing a considerable decrease in total blood fat of rabbits, during the actual course of a series of daily inhalations of benzine vapour, found that the decrease became smaller the longer the inhalations were continued, and that if the blood were examined from 1 to 2 months after the cessation of the experiment, the decrease had been converted to an increase. They also found a tendency to increase in the neutral fat and cholesterol ester— the easily transportable fat—and this they consider to be the chief indication of a protective reaction.

A tendency to decrease of the phosphatide and free cholesterol content of the blood, with a disturbance of the ratio between these substances, is regarded by Nikulin and Hetmann as an indication of the actual toxic action of the benzine. It should be noted, however, that Schachnowskaja (1935) found no characteristic or constant change in the cholesterol content of the blood of her experimental animals either in acute or chronic benzine poisoning.

TOXIC EFFECTS IN MAN

Although benzine vapour may, in circumstances of faulty mechanism, appear as such in exhaust gases from motor fuel, symptoms of poisoning in transport drivers from exhaust gases are, according to Flury and Zernik (1931), more likely to be due to the accompanying high concentration of carbon monoxide than to the benzine itself. The danger of poisoning is greater in garages and repair works from emptying benzine containers, cleaning and spraying cars in closed rooms, etc. The possibility of poisoning from constituents added to benzine to produce an "anti-knock" mixture (lead tetraethyl, aniline, nitrobenzene, iron carbonyl, etc.) cannot be overlooked, though with regard to iron carbonyl in "Motalin", it was stated in 1926 that after 2 years' intensive use of this substance no cases of poisoning had been reported.

The middle and heavy benzines are used as cleaning materials; the danger of their use in open vessels is specially emphasized by Segitz (1930), who points out that benzol and unsaturated compounds which increase the solvent power of the benzine also increase its toxicity.

Cleaning and repairing storage tanks, which have contained benzine, has been responsible for most of the acute cases of poisoning recorded, the early cases of Foulerton (1886), Leidy (1889), Siemon (1896), the later cases of Floret (1926), Zangger (1933), and a case reported to the Home Office in 1935. A series of cases of poisoning in a factory making waterproofs were described by Frumina and Fainstein in 1934. Acute poisoning from the use of a parasiticide spray consisting of 10 per cent methyl salicylate dissolved in benzine was reported by Judica-Cordiglia (1932), but he states that it is not certain that the effects were due entirely to benzine.

Acute Poisoning

Both in fatal and non-fatal cases of acute benzine poisoning the predominant symptoms are those of a narcotic poison with an added irritative effect. According to von Jaksch (1897) benzine was regarded by earlier investigators as a stronger narcotic than petrol. The lethal concentration for man, according to Flury (1928), is 30,000 p.p.m. for 5 to 10 minutes. Toxic effects are produced by 20,000 p.p.m. for $\frac{1}{2}$ to 1 hour, while 15,000 p.p.m. can be tolerated for $\frac{1}{2}$ to 1 hour.

Fatal Cases

Although acute benzine poisoning is generally less common than acute benzene poisoning, the actual percentage of fatalities is higher, indicating that severe exposure to benzine carries a very definite and serious hazard.

Among the fatal cases reported by early workers (eleven were recorded by Kobert in 1893) the majority were due to taking benzine by the mouth. One such case (Falck, 1892) was fatal after 10 minutes. According to Siwe (1932) there is little difference between the effects of benzine and petroleum when thus ingested. Children are specially liable to develop nervous symptoms.

Symptoms

The usual progress of a fatal case, which as a rule occurs after entering tanks filled with benzine vapour or filling cans with benzine in closed rooms, as in one of Floret's (1926) two cases, is loss of consciousness, irregular respiration, cyanosis, coldness of the skin, sometimes muscular twitchings going on to convulsions, disappearance of reflexes, paralysis of bladder and rectum, and fatal coma with cardiac collapse. In some cases the symptoms of involvement of the nervous system specially predominate. In Floret's case, for example, the clinical picture closely resembled that of meningitis. After recovery of consciousness the man complained of pain in the head and neck and stiffness of the neck, the muscle reflexes were increased to the point of clonus, and there was much restlessness and twitching of the muscles. Vomiting and renewed unconsciousness preceded death 5 weeks later.

Vomiting, with traces of blood, was a feature of the fatal case recorded by Zangger (1933). Smithies (1927) reported a case of gastroduodenal haemorrhage, the blood being haemolysed, in a worker using benzine and xylol. Necrosis of areas of the skin and mucous membranes was present in Floret's (1926) case, while multiple haemorrhages of the lungs were recorded in two fatal cases by Jaffe (1914). According to Burgl (1906), the blood is dark and fluid after death.

Non-fatal Cases
  Symptoms

One of the earliest cases of acute poisoning from inhalation of benzine fumes was that recorded by Foulerton (1886) of a workman who became unconscious after entering a reservoir. In this case, as in many others, "intoxication" with psychic disturbance was a conspicuous feature. Return to consciousness was accompanied by hysterical laughter and muscle twitching. Paralysis of the legs, bladder and rectum (Leidy, 1889), cyanosis (Foulerton, Leidy), and disturbances of respiration and heart action (Foulerton, Floret), headache, dizziness, nausea, abdominal pain, blurred vision (Hamilton, 1934), tingling of the limbs, drowsiness and giddiness (Factory Department, 1938) are other symptoms observed in severe acute poisoning. In slight cases of benzine "intoxication", according to Kobert (1906), there is often euphoria accompanied by slight disturbances of sight and hearing. Two cases of acute "gassing" by benzine, in which there was unconsciousness, were reported to the Factory Department in 1928 and 1929, respectively.

  Sequelae

Epileptiform condition. Severe epileptiform attacks with reduction of reflex irritability, persisted for a year after the initial exposure in one of Floret's cases, and were also recorded by Stiefler (1928).

Motor paralysis. Dorner (1915) records this as a residual symptom.

Anaemia. Acute benzine poisoning was followed by anaemia (Koelsch, 1926), which was also present after recovery from the acute symptoms (together with dyspnoea and lassitude) in a case reported to the Home Office in 1935.

Albuminuria. This was observed by Koelsch and also by Floret.

Inflammatory lesions in the lungs. These were reported by Flury and Zernik (1931).

Vertigo and nystagmus. Ruttin (1936) suggests that this may indicate some injury of the middle ear.

Chronic Poisoning

The symptoms of chronic benzine poisoning relate chiefly to the nervous system, a special feature observed being the occurrence of functional neuroses. The blood changes are apparently much slighter and less characteristic than those produced by benzol. Feil (1932) states that it is rare to find any variations of the blood picture, especially with those varieties of benzine which contain little or no benzol, but the results of several investigators, notably Rabinowitch (1925), Frumina and Fainstein (1934), and Macciotta (1942), appear to show that anaemia is the most constant change, with no leucopenia as in benzol poisoning, but a slight leucocytosis with some morphological changes in the leucocytes, or a relative lymphocytosis.

Symptoms
  Disturbance of general health

Giddiness, black spots before the eyes, insomnia, headache, somnolence, asthma, and gastro-intestinal disturbances (loss of appetite, eructations, burning sensation in the stomach, sometimes vomiting) have been described by a number of authors, including Chambovet (1921), Spencer (1922), Frumina and Fainstein (1934), etc. Many of the 400 painters examined by Harris (1918) stated that benzine caused more discomfort than any of the other solvents used and that its effects, giddiness, dizziness, weakness, were more lasting.

Irritation of mucous membranes

Laryngeal and respiratory irritation are mentioned by Feil (1932), and conjunctivitis was observed in 12 cases by Spencer (1922). An increased incidence of bronchitis, respiratory catarrh, conjunctival and skin lesions in workers continuously exposed to benzine was reported by Vigdortschik (1933). It has been suggested by Bartosch (1936) and Corelli (1937) that contact of benzine with the skin may cause sensitisation to other agents.

Nervous disturbances

Both psychic and motor and sensory changes have been reported, though the latter appear to be regarded by most authors as hysterical or functional in origin rather than organic. Koelsch (1926) states that long-continued exposure produces apathy, mental confusion, forgetfulness and change of character, while in one of Dorendorf's (1901) cases there was difficulty of speech, loss of memory and lack of concentration. Hysterical anaesthesia was observed by Werbow, Aschkewicz and Stopjanowskaja (1925), and hysterical and neuropsychic disturbances in golosh-workers were recorded by Rawkin and Kilkow (quoted by Schachnowskaja, 1935). Frumina and Fainstein (1934) in their investigation of 88 men in a rubber waterproof factory found that the incidence of functional neuroses was higher in those parts of the factory where the concentrations of benzine vapour were highest (45·4 per cent in the higher and 34·1 per cent in the lower concentrations).

The experimental results obtained by Schachnowskaja (1935) on the increased permeability of the cerebrospinal membrane in animals exposed to benzine have led her to conclude that this phenomenon is to be correlated with the nervous symptoms produced.

Motor and sensory changes were prominent in the cases recorded by Dorendorf (1901) and Haden (1919). Heaviness in the limbs with pain and actual loss of motor power in the hand, arm and leg, with tremors of the tongue, hand and eyelids, nystagmus, sensations of cold, and paraesthesia in the hand and arm, and increased knee jerks were special features of Dorendorf's cases.

Addiction

Lazarew (1929b), Lewin (1932), and Schwarz (1933) all mention that benzine may lead to addiction, but no actual accounts of such cases have been found.

Effect on the Blood

Changes similar to those occurring in benzol poisoning, though slighter and less characteristic, have been recorded by Looft (1930), viz. leucopenia and relative lymphocytosis, while Rabinowitch (1925) in an examination of 196 female workers in a shoe factory in Russia described the occurrence of a lymphopenia, a displacement of the Arneth count to the left, eosinophilia and polychromasia, which were interpreted as signs of irritation of the bone-marrow. Frumina and Fainstein (1934) on the other hand noted a leucocytosis and lymphocytosis with a reduction of neutrophils.

In most of the cases recorded, however, the main site of injury appears to be the red cells rather than the white. Thus Frumina and Fainstein found the red cells and haemoglobin reduced in most of their 88 workers in a rubber waterproof factory, while Werbow and co-workers also noted anaemia as the chief change in 10 women examined by them. Gram (1933) regards anaemia as the prevailing syndrome in occupational benzine poisoning. Whether this anaemia is hyperchromic in character seems uncertain; writers of articles, such as Simonin (1934), often assert that the colour index is raised in chronic benzine poisoning, but it was not so in Frumina and Fainstein's investigation, nor in that of Macciotta (1942). In the latter's series of 137 operatives (102 women and 35 men) exposed to benzine, chiefly in the rubber industry, the anaemia was of hypochromic type and was more pronounced in the women than in the men, as also was the tendency to leucocytosis. Macciotta, however, gives no data as to the constitution of the benzine used, or as to its possible contamination with benzene or its homologues. Moreover, consideration of the individual protocols shows that the average value of 8,066 total white cells per c.mm. for women and 7,928 for men includes several cases of slight leucopenia (one case among the men had only 3,450 per c.mm., and 12 of the women between 4,000 and 5,000). The same holds for his findings of "significant" lymphocytosis accompanied at times by monocytosis, and of "slight" eosinophilia. In only 13 of the 102 women and 4 of the 35 men is the lymphocyte percentage above 40, while an eosinophilia above 5 per cent occurs in only 14 women and in 2 men.

Morphological changes in the red cells have been infrequently observed—poikilocytosis and the presence of erythro-karyocytes (Looft, 1930).

Morphological changes in the white cells are also not frequently recorded but are mentioned by Frumina and Fainstein and Rabinowitch. Increase of bilirubin in the blood was noted by Schustrow and Salistowskaja (1926).

Effect on Fat Metabolism

A group of Russian workers, notably Schustrow and Salistowskaja (1926) and Schustrow and Letawet (1927), have suggested that benzine, by virtue of its lipoid-solvent capacity has a special action on fat metabolism. They claim that considerable loss of fat occurs in benzine workers, and that adipose workers are more sensitive to its action than the thin type. In support of this view they have stated on the basis of animal experiments that the blood shows a decrease of fatty substances (total fat) in benzine poisoning.

Investigations carried out by Nikulin and Hetmann (1933) and by Kornetow (1929) do not confirm this view. Nikulin and Hetmann found variations in the fat and lipoid content of the blood in workers exposed to benzine, but the changes in the total fat, neutral fat and cholesterol were in the nature of an increase and are regarded by these authors as a protective reaction. The blood examinations were carried out on 89 men and 53 women working in a rubber factory, and divided into groups according to the severity of their exposure to benzine. Thus one group was exposed to 2·2 to 2·9 mg. per l., another to concentrations up to 9 mg. per l. It was found that the changes in the blood fat were more pronounced in the latter group. These changes were: (1) increase of whole blood fat; (2) increase in neutral fat, cholesterol ester and total cholesterol; (3) decrease in phosphates and free cholesterol; (4) change in ratio between cholesterol ester and phosphatide. These changes, especially the increase in transportable fat, are regarded as a protective reaction of the organism, a mobilization of fat for the neutralization of the toxic effect of benzine.

Gelman (1932) states that the feeding of fat increases the resistance of the organism to benzine poisoning.

HYDROCARBONS

CONCENTRATION OF VAPOUR

Producing Acute Poisoning

Table 19 shows figures for effective concentrations of benzine vapour which have been estimated by the U.S. Bureau of Mines (1921) and by Lehmann (1912) in parts per million and mg. per litre of air respectively.

TABLE 19
Toxic concentrations of benzine vapour

Effect	Concentration	Author
Perceptible odour	(p.p.m.) 300	U.S. Bureau of Mines
Dangerous by inhalation for short periods	11,000–22,000	
Rapidly fatal	24,000–30,000	
Fatal immediately or after ¼–1 hr.	(mg./l.) 30–40	Lehmann
Dangerous to life after ½–1 hr.	25–30	
Tolerated for ½–1 hr. without immediate or later effect	10–20	
Toxic effects after several hours' inhalation	5–10	
Tolerated for 6 hr. without symptoms	10	

Producing Chronic Poisoning

According to Lazarew (1929b) the highest possible concentration of Russian Baku benzine in air at room temperature is 325·6 mg. per l., the concentrations in one of the few estimations which appear to have been made (that of Frumina and Fainstein, 1934) were very much less than this. In one workshop where rubber was dissolved in benzine and where 14,000 kg. of benzine was used in 24 hours the amount in the air reached 7·6 mg. per l., the highest concentration observed. In other parts of the factory the concentrations varied from 2·9 mg., the most usual, to 3 mg. per l., where benzine was poured into the apparatus. The latter concentrations persisted in spite of exhaust ventilation and were accompanied by symptoms of chronic poisoning in a large number of the workmen examined.

12. White Spirit

White spirit, or mineral spirit, is a petroleum distillate, chiefly used as a turpentine substitute.

PROPERTIES

A clear, colourless fluid.

British Standard Specification, No. 245 (1936) for paints. B.R. below 150° C. 10 per cent maximum; below 190° C. 80 per cent minimum. Fl.P. 25° C. Neutral, free from grease and objectionable sulphur compounds.

Its solvent properties vary with its source, the Roumanian product having the greatest general solvent power and the least odour.

USES

(1) As a solvent for paint.
(2) As a dry-cleaning agent especially used by larger firms equipped for vacuum distillation (Brown, 1933).

TOXIC EFFECTS IN ANIMALS

No experiments on animals with white spirit alone appear to have been carried out.

TOXIC EFFECTS IN MAN

Acute Poisoning

Simonin (1934) remarks that the white spirit used by painters gives off little vapour, and except for its effect upon the skin, is not dangerous; but since 1929 two cases of intoxication, one accompanied by unconsciousness, have been attributed by the Home Office to the use of white spirit. The first case (1929) was that of a workman painting in a confined space with a red lead paint containing oil of terebene (17·29 per cent) and white spirit (2·5 per cent); he collapsed and became unconscious. The second case occurred in a dry-cleaning works, where the man bent over a tank containing white spirit while moving it; he suffered from giddiness and vomiting, but was not unconscious.

No other cases of poisoning from the fumes of white spirit seem to have been recorded, but the symptoms of acute poisoning from ingestion were observed by Dubois (1885). These included violent delirium with clonic convulsions, a burning sensation in the throat and stomach, salivation and profuse sweating. Rapid recovery took place with no residual symptoms except headache.

Chronic Poisoning

An inquiry carried out by the Industrial Paints Committee (1920) in works where white spirit was used alone or with turpentine as a paint thinner revealed no evidence of toxic symptoms from its use. In one case only, where the white spirit used for painting heating radiators by a dipping process produced what were described as "very strong fumes from tank", a man complained of the fumes and gave up the work, but no colic, anaemia, or symptoms of kidney injury were observed.

The only recorded toxic effect attributed in the literature to chronic poisoning by white spirit is that by Hicguet (1930) of a "toxic neuritis", producing deafness, but there was no definite evidence in this case that the injury was actually due to the use of white spirit. The case was that of a painter, aged 47, who was using a paint in which white spirit had been substituted for turpentine. He complained of sudden deafness with noises in the head, but made a complete recovery after treatment with iodides. Hicguet states that as there was no clinical evidence of lead poisoning, he attributed the condition to white spirit.

An examination by the Home Office of three workers using white spirit in 1935 revealed no evidence of any ill-effect, with the exception of some dermatitis of the wrists and forearms in one, a worker of 25 years' standing.

13. *cyclo*Hexane

(*Hexamethylene, Naphthene, Hexahydrobenzene*)

$$C_6H_{12}, \text{ i.e. } CH_2 < \genfrac{}{}{0pt}{}{CH_2 \cdot CH_2}{CH_2 \cdot CH_2} > CH_2$$

*cyclo*Hexane occurs in Caucasian petroleum and is manufactured by the catalytic hydrogenation of benzene.

PROPERTIES

A colourless mobile liquid with an odour less pungent than benzene, rather similar to that of carbon tetrachloride.

B.R. 80–84° C. Sp.Gr. 0·774. Fl.P. —17° C. Insoluble in water at 20° C. Vapour pressure at 25° C. 97 mm. Hg. Evaporation rate at 45° C. 9·4 mg./min./sq. cm. Inflammable in air between the limits of 1·3 and 8·3 per cent by volume.

Good solvent for rubber, bitumen, caoutchouc and paraffin wax, but not for cellulose esters.

USES

(1) Chiefly as a degreasing agent, a solvent for rubber and a paint thinner.

(2) As a fungicide *cyclo*hexane is weak compared with dekalin or dipentene (7 per cent concentration as compared with 3 to 4 per cent for dekalin and 0·5 to 1 per cent for dipentene (Martin and Salmon, 1934)).

TOXICITY

*cyclo*Hexane is a narcotic in high concentrations, causing death from respiratory paralysis; it is less toxic in this respect than its derivatives, *cyclo*hexanol, *cyclo*hexanone, methyl*cyclo*hexanol and methyl*cyclo*hexanone. Judging from animal experiments, it has no severe or specific chronic toxic action, and no ill-effect has been reported in human beings.

Effect on ciliated epithelium. From an estimation of its toxicity by the inhibition of ciliated epithelium of the frog pharynx, cultivated *in vitro*, *cyclo*hexane appears to be less toxic than either benzene or *cyclo*hexanol (Umeda, 1928). According to Umeda, the decrease in the toxicity of the series benzene, *cyclo*hexene, *cyclo*hexane, depends on the increasing amount of hydrogen in the molecule.

Maximum Permissible Concentration

Cook (1945) suggests 400 p.p.m. (1,000 mg. per c.m.) as the maximum allowable concentration for prolonged exposure. 1·46 to 2·65 mg. per l. (434 to 786 p.p.m.) has also been suggested.

TOXIC EFFECTS IN ANIMALS

The opinion of Lewin (1929), that *cyclo*hexane is more lethal than benzene for animals, is not substantiated by the work of Lazarew (1929a), Sato (1928), Flury and Zernik (1931) and Treon, Crutchfield and Kitzmiller (1943), who, in reviewing the work of earlier authorities and also recording their own investigations on the oral, cutaneous and inhalation methods of administration, conclude that by the oral and cutaneous route, and by inhalation of high concentrations, *cyclo*hexane is less toxic than the oxygenated or methylated compounds, *cyclo*hexanol, *cyclo*hexanone, methyl*cyclo*hexanol and methyl*cyclo*hexanone but more toxic with prolonged exposure to low concentration than the methylated compounds. They emphasized the fact that if animal experiments can be applied to human beings, *cyclo*hexane should be a much safer solvent than benzene from the point of view of blood injury since it has no effect on the haemopoietic system of animals.

Acute Poisoning

Lethal and Narcotic Concentrations

By oral administration. In the study by Treon *et al.* (1943) *cyclo*hexane proved less toxic than any of its derivatives, the lethal dose for rabbits, with death taking place only after several hours or even days, being 5·5 to 6 g. per kg. Death was preceded by severe diarrhoea, but not by narcosis or convulsions.

By cutaneous application. Application of more than 180·2 g. per kg. was neither lethal, nor narcotic, and produced no diarrhoea (Treon *et al.*, 1943). This was in contrast to the application of some of the derivatives, e.g. *cyclo*hexanol, which was lethal in a dosage between 12·4 and 22·7 g. per kg., and in smaller dosage caused both tremors and narcosis. Irritation and thickening with fissure and bleeding of the skin followed the application of 5 ml. of *cyclo*hexane at intervals of 5 minutes for 5 days. These lesions healed within a week of cessation of the applications.

By intraperitoneal, intravenous and subcutaneous injection. The lethal dose by intraperitoneal injection for mice was found by Sato (1928) to be 0·4 ml., and the intravenous dose for rabbits 0·77 g. per kg. body weight. By subcutaneous injection for rabbits the lethal dose, according to Launoy and Lèvy-Bruhl (1920), is 12 to 15 ml. The narcotic dose by intraperitoneal injection, according to Sato, is 0·03 ml. per kg. for mice, and 0·8 ml. for rabbits.

By inhalation. There is some discrepancy among several groups of workers as to the lethal dose of *cyclo*hexane by inhalation. Henderson and Johnston (1931) in particular give much higher values than any other authority (386 mg. per l. just lethal for mice), while Treon *et al.* (1943) are much closer to the value given by Lazarew in 1929 (60 to 70 mg. per l.). Treon *et al.*, using rabbits and monkeys, found that concentrations over 25·1 mg. per l. were fatal on repeated inhalation. The narcotic dose shows a similar discrepancy, Henderson and Johnston's value of 132 to 141 mg. per l. being much higher than Lazarew's 50 mg., and Treon's and his co-workers' 25·1 mg. per l.

Table 20 shows the results of some of the principal investigations.

TABLE 20

Lethal and narcotic doses of cyclo*hexane by inhalation*

Authority	Lethal dose (mg./l.)	Narcotic dose (mg./l.)
Lazarew (1929a)	60–70	50
Flury and Zernik (1931)	—	62·5
Henderson and Johnston (1931)	386	132–141
Lehmann and Flury (1938)	150	—
Treon *et al.* (1943) ..	Above 25·1	25·1

The only suggested explanation of the discrepancy is that made by Henderson and Johnston, that accumulation of CO_2 may have been a factor in Lazarew's experiments.

Nature of the Effects

There is some difference of opinion on one aspect of the acute toxic effect of *cyclo*hexane, viz. its irritative, as distinct from its paralytic, effect on the central nervous system. According to Krawkow, whose work is freely cited by Lazarew (1929a) and Filippi (1914), *cyclo*hexane, like methyl*cyclo*hexane, is among those substances of the *cyclo*paraffin series, which produce the characteristic effect on animals called by Führer (1921a) "*Streckkrämpfe*"—a general tetanic spasm occurring suddenly and leading to sudden death. This condition is not characteristic of poisoning by pentane, octane or ethyl*cyclo*hexane. Sato (1928) on the other hand states emphatically, that this is not the case, and that *cyclo*hexane is a purely paralytic agent, producing no clonic spasm of any kind. He believes that Filippi's results are due to the fact that his *cyclo*hexane was contaminated by *cyclo*hexene (tetrahydrobenzene) or benzene itself. This difference of opinion may be explained by the results of the investigations of Treon et al. (1943). They found that convulsions were produced only by inhalation of *cyclo*hexane at high concentrations, and not at all by cutaneous or oral administration, even in narcotic or lethal dosage. On the other hand, methyl *cyclo*hexanol and methyl*cyclo*hexanone produced convulsions with both oral and cutaneous administration, but not with inhalation. The suggested explanation is based on the difference in rate of absorption and in the variation of the threshold of concentration of the solvent in the tissues, or the total spread of concentrations that constitutes an effective stimulus for the induction of convulsions.

Symptoms

Disturbances of equilibrium, staggering, stupor and final paralysis, followed by death from respiratory paralysis, have been noted by most observers as the result of acute poisoning from both injection and inhalation of *cyclo*hexane. Severe diarrhoea and loss of weight were recorded by Treon et al. in animals receiving lethal doses by all methods of administration, convulsions and rarely, opisthotonos only during exposure to high concentrations by inhalation (89·6, 62·6 and 42·4 mg. per l.). Light narcosis and paresis of the legs occurred during exposure to concentrations of 25·1 mg. per l., but no other noteworthy effects in animals exposed to non-lethal concentrations.

Metabolic Effects

A temporary decrease in the inorganic sulphate content of the urine and a temporary increase of the glucuronic acid output are stated by Treon et al. to be a useful measure of the gastro-enteric absorption of *cyclo*hexane and of its derivatives. In the case of *cyclo*hexane, however, the increase in glucuronic acid amounted to only about 4 to 5 per cent as compared with 45 to 50 per cent for *cyclo*hexanol, *cyclo*hexanone and methyl*cyclo*hexanol, and 92 per cent for methyl*cyclo*hexanone.

Lesions of the Internal Organs

The post-mortem appearances in animals receiving lethal doses of *cyclo*hexane by any route are, according to Treon and co-workers, those of a non-specific acute toxic process, death resulting from general cellular and vascular damage and inflammatory response. Extensive cellular degeneration of the liver with a toxic glomerulo-nephritis were observed, but these lesions were not so severe as with the oxygenated derivatives.

Chronic Poisoning

Effect of Repeated Oral, Subcutaneous and Cutaneous Administration

Toxic changes of a non-specific nature (Treon et al., 1943), a fall in blood pressure (Sato, 1928), increase in pulse and respiration rate (Sato, 1928) are the chief effects noted with non-lethal dosage.

Blood changes. Some observations by Launoy and Lévy-Bruhl (1920) on the effects of repeated injections of *cyclo*hexane into animals suggest that *cyclo*hexane has no significant effect on the bone-marrow.

Effect of Repeated Inhalation

No noteworthy effects were observed by Treon and co-workers in animals exposed to non-lethal concentrations (below 25 mg. per l.). Small increases in the rectal temperature were noted following exposure to fairly high non-lethal concentrations, but still lower levels (11·23 mg. per l. or lower) produced subnormal temperatures.

Blood changes. Treon *et al.* were unable to detect any specific injury of the cellular elements of the peripheral blood in animals subjected to repeated inhalation of non-lethal concentrations.

Metabolic Effects

Urinary sulphates. No abnormal variations in the rate of excretion of organic sulphates were observed by Treon and co-workers either during or after exposure to concentrations ranging from 1·46 to 25·1 mg. per l.

Glucuronic acid. The average daily output of glucuronic acid during exposure to 1·47 mg. per l. daily for 26 weeks was somewhat increased, but returned to the normal level during 2 months following exposure.

Glutathione content of the blood. According to Di Prisco (1932) *cyclo*hexane is amongst those substances which diminish the glutathione content of the blood in animals, in contrast to the true narcotics, which increase it.

TOXIC EFFECTS IN MAN

None has been recorded. An investigation by the Home Office of a factory where *cyclo*hexane was used as a degreasing agent for nickel sheets revealed no ill-effects of any kind amongst the workers.

14. Methyl*cyclo*hexane

(*Hexahydrotoluene, Heptanaphthene*)

$$C_6H_{11} \cdot CH_3, \text{ i.e. } CH_3 \cdot CH \begin{array}{c} CH_2 \cdot CH_2 \\ CH_2 \cdot CH_2 \end{array} CH_2$$

PROPERTIES

A colourless liquid present in the distillates of certain petroleums, mostly Russian and Galician, chiefly in the fractions boiling between 98 and 100°. It is also produced by the catalytic reduction of toluene, or by the reaction at high temperature of benzene and methane (Timmermanns and Martin, 1926).

B.P. 101·2° C. Sp.Gr. 0·733–0·760. Vapour pressure 43 mm. Hg at 25° C. Evaporation rate 6·7 mg./min./sq. cm. Insoluble in water at 20° C.

USES

Similar to those of *cyclo*hexane.

TOXICITY

Like *cyclo*hexane, but in a greater degree, methyl*cyclo*hexane is a narcotic in high concentrations, causing death to animals from respiratory paralysis, with also a convulsive effect. It is scarcely toxic from prolonged exposure to sublethal concentrations; in this respect even less so than *cyclo*hexane. In human beings no toxic effects from its industrial use have been recorded.

Maximum Permissible Concentration

Cook (1945) suggests 1,000 p.p.m. (4,000 mg. per c.m.) as the maximum allowable concentration for prolonged exposure. 4·57 to 11·35 mg. per l. (1,162 to 2,886 p.p.m.) has been found to be safe.

TOXIC EFFECTS IN ANIMALS

From animal experiments it appears that in all methods of administration, methyl*cyclo*hexane is more toxic than *cyclo*hexane, but less toxic than the oxygenated compounds, *cyclo*hexanol and *cyclo*hexanone, and still less than the methylated oxygenated compounds, methyl*cyclo*hexanol and methyl*cyclo*hexanone.

Lethal and Narcotic Concentrations

By oral administration. The minimum lethal dose for rabbits lies between 4·0 and 4·5 g. per kg., death occurring after 5½ hours. Death was preceded by diarrhoea, but this was less regular than with *cyclo*hexane.

By cutaneous application. Repeated applications of 60 ml. in 5 ml. portions at 5 minute intervals on 6 successive days induced local irritation, thickening and ulceration, but failed to cause death.

By intraperitoneal injection. Stupor, insensibility, marked hypothermia and death in 5 or 6 hours were stated by Filippi (1914) to follow intraperitoneal injection in mice, but no convulsions. Sato (1928) observed laboured respiration, paresis, and death 17 hours after the injection of 3·1 g. per kg. into the thoracic lymph sac of the frog.

By inhalation. The lethal concentration by inhalation for rats is given as 40 to 50 mg. per l., as compared with that of 60 to 70 mg. per l. for *cyclo*hexane. The narcotic dose given by Lazarew is 30 to 40 mg. per l., and the various stages of narcosis in rabbits is given by Treon *et al.* as follows: slight lethargy 21·9 mg. per l., lethargy 28·8 mg. per l., light narcosis 39·5 mg. per l., severe narcosis 59·9 mg. per l.

Nature of Effects

As in the case of *cyclo*hexane the occurrence of convulsions following oral or parenteral administration of methyl*cyclo*hexane has been disputed, Lazarew (1929a) stating that they appear with inhalations, and Sato (1928) that both with frogs' heart preparations, and with intraperitoneal injections into mice, the effect is purely paralytic. In the experiments of Treon *et al.* (1943) no convulsions followed administration of methyl*cyclo*hexane orally, in lethal or sublethal dosage, but occurred following inhalation at a concentration of 39·55 mg. per l.

Symptoms

Exposure to lethal concentrations caused conjunctival congestion with mucoid secretion and lachrymation, salivation, coughing, sneezing, laboured breathing, loss of weight, and diarrhoea, with convulsions at the highest levels.

Metabolic Effects

The effect from inhalation of high concentrations on the conjugation of urinary sulphates was similar, but in even less degree, to that of *cyclo*hexane, while the daily output of glucuronic acid was increased during intense exposure.

Lesions of the Internal Organs

These, as in the case of *cyclo*hexane, were not specific, but were the expression of general vascular injury and of inflammatory response.

Chronic Poisoning

Repeated inhalations of non-lethal concentrations caused no obvious illness in the animals investigated by Treon *et al.* During exposure to concentrations lower than 28·8 mg. per l., there was a slight fall in the daily rectal temperature, but above this level, there was a slight rise. There was no loss of weight, but rather a rise during exposure to non-lethal concentrations. There were no significant blood changes.

TOXIC EFFECTS IN MAN

No toxic effects in man have been reported.

15. Turpentine
$C_{10}H_{16}$

COMPOSITION AND PROPERTIES

Turpentine is a volatile oil obtained from various species of pine, and according to Kobert (1906) is a mixture of terpenes all having the formula $C_{10}H_{16}$. The following grades of turpentine are described by McCord (1926)*:—

(1) Gum spirit of turpentine, produced by tapping the living tree.

(2) Steam-distilled wood turpentine, distilled with steam from the oleoresin within or extracted from the wood.

(3) Destructively distilled wood turpentine, produced by the destructive distillation of the wood, including brush and stumps.

(4) Wood turpentine, a mixture of the last two, containing methyl alcohol, formaldehyde, methyl acetate, phenols, which are considered by McCord and other workers to be chiefly responsible for the well-known turpentine dermatitis.

Spirit of turpentine is a colourless, limpid liquid, sometimes yellowish or greenish, with a characteristic penetrating odour and burning bitter taste.

B.R. 155–180° C. Sp.Gr. 0·863–0·875. Fl.P. 33–36° C.

The pure "gum spirit" is composed almost exclusively of the hydrocarbon pinene ($C_{10}H_{16}$) which may be laevo-rotatory (terebenthene) or dextro-rotatory (australene). Commercial turpentine is often adulterated by naphtha, pine oil, benzene and its homologues, spirit or oil of resin (Durrans, 1933, 1935).

Turpentine is a good solvent for fats, resins, rubber, asphalt, bitumen, phosphorus and sulphur.

USES

(1) Chiefly in the paint industry as a solvent or thinner.

(2) In the manufacture of polishing and cleansing creams, especially shoe polishes (Wollenberg, 1927).

(3) In the textile industry for cleaning fabrics.

(4) In the printing industry for cleaning type and rollers of printing machines.

(5) In the rubber industry.

(6) For oil extraction.

(7) For decolorizing ivory, etc.

(8) For the manufacture of synthetic camphor and terpineol.

---

* Compare British Standards Specification, No. 244 (1936), also A.S.T.M. Specification D, 13–34 (American Society for Testing Materials, Philadelphia).

HYDROCARBONS

ESTIMATION IN AIR

A method described by Ficklen (1940) depends upon the use of vanillin in concentrated hydrochloric acid. Yates and Levinson (1943) have described a simple colorimetric method in which the light-deep orange colour produced when air containing turpentine is passed through concentrated sulphuric acid is matched by appropriate standard solutions. This method is not specific for turpentine, variations of the orange-yellow colour being given by *iso*propyl alcohol, gasoline, and methyl ethyl ketone.

TOXICITY

Whether taken by the mouth or inhaled as vapour or applied to the skin and mucous membranes, turpentine has a stimulating, followed by a paralysing, effect on the central nervous system, and an injurious effect on the liver and kidneys. In animals these effects are not produced by exposure to small, continued dosage, and they have not been indubitably confirmed in man.

Maximum Permissible Concentration

Basing their estimation on the atmospheric concentration which is tolerated without discomfort, Nelson, Ege, Ross, Woodman and Silverman (1943) suggest 100 p.p.m. (500 mg. per l.) as the maximum safe level for continued exposure.

TOXIC EFFECTS IN ANIMALS

Absorption and Elimination

Turpentine can be absorbed by the skin and mucous membranes as well as by the respiratory route. Poisoning by skin absorption has been recorded by Ridder (1923). Elimination takes place, partly unchanged, through the lungs but the greater elimination is through the kidneys, partly unchanged, partly combined with glycuronic acid. After absorption the urine has an atypical odour, not unlike that of violets.

Acute Poisoning

In general, turpentine in doses sufficient to produce acute symptoms acts as a local irritant to mucous membranes and as a central nervous depressant with initial excitation; it also causes injuries to the kidneys and liver.

Lethal dose. For cats Lehmann (1914) found 16 mg. per l. (2,900 p.p.m.) the minimal fatal dose by inhalation, death occurring after $\frac{3}{4}$ to 1 hour.

Convulsive and paralytic dose. In rabbits convulsions followed by paralysis occurred with concentrations higher than 4 mg. per l. (Eulenberg, 1865). In cats convulsions began at 4 mg. and became tetanic and were followed by paralysis at 8 mg. per l. after $2\frac{1}{2}$ to 3 hours (Lehmann, 1914). Geppert (1926) found that 10 mg. per l. after $\frac{1}{2}$ hour produced excitation and convulsions followed by paralysis; 10 mg. per l. was also the dose found by Goadby (1920) to produce disturbance of equilibrium, divergent pupils and opisthotonos in guinea-pigs.

Symptoms

The first effect of inhalation of turpentine is irritation of mucous membranes as shown by salivation and lachrymation; after a few hours, slight convulsions and disturbances of equilibrium ensue, with staggering and inability to walk straight, later more severe convulsions, tetanic or opisthotonic, then paralysis, which in the case of higher concentrations leads to death. Goadby also observed a tendency to diarrhoea.

Post-mortem Findings

Goadby found the following appearances in some of his animals: lesions in the kidney, especially of the tubules, consisting of congestion, a hyaline appearance and small haemorrhages; the lungs showed congestion and local patches of haemorrhage; in the intestines there were small ulcers in the ileum and melaena was present.

Chronic Poisoning

Chronic effects in animals have been found to be practically negligible. The inhalation by guinea-pigs of small amounts of turpentine vapour over long periods (715 p.p.m., or 4 mg. per l.) was found by Smyth and Smyth (1928) to produce no symptoms. The turpentine used in these experiments was steam-distilled (as used in a lacquer-manufacturing plant), and Smyth and Smyth state that their results agree with some unpublished results of experiments with gum spirit turpentine. They agree also with the results of Gardner (1925), who found pure gum spirits of turpentine or highly refined steam-distilled wood turpentine non-toxic to animals. These animals showed, post mortem, very slight changes in the liver and moderate scattered tubular degeneration of the kidneys. Similar results were obtained by Chapman (1941) with rats exposed to concentrations of turpentine vapour simulating those possible in industrial conditions (probably 5 and 10 mg. per l.). No histological evidence of any definite form was found.

Effect on the Blood

Inhalations of non-convulsive doses of turpentine, according to Smyth and Smyth, produced no variations in the blood picture. With subcutaneous injections, however, Bickert (1931–32), who studied not only the effect on the blood picture but on the immunity reactions, found the effect of turpentine in direct contrast to that of benzol; it produced a very definite leucocytosis (up to 25 per cent greater than the controls) and a considerable increase in the haemolysin titre (400 per cent higher after 6 weeks) of animals immunized with sheep cells.

TOXIC EFFECTS IN MAN

It is well known that turpentine when ingested is a violent irritant poison, which according to Taylor (1928) "may irritate the stomach as a primary effect, and after absorption has a specific influence on the kidneys, causing irritation which may lead to inflammation." Death occurred in a case reported by Maitland (1931) in an adult male who drank 6 oz. of turpentine, while severe symptoms—pain in epigastrium, giddiness, nausea, vomiting and oliguria—followed the ingestion of 40 g. in an anticolic mixture (Werner, 1932), and vomiting, diarrhoea and a violet odour of the urine lasting 8 days were recorded by Lodemann (1920) after 15 to 20 g.

Apparently these effects can be produced also by inhalation. Death from the industrial use of turpentine has been recorded by Schaefer (1901), and Drescher (1906), of a man employed in varnishing the interior of a box. Von Tappeiner (1916) has also stated that death has resulted from sleeping in a room freshly polished with turpentine. Vertigo, faintness, inco-ordination and spasmodic contractions of the arms in a worker employed in lacquering the inside of a cylinder with a varnish composed of equal parts of turpentine and petroleum distillate is attributed by Hagen (1939) to the turpentine component.

Acute Poisoning

Symptoms

Irritation of skin and mucous membranes. Burning of the eyes is experienced with comparatively low concentrations, e.g. 4 mg. per l. (Lehmann, 1914), while Hamilton (1934) states that reddening of the skin and mucous membranes with swelling and some exudation may occur. According to Nelson *et al.* (1943), irritation of the eyes and nose occurs at concentrations of 175 p.p.m. (less than 1 mg. per l.), and of the throat at 125 p.p.m. In one case recorded by Adler-Herzmark (1928), the inhalation of fumes from hot turpentine was responsible for an acute oedema of the glottis for which tracheotomy had to be performed. Nausea and vomiting and pain in the stomach were prominent symptoms in four cases recorded by Adler-Herzmark and abdominal discomfort was complained of in one case reported to the Home Office in 1931, but according to the Report of the Industrial Paints Committee (1920) true colic due to the use of turpentine in paint, apart from the simultaneous use of lead, is rare.

Nervous symptoms. Headache and giddiness are the commonest symptoms of slight intoxication; Goadby (1920) experienced them himself in the course of his animal experiments, and a few cases of dizziness occurred in workmen examined by the Home Office (1928 and 1931). Transient symptoms of respiratory malaise, headache, giddiness and a slight feeling of drunkenness were observed by Heim de Balsac, Agasse-Lafont and Feil (1923). In more severe cases mental confusion and excitement may be produced with vertigo, buzzing in the head and visual disturbances.

Kidney injury. A number of cases of haematuria with scanty urine and difficulty of urination have been recorded both from inhalation of paint fumes and from external application or contact with turpentine, but it appears uncertain whether these symptoms represent a true toxic nephritis. An interesting feature is pointed out by Braverman (1933)—the apparent aggravation, by exposure to paint containing turpentine, of retention of urine due to an enlarged prostate. Since, however, the paint in question also contained lead, it was not possible to attribute the symptoms definitely to turpentine. The same reservation applies to the observations of Harris (1918) who, in his investigation of a number of painters, noted many cases of painful and frequent urination, a few showing blood in the stools and the urine, and several, dark-coloured urine with a peculiar odour. Harris did not definitely distinguish between the paint solvents used, which included benzine, benzol, acetone, and wood alcohol, as well as turpentine.

Several cases of apparent injury to the kidneys from external application of turpentine have been observed. Ridder (1923) describes the case of a man whose urine, a week after the onset of severe inflammation of the foot due to the application of turpentine, contained albumin, red and white blood corpuscles and a few casts, while de Jongh (quoted by Pohl, 1923) observed haematuria and strangury in a man who had spilt turpentine on his hands and feet. The urine contained combined glycuronic acid but no casts. Hamilton (1925) mentions the case of a young woman using hot shoe polish containing turpentine, who complained of frequent urination, and whose urine was whitish and had a curious odour.

The cases of haematuria reported by Wilks (1930) and by McCorkle (1929) are more directly related to the use of turpentine as a solvent. In Wilks' cases the occurrence amounted to an outbreak of haematuria, for no less than seven seamen were affected within a week. No other symptoms except slight pain on micturition and slight frequency in two or three cases were present, but the urine was sweet-smelling in several cases. The turpentine used was analysed and found to be good and normal and the outbreak of poisoning was considered probably to be due to insufficient mixing of the turpentine with the "flatting", so that the vapour was released more rapidly than it was removed by ventilation. All the symptoms rapidly cleared when the men were removed from exposure, and Wilks considered it unlikely, in the light of their subsequent good health, that any permanent damage had been done to the kidneys. Chapman (1941), however, quotes two cases of fatal extensive glomerulo-nephritis, one an indoors painter, the other a girl who had slept in an ill-ventilated, newly painted room.

Of the two cases recorded by McCorkle (1929), both in painters, the urine in one showed 1·5 g. per l. of albumin, and in both was scanty, dark red and had an odour of violets. This odour, which was observed by Reinhard as early as 1887, is, according to Poulsson (1930), probably not due to a new compound formed in the body during the excretion of turpentine, since it can be produced merely by shaking normal urine with turpentine. Flury and Zernik state that turpentine inhaled by the lungs is excreted partly by the lungs, but chiefly by the urine, partly unchanged, partly in combination with glycuronic acid.

Chronic Poisoning

The question of the occurrence of chronic poisoning from turpentine inhalation, especially in painters using lead and turpentine simultaneously, has in the past given rise to much controversy. The opinion that turpentine might produce more serious effects than lead in paints was first formulated by Leclaire (1861), who stated "it is not lead, but turpentine which ought to be permanently excluded from the painting industry."

The whole question was discussed in London in great detail by the Industrial Paints Committee and the evidence given by Goadby to the Committee was published by the Departmental Committee in 1923. It was then concluded that the symptoms of abdominal pain, colic, headache and anaemia, affecting both red cells and haemoglobin, very similar to those resulting from acute lead poisoning, might result from the inhalation of turpentine in considerable concentration and in unventilated surroundings, but that a definite distinction between lead and turpentine as the causative agent could not be made with certainty. There was, however, no clinical evidence of chronic turpentine poisoning in man.

With regard to the incidence of mortality from Bright's disease in painters and lead manufacturers respectively, Goadby's figures appeared to show that the incidence in painters was higher and that the inhalation of turpentine was a factor in the comparison. Further statistical information provided by Stevenson, however, revealed possible fallacies in Goadby's interpretation, and it was finally concluded that the assumption that turpentine was the cause of a large proportion of the excess Bright's disease among painters was not confirmed. Furthermore, inquiries in works where turpentine was used without lead (Appendix IV, 1923 Report) failed to elicit evidence of any chronic toxic effects from the use of turpentine.

In a review by Chapman (1941) of 3,705 persons suffering from chronic Bright's disease, only 1·8 per cent were known to have been exposed to turpentine or paint, and post-mortem examination of 8 of these showed glomerular nephritis in one painter only; the rest showed vascular nephritis. These facts, according to Chapman, do not suggest that exposure to turpentine predisposes the human kidney to nephritis.

This comparative harmlessness of turpentine as a paint thinner was further confirmed by Heim de Balsac, Agasse-Lafont and Feil (1923). They examined two groups of painters, 21 who had never come into contact with lead, but had used only white zinc and turpentine over periods of from 6 to 30 years, and 14 who had used lead at some period before entering on their present, lead-free, work. No symptoms of disturbance of health were found in the group using turpentine without lead. The possible occurrence of nephritis was specially ruled out by examinations of urine and blood urea, all of which gave normal results. The blood pressure was slightly raised in 9 per cent of this group, as compared with 14 per cent in the group which had at one time used lead, and with 30 per cent in a group using lead at the time of an earlier investigation by Heim de Balsac, Agasse-Lafont and Feil (1922).

The whole question of possible chronic injury to health from turpentine fumes was further investigated in 1931 by the Compensation for Industrial Diseases Committee, with the result that turpentine was not added to the Schedule of Compensation. Legge and Goadby stated that they associated exposure to turpentine vapour with dizziness, drowsiness, nausea, abdominal pain or colic, but that these symptoms were of short duration, and Legge said he did not believe that turpentine vapour ever gave rise to chronic poisoning. Goadby, however, considered that it might produce a tendency to high blood-pressure or gout, though cardiac symptoms were not found referable to turpentine.

Effect on the Skin

Turpentine, whether applied directly to the skin or by exposure to its vapour, may produce severe dermatitis. In an examination of a large number of painters, polishers and paint factory workers in Finland, amongst whom the incidence of dermatitis was 10·7 per cent, Pirila (1947) concluded that, of all the paint ingredients used, turpentine was chiefly responsible for this incidence. The turpentines used by these workers were sulphate and kiln turpentines with a high content of high boiling point fractions (carotene, limonene, dipentene).

Effect on the Blood

In the few examinations which have been made of the blood of workers exposed to turpentine the changes have been very slight, the chief feature being a leucocytosis. In the investigation by Heim de Balsac and Agasse-Lafont (1910), the red cells showed no change, the white cells a slight polymorphonuclear leucocytosis but no lymphocytosis. In an examination of three workers by the Home Office in 1931, two showed slight anaemia (reduction of the red cells and haemoglobin) and a slight leucocytosis. There were also some abnormalities of both red and white cells—occasional normoblasts in one case, punctate basophilia in another, while a few of the white cells showed toxic nuclei with the occasional presence of a lymphoblast. These workers had been using a shoe polish containing 25 per cent white spirit and 50 per cent of turpentine and had complained of slight symptoms of shakiness, liveliness followed by depression, abdominal discomfort and loss of appetite.

16. Dipentene
(*Limonene*)

$C_{10}H_{16}$

A dehydration product of terpineol occurring widely in volatile essential oils.

PROPERTIES

A colourless liquid with a faint lemon odour.

Pure dipentene—B.P. 175–176° C. Sp.Gr. 0·844.
Technical ,,   —B.P. 170–178° C. Sp.Gr. 0·845–0·854.

Dipentene is somewhat similar in properties to turpentine but oxidizes at only one seventh the rate.

Good solvent for ester gum, coumarone, waxes, some bakelites and glyptals, and for rubber; does not dissolve cellulose acetate or nitrate.

USES

(1) Chiefly as a paint and varnish thinner; it is a good solvent for most synthetic resin varnishes and retards "skinning" in the can (Schantz, 1932).

(2) As an insecticide; according to Martin and Salmon (1934) it is also fungicidal in a spray in low concentrations (0·5 to 1 per cent).

TOXIC EFFECTS

No toxic effects in either animals or human beings are recorded. The only indication that it may possess toxic properties appears from the work of Umeda (1928) on the inhibiting effect of certain solvents on the ciliary epithelium of the frog pharynx, cultured *in vitro*. From this point of view dipentene was found to be more toxic than *cyclo*hexene, which was in its turn more toxic than *cyclo*hexane in another series of experiments. Umeda ascribes the increase in toxicity to the increase in the number of side-chains in the molecule. It is included by Schwartz, Tulipan and Peck (1947) among those solvents regarded as skin irritants.

17. *cyclo*Pentadiene
(*Cyclopentylene*, $\triangle^{1:3}$-*Pentamethylene, Pentol*)

$$C_5H_6, \text{ i.e. } \begin{array}{c} CH=CH \\ | \\ CH=CH \end{array} \!\!> CH_2$$

PROPERTIES

A colourless liquid, insoluble in water but miscible with alcohol or ether. B.P. 41–42°C. It is one of the many compounds formed during the pyrolysis of hydrocarbons at high temperatures; its presence has been reported in coal tar fractions, and especially in benzol forerunnings. Calculated amount in coal 0·1–0·2 lb. of *cyclo*pentadiene per ton of coal carbonized. It is also obtained from cracking of the lower paraffins, other than methane, at 850°C. At ordinary temperatures *cyclo*pentadiene polymerizes spontaneously to di*cyclo*pentadiene.

HYDROCARBONS 107

TOXICITY

*cyclo*Pentadiene is a narcotic, with some irritant effect on the internal organs. According to some investigations made by Browning (1948, unpublished) of a few cases, it may have a slight "benzene-like" effect on the blood.

TOXIC EFFECTS IN ANIMALS

Effect of Inhalation

Elfstrand (1900, cited by von Oettingen, 1940) states that inhalation of the vapour causes complete narcosis in 10 minutes, with an initial phase of motor excitation; sometimes convulsions occur during narcosis, and finally, respiratory paralysis. Recovery, if it takes place, is slow.

Effect of Subcutaneous Injection

Elfstrand found that in rabbits, the narcotic dose, 3·0 ml., was followed by fatal convulsions.

Injury to Internal Organs

The pleural and pericardial cavities contained exudates, and the kidneys were hyperaemic.

TOXIC EFFECTS IN MAN

No toxic effects are reported in the literature, but the following details are the unpublished results of an investigation made by Browning (1948, unpublished) of nine men engaged in cracking gas oil where an analysis of the air had revealed the presence of *cyclo*pentadiene.

Symptoms

All the men complained of lassitude, sleeplessness at night and irritability and some also of loss of appetite and nausea.

Effect on the Blood

Blood changes were present in the men who had the greatest potential exposure to *cyclo*pentadiene: anaemia 3 cases, neutropenia 2 cases, slight leucopenia 1 case, eosinophilia 1 case.

18. Di*cyclo*pentadiene
$C_{10}H_{12}$

PROPERTIES

Present in the higher boiling fraction of light oil, particularly after standing, or in coal tar fractions. It is always found in *cyclo*pentadiene when the latter is allowed to stand.

B.P. 170° C.

TOXICITY

No animal experiments have been made and no toxic effects in man have been recorded.

BIBLIOGRAPHY

ABRAMOWSKAJA and TER (1924). (Cited by Flury and Zernik in Schädliche Gase, Dämpfe, Nebel, Rauch- und Staubarten. Springer, Berlin 1931).
ADLER-HERZMARK, J. (1928). Seltener Fall von Terpentinvergiftung. *Zbl. GewHyg.*, **15**, 65.
ADLER-HERZMARK, J. (1930). Benzolvergiftung. *Wien. med. Wschr.*, **80**, 368.
ADLER-HERZMARK, J. (1933). Periodische Untersuchung von Wiener Arbeitern, die mit benzol-toluol-xylolhältigen Materialien beschäftigt sind. *Arch. Gewerbepath. Gewerbehyg.*, **4**, 486.
ADLER-HERZMARK, J. and SELINGER, A. (1931). Untersuchungen von Wiener Arbeitern, die mit benzol-, toluol- und xylolhältigen Materialien beschäftigt sind. *Arch. Gewerbepath. Gewerbehyg.*, **1**, 763.
ALBRECHT, K. (1932). Unter dem Bilde eines Hirntumors verlaufende chronische Benzolintoxikation. *Mschr. Psychiat. Neurol.*, **82**, 108.
AMBROSIO, L. (1941). La resistenza globulare nell' intossicazione acuta eronica da benzina. *Folia med. Napoli*, **27**, 537.
AMBROSIO, L. (1942). Behaviour of blood enzymes in poisoning by solvents. *Biochim. Terap. sper.*, **29**, 335.
AMENOSSOW, M. M. and BLINKOW, S. M. (1929). Experimentelle Erforschung der Wirkung des Benzols auf das zentral Nervensystem. *Arb. Baku Inst. Berufskrank.*, **1**, 212. (Cited by Mitnik and Genkin in *Arch. Gewerbepath. Gewerbehyg.*, **2**, 457).
ANSWER TO LETTER, Journal of the American Medical Association (1935). Use of interferometer in determining benzene fumes. *J. Amer. med. Ass.*, **105**, 1454.
APITZ, K. and HÜHN, U. (1942). Über die Thrombopenie bei Benzolvergiftung der Ratte. *Z. ges. exp. Med.*, **111**, 540.
ARDAGH, E. G. R. and BOWMAN, W. H. (1935). Removal of thiophen from benzene by the action of acidified hypochlorite solution. *J. Soc. Chem. Ind.*, **54**, 267T.
ASKEY, J. M. (1928). Aplastic anemia due to benzol poisoning. *Calif. West. Med.*, **29**, 262.
AVERILL, C. (1889). Benzole poisoning. *Brit. med. J.*, i, 709.
BABSKY, E. and LEITES, R. (1931). Über die Bildung eines bedingten Reflexes bei Benzinvergiftung. *Arch. exp. Path. Pharmak.*, **161**, 1.
BADINAND, A., PAUFIQUE, L. and RODIER, J. (1947). Intoxication expérimentale par la tétraline. *Arch. Mal. prof.*, **8**, 124.
BAMESREITER, O. (1932). Neue Versuche über die quantitative Giftigkeit von Benzol und Benzindämpfen. *Arch. Hyg., Berl.*, **108**, 129.
BARKER, L. F. and GIBBES, J. H. (1913). On the treatment of leukaemia with benzol. *Arch. exp. Path. Pharmak.*, **181**, 176.
BARTOSCH (1936). Über die Freisetzung von Histamin durch chemisch bekannte Substanzen. *Arch. exp. Path. Pharmak.*, **181**, 176.
BARZILAI, G. (1933). Alterazioni del genitale femminile nel benzolismo sperimentale. *Folia gynaec., Genova*, **30**, 669.
BATCHELOR, J. J. (1927). Relative toxicity of benzol and its higher homologues. Study conducted for Committee on Benzol of National Safety Council. *Amer. J. Hyg.*, **7**, 276.
BEINHAUER, F. (1896). Über Benzolvergiftung, sowie über die Betheiligung der Medicinalbeatem bei Begutachtung von Neuanlagen und Veränderungen gewerblicher Anlagen. *Münch. med. Wschr.*, **43**, 902.
BEISELE, P. (1912). Ein Beitrag zur Kasuistik der Benzoldämpfvergiftung. *Münch. med. Wschr.*, **49**, 2286.
BENÉCH (1897). Rec. Mém. méd. chir. pharm mil., Paris III, s.35, 81. (Cited in Chevallier, P. and Moutier, F. 1947, La gastrite aiguë de intoxiqués benzoliques récents. *Sang*, **18**, 102).
BENZOL POISONING COMMITTEE (1922). See National Safety Council, Final Report on Benzol.
BERNARD, J. (1942). La lymphocytose benzénique. *Sang*, **15**, 501.
BEYER, G. (1933). Chronische Benzolvergiftung bei Kaninchen, *Z. ges. exp. Med.*, **91**, 410.
BICKERT, F. W. (1931–2). Untersuchungen über den Einfluss gewerblicher Gifte auf die Immunkörperbildung. *Arch. Hyg., Berl.*, **107**, 1.
BILLINGS, F. (1913). Benzol in the treatment of leukaemia. *J. Amer. med. Ass.*, **60**, 495.
BINDER, A. (1921). Zur akuten tödlichen Vergiftung mit Benzoldämpfen. *Mschr. Unfallheilk.*, **28**, 202.
BLOOMFIELD, J. J. (1928). Benzol poisoning as a possible hazard in chemical laboratories. *Publ. Hlth. Rep. Wash.*, **43**, 1895.
BÖHME, A. and KÖSTER, R. (1917). Klinische und experimentelle Beobachtungen über Benzinvergiftung. *Arch. exp. Path. Pharmak.*, **81**, 1.
BOWDITCH, M., DRINKER, C. K., DRINKER, P., HAGGARD, H. H. and HAMILTON, A. (1940). Code for safe concentrations of certain common toxic substances used in industry. *J. industr. Hyg.*, **22**, 251.
BOWERS, V. H. (1947). Reaction of human blood-forming tissues to chronic benzene exposure. *Brit. J. industr. Med.*, **4**, 87.
BRACHMANN, J. (1937). Gesundheitsschädigungen in Tiefdruckbetrieben. *Arch. Hyg., Berl.*, **118**, 328.
BRANDINO, G. (1922). Osservazioni istologische nell' intossicazione acute e croniche da benzolo. *Gass. med. lombarda*, **81**, 141.

BRAVERMAN, A. (1933). Retention of urine due to hypertrophy of prostate; two recrudescences, effects of paint; recovery. *Med. J. Rec.*, **137**, 287.
BREZINA, E. (1921). Internationale Übersicht über Gewerbekrankheiten. Springer, Berlin.
BREZINA, E. (1929). Internationale Übersicht über Gewerbekrankheiten, nach den Berichten der Gewerbeaufsichtsbehörden der Kulturländer über die Jahre 1920 bis 1926. Springer, Berlin.
BREZINA, E. and SCHMIDT, W. (1934). Versuche über die Wirkung von Absaugevorrichtungen. *Arch. Gewerbepath. Gewerbehyg.*, **5**, 382.
BRIGANTI, A. and AMBROSIO, L. (1941). Modificazioni anatomo-isopatologiche nella intossicazione sperimentale da benzina. *Rass. Med. Lav. industr.*, **11**, 577.
BRINDEAU, A (1931a). Deux cas d'anémie pernicieuse par intoxication benzolique. *Ann. Hyg. publ., Paris, N.S.*, **9**, 99.
BRINDEAU, A. (1931b). Gravidät und chronische Benzol Vergiftung. (Abstr. in *Dtsch. med. Wschr.*, **57**, 1520).
BROCHER, J. E. W. (1929). Beitrag zur Panmyelopathia atrophicans und zur Frage der Benzolintoxikation in Druckereien. *Zbl. inn. Med.*, **50**, 1186.
BROWN, W. (1933). Dry-cleaning solvents. *J. Soc. Dy. Col., Bradford*, **49**, 42.
BRÜCKEN (1923). Über chronische Benzolvergiftung. *Dtsch. med. Wschr.*, **49**, 1120.
BRÜLLOWA, L. P., BRUSSILOWSKAJA, A. S., LAZAREW, N. W., LÜBIMOWA, M.P. and STALSKAJA, D. I. (1930). Das Blut bei der experimentellen Benzinvergiftung. *Arch. Hyg., Berl.*, **104**, 226.
BRÜLLOWA, L. P. and LUBIMOVA, M.P. (1928). *Gigiena Truda*, **11**, 35. (Cited by Engelhardt, W. E., 1931, in *Arch. Gewerbepath. Gewerbehyg.*, **2**, 479).
BUCHMANN, E. (1911). Zur Frage der akuten Benzolvergiftung. *Klin. Wschr.*, **48**, 936.
BURGL, G. (1906). Über tödliche innere Benzinvergiftung und insbesondere den Sektionsbefund bei derselben. *Münch. med. Wschr.*, **53**, 412.
CACCURI, S. (1940). L'apparato cardiovascolare nell' intossicazione acuta sperimentale da benzolo (ricerche sul comportamento dell-elettro-cardiogramma e della pressione arteriosa). *Rass. Med. Lav. industr.*, **11**, 403.
CACCURI, S. (1946). Su alcuni casi di "benzolismo latente". *Rif. med.*, **60**, 416.
CAMERON, G. R., PATERSON, J. L. H., DE SARAM, G. S. W. and THOMAS, J. C. (1938). Toxicity of some methyl derivatives of benzene with special reference to pseudocumene and heavy coal tar naphtha. *J. Path. Bact.*, **46**, 95.
CAMP, W. E. and BAUMGARTNER, E. H. (1915). Inflammatory reactions in rabbits with a severe leucopenia. *J. exp. Med.*, **22**, 174.
CARDANI, A. (1942a). Tetralina. *Med. d. Lavoro*, **33**, 145.
CARDANI, A. (1942b). Studio sperimentale sulla tossicita della tetralina e della decalina. *Med. d. Lavoro*, **33**, 169.
CARTER, G. (1928). Fatal case of accidental poisoning by benzol vapour. *Brit. med. J.*, **ii**, 794.
CAVAGLIANO, B. (1932). La riserva alcalina nel benzolismo cronico sperimentale. *Med. d. Lavoro*, **23**, 238.
CESARO, A. N. (1941). La resistanza globuolaire nell' intossicazione sperimentale acuta da tetralina. *Folio med., Napoli*, **27**, 65.
CESARO, A. N. (1946). Is percutaneous absorption of benzene possible? *Med. d. Lavoro*, **37**, 151.
CHAMBOVET, L. (1921). Intoxications par le benzène. Étude toxicologique, clinique et médicolegale. Thèse, Lyon.
CHAPMAN, E. M. (1941). Observations on the effect of paint on the kidneys, with particular reference to the rôle of turpentine. *J. industr. Hyg.*, **23**, 277.
CHASSEVANT, A. and GARNIER, M. (1903). Toxicité de benzène et de quelques hydrocarbures aromatiques homologues. *C. R. Soc. Biol., Paris*, **55**, 1255.
CHEMICAL WORKS REGULATIONS (1922). Regulation 4b. Statutory Rule and Order No. 731. H.M. Stationery Office, London.
CHEVALLIER, P. and DÉSOILLE, P. (1947). Les groupes sanguines des benzoliques. *Sang*, **18**, 126.
CHEVALLIER, P. and MOUTIER, F. (1947). La gastrite aiguë des intoxiqués benzoliques récents. *Sang*, **18**, 102.
CHIEF INSPECTOR OF FACTORIES AND WORKSHOPS (1918). Annual Report. Industrial poisoning, p.65. H.M. Stationery Office, London, 1919.
CHIEF INSPECTOR OF FACTORIES AND WORKSHOPS (1947). Annual Report. Industrial diseases, p.84. H.M. Stationery Office, London.
CLIMENKO, D. R. and MACLEOD, J. (1942). Leucotoxic action of benzol. *J. industr. Hyg.*, **24**, 289.
COLEMAN, J. B. and BILHAM, P. (1922). Properties and composition of dekalin. *Chem. Age (N.Y.)*, **7**, 554.
COLEMAN, W. (1934). Hydrogenated naphthalene against clothes moths. *J. econ. Ent.*, **27**, 860.
COMPENSATION FOR INDUSTRIAL DISEASES. Report of Departmental Committee appointed to enquire and report as to certain proposed extensions of Schedule of Industrial Diseases. H.M. Stationery Office, London, 1931.
CONGRESS OF AMERICAN HYGIENISTS (1949). Transactions of American Conference of Governmental Industrial Hygienists.
COOK, W. A. (1945). Maximum allowable concentrations of industrial atmospheric contaminants. *Industr. Med.*, **14**, 936.

Cook, W. A. and Ficklen, J. B. (1935). Determination of benzene in air. *J. industr. Hyg.*, **17**, 41.
Corelli, F. (1937). Das Phänomen des Wiederauflackerns allergischer Erscheinungen nach Histamin. *Klin. Wschr.*, **16**, **ii**, 1546.
Cronin, H. J. (1924). Benzol poisoning in the rubber industry. *Boston med. surg. J.*, **191**, 1164.
Dameshek, W. (1929). Benzene poisoning and agranulocytosis. *J. Amer. med. Ass.*, **93**, 712.
Danysz (1942). Quelques résultats hématologiques constatés chez des ouvriers travaillant dans le benzène. *Sang*, **15**, 348.
Dautrebande, L. (1933). La paralysie périphérique du système vasomoteur par le benzol. La syncope adrénalino-benzolique. *Arch. int. Pharmacodyn.*, **44**, 394.
Dautrebande, L. (1935a). L'action sur le système vasomoteur de certains solvants volatils utilisés dans l'industrie (benzol, éther de pétrole, acétate d'amyle, vernis cellulosique). *Médecine*, **16**, 202.
Dautrebande, L. (1935b). La paralysie du système vasomoteur par les solvants volatils industriels. *Pr. méd.*, **43**, 1081.
Dautrebande, L. and Waucomont, R. (1933). L'action du benzol sur les organes isolés. *C. R. Soc. Biol., Paris*, **112**, 698.
Delarue, R. (1919). De l'intoxication chronique par le benzol. Thèse, Paris.
Delore, P. and Borgomano, J. (1928). Leucémie aiguë au cours de l'intoxication benzénique. Sur l'origine toxique de certains leucémies aiguës et leurs relations avec les anémies graves. *J. Méd. Lyon*, **9**, 227.
Department of Scientific and Industrial Research (1939). Methods for the detection of toxic gases in industry. Leaflet No. 4: Benzene vapour. H.M. Stationery Office, London.
Dimmel, H. (1932). Vergiftungen mit aromatischen Substanzen. *Wien. med. Wschr.*, **82**, 526.
Dimmel, H. (1933). Zur Klinik der chronischen Benzolvergiftung. *Arch. Gewerbepath. Gewerbehyg.*, **4**, 414.
Dolin, B. H. (1943). Determination of benzene in presence of xylene, toluene, etc. New York State Dept. of Labor, *Industr. Bull.* **22**, 373.
Dorendorf (1901). Benzinvergiftung als gewerbliche Erkrankung. *Z. klin. Med.*, **43**, 42.
Dorner, G. (1915). Akute Benzinvergiftung mit nachfolgender spinaler Erkrankung. *Dtsch. Z. Nervenheilk.*, **54**, 66.
Drescher (1906). Tödliche Vergiftung durch Inhalation von Terpentinöldampfen. *Z. Med-Beamt.*, **19**, 131.
Drinker, P., Yaglou, C. P. and Warren, M. F. (1943). The threshold toxicity of gasoline. *J. industr. Hyg.*, **15**, 225.
Dubois, R. (1885). Note sur la vaseline et son emploi dans l'alimentation. *Gaz. Hôp., Paris*, **58**, 1067.
Duke, W. W. (1913). Causes in variations of the platelet count. *Arch. intern. Med.*, **11**, 100.
Dumas, G. (1934). Les intoxications par les produits benzéniques, benzène, benzols et benzines. Thèse, Nancy.
Dunn, J. F. and Brockett, F. S. (1948). Photosensitising properties of some petroleum solvents. *Industr. Med.*, **17**, 303.
Durrans, T. H. (1933). Solvents. 3rd ed. Chapman & Hall, London.
Durrans, T. H. (1935). Solvents and plasticisers. *J. Oil Col. Chem. Ass.*, **18**, 340.
Durrans, T. H. (1935). Review of the technical aspects of industrial solvents. *Chem. and Ind.*, **13**, 585.
Durrans, T. H. (1944). Solvents. 5th ed. Chapman & Hall, London.
Duvoir (1922). À propos des intoxications par le benzène et les benzoles. *Bull. Soc. méd. Hôp., Paris*, **46**, 1541.
Duvoir (1928). Les maladies professionnelles causées par la manipulation des hydrocarbures et de leurs principaux dérivés. *Pr. méd.*, **36**, (ii), 1365.
Duvoir, M. and Dérobert, L. (1942). L'eosinophilie des benzéniques. *Sang*, **15**, 241.
Duvoir, M. and Dérobert, L. (1946). Les réactions hématologiques exceptionelles du benzolisme chronique. *Rec. trav. Inst. nat. Hyg.*, **2**, 550.
Duvoir, M. and Fabre, R. (1946). Recherches sur l'évolution comparée du benzène dans le sang. *Rec. trav. Inst. nat. Hyg.*, **2**, 518.
Dworetzky, A. (1914). Rätselhafte Massenvergiftungen in russischen Fabriken. *Münch. med. Wschr.*, **61**, 1306.
Elfstrand, M. (1900). Beobachtungen über die Wirkung einiger aliphatischer Kohlenwasserstoffe. *Arch. exp. Path. Pharmak.*, **43**, 435.
Elliott, M. S. (1913). Poisoning by petroleum spirits. *Nav. med. Bull., Wash.*, **7**, 416.
Ellis, C. and Meigs, J. V. (1921). Gasoline and other motor fuels. Van Nostrand, New York.
Engelhardt, W. E. (1931). Benzolvergiftung, chronische, gewerbliche. *Samml. Vergiftungsf.*, **2**, 23, C6.
Engelhardt, W. E. (1931). Vergleichende Tierversuche über die Blutwirkung von Benzin und Benzol. *Arch. Gewerbepath. Gewerbehyg.*, **2**, 479.
Engelhardt, W. E. (1935). Versuche über die akut narkotische Wirkung aliphatischer und aromatischer Kohlenwasserstoffe. *Arch. Hyg., Berl.*, **114**, 249.

Espeut, G. and Salinger, J. (1930). Veränderungen des Blutbildes bei Kraftfahrern (Benzolschädigung?) *Dtsch. med. Wschr.*, **56**, 360.
des Essarts, J. Q. (1932). Notes sur quelques cas d'intoxication due à l'emploi de préparations dites bitume-mastic. *Arch. Med. Pharm. nav.*, **122**, 235.
Estler, W. (1935). Versuche über die akut narkotische Wirkung aliphatischer und aromatischer Kohlenwasserstoffe. II. Die Wirkung wiederholter Einatmungen verschiedener Konzentrationen von Benzin, Benzol, Toluol und Xylol auf weisse Mäuse. *Arch. Hyg., Berl.*, **114**, 261.
Eulenberg, H. (1865). Die Lehre von den schädlichen und giftigen Gasen . . . mit besonderer Berücksichtigung der offentlichen Gesundheitspflege und gerichtlichen Medicin. Braunschweig, Vieweg.
Falck, F. (1892). Tödliche Benzinvergiftung. *Vjschr. gerichtl. Med.*, **3**, 399.
Falconer, E. H. (1931). Benzol poisoning. *Calif. West. Med.*, **35**, 365.
Farber, M. (1933). Das Verhalten des Blutbildes unter dem Einfluss von Xylol. *Beitr. path. Anat.*, **91**, 554.
Fauré-Beaulieu, M. and Lévy-Bruhl, M. (1922). L'intoxication benzolique professionnelle: anémie grave avec purpura hémorragique; syndrome medullaire fruste. *Bull. Soc. méd. Hôp., Paris, 3me sér.*, **46**, 1466, 1543.
Feil, A. (1932). Les intoxications professionelles par le benzine et le benzol. *Pr. méd.*, **40**, 1973.
Feil, A. (1933). Le benzolisme professionnel. *Pr. méd.*, **41**, 129.
Ferguson, T., Harvey, W. H. and Hamilton, T. D. (1933). Enquiry into the relative toxicity of benzene and toluene. *J. Hyg., Camb.*, **33**, 547.
Ficklen, J. B. (1940). Manual of industrial health hazards. Service to Industry, Connecticut.
Ficklen, J. B. and Cook, W. H. (1933). Empfindlichkeit der Persalpetersäurereaktion bei der Bestimmung von Benzol. *Z. anorg. Chem.*, **211**, 141.
Filippi, E. (1914). Azione fisiologica e compartamento di alcuni derivati del benzene in confronto con quelli di cicloesano. *Arch. Farmacol. sper.*, **18**, 178.
Fischbach, H. and Terbrüggen, A. (1938). Über die Wirkung von Vitamin C, thyreotropem Hormon und Thyroxin auf das Leberglykogen und die Schilddrüse sowie ihre gegenseitige Beeinflussung. *Virchows Arch.*, **301**, 186.
Fischer, H. (1932). Danger of benzene poisoning in the rubber factories. *Zbl. GewHyg.*, **19**, 87.
Flandin C. and Roberti, J. (1922). Purpura hémorragique mortel du à une intoxication professionelle par les vapeurs de benzol. *Bull. Soc. méd. Hôp., Paris, 3me sér.*, **46**, 58.
Floret, F. (1926). Neuere Beobachtungen über gewerbliche Schädigungen durch Kohlenwasserstoffe. *Zbl. GewHyg.*, **13**, 7.
Flury, F. (1928). Moderne gewerbliche Vergiftungen in pharmakologischer toxikologischer Hinsicht. *Arch. exp. Path. Pharmak.*, **138**, 65.
Flury, F. and Zernik, F. (1931). Schädliche Gase, Dämpfe, Nebel, Rauch- und Staubarten. Springer, Berlin.
Fontana, G. (1921). Nuove ricerche sul sangue e sugli organi ematopoietici nell' intossicazione benzolica. *G. Clin. med.* **2**, 93.
Forssman, S. and Frykholm, K. O. (1947). Benzene poisoning; II. Examination of workers exposed to benzene, with reference to the presence of ester sulfate, muconic acid, urochrome A, and polyphenols in the urine, together with vitamin C deficiency: prophylactic measures. *Acta med. scand.*, **128**, 256.
Foulerton, A. G. R. (1886). Poisoning by benzoline vapour. *Lancet*, **ii**, 865.
Frada, P. (1937). Azione dell' acido ascorbico sul tasso glicemico del coniglio. *Biochim. Terap. sper.*, **24**, 125.
Friemann, A. (1936). Zur Diagnose der chronischen Benzolvergiftung. *Arch. Gewerbepath. Gewerbehyg.*, **7**, 278.
Frumina, L. M. and Fainstein, S. S. (1934). Zur Benzintoxikologie. *Zbl. GewHyg.*, **21**, 161.
Führer, H. (1921a). Die narkotische Wirkung des Benzins und seiner Bestandteile (Pentan, Hexan, Heptan, Octan). *Biochem. Z.*, **115**, 235.
Führer, H. (1921b). Die Wirkungsstärke der Narkotica. *Biochem. Z.*, **120**, 143.
Führer, H. and Pietrusky, F. (1934). Butylazetat-Toluol-Vergiftung, chronische, berufliche. Spätfolgen? *Samml. Vergiftungsf.*, **5**, 1, B40.
Gadaskin, I. D. (1928). Über die Umwandlung des Benzols im Organismus und Methoden zur Bestimmung desselben. *Biochem. Z.*, **198**, 149.
Gaede, D. (1944). Über die Wirkung einiger gesättigter aliphatischer und aromatischer Kohlenwasserstoffe auf die Körpertemperatur. *Arch. Gewerbepath. Gewerbehyg.*, **12**, 308.
Galewsky, G. (1922). Über Dermatitiden durch Tetralin. *Dermat. Woch.*, **273**, 161. (cited by Cardani, 1942. *Med. d. Lavoro*, **33**, 145, 169).
Garan, P. S. (1938). Histamin-freisetzung in der Lunge durch Reizstoffbeatmung. *Arch. exp. Path. Pharmak.*, **188**, 250.
Gardner, H. A. (1925). Physiological effects of vapors from a few solvents used in paints, varnishes and lacquers. *Paint Manuf. Ass. U.S., Proc. Scientific Sec. Educ. Bur. Circ. No.* 250.
Garnier, J. and Cordier, R. (1942). Sur deux cas de réaction lymphomonocytose, importante chez des ouvriers en contact quotidien avec les émanations benzoliques. *Sang*, **15**, 496.
Gelman (1932). Über die Frage der Empfanglichkeit des Organismus gegen gewerbliche Gifte der Benzingruppe. VI Internationaler Kongress für Unfallheilkunde und Arbeitsmedizin reported by Engelhardt, W. E., *Zbl. GewHyg.*, **19**, 17.

GENHARD, A. (1910). Gesellschaft der Aerzte in Zürich. Über Benzolvergiftung. *CorrespBl. schweiz. Arz.*, **4**, 387.
GEPPERT, J. (1926). Zur Frage von Gesundheitsschädlichen Bohnerwachs. *Dtsch. med. Wschr.*, **52**, 1080.
GLADYCHERSKAJA, V. A. (1936). Effect of benzene and benzine on the respiratory center of a frog. *J. industr. Hyg.*, **18**, abstr. sec., 175.
GLIBERT, D. (1935). Les méfaits de l'héliogravure. *Brux.-méd.*, **16**, 194.
GLOOR, W. (1929). Die klinische Bedeutung der qualitativen Veränderung der Leukocyten. Thieme, Leipzig.
GOADBY, K. (1920). Evidence to Departmental Committee on Use of Lead in Painting. Home Office; Report of the Departmental Committee on Industrial Paint, pp. 25–34. H.M. Stationery Office, London, 1923.
GOLDMANN, H. (1930). Eine neue gewerbliche retrobulbär Neuritis. *Klin. Mbl. Augenheilk.*, **84**, 761.
GOLDWATER, L. J. and TEWKSBURY, M. P. (1941). Recovery following exposure to benzene (benzol). *J. industr. Hyg.*, **23**, 217.
GOODERHAM, W. J. (1935). Analysis of benzoles. *J. Soc. chem. Ind.*, **54**, 297 T.
GOUNELLE, H. and DUMAS, G. (1935). Manifestations prolongées de l'intoxication benzolique. Persistence de troubles morbides dix-huit mois plus tard. *Sang*, **9**, 204.
GRAM, B. (1933). Di un caso di benzinismo professionale con prevalente sindrome anemica. *Gazz. Osp. Clin.*, **54**, 643.
GREENBURG, L. M., MAYERS, M. R., HEIMANN, H. and MOSKOWITZ, S. (1942). Effects of exposure to toluene in industry. *J. Amer. med. Ass.*, **118**, 573.
GREENBURG, L. M. and MOSKOWITZ, S. (1945). Safe use of solvents for synthetic rubber. *Industr. Med.*, **14**, 359.
GROSS, E. and KUSS, E. (1931). Über die Dosierung von Dämpfen in chronischen Inhalationsversuchen. *Zbl. GewHyg.*, **18**, 95.
GUEFFROY, W. and LUCE, F. (1937). Untersuchungen über gewerbliche Exposition gegenüber den Dämpfen des Benzols und seiner Homologe. *Arch. Gewerbepath. Gewerbehyg.*, **8**, 426.
HADEN, R. L. (1919). Benzine poisoning with report of a case. *Johns Hopk. Hosp. Bull.*, **30**, 309.
HAGEN, J. (1939). Eine akute Terpentinöl-Vergiftung. *Samml. Vergiftungsf.*, **10**, 191, A820.
HAMILTON, A. (1925). Industrial poisons in the United States. Macmillan, New York.
HAMILTON, A. (1928). Lessening menace of benzol poisoning in American industry. *J. industr. Hyg.*, **10**, 227.
HAMILTON, A. (1928). Protection against industrial poisoning; prevention of slow, chronic poisoning from gas. Chemistry in Medicine. Chemical Foundation Inc., New York.
HAMILTON, A. (1931). Benzene (benzol) poisoning. *Arch. Path. Lab. Med.*, **11**, 434, 601.
HAMILTON, A. (1934). Industrial toxicology. Harper, New York.
HAMILTON, A. and JOHNSTONE, R. T. (1945). Industrial toxicology. Oxford Univ. Pr., New York.
HAMILTON-PATERSON, J. L. and BROWNING, E. (1944). Toxic effects in women exposed to industrial rubber solutions. *Brit. med. J.*, i, 349.
HARBECK, E. and LUNGE, G. (1898). Quantitative Scheidung von Äthylene und Benzoldampf. *Z. anorg. Chem.*, **16**, 26.
HARRINGTON, T. F. (1917). Industrial benzol poisoning in Massachussetts. *Boston med. Surg. J.*, **177**, 203.
HARRIS, L. J. (1918). Clinical study of frequency of lead, turpentine and benzine poisoning in 400 painters. *Arch. intern. Med.*, **22**, 129.
HAYHURST, E. R. and NEISWANDER, B. E. (1931). Case of chronic benzene poisoning. *J. Amer. med. Assoc.*, **96**, 269.
HEATON, N. (1923). Some hydrogenation products of benzene and naphthalene. *J. Oil Col. Chem. Ass.*, **6**, 93.
HEFFTER, A. (1915). Über die akute Vergiftung durch Benzoldämpfen. *Dtsch. med. Wschr.*, **41**, 182.
HEGLER, C. (1933). Benzol- und Phosgenvergiftungen. 1. Benzol. *Med. Welt*, **7**, 10.
HEIM DE BALSAC, F. and AGASSE-LAFONT, E. (1910). Réactions hématiques du benzénisme professionel. Association française pour l'avancement des Sciences. Congrès de Toulouse.
HEIM DE BALSAC, F., AGASSE-LAFONT, E. and FEIL, A. (1922). Enquête sur les manifestations morbides présentées par les ouvriers peintres en voitures. *Progr. méd.*, Paris, **49**, 157.
HEIM DE BALSAC, F., AGASSE-LAFONT, E. and FEIL, A. (1922). Manifestations morbides chez les ouvriers maniant le celluloid et ses solvents. *Paris méd.*, i, 477.
HEIM DE BALSAC, F., AGASSE-LAFONT, E. and FEIL, A. (1922). Morbid disturbances due to handling of celluloid. *J. Amer. med. Ass.*, **79**, 146.
HEIM DE BALSAC, F., AGASSE-LAFONT, E. and FEIL, A. (1923). L'essence de térébenthine a t'elle un rôle dans la pathologie professionelle des peintres? *Pr. méd.*, **31**, 537.
HEIM DE BALSAC, F., and AGASSE-LAFONT, E. (1933). Intoxications mortelles ou de gravité variable, en série, par emploi d'un adhesif solubilisé par le benzène; indications prophylactiques. *Bull. Acad. Méd.*, Paris, **110**, 31.
HEITZMANN, O. (1931). Vergleichende pathologische Anatomie der experimentellen Benzol- und Benzinvergiftung. *Arch. Gewerbepath. Gewerbehyg.*, **2**, 515.

HEKTOEN, L. (1916a). Effect of benzene on the production of antibodies. *J. infect. Dis.*, **19**, 69.
HEKTOEN, L. (1916b). Effect of toluene on the production of antibodies. *J. infect. Dis.*, **19**, 737.
HELMER, K. J. (1944). Accumulated cases of chronic benzene poisoning in the rubber industry. *Acta med. scand.*, **118**, 354.
HENDERSON, V. E. and JOHNSTON, J. F. A. (1931). Anesthetic potency in the cyclo-hydrocarbon series. *J. Pharmacol.*, **43**, 89.
HENDERSON, Y. and HAGGARD, H. W. (1927). Noxious gases and the principles of respiration influencing their action. Amer. Chem. Soc. Chemical Catalog Co., New York.
HETZER, W. (1922). Akut entstandene Pylorusstenose nach Benzolvergiftung. *Dtsch. med. Wschr.*, **48**, 627.
HICGUET, G. (1930). Un cas de surdité nerveuse; toxi-névrite due au White Spirit. *J. Neurol. Psychiat.*, **30**, 89.
HIRSCH, S. (1932). Über chronische Xylolvergiftung, insbesondere über die Einwirkung des Xylols auf Herz und Gefasse. *Verhandl. d. deutsch. Gesellschaft f. inn. Med. Kong.*, **44**, 483.
HOGAN, J. F. and SHRADER, J. H. (1923). Benzol poisoning. *Amer. J. publ. Hlth.*, **13**, 279.
HULTGREN, G. (1926). Action du benzol sur le teneur du sang en thrombocytes, leucocytes et érythrocytes. *C. R. Soc. Biol.*, Paris, **95**, 1060.
HUNT, E. and WEISKOTTEN, H. G. (1930). Value of the Arneth count in determining the age of neutrophile (amphophile) leucocytes (rabbit). VIII. Action of benzol. *Amer. J. Path.*, **6**, 175.
HUNTER, F. T. (1939). Chronic exposure to benzene; clinical effects. *J. industr. Hyg.*, **21**, 331.
HUNTER, F. T. and HANFLIG, S. S. (1927). Chronic benzol poisoning. *Boston med. surg. J.*, **197**, 292.
HURWITZ, S. H. and DRINKER, C. K. (1915). Factors of coagulation in the experimental aplastic anaemia of benzol poisoning with special reference to the origin of prothrombin. *J. exp. Med.*, **21**, 401.
I. G. ELBERFELD TOXICOLOGY INDEX (1931). Report of LK Division of I.G. Ludwigshafen.
I. G. FARBENINDUSTRIE AKTIENGESELLSCHAFT (1930). Lösungsmittel und Weichmachungsmittel, p.134. Tabelle der Verdunstungszeiten und der Siedegrenzen. Frankfurt-am-Main.
INDUSTRIAL PAINTS COMMITTEE (1920). Report of Departmental Committee appointed to examine danger of lead paints to workers, etc. H.M. Stationery Office, London, 1923.
INDUSTRIAL PAINTS COMMITTEE (1920). Departmental Committee appointed to examine danger of lead paints to workers, etc. Minutes of evidence and appendices to Report. H.M. Stationery Office, London, 1923.
ISRAEL, L. (1934). Étude clinique et expérimentale des hémopathies benzoliques et d'une aleucie vraie (absence de globules blancs dans le sang). *Strasbourg méd.*, **94**, 569.
JACKSON, H. jr., PARKER, F., jr. and LEMON, H. M. (1940). Agnogenic myeloid metaplasia of the spleen. *New Engl. J. Med.*, **222**, 985.
JAFFÉ, M. (1909). Über die Aufspaltung des Benzolrings im Organismus. *Hoppe-Seyl. Z.*, **62**, 58.
JAFFÉ, R. (1914). Über Benzinvergiftung nach Sektionsergebnissen und Tierversuchen. *Münch. med. Wschr.*, **61**, 175.
JAKOBSEN, J. (1939). Spray painting hazards: a clinical and experimental haematological study with special reference to changes in the differential cell count. Oxford Univ. Pr., London.
VON JAKSCH, R. (1897). Die Vergiftungen. Spezielle Pathologie u. Therapie. Vol. **1**, 383. Ed. by H. Northnagel. Cited by Chapman, E. M. (1941).
JEPHCOTT, C. M. and BULMER, F. M. R. (1939). Urinary sulfate test in the supervision of workers exposed to benzene. *J. industr. Hyg.*, **21**, 132.
JOACHIMOGLU (1915). Über den Nachweis des Benzols in Organen und seine Verteilung in Organismus. *Biochem. Z.*, **70**, 93.
JONGH, de. Quoted by Pohl, J. (1923), Spezielle Pathologie und Therapie innere Krankheiten, *Hft.* **9**, 1168.
JOST, H. (1932). Harnuntersuchungen bei chronischer Schädigung durch Benzol und Benzolderivate. *Arch. Gewerbepath. Gewerbehyg.*, **3**, 791.
JUDICA-CORDIGLIA, G. (1932). Benzinvergiftung durch ein Fliegenvertilgungsmittel. *Samml. Vergiftungsf.* **3**, 21, A187.
KAMMER, A. G., ISENBERG, N. and BERG, M.E. (1938). Medical supervision of benzene plant workers. *J. Amer. med. Assoc.*, **111**, 1452.
KANITZ, H. R., LOHMEYER, A. and SCHOLZ, J. (1935). Über die Wirkungen von Tetralin, 5-Tetralol und 5-Tetralon auf Körpertemperatur und Stoffwechsel. *Arch. Hyg.*, Berl., **113**, 234.
KEYES, D. B. (1925). Solvents and automobile lacquers. *Industr. Engng. Chem.*, **17**, 558.
KIRÁLYFI, G. (1913). Weitere Beiträge zur therapeutischen Verwendung des Benzoles. *Wien. klin. Wschr.*, **26**, 1062.
KLEMPERER, G. and HIRSCHFELD, H. (1913). Weitere Mitteilungen über die Behandlung der Blutkrankheiten mit Thorium X. *Ther. Gegenw.*, **54**, 57.
KLINE, B. S. and WINTERNITZ, M.C. (1913). Studies upon experimental pneumonia in rabbits. V. The rôle of the leucocyte in experimental pneumonia. The relation of the number of organisms injected to the mortality. *J. exp. Med.*, **18**, 50.
KOBERT, R. (1893). Lehrbuch der Intoxikationen. Stuttgart.
KOBERT, R. (1906). Lehrbuch der Intoxikationen. 2 Aufl. Stuttgart.
KOELSCH, F. (1926). Vergiftungen. Aliphatische Verbindungen. *Hb. soz. Hyg.*, **2**, 390.

KOLESNIKOW (1932). (Cited by LEWIN, I. E., 1932, Zur Frage der pathologischen Veränderung und der Funktionsfahigkeit, *Arch. Gewerbepath. Gewerbehyg.*, **3**, 340).

KOPPENHÖFER, G. F. (1935). Morphologische und chemische Untersuchungen bei einem Fall einer tödlichen akuten Benzolvergiftung. *Arch. Gewerbepath. Gewerbehyg.*, **6**, 416.

KORANYI, A. (1912). Die Beeinflussung der Leukämie durch Benzol. *Berl. klin. Wschr.*, **49**, 1357.

KORNETOW (1929). Arb. Baku Inst. Berufskrank. I. (Cited by Nikulin and Hetmann, 1933 in *Arch. Gewerbepath, Gewerbehyg.*, **4**, 653).

KORVIN, E. (1933). Über das Auftreten von Epilepsie bei chronischer Benzolvergiftung. *Dtsch. med. Wschr.*, **59**, 816.

KRACKE, R. R. (1932). Experimental production of agranulocytosis. *Amer. J. clin. Path.*, **2**, 11.

KRACKE, R. R. and PARKER, F. P. (1934). Etiology of granulopenia (agranulocytosis); with particular reference to the drugs containing the benzene ring. *J. Lab. clin. Med.*, **19**, 799.

KRACKE, R. R. and PARKER, F. P. (1935). Relationship of drug therapy to agranulocytosis. *J. Amer. med. Ass.*, **105**, 960.

KRANENBERG, W. R. H. and PEETERS, H. (1928). Chronische Benzolvergiftung. *Zbl. GewHyg.*, **15**, 358.

KRAWKOW (1916). *Russki Wratsch.*, **15**, 338. (Quoted by Lazarew, N. W., 1929, Toxic effects of cyclohexane. *Arch. exp. Path. Pharmak.*, **143**, 223).

KUNTZEN, H. (1932). Thrombosis, embolism and chronic poisoning from benzine and benzene mixtures. *Z. ärztl. Fortbild.*, **29**, 663.

LAIGNEL-LAVASTINE, LÉVY, R. and DESOILLE, H. (1928). Un cas mortel d'anémie aplastique hémorragique par intoxication benzénique professionelle. *Bull. Soc. méd. Hôp., Paris*, **52**, 1264.

LANDÉ, K. and KALINOWSKY, L. (1928). Zur Klinik der gewerblichen Berufserkrankungen durch Benzol. *Med. Klinik*, **24**, 655.

LANGELEZ, A., PEREMANS, G. and BASTENIER, H. (1940). À propos de l'action toxique du toluol dans l'industrie. *Rev. du Travail. (Brux.)*, **41**, 965.

LANGLOIS, J. P. and DESBOUIS, G. (1907). Des effets des vapeurs hydrocarbonées sur le sang. *J. Physiol. Path. gén.*, **9**, 253.

LARIONOW, L. T. (1932). Über die Wirkung des Benzols auf die Gewebskulturen. *Arch. exp. Zellforsch.*, **13**, 445.

LARIONOW, T. and LAZAREW, N. W. (1931). Experimentelle Untersuchungen über die Wirkung von Einatmung kleiner Benzin- und Benzolmengen. *Klin. Wschr.*, **10**, 356.

LAUNOY, L. and LÉVY-BRUHL, M. (1920). De l'action comparée du benzène et du cyclohexane sur les organes hématopoiétiques. *C. R. Soc. Biol., Paris*, **83**, 215.

LAZAREW, N. W. (1929a). Über die Giftigkeit verschiedener Kohlenwasserstoffdämpfe. *Arch. exp. Path. Pharmak.*, **143**, 223.

LAZAREW, N. W. (1929b). Zur Toxicologie des Benzins. *Arch. Hyg., Berl.*, **102**, 227.

LAZAREW, N. W., BRUSSILOWSKAJA, A. J. and LAWROW, J. N. (1931a). Quantitative Bestimmung einiger fluchtige Stoffe im Blut. *Biochem. Z.*, **240**, 12.

LAZAREW, N. W., BRUSSILOWSKAJA, A. J. and LAWROW, J. N. (1931b). Quantitative Untersuchungen über die Resorption einiger organischer Gifte durch die Haut ins Blut. *Arch. Gewerbepath. Gewerbehyg.*, **2**, 641.

LAZAREW, N. W., BRUSSILOWSKAJA, A. J., LAWROW, J. N. and LIFSCHITZ, F. P. (1931c). Über die Durchlässigkeit der Haut für Benzin und Benzol. *Arch. Hyg., Berl.*, **106**, 112.

LAZAREW, N. W., BRULLOWA, L. P., KREMNEWA, S. N., LARIONOW, L. T., LUBIMOWA, M. P. and STALSKAJA, D. J. (1931d). Experimentelle Untersuchungen über die Gewöhnung von Benzin. *Arch. exp. Path. Pharmak.*, **159**, 345.

LECHELLE, COSTE, THIEFFREY and CUADRANO (1940). Les modalités cliniques de l'intoxication benzolique, leur pronostic, leur prophylaxie. *Bull. Soc. méd. Hôp., Paris*, **56**, 353.

LECLAIRE, E. J. (1861). Recherches concernant l'influence que peut avoir l'essence de térébenthine sur la santé des ouvriers peintres en bâtiments et des personnes qui habitent un appartement nouvellement peint. Bouchard-Huzard, Paris.

LEDERER, E. (1932). Chronische Benzolvergiftung unter Morbus Gaucherähnlichem Bilde? *Arch. Gewerbepath. Gewerbehyg.*, **3**, 535.

LEGGE, T. M. (1916). Special discussion on the origin, symptoms, pathology, treatment and prophylaxis of toxic jaundice observed in munition workers. *Proc. R. Soc. Med.*, **10**, *General Reports*, 33.

LEGGE, T. M. (1918). See Chief Inspector of Factories, Annual Report, 1918.

LEGGE, T. M. (1919–20). Chronic benzol poisoning. *J. industr. Hyg.*, **1**, 539.

LEGLUDIC, H. and TURLAIS, C. (1914). Recherches sur la toxicité du petrole et quelques-unes de ses actions physiologiques. *Ann. Hyg. publ., Paris*, 4me sér., **21**, 385.

LEHMANN, K. B. (1910). Quantitative Untersuchungen über die Aufnahme von Benzol durch Tier und Mensch aus der Luft. *Arch. Hyg., Berl.*, **72**, 307.

LEHMANN, K. B. (1912). Experimentelle Studien über den Einfluss technisch und hygienisch wichtiger Gase und Dämpfe auf den Organismus. Die Kohlenwasserstoffe: Benzol, Toluol, Xylol, Leichtbenzin und Schwerbenzin. *Arch. Hyg., Berl.*, **75**, 1.

LEHMANN, K. B. (1914). Experimentelle Studien über den Einfluss technisch und hygienisch wichtiger Gase und Dämpfe auf den Organismus. Vergleichende Untersuchungen über die Giftigkeit von Terapin (Sangajol) und Terpentin. *Arch. Hyg., Berl.*, **83**, 239.

LEHMANN, K. B. and FLURY, F. (1938). Toxikologie und Hygiene der technischen Loesungsmittel. Berlin.
LEHMANN, K. B. and FLURY, F. (1943). Toxicology and hygiene of industrial solvents. (Tr. by E. King and H. F. Smyth, jr.). Williams and Wilkins, Baltimore.
LEIDY, J. (1889). Case of petroleum ether poisoning. *Therap. Gaz.*, p. 443.
LE NOIR and CLAUDE (1897). Sur un cas de purpura attribué à l'intoxication par la benzine. *Bull. Soc. méd. Hôp., Paris*, **14**, 1251.
LESCHKE, E. (1932). Fortschritte in der Erkennung und Behandlung der wichtigsten Vergiftungen. *Münch. med. Wschr.*, **79**, 751, 1755, 1786.
LESTCHINSKAJA, O. (1932–33). Zur relativen Giftigkeit verschiedener Benzine. *Arch. Gewerbepath. Gewerbehyg.*, **4**, 508.
LÉVY, R. (1935). Les hémopathies benzéniques. *Médecine*, **16**, 245.
LEWIN, I. E. (1928). Involution und Regeneration des Thymus unter den Einfluss von Benzol. *Virchows Arch.*, **218**, 1.
LEWIN, I. E. (1932). Zur Frage der pathologischen Veränderungen und der Funktionsfähigkeit des Reticulo-endothelsystems bei Vergiftung mit Benzindämpfen. *Arch. Gewerbepath. Gewerbehyg.*, **3**, 340.
LEWIN, L. (1888). Über Allgemeine- und Hautvergiftung durch Petroleum. *Virchows Arch.*, **112**, 35.
LEWIN, L. (1907). Die akute tödliche Vergiftung durch Benzoldampf. *Münch. med. Wschr.*, **54**, 2377.
LEWIN, L. (1920). Über giftige Extraktionsmittel für Fette, Wachse, Harze, und andere ähnliche wasserunlösliche Stoffe. *Z. Dtsch. Öl- u. Fettindustr.*, **40**, 421.
LEWIN, L. (1929). Gifte und Vergiftungen. 4th Aufl., p. 384. Springer, Berlin.
LIFSCHITZ, I. I. (1935). De l'influence des conditions de température sur la toxicité de la benzine. *La Médicine du Travail*, **7**, 41.
LIGNAC, G. O. E. (1932). Die Benzolleukämie bei Menschen und weissen Mäusen. *Krankheitsforschung*, **9**, 403.
LITZNER, S. (1932). Erkrankungen durch Benzol und seine Homologen. *Ergebn. ges. Med.*, **17**, 367.
LITZNER, S. and EDLICH, W. (1934). Toluol-Vergiftungen, chronische, gewerbliche. *Samml. Vergiftungsf.*, **5**, 9, A398.
LODEMANN (1920). Vergiftung durch Genuss von Terpentinöl. *Med. Klinik*, **16**, 340.
LOEPER, M. (1941). Benzol et foie. *Progr. méd., Paris*, **69**, 729.
LOEPER, FABRE and BORREAU (1946). Leucémie benzénique de 15 mois avec benzol dans le sang. *Progr. méd., Paris*, **74**, 581.
LOOFT, A. (1930). Blood changes in benzine poisoning. *Med. Rev., Bergen*, **47**, 1.
LÖWY, J. (1926). Die Berufskrankheiten der Ärzte. *Med. Klinik*, **22**, 567.
LUIG, B. (1913). Beiträge zur Schwefelkohlenstoff- und Benzolvergiftung in akuten und chronischer Versuchen. Dissertation, Würzburg.
LUSZCZAK, A. (1935). Die Bestimmung von Xylol und Xylol-Toluol Dämpfgemischen in der Raumluft. Grassgerben, R. (ed.). Abhandlungen aus dem Gesamtgebiete der Hygiene. Urban and Schwarzenberg, Berlin.
LÜTHY, F. (1940). Polynévrite due à une intoxication par l'inertol. *Rev. des Acc. du Trav. et Malad. profess.*, **34**, 256.
MABILLE (1896). Note sur l'ivresse pétrolique. *Ann. Hyg. publ., Paris*, 3me sér., **35**, 360.
MACCIOTTA, A. (1942). Crasi sanguigne e benzinismo de lavatori. *Rass. Med. Lav. industr.*, **13**, 374.
MCCLURE, R. D. (1916). Transfusion in benzole poisoning. *J. Amer. med. Ass.*, **67**, 793.
MCCORD, C. P. (1926). Occupational dermatitis from wood turpentine. *J. Amer. med. Ass.*, **86**, 1979.
MCCORD, C. P. (1929). Present state of benzene (benzol) poisoning. *J. Amer. med. Ass.*, **93**, 280.
MCCORD, C. P., COX, N. and O'BOYLE, C. (1932). New investigation of the toxicity of benzene and benzene impurities. Industrial Health Conservancy Laboratories, Cincinnati.
MCCORKLE, W. E. (1929). Acute nephritis due to turpentine poisoning. *Hahnemann Monthly*, **64**, 609.
MAITLAND, F. P. (1931). Toxicity and fatal dose of turpentine. *Brit. med. J.*, ii, 77.
MALLORY, T. B., GALL, F. A. and BRICKLEY, W. J. (1939). Chronic exposure to benzene (Benzol). III. Pathologic results. *J. industr. Hyg.*, **21**, 355.
MARIAN-WOLFEN, T. (1925). Unfallgefahren in Benzolbehältern und deren Verhütung. *Zbl. GewHyg.*, **12**, 107.
MARTIN, H. and SALMON, E. S. (1934). Fungicidal properties of certain spray fluids. *J. agric. Sci.*, **24**, 469.
MARTLAND, H. S. (1917). Cited by Hamilton, A., 1931, in *Arch. Path. Lab. Med.*, **11**, 434, 601.
MATSUSHITA, K. (1935). Pathologische-histologische Studien über das Zentralnervensystem bei experimenteller Petroleumvergiftung. *Nagasaki Igakkwai Zassi*, **13**, 967. (*Zbl. GewHyg.*, **22**, abstr. sec., 190.).
MATVEEV, A. P., PRONIN, Yu. B. and FROST, O. I. (1930). *Zhur. Prikladnoi Khim.* **3**, 1223. Method for continuous analysis of benzene vapors in air. *Chem. Abs.*, **25**, 3273.

Mauro, G. (1925). Avvelenamento sperimentale da benzolo, prima sintomi. *Med. d. Lavoro*, **16**, 168.
Mazel, P., Picard, D. and Bourret, J. (1944). Le mononucléoses est-elle une forme " actuelle" de la myelotoxicose benzolique. *Arch. mal. Profess.*, **6**, 18.
Meda, G. (1922). Il benzolismo professionale. *Med. d. Lavoro*, **13**, 264.
Mendel, L. B. and Rose, W. C. (1911). Experimental studies on creatine and creatinine : II. Inanition and the creatine content of muscle. *J. biol. Chem.*, **10**, 255.
Merklen, P. and Israël, L. (1934). L'intoxication par le benzol; aleucie hémorragique. *Sang*, **8**, 700.
Merle, F. (1928). Les maladies professionelles causées par la manipulation des hydrocarbones et de leurs principaux derivés. *Pr. méd.*, **36**, 1366.
Meyer, A. (1937). Chronische Benzolvergiftung und Vitamin C. *Z. Vitaminforsch.*, **6**, 83.
Meyer, L. M. and Ginsberg, V. (1942). Aplastic anemia. *J. industr. Hyg.*, **24**, 37.
Meyer, S. (1928). Changes in the blood as reflecting industrial damage. *J. industr. Hyg.*, **10**, 29.
Meyer, S. (1931). Über Blutveränderungen bei gewerblichen Schädigungen. *Arch. Gewerbepath. Gewerbehyg.*, **2**, 516, 553.
Mgebrow, L. (1930). Materialen zum Studium der Wirkung einiger Destillations-produkte des Bakuer Naptha auf den Tierschen Organismus. *Virchows Arch.*, **278**, 610.
Milton, R. (1945). Absorptiometric method for estimation of atmospheric benzene (in the presence of xylene and toluene). *Brit. J. industr. Med.*, **2**, 36.
Ministry of Health, Berlin (1920). Quoted by Heaton, N. (1923). *J. Oil Col. Chem. Ass.*, **6**, 93.
Mitnik, P. and Genkin, S. (1931). Zur Klinik der chronischen Benzolvergiftung. *Arch. Gewerbepath. Gewerbehyg.*, **2**, 457.
Muto, T. (1931), Beiträge zur Kenntnis der experimentellen Benzolvergiftung. *J. orient. Med.*, **15**, 97.
Myers, J. and Jenkins, T. (1913). Benzol in the treatment of leukemia. Report of Proc. Nat. Assoc. Study and Prevention of Tuberculosis. 9th Annual Meeting, May 8–9, Washington. (Abstr. *J. Amer. med. Ass.*, **60**, 1575.)
Naegeli, O. (1931). Blutkrankheiten und Blutdiagnostik. 5th Aufl. Springer, Berlin.
Nahum, L. H. and Hoff, H. E. (1934). Mechanism of sudden death in experimental acute benzol poisoning. *J. Pharmacol.*, **50**, 336.
National Safety Council, The Chemical and Rubber Sections. Final Report on Benzol. Published by National Bureau of Casualty and Surety Underwriters, 1926. See also Winslow, C. E. A. (1927).
National Safety Council. Final Report of the Committee on Spray Coating. See also Smyth, H. F. and Smyth, H. F., jr. (1928).
Nelken, F. (1931). Untersuchungen über Xylolschädigungen in Berliner Tiefdruckbetrieben. *Zbl. GewHyg.*, **18**, 182.
Nelson, K. W., Ege, J. F., Ross, M., Woodman, L. E. and Silverman, L. (1943). Sensory response to certain industrial solvent vapors. *J. industr. Hyg.*, **25**, 282.
Nencki, M. and Giacosa, P. (1880). Über die Oxydation der aromatischen Kohlenwasserstoffe im Thierkörper. *Hoppe-Seyl. Z.*, **4**, 325.
Neumann, W. (1915). Experimentelles zur Wirkung des Benzols. *Dtsch. med. Wschr.*, **41**, 394.
Newton, C. R. (1920). Industrial blood poisons. *J. Amer. med. Ass.*, **74**, 1149.
New York State Department of Labor (1927). Chronic benzol poisoning among women industrial workers. *Spec. Bull.* No. 150. Bureau of Women in Industry.
Nick, H. (1922). Erfolgreiche Behandlung einer schweren akuten Benzolvergiftung durch Lecithin-emulsion. *Klin. Wschr.*, **1**, 68.
Nicolajew, N. M. and Schparo, L. A. (1929). Studien über Benzolwirkung auf den tierischen Organismus. *Virchows Arch.*, **272**, 123.
Nikulin, M. and Hetmann, Z. (1933). Zur Frage der Benzinwirkung auf Blutzelle und Lipoide. *Arch. Gewerbepath. Gewerbehyg.*, **4**, 653.
Nikulina, M. and Titowa, A. (1934). Zur Frage der. Thrombopenie, als eines der frühesten Symptome der chronischen Benzolintoxikationen. *Arch. Gewerbepath. Gewerbehyg.*, **5**, 201.
Nunn, J. A. and Martin, F. M. (1934). Gasolin (Petroläther)- und Kerosin (Petroleum)-Vergiftungen bei Kindern. *Samml. Vergiftungsf.*, **5**, 183, A459.
Oettel, H. (1936). Einwirkung organische Flüssigkeiten auf die Haut. *Arch. exp. Path. Pharmak.*, **183**, 641.
von Oettingen, W. F. See Delarue, R. (1919). Thèse, Paris.
von Oettingen, W. F. (1940). Toxicity and potential dangers of aliphatic and aromatic hydrocarbons. *Publ. Hlth. Bull., Wash.*, No. 255, 38.
von Oettingen, W. F., Neal, P. A. and Donahue, D. D. (1942a). Toxicity and potential dangers of toluene: a preliminary report. *J. Amer. med. Ass.*, **118**, 579.
von Oettingen, W. F., Neal, P. A., Donahue, D. D., Svirbely, J. L., Baernstein, H. D., Monaco, A. R., Valaer, P. J. and Mitchell, J. L. (1942b). Toxicity and potential dangers of toluene, with special reference to its maximum permissible concentration. *Publ. Hlth. Bull., Wash.*, No. 279.
Oppenheim, M. (1930). Hautschädigungen durch die Arbeit mit einer Benzol-Vergussmasselösung in einer Minenzünderfabrik. *Wien. klin. Wschr.*, **43**, 249.

Orzechowski, G. (1929). Chronische Benzolvergiftung und Knochenmark. *Virchows Arch.*, **271**, 191.
Panse, F. and Bender, W. (1934). Toluol-Xylol Vergiftung: chronische (Psychose) bei einem Tiefdruckarbeiter. *Samml. Vergiftungsf.*, **5**, 179, A458.
Pappenheim, A. (1913). Zur Benzolbehandlung der Leukämie und sonstiger Blutkrankheiten. *Wien. klin. Wschr.*, **26**, 48.
Patty, F. A. ed. (1949). Industrial hygiene and toxicology. Vols. I and II. Interscience Publishers, New York.
Paul, W. D., Friedlander, V. A. and McCord, C. P. (1927). Baso-philic material in benzol poisoning. *J. industr. Hyg.*, **9**, 193.
Penni, G. and de Steffanis, C. (1940). Ricerche sull'intossicazione cronica da vapori di benzina. *Rass. med. Lav. industr.*, **11**, 516.
Pennsylvania Department of Labor and Industry (1926). Special Bull. No. 16 quoted by Smyth, H. F. and Smyth, H. F., jr. (1928). *J. industr. Hyg.*, **10**, 163.
Peronnet, M. (1934). Recherches sur la dosage du benzène en toxicologie. *J. Pharm. Chim., Paris*, **20**, 145, 195, 244.
Perrault, M., Dérobert, L. and Tiret (1944). Hémopathie benzolique retardée: dosage du benzène dans le moelle osseuse. *Arch. Mal. prof.*, **6**, 239.
Petri, E. (1930). In Henke and Lubarsch: Handbuch der spez. pathologischen Anatomie und Histologie. Pathologische Anatomie und Histologie der Vergiftungen. *Hft.* **10**, 309.
Petrie, A. S. (1908). Toxic effects of petrol fumes. *Brit. med. J.*, **i**, 987.
Petrini, M. (1941). Intossicazione acuta e sub-acuta de benzina e da benzolo. *Rass. Med. Lav. industr.*, **12**, 453.
Pfeil (1932). Auspuffgase von Benzolmaschinen. *Zbl. GewHyg.*, **19**, abstr. sec., 236.
Pirila, V. (1947). Occupational disease of the skin among paint factory workers, painters, polishers, and varnishers in Finland. *Acta derm.-venereol., Stockh.*, **27**, Suppl. 16.
Plummer, S. W. (1913). Case of petrol intoxication. *Brit. med. J.*, **i**, 661.
Pohl, J. (1923). Gifte der aliphatischen, der Benzol- und Naphthalinreihe. In: Spezielle Pathologie und Therapie innere Krankheiten, *Hft.* **9**, 1168. Edited by Kraus, F. and Brugsch, T. Urban and Schwarzenberg.
Pohl, J. and Rawicz, M. (1919). Über das Schicksal des Tetrahydronaphthalins (Tetralins) im Tierkörper. *Hoppe-Seyl. Z.*, **104**, 95.
Poincaré, L. (1885). Recherches expérimentales sur les effets d'un air chargé de vapeurs de pétrole. *Ann. Hyg. publ., Paris*, 3me sér., **13**, 312.
Porrini. *Gaz. degli osped. et delle clin.* No. 80. (Quoted by Legludic, H. and Turlais, C. (1914). Intoxication par les dissolvants des vernis. *Ann. Hyg. publ., Paris*, 4me sér., **21**, 385.)
Poulsson, E. (1930). Lehrbuch der Pharmakologie, für Ärzte und Studierende. 9 Aufl. Hirzel, Leipzig.
Price, J. P. (1933). Petroleumvergiftung bei Kindern. *Samml. Vergiftungsf.*, **4**, 245, A385.
di Prisco, L. (1932). Glutathione content of blood in experimental chronic intoxication with solvents. *Minerva med.*, **2**, 423. (Abstr. in *Chem Abs.*, 1932, **26**, 718.)
Pugliese, A. (1922). Richerche comparative sulla tossicita di alcuni campioni di benzolo. *R. C. Ist. Lombardo*, **55**, 404.
Pugliese, A. (1922). L'azione tossica della benzin, del etere di petrolio a del toluolo. *R. C. Ist. Lombardo*, **55**, 443.
Pulford, D. S. (1931). Benzol poisoning: report of a case. *Calif. West. Med.*, **35**, 361.
Quarterly Safety Summary, **1**, No. 1, 3. Fire and explosion. Dangers in the manufacture of benzole and their prevention. Association of British Chemical Manufacturers, 1930.
Quarterly Safety Summary, **2**, No. 6., 25. Fire and explosion. Limits of inflammability of gases and vapours. Association of British Chemical Manufacturers, 1931.
Quarterly Safety Summary, **3**, No. 9, 1. Benzole. Fatality due to ignition of benzole. Association of British Chemical Manufacturers, 1932.
Rabe, R. F. and Hirschland, F. H. (1920). Partial provings of benzol, iodine and kalichromium. *N. Y. Homeopathic Coll. J.*, **13**, 499.
Rabinowitch (1925). Klinik chronischer Intoxikation durch Benzindämpfe. *Gigiena Truda*, **9**, 40 (Cited by Engelhardt, W. E., 1931 in *Arch. Gewerbepath. Gewerbehyg.*, **2**, 479.)
Rachet, J., Lumière, and Dérobert, L. (1944). Hémopathie benzolique: dosage du benzène dans la moelle osseuse. *Arch. Mal. prof.*, **6**, 316.
Rambousek, J. (1913). Industrial poisoning from fumes, gases, and poisons of manufacturing processes, tr. and ed. by T. M. Legge. Arnold, London.
Rawkin and Kulkow. See Schachnowskaja, S.B. (1935). *Arch. Gewerbepath. Gewerbehyg.*, 6, 144.
Reifschneider, C. A. (1922). Benzol poisoning, its occurrence and prevention. National Safety Council—11th Annual Congress, Proc., Detroit. p.249.
Reinhard (1887). Ein Fall von Terpentinintoxikation in Folge Einathmens von Terpentinöl. *Dtsch. med. Wschr.*, **13**, 256.
Reznikoff, P. and Fullarton, R. (1933). Action of benzol on granulocytes. *Folia haemat., Lpz.*, **50**, 454.
Rich, A. R. and McKee, C. M. (1934). A study of the character and degree of protection afforded by the immune state independently of the leucocytes. *Bull. Johns Hopk. Hosp.*, **54**, 277.

Ridder (1923). Terpentinölvergiftung mit Nierenschädigung durch äusserliche Anwendung des Oels. *Dtsch. med. Wschr.*, **49**, 1369.
Rivet, L. and Guédé, M. (1928). L'intoxication benzénique mortelle. *Bull. Soc. méd. Hôp., Paris*, **52**, 1234.
Robinson, F. J. and Climenko, D. R. (1941). Effects of inhalation of benzene vapors on red blood cells of rabbits. *J. industr. Hyg.*, **23**, 232.
Rockemann, W. (1922). Über Tetralinharn. *Arch. exp. Path. Pharmak.*, **92**, 52.
Rohner, F. J., Baldridge, C. W. and Hansmann, G. H. (1926). Chronic benzene poisoning: report of a case with necropsy findings. *Arch. Path. Lab. Med.*, **1**, 220.
Ronchetti, V. (1922). Due casi di anemia perniciosa da benzolo in operaie di una fabbrica di impermeabli. *Atti Soc. lombarda Sci. med. biol.*, **11**, 322.
Rosenblath (1902). Neurasthenie hervorgerufen durch Einathmung von Xylol-Dämpfen. *Arztliche Sachverständigen Zeitung* **8**, 197.
Rosenthal-Deussen, E. (1931). Vergiftungen durch ein Anstrichmittel (Inertol). *Arch. Gewerbepath. Gewerbehyg.*, **2**, 92.
Roubinet, R. (1939). Occupational benzene poisoning. Rôle of individual predisposition and lack of Vitamin C. Thèse. Peyronnet et Cie., Paris. (Abstract in *J. industr. Hyg.*, (1941), **23**, 34.)
Rusk, G. Y. (1914). Studies on the locus of antibody formation. II. The effect of benzol intoxication and consequent leucopenia on the formation of artificial hemolysins and precipitins. *Univ. Calif. Publ. Path.*, **2**, No. 16.
Ruttin, E. (1936). Ohrbefunde bei Benzindampfintoxication. *Acta oto-laryng.*, **23**, 410.
Sack, G. (1941). Ein Fall von Toluol-Vergiftung. *Samml. Vergiftungsf.*, **10**, 41, B98.
St. George's Hospital Reports, Lond. (1877–8). Two cases: benzol poisoning by swallowing.
Saita, G. (1945). Myélose aplastique provoquée par le benzol. *Med. d. Lavoro*, **36**, 143.
Saita, G. and Dompe, M. (1947). Sul rischio benzolico nei principali stabilimenti rotocalcografici di Milano. *Med. d. Lavoro*, **38**, 269.
Saenger (1914). Auftreten einer zirkumskripten Myelitis nach Einatmen von Benzoldämpfen. *Münch. med. Wschr.*, **61**, 385.
Safety Circular No. 29. Two unusual cases of fire. Association of British Chemical Manufacturers, 1929.
Safety Circular No. 43. Fire at a benzole plant. Association of British Chemical Manufacturers, 1931.
Safety Circular No. 51. Benzole. Association of British Chemical Manufacturers, 1931.
Santesson, C. G. (1897). Über chronische Vergiftungen mit Steinkohlen-Theerbenzen. *Arch. Hyg., Berl.*, **31**, 336.
Sartorius, F. and Sudhues, M. (1933). Studien bei experimenteller chronischer Benzolvergiftung. *Arch. Hyg., Berl.*, **110**, 254.
Sato, K. (1928). Über die pharmakologischen Wirkungen der hydro-aromatischen Verbindungen: Cyclohexen, Cyclohexan und Cyclohexanol. *Jap. J. med. Sci., Pharmacol.*, **3**, 1.
Säverin (1940). Tödlicher Unfall beim Streichen eines Wasserbehälters. *Reichsarbeitsblatt*, **20**, III, 224, (abstr. in *Z. GewHyg.*, (1941), **28**, 29).
Sayers, R. R. and Dallavalle, J. M. (1935). Prevention of occupational diseases other than those that are caused by toxic dust. *Mech. Engng.*, **57**, 230.
Schachnowskaja, S. B. (1935). Über die Durchlassigkeit der Blutliquorschranke und Blutveränderungen bei experimenteller Benzinvergiftung. *Arch. Gewerbepath. Gewerbehyg.*, **6**, 144.
Schaefer (1901). *Jahresberichte d. Gewerbe-Aufsichtsbeamten, Berlin*, **3**, 19.
Schaefer, E. (1909). Verwendungs und schädliche Wirkung einiger Kohlenwasser- und anderen Kohlenstoffverbindungen. Hamb. Gew. Insp., Arbeit und Sonderberichte.
Schantz, J. M. (1932). Dipentene as a paint or varnish thinner. *Paint Oil Chem. Rev.*, **93**, No. 22, 13.
Schiff, F. (1914). Einfluss der Benzols auf die aktive Anaphylaxie des Meerschweinchens. (Über Anaphylaxie). *Z. ImmunForsch.*, **23**, 61.
Schilling, V. (1935). Wissenschaftliche Kongresse und Vereine: 3 Sitzung. Akute Blutkrankheiten des myeloischen Systems. *Münch. med. Wschr.*, **82**, 767.
Schillowa, A. (1933). Veränderungen bei chronischer Benzolvergiftung. *Folia haemat., Lpz.*, **49**, 447.
Schmidtmann, M. (1930). Experimentelle Untersuchungen über die Wirkung von Einatmung kleiner Benzin- und Benzolmengen auf Atmungsorgane und Gesamtorganismus. *Klin. Wschr.*, **9**, 2106.
Schmiedeberg, O. (1881). Über Spaltungen und Synthesen im Thierkörper. *Arch. exp. Path. Pharmak.*, **14**, 379.
Schneider, H. (1930). Zur Klinik und Therapie der chronischen gewerblichen "Benzolvergiftung". *Med. Klinik*, **26**, 1112.
Schnitzer, R. J. and Goddard, I. G. (1943). Influence of benzene-poisoning upon streptococcal infections in rabbits: I. Benzene-poisoning and natural resistance to intracutaneous streptococcal infection. *J. Immunol.*, **46**, 133.
Schrenk, H. H., Pearce, S. J. and Yant, W. P. (1935). Microcolorimetric method for the determination of benzene. *U.S. Bur. Mines Rept. Investigation* No. 3287.

SCHRENK, H. H., YANT, W. P., PEARCE, S. J., PATTY, F. A. and SAYERS, R. R. (1941). Absorption, distribution and elimination of benzene by body tissues and fluids of dogs exposed to benzene vapor. *J. industr. Hyg.*, **23**, 20.
SCHULTE, H. F. (1945). Report on the quartz crystal industry. *Industr. Med.*, **14**, 68.
SCHUSTROW, N. and LETAWET, K. (1927). Die Bedeutung der Fettsubstanzen bei der Benzinintoxikation. *Dtsch. Arch. klin. Med.*, **154**, 180.
SCHUSTROW, N. and SALISTOWSKAJA, E. (1926). Das Blut bei Benzinintoxikation. *Dtsch. Arch. klin. Med.*, **150**, 271.
SCHUSTROW, N. and SALISTOWSKAJA, E. (1926). Die Benzinangewöhnung. *Dtsch. Arch. Klin. Med.*, **150**, 277.
SCHWARTZ, E. and TELEKY, L. (1941). Some facts and reflections on the problem of poisoning by benzene and its homologs. *J. industr. Hyg.*, **23**, 1.
SCHWARTZ, L., TULIPAN, L. and PECK, S. M. (1947). Occupational diseases of the skin. Kimpton, London.
SCHWARTZ, S. M. (1931). Über den Einfluss der akuten und der chronischen Benzinvergiftung auf der Tierorganismus. *Fortschr. Med.*, **49**, 215.
SCHWARZ, H. G. (1933). Benzin-vergiftung chronische, medizinale. *Samml. Vergiftungsf.*, **4**, 247, A386.
SECCHI, P. (1914). Ricerche ematologiche nelle intossicazioni acute e croniche da benzolo. *Rif. med.*, **30**, 995.
SECRETARY FOR MINES (1934). Annual Report of the Secretary for Mines, p.16. Fuel treatment and utilisation. Production of light oils. H.M. Stationery Office.
SEGHINI, C. (1941). Intossicazione di tolulo. *Med. d. Lavoro*, **32**, 179.
SEGITZ, A. F. G. (1930). Über Benzin und Tetrachlorkohlenstoff in chemischen Reinigungsanstalten. *Zbl. GewHyg.*, **17**, 298.
SELLING, L. (1911). Benzols als leucotoxin. *Beitr. path. Anat.*, **51**, 576.
SELLING, L. (1916). Benzol as a leucotoxin. *Johns Hopk. Hosp. Rep.*, **17**, 83.
SEYFRIED, H. (1942). Über Benzolschäden. *Wien. klin. Wschr.*, **55**, 399.
SEYFRIED, H. (1942). Zur Pathogenese und Therapie beruflich erworbener Schäden und Beschwerden durch chronische Einwirkung von Benzolkörper. *Arch. Gewerbepath. Gewerbehyg.*, **11**, 588.
SHARP (1889). La toxicité du petrole et des maladies professionelles des ouvriers pétroliers. *Ann. Hyg. publ., Paris*, 3me sér., **22**, 550.
SIEMON, O. (1896). Ein Vergiftungsfall nach Einatmung grosser Mengen von Benzin. *Mschr. Unfallheilk.*, **11**, 366.
SILBERBERG, M. (1928). Das Verhalten des aleukocytären und vital gespeicherten Körpers gegenüber der septischen Allegemeininfektion als Beitrag zur Entzündungs- und Monocytenlehre. *Virchows Arch.*, **267**, 483.
SIMONDS, J. P. and JONES, H. M. (1915). Effect of injections of benzol upon the production of antibodies. *J. med. Res.*, **33**, 197.
SIMONIN, C. (1903). L'intoxication par ingestion accidentelle de benzine. *Bull. Soc. méd. Hôp., Paris*, **20**, 199.
SIMONIN, C. (1934). Considérations toxicologiques et médico-légales sur le benzolisme et le pétrolisme professionnels. *Paris méd.*, ii, 408.
SIWE, S. A. (1932–33). Zur Frage nach den Symptomen bei akuter Vergiftung mit flüssigen Kohlenwasserstoffen der aliphatischen Reihe. *Mschr. Kinderheilk.*, **55**, 146.
SKLAWUNOS, T. G. (1925). Experimentell-histologische Studien über Entzündung bei "möglichst" leukozytenfrei gemachten Kaninchen. *Krankeitsforschung.*, **1**, 507.
SMITH, A. R. (1928). Chronic benzol poisoning among women industrial workers: a study of the women exposed to benzol fumes in six factories. *J. industr. Hyg.*, **10**, 73.
SMITH, A. R. (1943). Fatality after a year's freedom from exposure to benzene. *Industr. Hyg. Bull.*, **22**, 329.
SMITHIES, F. (1927). Gastro-duodenal hemorrhage. *Ann. int. Med.*, **1**, 637.
SMYTH, H. F., jr. (1929). Determination of small amounts of benzene vapors in air. *J. industr. Hyg.*, **11**, 338.
SMYTH, H. F. (1931). Toxicity of certain benzine derivatives and related compounds. *J. industr. Hyg.*, **13**, 87.
SMYTH, H. F., jr. (1931). Note on the determination of small amounts of benzene vapor in air. *J. industr. Hyg.*, **13**, 227.
SMYTH, H. F. and SMYTH, H. F., jr. (1928). Spray painting hazards as determined by the Pennsylvania and the National Safety Council Surveys. *J. industr. Hyg.*, **10**, 163.
SMYTH, H. F. and SMYTH, H. F., jr. (1928). Inhalation experiments with certain lacquer solvents. *J. industr. Hyg.*, **10**, 261.
SORRENTINI, E. (1930). Considerazioni su di un caso di benzolismo cronica. *Med. d. Lavoro*, **21**, 207.
SPECIAL ARTICLE (1934). Aplastic anaemia: inquest on cellulose sprayer. *Lancet*, i, 1025.
SPENCER, O. M. (1922). Effect of gasoline fumes on dispensary attendance and output in a group of workers. *Publ. Hlth. Rep., Wash.*, **37**, 2291.
STARR, E. B. (1922). Poisoning by benzol-carbon tetrachloride cement. *J. industr. Hyg.*, **4**, 203.

STERNER, J. H. (1941). Study of hazards in spray painting with gasoline as a diluent. *J. industr. Hyg.*, **23**, 437.
STIEFLER, G. (1928). Epilepsie nach Benzinvergiftung. *Wien. med. Wschr.*, **78**, 938.
STOCKE, A. (1929). Akute Xylol- und Toluolvergiftungen beim Tierdruckfähren. *Zbl. GewHyg.*, **16**, 355.
STOCKE, A. (1931). Gewerbemedizinische Erfährungen mit dem Anstrichmittel "Inertol 49". *Arch. Gewerbepath. Gewerbehyg.*, **2**, 99.
SURY-BIENZ (1888). Tödliche Benzoldampfvergiftung. *Vjschr. gerichtl. Med.*, **49**, 138.
SVIRBELY, J. L. (1946). Appraisal and potential dangers of petroleum solvents with special reference to particle size. *Industr. Med.*, **15**, 483.
SVIRBELY, J. L., DUNN, R. C. and VON OETTINGEN, W. F. (1943). The acute toxicity of vapors of certain solvents containing appreciable amounts of benzene and toluene. *J. industr. Hyg.*, **25**, 366.
SVIRBELY, J. L., DUNN, R. C. and VON OETTINGEN, W. F. (1944). Chronic toxicity of benzene mixtures. *J. industr. Hyg.*, **26**, 37.
SWEENEY, J. S. (1928). Chronic aplastic anemia and symptomatic hemorrhagic purpura due to benzol poisoning. *Amer. J. med. Sci.*, **175**, 317.
VON TAPPEINER, H. (1916). Lehrbuch der Arzneimittellehre und Arzneiverordnungslehre unter besonderer Berücksichtigung der deutschen und österreichischen Pharmakopoe. 11th Aufl. Vogel, Leipzig.
TAUSZ, J. (1924). Determination of benzol content of coke oven gas. *Mitt. chem. tech. Inst. tech. Hochsch. Karlsruhe*, **1**, 19.
TAYLOR, A. S. (1928). Principles and practices of medical jurisprudence. Vol. 2. 8th edn., ed. by Sidney Smith. Churchill, London.
TELEKY, L. and WEINER, E. (1924). Über Benzolvergiftung. *Klin. Wschr.*, **3**, 226.
THEIS, B. and BENEDICT, S. R. (1924). Determination of phenols in the blood. *J. biol. Chem.*, **61**, 67.
THIERFELDER, H. and KLENK, E. (1924). Weitere Untersuchungen über das Verhalten fettaromatischer Verbindungen im Tierkörper. *Hoppe-Seyl. Z.*, **141**, 13.
TIMMERMANS, J. and MARTIN, F. (1926). Étude de vingt hydrocarbures et dérivés halogénés. *J. chim. phys.*, **23**, 747.
TOLLENS, C. (1909). Quantitative Bestimmung der Glukonsaure im Urin. *Hoppe-Seyl. Z.*, **61**, 95.
TREON, J. F., CRUTCHFIELD, W. E., jr. and KITZMILLER, K. V. (1943). Physiological response of rabbits to cyclohexane, methylcyclohexane and certain derivatives of these compounds. *J. industr. Hyg.*, **25**, 199, 323.
UMEDA, T. (1928). Influence of varying chemical structure of some chemicals upon the movement of ciliated epithelium. *Acta derm., Kyoto*, **11**, 481.
UNDERHILL, F. P. and HARRIS, B. R. (1923). Influence of benzol upon certain aspects of metabolism. *J. industr. Hyg.*, **4**, 491.
U.S. BUREAU OF MINES (1921). Gas masks for gases met in fighting fires. Technical Report No. 248 by A. C. Fieldner, S. H. Katz and S. W. Kinney.
U.S. BUREAU OF STANDARDS. Technologic Papers of the Bureau of Standards. No. 131. Application of the interferometer to gas analysis. J. D. Edwards. Government Printing Office, Washington.
U.S. PUBLIC HEALTH SERVICE (1942). See von Oettingen *et al.*, 1942b.
VERHOOGEN, R. (1934). Anémie de type aplastique et leucopénie extrème au cours d'une intoxication mortelle par le xylol. *Brux. méd.*, **14**, (ii), 884.
VIGDORTSCHIK, N. A. (1933). Zur Frage der chronischen Benzinwirkung auf den Organismus. *Zbl. GewHyg.*, **10**, 219.
WALLBACH, G. (1929). Untersuchungen über die unterschiedliche Wirkung einiger leukocytenvermindernder Substanzen. *Z. ges. exp. Med.*, **68**, 621.
WALLBACH, G. (1931). Experimentelle Untersuchungen über die Beeinflussung der Wirkung leukocytenvermindernder Substanzen. *Folia haemat.*, **43**, 340.
WALLBACH, G. (1933). Über Einwirkungen von Benzol und von Thorium auf einige Infektionsprogresse. *Folia haemat.*, **49**, 241.
WARD, A. L. and KURTZ, S. S. (1938). Refraction, dispersion and related properties of pure hydrocarbons. *Industr. Engng. Chem.*, **30**, 559.
WARING, J. I. (1933). Pneumonia in kerosene poisoning. *Amer. J. med. Sci.*, **185**, 325.
WEIL, E. P. (1932). La leucémie post-benzolique. *Bull. Soc. méd. Hôp.*, **46**, 750.
WEIL, E. P. (1933). Les hématopathies benzoliques. *Paris méd.*, **89**, 112.
WEIL, E. P. (1935). Manifestations prolongées de l'intoxication benzolique. *Sang*, **9**, 206.
WEISKOTTEN, H. G. (1930). Normal life span of neutrophile (amphophile) leucocyte (rabbit): action of benzol. *Amer. J. Path.*, **6**, 183.
WEISKOTTEN, H. G., GIBBS, C. B. F., BOGGS, E. O. and TEMPLETON, E. R. (1920). Action of benzol. VI. Benzol vapor leucopenia (rabbit). *J. med. Res.*, **41**, 425.
WEISKOTTEN, H. G., SCHWARTZ, S. C. and STEENSLAND, H. S. (1915). Action of benzol. I. On the significance of myeloid metaplasia of the spleen. *J. med. Res.*, **33**, 127.
WEISKOTTEN, H. G., SCHWARTZ, S. C. and STEENSLAND, H. S. (1916). Action of benzol. II. The deuterophase of the diphasic leucopenia and antigen-antibody reaction. *J. med. Res.*, **35**, 63.

WEISKOTTEN, H. G. and STEENSLAND, H. S. (1919). Action of benzol. V. The diphasic leucopenia as a polynuclear amphophile phenomenon (rabbit). *J. med. Res.*, **39**, 485.
WERBOW, ASCHKEWICZ and STOPJANOWSKAJA (1925). Über die Veränderungen des morphologischen und physikalchemischen Blutbildes unter dem Einfluss von Benzindampfen. *Gigiena Truda*, No. 8, 18. (Cited by Lewin, I. E., 1932, in: *Arch. Gewerbepath. Gewerbehyg.*, **3**, 341.)
WERNER, F. F. (1932). Terpentinöl-Vergiftung, medizinale, durch das Gallensteinmittel Anticolicum. *Samml. Vergiftungsf.*, **3**, 157, A242.
WERNER, H. W., DUNN, R. C. and VON OETTINGEN, W. F. (1944). Acute effects of cumene vapors in mice. *J. industr. Hyg.*, **26**, 264.
WHITE, W. C. and GAMMON, A. M. (1914). Influence of benzol inhalations on experimental pulmonary tuberculosis in rabbits. *Trans. Ass. Amer. Phys.*, **29**, 332.
WIECYK (1888). L'influence des émanations de pétrole sur la santé. *Ann. Hyg. publ.*, Paris, 3me sér., **19**, 176.
WILKS, H. (1930). Some cases of haematuria caused by turpentine poisoning. *J. R. Nav. med. Serv.*, **16**, 53.
WINSLOW, C. E. A. (1927). Survey of the National Safety Council study of benzol poisoning. *J. industr. Hyg.*, **9**, 61.
WINTERNITZ, M. C. and HIRSCHFELDER, A. D. (1913). Studies on experimental pneumonia in rabbits. *J. exp. Med.*, **17**, 657.
WOCJIECHOWSKI, A. (1910). Studien über die Giftigkeit verschiedener Händelssorten des Benzols in Gasform. Dissertation, Würzburg.
WOLLENBERG, A. (1927). Verletzungen der äusseren Augenhäute durch terpentinhältige Schuhputzmittel. *Klin. Mbl. Augenheilk.*, **78**, 410.
WORONOW, A. (1929). Über die morphologische Veränderungen des Blutes unter dem Einfluss des Benzols und der bluterzeugenden Organe unter dem Einfluss des Benzols und dessen Abkömmlinge. *Virchows Arch.*, **271**, 173.
WURM, E. (1931). Die gebrauchlichsten Lösungsmittel in der Gummiindustrie: ihre Gefähren und der Verhütung. *Arch. Gewerbepath. Gewerbehyg.*, **2**, 776.
WYSS, M. O. (1910). Gesellschaft der Aerzte in Zürich. Über Benzolvergiftung. *CorrespBl. schweiz. Ärz.*, **4**, 387.
YANT, W.P., SCHRENK, H. H., WAITE, C. P. and PATTY, F. A. (1930). Acute response of guineapigs to vapors of some new commercial organic compounds: ethyl benzene. *Publ. Hlth. Rep., Wash.*, **45**, 1241, pt. I.
YANT, W. P., SCHRENK, H. H., SAYERS, R. R., HORVATH, A. A. and REINHARD, W. H. (1936). Urine sulfate determinations as measure of benzene exposure. *J. industr. Hyg.*, **18**, 69.
YATES, M. T. and LEVINSON, S. (1943). Determination of turpentine in air. *Nav. med. Bull., Wash.*, **41**, 1138.
ZANGGER, H. (1933). Arbeitsunfälle und Arbeitsgefährung bei den Arbeit im Innern von geschlossenen Behältern. *Arch. Gewerbepath. Gewerbehyg.*, **4**, 117.
ZERNIK, F. (1933). Neuere Erkenntnisse auf dem Gebiete der schädlichen Gase und Dämpfe. *Ergebn. Hyg. Bakt.*, **14**, 139. 202.
ZIEL (1925). Zur Benzolvergiftung. *Med. Klinik*, **21**, 93.

CHAPTER II
CHLORINATED HYDROCARBONS
1. Methylene Dichloride
(*Dichloromethane, Methylene Chloride*)

CH_2Cl_2

PROPERTIES

A COLOURLESS, non-inflammable, highly volatile liquid (evaporation rate 1·8 compared with that of ether, 1).

B.P. 40–42° C. Sp.Gr. 1·346.

Solvent for cellulose esters, fats, oils, resins and rubber. Commercial methylene dichloride frequently contains traces of methyl chloride, which lower its boiling point, and also chloroform (10 to 15 per cent) as well as small amounts of carbon tetrachloride. Methylene dichloride is often referred to commercially as methylene chloride. The pure substance, used sometimes as an anaesthetic in Germany under the name of "Solaesthin," is, according to Müller (1925), less toxic than the commercial product, both as regards respiratory paralysis and the effect upon internal organs.

USES

(1) Chiefly used as a paint remover.
(2) Also used in the artificial silk industry as a "stretching" solvent.
(3) As an anaesthetic.

TOXICITY

Methylene dichloride is a narcotic, less pronounced than chloroform, ($2\frac{1}{2}$ times weaker according to Hellwig, 1922), but more irritating to the respiratory passages and more excitatory. It was first used as a general anaesthetic by Richardson in 1867 and was employed up to 1880 as an alternative to chloroform. During that period ten fatal cases under general anaesthesia were recorded, which have since been misquoted (by Carozzi (1936) and others) as due to its industrial use.

Methylene dichloride has been found unsuitable by Hellwig (1922) for complete anaesthetization, since it produces an alarming excitation stage with laboured breathing, cyanosis, dilated pupils and a rapid, weak pulse, but Bourne and Stehle (1923) have recommended it for the induction of anaesthesia before the use of less agreeable anaesthetics, for analgesia in labour pains, minor operations, etc.

The lesions of the liver and kidneys which have been observed by some investigators after exposure to commercial methylene dichloride are probably due to the impurities contained in it. From its industrial use, only slight acute effects of a rather indefinite nature, and no chronic effects, have been observed.

Maximum Permissible Concentration

The maximum allowable concentration recommended for 8 hours daily exposure is 500 p.p.m. (Cook, 1945; Heppel, Neal, Perrin, Orr and Porterfield, 1944b).

TOXIC EFFECTS IN ANIMALS

Methylene dichloride is relatively well tolerated by animals, especially by inhalation.

Acute Poisoning

Lethal and Narcotic Concentrations

Lethal dose

By intravenous injection. For the dog, 200 mg. per kg. body weight, compared with 90 mg. for chloroform.

By subcutaneous injection. For the rabbit, 2·7 g. per kg. body weight, compared with 0·99 g. for chloroform (Barsoum and Saad, 1934).

By inhalation. For the mouse, 50 mg. per l. (14,500 p.p.m.), compared with 30 to 40 mg. per l. for chloroform (Lazarew, 1929). More recent results by Pantelitsch (1933) agree closely with this value; he gives 45 mg. per l. after 110 minutes, 50 mg. per l. after 100 minutes.

Narcotic dose

For light narcosis, 30 to 35 mg. per l. (8,700 to 10,000 p.p.m.); for light to moderate narcosis, Heppel *et al.* (1944b) give 10,000 p.p.m.; for narcosis involving loss of reflexes, 35 mg. per l. (8,700 p.p.m.), compared with chloroform, 20 mg. per l. (4,100 p.p.m.) (Lazarew, 1929).

Symptoms

Most of the early workers observed excitation as a preliminary phase of narcosis. The early experiments of Regnauld and Villejean (1884) on dogs gave a typical picture of the effects as follows: after ½ minute, agitation; after 1½ minutes, dilated pupils and nystagmus; after 2 minutes, abolition of corneal and palpebral reflexes and general insensibility; after 3 minutes, clonic movements; during recovery, epileptiform attacks; normal in 22 to 30 minutes.

Chronic Poisoning

The effects of repeated exposure to concentrations of 5,000 p.p.m. (17 mg. per l.) for 7 hours a day, 5 days a week for 6 months, have been investigated by Heppel *et al.* (1944b). They have found that such exposure produced no evident toxic reaction, no change in the blood picture and no evidence of liver damage. It was only when light to moderate narcosis was produced by exposure to 10,000 p.p.m. that moderate fatty degeneration of the liver was found in some of the exposed animals.

TOXIC EFFECTS IN MAN

The opinion of most authorities on the possibility of industrial poisoning by methylene dichloride is summed up by Zernik (1933), who states that it is practically harmless under conditions of good ventilation and if it is free from other products such as petroleum distillates and other chlorinated hydrocarbons. With this opinion Collier (1936) seems inclined to disagree; he reports several cases of acute intoxication from its use as a paint remover. The symptoms in the two most severely affected patients who had to leave work were, however, largely subjective, and the evidence was rendered to some extent indefinite by the presence of intercurrent affections, lead poisoning in the one and duodenal ulcer in the other. The first complained of irregular but severe pains in legs and arms, hot flushes, headache, vertigo, stupidity whilst at work, difficulty in reading on account of eyesight not being clear, anorexia, precordial pain, rapid pulse, shortness of breath, fatigue on exertion and attacks of rapid beating of the heart. No objective abnormalities were present with the exception of haematological changes including slight anaemia and punctate basophilia, attributable to his previous exposure to lead. In the second case the subjective symptoms were very slight, e.g. drowsiness, pains in the head, irritability and tingling in the hands and feet, and a diagnosis of gastro-duodenal ulcer was made on admission to hospital.

Collier's statement that the effects attributable to methylene dichloride include "possibly some degree of chronic anaemia" is not borne out by that of Beyer and Gerbis (1931) that its chronic use produces no anaemia.

A few cases of intoxication, though not severe in degree, have been reported to the Factory Department. In one of them (1944), where nausea and a choking sensation were followed by collapse, loss of consciousness and vomiting lasting for 4 days, the solvent to which the man was exposed contained 40 per cent dichloroethylene as well as 45 per cent methylene dichloride. In another case (1945), a female operative using methylene dichloride as a paint remover complained of nausea, vomited, and then fainted; a fellow-worker also vomited. It appears that if methylene chloride is deeply inhaled it produces a painful sensation at the back of the nose, or the "bursting headache" as in the case reported to the Home Office in 1935.

2. Chloroform
(*Trichloromethane*)
$CHCl_3$

PROPERTIES

A colourless, non-inflammable liquid with a characteristic odour. Solubility in water, 0·5 per cent (Durrans, 1950). Sp.Gr. 1·49–1·50. B.P. 61·2° C. Slowly becomes acid in presence of light and moisture; a small quantity of alcohol (2 per cent) is usually added to prevent this.

Solvent for cellulose acetate and benzoate, ethyl cellulose, and for rubber, guttapercha and most resins.

USES

Chloroform has not a wide use as an industrial solvent. It is, however, used to some extent as follows:

(1) As a lacquer solvent. Hamilton (1934) states that no instance has come to her knowledge of the use of chloroform in lacquer coatings, but it is mentioned in this connection by Hofmann and Reid (1929).

(2) In artificial silk manufacture owing to its high solvent power for cellulose acetate.

(3) In the sterilization of catgut but usually mixed with methanol.

(4) As a constituent of floor polishes.

(5) In the extraction of certain alkaloids.

(6) As an anaesthetic.

TOXICITY

The acute toxic narcotic effects of chloroform used as an anaesthetic have been recognized since Simpson in 1847 discovered that consciousness might be completely suspended by its action. Many cases of death from cardiac and respiratory failure have been reported in this connection. Delayed chloroform poisoning, in which its toxic effect is exerted on the liver with the appearance of jaundice, and on the kidneys with uraemia, leading to death, is also well known (Willcox, 1934). These effects, especially that of liver injury, are practically unknown from industrial poisoning.

Animal experiments have shown that its lethal narcotic effect is greater than that of carbon tetrachloride and less than that of tetrachloroethane, that its toxicity as judged by its effect on the isolated frog's heart bears the same relation to these two compounds (Barsoum and Saad, 1934; Mezey and Staub, 1935), and that it produces fatty degeneration of the liver in the same way as tetrachloroethane (Whipple and Sperry, 1909; Wells, 1908; Meerssemann, 1934).

Maximum Permissible Concentration

The maximum allowable concentration for repeated exposure recommended by the United States Public Health Service is 100 p.p.m. (Cook, 1945).

TOXIC EFFECTS IN ANIMALS
Acute Poisoning

Lethal and Narcotic Concentrations

Lethal dose

By oral administration. Barsoum and Saad (1934) found that 2·25 g. per kg. body weight is lethal for the dog compared with 4 g. for carbon tetrachloride and 0·7 g. for tetrachloroethane.

By intravenous injection. 90 mg. per kg. body weight is lethal for the dog compared with 125 mg. for carbon tetrachloride and 60 mg. for tetrachloroethane.

By subcutaneous injection. 0·9 g. per kg. body weight is lethal for the rabbit as compared with 0·5 g. tetrachloroethane; 2·61 g. of carbon tetrachloride is lethal only after 4 days.

By inhalation. The concentration given by a number of authors varies from 20 to 40 mg. per l. (4,100 to 8,200 p.p.m.) for the mouse (Führer, 1929; Lazarew, 1929; Meyer and Gottlieb-Billroth, 1921; Müller, 1925) to 60 mg. per l. for the rabbit (Lehmann, 1911) and 80 to 100 mg. per l. for the guinea-pig and dog (Wittgenstein, 1918). Cole (1927) found 0·2 ml. evaporated in a dish of 1 l. capacity fatal to rats in 11·2 minutes, 0·4 ml. in 5·4 minutes and 5 ml. or more in 1·5 minutes.

Narcotic dose

The lowest concentration inducing narcosis is about 20 mg. per l. for the mouse (Lazarew, 1929) and this concentration produces deep narcosis if prolonged for ½ hour. The results of most of the authors quoted above as to narcotic and lethal dosage for the mouse have been summarized by Flury and Zernik (1931) and are shown in Table 21.

TABLE 21
Effect of different concentrations of chloroform on the mouse

Concentration (mg./l.)	(p.p.m.)	Duration (hr.)	Effect
Up to 12	2,500	Up to 2	No decided effect
16·5	3,400	1½	After 1 hr. side position and light narcosis; recovery
20	4,100	½	After 20 min. side position; after 30 min. deep narcosis; recovery
		2	Death
20–30	4,100–6,150	¾–1	After ½–¾ hr. deep narcosis; occasionally death after 1–9 days
36	7,400	¼	Deep narcosis; recovery
		½	Deep narcosis; death after ½–1 day
45–55	9,200–11,250	1½	After 3–4 min. side position; after 4–5 min. light narcosis; after 6–9 min. deep narcosis; after ½–¾ hr. death

Symptoms

Local irritation of mucous membranes is followed by a stage of excitation, then by somnolence, stupor, light and then deep narcosis; finally death occurs from respiratory paralysis.

Lesions of the Internal Organs

Liver. It has been shown that both subcutaneous injection and prolonged inhalation of narcotic doses of chloroform may produce fatty degeneration with or without acute central necrosis of the liver in animals, the severity of the lesion varying not only with the amount administered, but with the individual susceptibility of the animal and its previous state of nutrition. Dogs are specially susceptible before they are fully grown, though during the first 3 weeks of life they are very resistant (Whipple and Sperry, 1909; Whipple, 1912), while starvation and a fat-rich diet also constitute predisposing factors (Davis and Whipple, 1919); a diet rich in sugar, other carbohydrates and milk is specially protective (Opie and Alford, 1915; Althausen and Thoenes, 1932).

The histological picture of liver injury and repair has been described in detail by Whipple and Sperry (1909), and contrasted by Meerssemann (1934) with the lesions produced by phosphorus poisoning. The essential lesion is a uniformly distributed central necrosis. Meerssemann has used special staining methods to show mitochondrial lesions, and found these more advanced with chloroform than with carbon tetrachloride poisoning. From the results of urinary tests for liver insufficiency, however, he concluded that the injury produced by chloroform poisoning is neither so severe nor so irremediable as that of phosphorus poisoning, being more definitely functional and capable of regression after exposure to the poison has ceased. This view conforms with the earlier observations of Whipple and Sperry (1909), that, when animals recover from the liver disturbance produced by chloroform, repair is rapid and the liver becomes normal in 2 or 3 weeks, and the degeneration and necrosis are not followed by cirrhosis.

Hypoglycaemia and a diminution of sugar tolerance have been shown by Althausen and Thoenes (1932) to accompany the acute liver necrosis produced by subcutaneous injection of 0.5 to 1 ml. of chloroform per kg. body weight in animals. This is explained as due to lack of deposition of glycogen in the liver. Regeneration after small doses was also observed by these authors and was accompanied by a rise in the blood sugar level.

Acquired tolerance to chloroform poisoning in animals anaesthetized with chloroform has been observed by Davis and Whipple (1919) and by Inman 1915). It was found that dogs which survived the initial injury by chloroform might be given it afterwards on successive or alternate days, for a week or two without increasing the original necrosis, though the liver might become intensely fatty and the animal much jaundiced and intoxicated. If, however, the animal were allowed to recover for 2 or 3 weeks, tolerance to a second dose was not shown. Nevertheless, Davis and Whipple (1919) do not agree with Wells (1908) that previous liver injury is a predisposing factor in chloroform liver injury.

Other organs. Moderate fatty degeneration of other organs may occur, and occasionally severe necrosis of the kidney tubules (Opie and Alford, 1915).

Chronic Poisoning

Repeated inhalation of small quantities of chloroform, 9 mg. per l. (1,800 p.p.m.) for 7 hours at intervals of 2 to 4 days, was found by Lehmann (1911) to cause death in rabbits 2 to 4 days later. Cats tolerated this dosage better, in some cases surviving 17 days. The only symptoms in these animals were vomiting and loss of weight.

ESTIMATION IN ANIMAL TISSUES

According to Cole (1926, 1927) chloroform is demonstrable in the brains of rats poisoned with it up to 25 days after death, in spite of considerable decay of the tissues. Chloroform is estimated both qualitatively and quantitatively by the pyridine test; the amount found varies directly with the dosage (Table 22).

TABLE 22
Chloroform concentration in brain tissue after inhalation

Dosage (inhaled)	Chloroform in brain tissue (ml./mg.)
0·2 ml. in litre dish	0·03780
5·0 ml. or more	0·09210
Anaesthesia (slight)	0·01190
Anaesthesia (deep)	0·03028

TOXIC EFFECTS IN MAN
Absorption and Excretion

According to Lehmann and Hasegewa (1910) 74 to 80 per cent of chloroform vapour in air is absorbed by human beings during the first half hour; later the absorption decreases to 60 per cent. This is twice as high as the rate of absorption in rabbits. No absorption takes place through the skin (Drescher, 1920). Excretion takes place chiefly through the lungs and a very small part through the urine; part is decomposed in the organism and is present in the urine in the form of chloride.

Acute Poisoning

Narcotic and Lethal Concentrations

The narcotic concentration for man varies with the individual but is usually about 70 to 80 mg. per 1. (14,000 to 16,000 p.p.m.). According to Flury and Zernik (1931) the interval between the narcotic and the toxic dose is only about 1 vol. per cent. Kohn Abrest (1924) states that "it is dangerous to remain in atmospheres containing more than 10 mg. per 1., and 150 mg. per 1. is promptly anaesthetic". The values for toxic concentrations according to duration are shown in Table 23 and are taken from Flury (1928).

TABLE 23
Toxic concentrations of chloroform

Concentration		Time of exposure (min.)	Effect
(mg./l.)	(p.p.m.)		
120	25,000	5–10	Lethal
75	15,000	30–60	Dangerous
24	5,000	30–60	Tolerated

A severe case of acute narcotic poisoning from the use of a floor polish containing carbon tetrachloride and chloroform was reported by Sartorius and Boedecker (1931).

Symptoms

Inhalation of concentrated chloroform vapour produces irritation of the mucous membranes of the eyes, mouth and nose, reflex cessation of breathing, feeling of suffocation with rapid pulse, often vomiting, tremor and mental disorientation. Sudden death from cardiac arrest may occur at this stage. In ordinary narcosis there is usually a preliminary stage of excitation, followed by loss of reflexes and sensation, and ultimately unconsciousness. At this stage the concentration of chloroform in the blood (about 0·25 g. per l., according to Flury and Zernik, 1931) is very close to the level which produces ventricular fibrillation, so that sudden heart stoppage may easily occur.

The after-effects consist of digestive disturbances, vomiting, sensation of pressure and pain in the liver region, jaundice, disturbances of cardiac activity and lastly, acute yellow atrophy of the liver. A typical case of this kind was described by Whipple and Sperry in 1909, in which death took place 4 days after the anaesthetic, and the liver lesions were found to correspond exactly with those found in chloroform-poisoned animals.

Chronic Poisoning

Chloroform addiction apparently gives rise to symptoms resembling those of chronic alcoholism—digestive disturbances, loss of appetite, vomiting, cachexia, general nervousness, insomnia, hallucinations, depression, mental disturbance (Flury and Zernik, 1931). With two exceptions, no such effects are recorded from the industrial use of chloroform; the Home Office (1930) reported the case of a female worker employed in sterilizing catgut, and who complained of slight drowsiness; Hofbauer described the case of a chemist's apprentice, engaged in decanting chloroform, who complained of vertigo, headache, and inability to stand upright (quoted in *Occupation and Health*, 1929).

3. Carbon Tetrachloride
(*Tetrachloromethane*)
CCl_4

PROPERTIES

A colourless, non-inflammable liquid of high specific gravity, with an odour somewhat similar to that of chloroform.

British Standard Specification, No. 575 (1934). B.R., up to 75·7°C. 2 per cent maximum; 75·7–76·7°C . 95 per cent minimum. Sp. Gr. 1·600–1·608. Sulphur compounds, 0·1 per cent maximum as carbon disulphide. No free chlorine. Residue 0·01 per cent maximum. Neutral.

COMMERCIAL PREPARATION

(1) By chlorination of carbon disulphide in presence of sulphur or iodine; this method yields an impure product, containing sulphur compounds in particular.

(2) By chlorination of ethylene, methane or methylene chloride.

(3) By the catalytic action of the electric arc on carbon and calcium chloride.

TOXIC IMPURITIES IN COMMERCIAL CARBON TETRACHLORIDE

According to Davies (1934) the most toxic impurities that may be present in commercial carbon tetrachloride are phosgene, hydrogen sulphide, free hydrochloric acid, organic sulphides and carbon disulphide.

Phosgene. The report of the formation of phosgene by heating carbon tetrachloride at high temperatures in presence of air appears to be well-established, both by experimental investigation and from the evidence provided by accidents from the use of fire extinguishers containing carbon tetrachloride. The danger of such accidents was first revealed in 1919 when two deaths occurred in Portsmouth Naval Yard, U.S.A., from spraying a fire extinguisher over a man whose clothes were ignited. Both men recovered from the narcotic effect of the fumes and from the burns, and showed no symptoms typical of delayed carbon tetrachloride poisoning, but died later from pneumonia.

Fieldner and his colleagues (Fieldner, Katz, Kinney and Longfellow, 1920; Fieldner and Katz, 1921; Fieldner, Katz and Kinney, 1921) then found that the gas produced from the fire extinguishers under experimental conditions consisted of 15 to 80 p.p.m. of phosgene, 2,000 to 5,840 parts of carbon tetrachloride and 60 to 236 parts of hydrochloric acid. Further confirmation of these facts has been provided by Fohlen (1922, 1924), Biesalski (1924), Hamilton (1933) and others. According to Biesalski the greatest yield of phosgene comes from carbon tetrachloride at 250° C. in the presence of ferrous chloride.

For the removal of such toxic decomposition products, the German firm manufacturing Excelsior Fire Apparatus recommends that rooms in which the extinguishers have been used should be sprayed with watery solutions of aliphatic, aromatic or heterocyclic bases. They state that these, even if a surplus is used, are less irritating than ammonia, which has also been used for this purpose. The Carbide and Carbon Chemicals Corporation, New York, recommend the mixture of carbon tetrachloride with other substances, such as ethylene chloride, while Boye (1935) suggests the use of extinguishers employing carbon tetrachloride in combination with solid carbon dioxide or stabilized aqueous emulsions of carbon tetrachloride.

In spite of the fact that a few deaths and several cases of acute intoxication have been recorded from the use of carbon tetrachloride fire extinguishers, the incidence of such poisoning would appear to be small in comparison with the extensive use of such apparatus, which has the special advantage of not conducting the current in fires due to short-circuiting in high tension plants. Olsen (1933), writing in reply to Hamilton's (1933) statement on the toxic decomposition products of carbon tetrachloride fire extinguishers, states on behalf of the Chemical Fire Extinguisher Association that no fatalities have been recorded by the numerous firms supplied by this Company, although 5,770,000 extinguishers and 7,500,000 refills were supplied between 1910 and 1930. Lehmann (1930) gives a comprehensive survey of the use of carbon tetrachloride in fire equipment.

In order to ascertain whether phosgene is formed when carbon tetrachloride is used for home dry-cleaning, Smyth and Smyth (1936) tested the phosgene content of the air in a kitchen where a gas stove pilot light was on, under conditions duplicating home dry-cleaning. They found no phosgene with carbon tetrachloride concentrations up to 5,000 p.p.m. and relative humidity of 70 per cent, and conclude that in home use of carbon tetrachloride under the conditions described, there is little possibility of a hazard due to the formation of phosgene.

Carbon disulphide. The dependence of the toxicity of commercial carbon tetrachloride, as prepared by modern methods, on its content of carbon disulphide is apparently not great. Tomb and Helmy (1933) point out that the lesions produced by carbon tetrachloride poisoning are identical with those produced by other closely allied chlorine substitution products of aliphatic hydrocarbons, e.g. chloroform, which are sulphur-free; and that these lesions are quite different from those produced by carbon disulphide. Møller (1933) also remarks that whereas at the beginning of this century commercial carbon tetrachloride contained more than 0·5 per cent of carbon disulphide, there is no possibility of intoxication due to admixture of carbon disulphide with that prepared by modern methods. It may be noted in this connection that in 1909, when a fatal case of carbon tetrachloride poisoning due to the use of a hair shampoo was reported, an analysis showed that it contained 1·5 per cent of carbon disulphide (Veley, 1909).

USES

(1) As a solvent in the rubber, chemical and paint industries, and in the extraction of fats from plant and animal substances.

(2) As a cleansing agent in the dry-cleaning industry.

(3) As a constituent of fire extinguishers.

(4) In machine shops and printing plants for the removal of grease, in combination with benzine to reduce fire hazard.

(5) As a dry hair shampoo; in France it was called *Lotion antiseptique* when used for this purpose, but was forbidden in Paris under a decree of March 1st, 1930 (Bordas, 1935).

(6) As a constituent of insecticide sprays.

(7) As a constituent of soap solutions especially in the textile industry.

(8) In the quartz-crystal industry.

TOXICITY

While experiments on animals have shown that in relatively large doses carbon tetrachloride is both a narcotic and a metabolic poison, and although numerous records of acute poisoning in human beings are to be found in the literature, several investigations refute the idea, once held, that carbon tetrachloride is one of the most dangerous of modern industrial solvents. The view now generally held is that, though it constitutes an industrial hazard with high potentiality for temporary disablement and severe discomfort because of the rapid and disagreeable effects which it produces, by the rapid regeneration of tissues following injury caused by exposure to it, it carries some degree of safeguard against permanent damage from repeated sub-lethal dosage.

Some knowledge of the toxic effects exerted on the internal organs by carbon tetrachloride has been gained from its oral administration as an anthelmintic, from accidental ingestion by human beings and from experimental oral administration to animals. While such effects are not entirely comparable with those produced by inhalation of the fumes during industrial exposure, or during the use of fire extinguishers, their observation has given a valuable indication of the pathological effects and symptoms likely to be produced by carbon tetrachloride in industry.

Dangerous Concentrations in Industry

The amount present in factory air which apparently can be regarded as the safety limit for acute poisoning is given as 1,000 p.p.m. by the National Safety

Council of the U.S.A., and as 1,600 p.p.m. by the Retail Credit Company of Atlanta (1931). Lehmann (1911) used concentrations of 730 to 2,400 p.p.m. in his experiments on the chronic poisoning of animals. The limit for amounts which may produce chronic effects, however, has been stated by Davies (1934) to be about 100 p.p.m., and on the basis of animal experiments using concentrations as low as 50 p.p.m., which approximate to the concentrations actually found in factories where carbon tetrachloride is used, Smyth and Smyth (1936) agree with this figure.

Later investigations by Heimann and Ford (1941) and Elkins (1942) suggest that even this is too high; the former consider the safe minimum should be less than 79 p.p.m., since this level induces mild narcotic and gastric symptoms. Greenburg and Moskowitz (1945) suggest 75 p.p.m. as the maximum allowable concentration in the synthetic rubber industry, although the Oregon State Board of Health specifies 50 p.p.m. (0·005 per cent). McGill (1946) states that this concentration is produced by the evaporation of half a pint of carbon tetrachloride in a room 50 ft. × 50 ft. × 15 ft.

Davies (1934) studied the effects of various concentrations when inhaled by men, and found that the odour could be detected at 0·5 mg. per l. (about 79 p.p.m.) and was very strong at 3 or 5 mg. per l. (476 to 790 p.p.m.). Symptoms of intoxication began at 2 mg. per l. (317 p.p.m.) with 30 minutes' exposure, and consisted of nausea, vomiting and headache, but no changes in the urine. At 2,382 p.p.m. dizziness, nausea, vomiting, throbbing in the head and sleepiness developed. Three men painting pure carbon tetrachloride on fabric in a closed room, but with exhaust ventilation 6 feet away from the table on which they worked, were nauseated, sleepy and giddy in 10 minutes. The air in this room contained 2,300 p.p.m. of carbon tetrachloride. Men exposed for 8 hours to 200 p.p.m. in a cement house, and to 370 p.p.m. while spraying cellulose cement, complained of fatigue and sleepiness, and one man had albuminuria.

Davies believes that a concentration of 100 p.p.m. is the limit of safety for the avoidance of chronic effects, and that individuals with predisposing conditions of ill-health, in which he includes obesity, under-nourishment, gastric disturbance, pulmonary disease, nephritis and possibly alcoholism, should not work even at this concentration.

The vapour concentrations actually existing in well-conducted plants manufacturing carbon tetrachloride and in some thirty American establishments using it have been estimated by Smyth and Smyth (1936); they found that as a rule the average concentration over 8 hours' exposure was not above 100 p.p.m. which they, like Davies, consider a safe concentration for continuous exposure. The plants examined included five of manufacturers of carbon tetrachloride, twenty-three dry-cleaners using five different types of machine—"closed"and so-called "open"—three fire extinguisher makers, and five using the substance in chemical processes. In fourteen work-places peak concentrations were found above 100 p.p.m., but in practically every case these peaks lasted not more than ½ minute each and were judged to total less than 30 minutes per day. Smyth and Smyth (1936) and Smyth, Smyth and Carpenter (1936) state that it is significant of the care being taken in the industry that the highest average daily exposures found were 117 p.p.m. for five men and 111 p.p.m. for two men. The only two peak exposures found to exceed 1,000 p.p.m. were 1,680 and 1,252 p.p.m., but several men worked with air helmets where the peak was 1,040 p.p.m. It is entirely possible and practical to use carbon tetrachloride for almost any purpose without exceeding a concentration of 100 p.p.m. Standard dry-cleaning apparatus on the market will keep the concentration much below this figure and can be maintained in condition without more than ordinary care. Other processes using carbon tetrachloride can be operated in a similarly safe manner, if plant managers realize that it is necessary, without unduly increasing expenses and in most cases with a saving of solvent.

In the manufacture of quartz-crystal oscillators, Schulte (1945) found concentrations of 174 p.p.m. during the operation of washing crystals after separation of blocks.

Dangerous Concentrations from the Use of Fire Extinguishers

The concentration of fumes from fire extinguishers used in a closed room may be so great that they produce narcosis. According to Gronow (1927) a concentration of 160 mg. per l. corresponds to a mixture of 1 l. of fluid evaporated in 10 c.m. of air; the experiments of Lamson, Gardner, Gustafson, Maire, McLean and Wells (1924) show that less than this amount is sufficient to produce narcosis in animals. Gronow states, however, that the vapour given off from 2 l. in a completely closed room of moderate size can be inhaled for several minutes without danger; more than 2 l. should not be used without fire helmets.

It will be seen, however, that narcosis is not by any means the only danger to be apprehended from the use of fire extinguishers; severe symptoms of organic injury may arise some hours after exposure when no immediate signs of narcosis have occurred.

It is pointed out by Gautier, Chatron and Seidmann (1933) that the toxicity of carbon tetrachloride is much increased if the fire extinguisher is sprayed shoulder high, and two further sources of danger are to be found in the formation of phosgene when the vapour is brought into contact with flames, burning wood or red-hot iron, and the possibility of absorption of the carbon tetrachloride through fresh burns when the spray is used for extinguishing burning clothing (Kionka, 1931). A detailed review of the dangers of and indications for the use of carbon tetrachloride as a fire extinguisher is given by Lehmann (1930).

Method of Measuring the Concentration

A 50 centimetre portable Zeiss interference refractometer or interferometer was used by Smyth and Smyth, calibrated in terms of refractive index difference per scale division, using the air pressure method described by Edwards (1917). Air samples were drawn by suction through a copper tube placed as close as possible to a workman's nose at the point of maximum working exposure, passed over anhydrone (magnesium perchlorate) and ascarite to dry and remove carbon dioxide and other acid vapours, then through the interferometer. Samples were taken at intervals of about 10 seconds to about 5 minutes, depending on the nature of the process and fluctuations in vapour production. A method for collection and analysis of carbon tetrachloride vapour by absorption on silica gel and subsequent digestion with alcoholic potassium hydroxide is described by Pernell (1944).

Estimation in the Blood

A method, based on the pyridine and alkali method of Daroga and Pollard (1941), has been modified by Habgood and Powell (1945) to give measurements of small amounts, between 2 and 0·02 mg., of carbon tetrachloride in blood. The blood sample is made up to about 100 ml. with water and the carbon tetrachloride removed by steam distillation into a receiver containing between 1·5 and 5 ml. of toluene, which increases the sensitivity of the estimation. For the final colorimetric analysis, 1 ml. of the toluene extract is added to 10 ml. of pyridine and 5 ml. of 20 per cent sodium hydroxide. After heating at 100° C. for 5 minutes and then cooling, the colour intensity of the pyridine layer, which is purplish red, is measured in a Pulfrich colorimeter. The method is applicable to the estimation of trichloroethylene in blood, the pyridine layer in this case giving an orange red colour.

Factors Predisposing to Susceptibility

Alcoholism. McMahon and Weiss (1929) reported a fatal result of carbon tetrachloride poisoning in an alcoholic. Minot and Cutler (1928), and Lamson and co-workers (1924) and Gardner, Grove, Gustafson, Maire, Thompson, Wells and Lamson (1925) found in both animals and human beings that the toxicity of a large dose of carbon tetrachloride was greatly increased by the simultaneous administration of alcohol, and Hammes (1941) observed that the total exposure necessary to cause illness was much greater in abstainers than in alcoholics.

Obesity. According to Sherman and Binder (1944), obese persons are particularly liable to carbon tetrachloride poisoning.

Exposure to other chlorinated hydrocarbons. Hammes (1941) states that susceptibility is increased by previous exposure to other chlorinated hydrocarbons. He quotes the case of a man, who during a period of 3 years, cleaned fur pelts with trichloroethylene, and who had his initial symptoms of nausea a few hours after his first exposure to carbon tetrachloride. This has not been found to be the case with animals (Barrett, Maclean and Cunningham, 1938).

Lack of dietary calcium. Minot (1927) has shown, in both animals and in human beings, that lack of calcium in the diet predisposes to an over-susceptibility to carbon tetrachloride, and Smyth and Smyth (1936) have confirmed the fact by showing that a high calcium diet increases the resistance of such unusually sensitive animals as guinea-pigs. Observations by Hassan and Salah (1935) and Emara (1935) show a difference of opinion as to the effect of calcium deficiency when carbon tetrachloride is used as an anthelmintic; the former reports that it has no relation to the deaths, the latter that it is an important factor.

TOXIC EFFECTS IN ANIMALS

While the acute effects of carbon tetrachloride intoxication on animals have been well established, and correlated with those occurring in human cases, until recently little work has been done on the effects on animals of chronic exposure to vapours of carbon tetrachloride under conditions simulating those found in industry.

Acute Poisoning

Effect of Ingestion and Subcutaneous Administrations

Lesions of the internal organs

The effects of carbon tetrachloride ingestion by animals have thrown much light on its toxic action on the liver and kidneys in man, since the essential tissue damage in animals is acute hepatic necrosis and acute nephritis.

Smillie and Pessôa (1923) reported that small doses by mouth produced fatty degeneration of the liver and kidneys in dogs; Rosenthal and Lillie (1931) found in one out of seven dogs the condition of lipaemia described in a human case by McMahon and Weiss (1939). Bollmann and Mann (1931) were able to reproduce the liver lesions in dogs; Dervillée and Castagnou (1934) found that oral administration to rabbits caused a strong hypoglycaemia due to insufficiency of hepatic parenchyma, and Lande and Dervillée (1934) produced liver changes resembling yellow atrophy in these animals with as little as 2 ml. given by the same route. Experiments by Cameron and Karunaratne (1936) indicate that the minimum amount which will produce toxic effects in the rat's liver is about 0·025 ml. per kg. body weight injected subcutaneously.

Effect on metabolism

Protein metabolism. The experiments of Takahashi (1929) on rabbits with either subcutaneous or oral administration of carbon tetrachloride have shown that protein metabolism is disturbed in a manner to be correlated with liver injury. Total nitrogen excretion in the urine was much increased; the urea was increased in absolute amount, but was decreased relatively on account of the increase in amino-acid and ammonia. Takahashi states also that the injury to the oxidative function of the liver for purine bases and uric acid gives rise to abnormal acid formation in the organism with resultant acidosis. These results should be compared with those of Chatron (1934) in a case of acute carbon tetrachloride poisoning in a human being (p. 142). A further indication that one of the first effects of acute carbon tetrachloride poisoning is disturbance of protein metabolism is provided by Rosin and Doljanski (1946). They noted, in young rats that had been given intraperitoneal injections of carbon tetrachloride, the rapid disappearance from the hepatic cells of special protein granules which stain red with methyl-green-pyronine mixtures.

Carbohydrate metabolism. A pronounced hypoglycaemia, with depletion of the glycogen stores of the liver, has been observed by Cutler (1932) in dogs given large doses of carbon tetrachloride (4 ml. per kg. body weight) and by Dervillée and Castagnou (1934) in rabbits given 2 ml. per kg. by stomach tube.

Effect of Inhalation

The acute narcotic effect of carbon tetrachloride when inhaled by animals was described as early as 1876 by Eulenberg and confirmed and amplified by later workers, up to Lehmann and Flury (1943). Preliminary irritation shown by conjunctivitis, twitching, convulsive movement, etc. was noted by most observers, and Rambousek (1911) found a greater irritative effect than with chloroform. According to Lehmann and Flury (1943), fatty degeneration and necrosis of the liver and degeneration and necrosis of the tubules of the kidney are produced by inhalation of doses causing acute intoxication.

Lethal and narcotic concentrations

Lethal dose. Immediate lethal effects of inhalation occur in the smaller animals but are not often observed in the larger ones. Pantelitsch (1933) found that 65 mg. per l. (10,300 p.p.m.) was fatal to mice in 100 minutes, 90 mg. per l. (14,000 p.p.m.) in 40 minutes; Lazarew (1929) found that the minimum lethal dose for mice was 65 to 70 mg. per l. (10,400 to 11,000 p.p.m.), whereas Reuss (1931) observed rapid recovery in cats from deep narcosis following inhalations of 92·6 mg. per l. for 70 minutes. With larger animals death apparently often takes place some days later, especially when relatively small exposures are repeated at intervals. Thus Davies (1934) exposed a rabbit to 3,200 p.p.m. for 3 hours daily for 3 days; it died on the fifth day. Of the cats studied by Reuss (1931) which recovered temporarily, 25 per cent died after 1 to 17 days.

Narcotic dose. The concentration necessary to produce narcosis differs according to the species of animal and the length of exposure, from 36 mg. per l. for 53 minutes (Fühner, 1923) or 40 to 50 mg. per l. for 2 hours (Lazarew, 1929) for mice, to 90 mg. per l. for 4½ hours for cats (Lehmann 1911) and 133 mg. per l. for 9 to 13 minutes for dogs (Lamson *et al.*, 1924).

The symptoms produced in dogs by various concentrations by Lamson and co-workers are fairly typical of the results obtained by the majority of workers (Table 24).

TABLE 24

Symptoms of carbon tetrachloride poisoning by inhalation in dogs

Stage	Concentration (mg./l.)	(p.p.m.)	Time	Effect
I	40	6,350	1 hr.	Restlessness, uneasiness, increased salivation
II	51	7,950	1 hr.	Great muscular restlessness and hypertonia, increased salivation, some unsteadiness
III	57	9,050	1 hr.	Loss of co-ordination and balance, very marked tremor of head
IV	73	11,600	1 hr.	Loss of equilibrium, excitement, tremor of head
V	133	21,000	9–13 min.	Quiet sleep, with cessation of all voluntary and involuntary muscular movements; narcosis may be continued for an hour without fatal results

With concentrated fumes, deep narcosis sets in after one or two minutes, and if continued for a few seconds respiratory paralysis will result. Fresh air and artificial respiration will still save life at this point.

Lesions of the organs and tissues

The characteristic lesions of the liver and kidneys found in many cases of poisoning in men and animals after ingestion of carbon tetrachloride have been reproduced in animals subjected to inhalation of acute concentrations. Thus Lehmann (1911) noted fatty degeneration of the liver and enlarged and fatty kidneys. Davies (1934), in addition to cloudy swelling and congestion of the kidneys and commencing central necrosis of the liver, observed signs of irritation of mucous membranes, congestion of lungs and bronchi (in one rabbit the lungs were mucopurulent), congestion of the gastric and intestinal mucosa, and conjunctivitis. In some experiments on mice, rabbits and guinea-pigs carried out at Porton in 1934 (unpublished work), microscopical examination showed changes in the liver similar to acute yellow atrophy—congestion, cell degeneration, minor fatty degeneration, commencing patchy necrosis, and in the kidneys, congestion, patchy degeneration and necrosis of the tubular epithelium and glomeruli, the lesions being mainly in the cortex. The blood urea rose in nearly all the animals. Individual animals undergoing the same exposure differed with regard to both the effect on the blood urea and on the organs.

In some experiments by Lande and Dervillée (1935), repeated every day for about a month, in which almost lethal concentrations were inhaled until convulsions began (3 or 4 minutes), the liver showed sub-acute diffuse hepatitis and periportal fibrosis, and the kidney some nephritis.

Effect on central nervous system

Biancalani (1934) has reported changes in the cerebral cortex of animals subjected to inhalations of carbon tetrachloride vapour aspirated under $1\frac{1}{2}$ atmospheres pressure for 2 minutes followed by aspiration of normal air for 2 minutes and exposure to the mixture for 15 minutes at a time. These changes consisted of degeneration in the form of chromatolysis, vacuolation in ganglion cells and hypertrophy and oedema of the neuroglia tissue.

Relative Toxicity of Carbon Tetrachloride and Chloroform

The results of most animal experiments indicate that the immediate lethal narcotic effect of inhalation of carbon tetrachloride is less than that of chloroform, but that its toxic action on the organism is more intense.* Thus, in the Porton experiments, rabbits and guinea-pigs exposed for 42 minutes to inhalations of 1 in 30 concentrations of carbon tetrachloride recovered, while two out of three rabbits similarly exposed to chloroform died. The degree of injury to the liver and kidneys due to carbon tetrachloride was, however, more severe than that of chloroform.

The data of Veley (1909) were based on the abolition and recovery of response of isolated muscle to solutions of carbon tetrachloride and chloroform. He concluded that though the action of carbon tetrachloride was less rapid than that of chloroform, it was more deadly, since the muscle recovered from the toxic effect of chloroform but was killed by carbon tetrachloride. It must be mentioned here, that Barsoum and Saad (1934) found no arrest of the isolated frog's heart with saturated solutions of carbon tetrachloride.

Chronic Poisoning

A very complete and extensive investigation of the effects of repeated inhalation of small doses of carbon tetrachloride has been carried out by Smyth and Smyth (1936) and the results have been closely correlated with the supposed safe concentration of the vapour for industrial workers.

The animals used in this experiment were monkeys, rats and guinea-pigs. Guinea-pigs, like rabbits (Lehmann, 1911), were found to be so much more susceptible to carbon tetrachloride that less importance was attached to the results obtained from them than to those from rats and monkeys. Rabbits were used in the experiments of Vallery-Radot, Mauric, Domart and Gauthier-Villars (1938) and these showed glomerulitis in 4 per cent, and liver changes of varying severity in 94 per cent. The concentrations used in Smyth's experiments were generally 50, 100, 200 and 400 p.p.m., and exposures were made for 8 hour periods, 4 to 6 days a week over periods up to $10\frac{1}{2}$ months, some of the animals receiving a total of 225 exposures.

---

* It should be noted that the relative narcotic potency is in opposition to the general rule, enunciated by Joachimoglu (1921a,c, 1925) and Lazarew (1929), that the narcotic effect of the chloro-compounds increases with the number of chlorine atoms. Lazarew points out, however, that the rule holds only for aqueous solutions of these substances, the toxicity of the vapours depending on their solubility coefficient. He states that in aqueous solution chloroform is less toxic, and in vapour form more toxic (i.e. narcotic) than carbon tetrachloride.

General Effects

None of the animals showed signs of any subjective symptoms or discomfort, nor was appetite decreased. Growth was affected very little, even by the higher concentrations, while 50 p.p.m. appeared actually to have a stimulating effect. Fertility was adversely affected by the higher concentrations, but was stimulated by the lower (50 p.p.m.). Infection could not be correlated with exposure. Blood counts were unaffected, though the spleen in many animals receiving long-continued or severe exposure indicated active red cell destruction. Apparently injury had not progressed to the stage where red cell production was unable to keep up with this destruction, for no cases of anaemia were found. It should be noted here that Takada (1932) found that repeated injections of carbon tetrachloride in animals produced anaemia and reticulocytosis, the former more marked in mature and the latter in young animals. Slightly raised icteric indices were found only with higher concentrations (400 p.p.m.). Albuminuria and bilirubinuria were not consistent findings.

Lesions of the Internal Organs

Liver. Guinea-pigs showed a much greater susceptibility than rats and monkeys, death occurring from combined rapidly developing hepatic and renal damage from concentrations of 100 p.p.m. or higher. While definite fatty degeneration and cirrhosis were produced by single heavy exposures to 400 p.p.m., and by comparatively few exposures to 200 p.p.m. or more, very little damage was produced by 100 p.p.m., and a striking feature of the investigation was the constant regeneration of liver cells occurring during the exposures, sufficient to maintain an adequately functioning liver for the whole 10 months' period of the experiment.

Concentrations of 400 p.p.m., however, were definitely and progressively harmful, producing fatty degeneration of the liver parenchyma. The regenerative process, when exposure was stopped at an early stage, would result in the entire acinus being made up of new cells replacing old degenerated ones, the new cells being more resistant to the material responsible for their formation than the old. With continued severe exposure, proliferation of interstitial cells and fibroblasts transformed the liver into a mass of small acini embedded in a matrix of fibrous tissue, producing cirrhosis, and with the contraction of this new fibrous tissue, the highly nodular "hobnail" liver. With cessation of exposure, regeneration also occurred and healed cirrhosis resulted, with a multi-lobulated liver consisting of multiple small but normally functioning acini, separated by comparatively narrow bands of contracted fibrous tissue. Disturbance of liver function of animals exposed to carbon tetrachloride has been described by McCord, Sterner, Kline and Williams (1946) and by Maclagan (1944) using the thymol turbidity test. This reaction, which is said to be due to a variation in the phospholipid metabolism, is the earliest demonstrable effect of carbon tetrachloride poisoning, and was negative after 7 days of non-exposure.

Kidneys. Kidney damage, in the form of congestion and cloudy or granular swelling, followed by degeneration, was evident after fifty-two exposures to 50 p.p.m. and eighteen to twenty to the higher concentrations. It was not, however, extreme, and the kidneys were able to function satisfactorily even in the animals subjected to 400 p.p.m. It appeared that cessation of exposure might lead to complete regeneration, fibrosis not resulting as easily as in the liver.

Adrenals. Fatty degeneration and necrosis occurred in guinea-pigs, but there was very little change in rats.

Nerve tissue. No examinations of central nervous tissues were undertaken, but changes in peripheral nerves (sciatic and optic) indicating nerve damage (fatty degeneration) were seen in some animals.

Ocular muscle. In some animals this also showed evidence of degeneration after long exposure and high concentrations.

Effect of Diet

A low protein, high carbohydrate diet is said to exert a protective action in animals inhaling carbon tetrachloride vapour (Davis, 1924; Bollmann and Mann 1931; Drill, Loomis and Belford, 1947). This is in contrast to chloroform and ethylene dichloride, where animals on a high protein diet show less liver damage than on low protein.

According to György, Seifter, Tomarelli and Goldblatt (1946) and Shaffer and Critchfield (1945), the effect of methionine on animals is not of the beneficial character found in human beings as reported by some authorities (Beattie, Herbert, Wetchel and Steele, 1944; Eddy, 1945). György and co-workers found it without effect on the liver damage in rats, though it had a favourable one in prolonging survival, and also on the occurrence of nephrosis. Shaffer and Critchfield (1945) were unable to explain the good results which may follow methionine therapy in acute carbon tetrachloride poisoning in human beings. They endeavoured to find out if inhalation of carbon tetrachloride produced a specific demand for sulphur-containing amino-acids in animals receiving adequate protein. They therefore exposed rats on a diet deficient in these amino-acids to successive daily exposures of 1,000 p.p.m. of carbon tetrachloride vapour, and compared them with similarly fed non-exposed controls. Although the livers of the exposed animals showed gross fatty infiltration, neither the sulphur content nor the total nitrogen was depleted.

Regeneration in Damaged Organs

While the lesions observed in the organs, especially the liver and kidney, of animals exposed to such concentration resemble those described by other workers who administered larger doses, a special feature of chronic poisoning appears to be the regeneration of damaged cells both in the liver and kidney. The regenerated cells or entirely new cells formed are more resistant to the carbon tetrachloride than were the original cells, as shown by the effects on the liver of anaesthetic concentrations (20,000 p.p.m.). It may be noted here that the experiments of Cameron and Karunaratne (1936), using subcutaneous injections, have shown the same capacity for liver cell regeneration after carbon tetrachloride, provided that time for recovery after each dose is allowed. If sufficient intervals are not allowed, however, a cirrhotic stage develops in which a permanent perilobular fibrosis occurs. It is concluded by Smyth and Smyth (1936) from their results, that in most cases men who are continuously exposed to concentrations of 100 p.p.m. or less of carbon tetrachloride increase their resistance rather than their susceptibility, in strong contrast to benzol, where injury is usually progressive after exposure has stopped. This view is confirmed by the investigation on human beings by Stewart and Witts (1944).

This study emphasizes the following features of chronic carbon tetrachloride exposure in animals: (1) acclimatization with regeneration of affected organs; (2) recovery on cessation of exposure; (3) development of increased resistance with continuing exposure.

TOXIC EFFECTS IN MAN

Acute Poisoning

As in the case of animals subjected to inhalation of carbon tetrachloride, in man immediate mortality due to the acute narcotic effect of the fumes is of comparatively rare occurrence, but many cases have been recorded in which death has taken place at an interval of 3 to 12 days after severe exposure, with symptoms of involvement of the liver or kidneys or both.

Symptoms

The delay in the appearance of acute symptoms has been very characteristic in several cases which have eventually recovered, and the initial symptoms, anorexia, nausea and vomiting, have usually been referred to the gastro-intestinal system. According to Davies (1934) this reaction is of central nervous origin, not a local gastric disturbance. Motor and psychic phenomena, ending in Jacksonian epilepsy, are regarded by Tietze (1933) as characteristic features of carbon tetrachloride intoxication, but in the case which he reports and on which he bases his evidence the patient was exposed to a mixture of methyl bromide and carbon tetrachloride. A case showing epileptiform convulsions with unconsciousness has been described by Hagen (1939). The attacks of convulsions lasted for 2 days after removal to hospital.

Some examples of poisoning from the use of carbon tetrachloride as an anthelmintic have been reported (Hall, 1921; Docherty and Burgess, 1922; Docherty and Nicholls, 1923; Minot, 1927; Lattes, 1934; etc.). In Lattes' case death took place 4 days after administration of three capsules of a remedy for ankylostomiasis, corresponding to 3·6 g. of carbon tetrachloride. The symptoms began with nausea, vomiting, diarrhoea, and went on to unconsciousness. There was swelling of the face and body, punctiform haemorrhages in the eyelids and sclera, palate and pharynx, complete anuria and icterus. A catheter specimen of the urine contained 5 per cent of albumin, renal cells, granular casts and a few red cells.

Most of the reported cases of acute poisoning from ingestion of carbon tetrachloride appear to show that the greatest damage occurs in the kidneys and liver. Diminished renal function, with an increase of the non-protein nitrogen in the blood, was a special feature in a case of acute poisoning from the accidental ingestion of 4 to 5 oz. of carbon tetrachloride recorded by Lehnherr (1935). The liver showed especially severe injury in the case recorded by McMahon and Weiss (1929). The symptoms in this case began with jaundice, abdominal pain, nausea, vomiting, and went on to coma and death after 5 days. Post-mortem examination showed the liver to be enlarged, soft, smooth and yellow, with congested red areas against the fatty yellow background. Histologically the liver showed extreme fatty degeneration and necrosis. A curious feature of this case was the high fat content of the blood, a finding which McMahon and Weiss explain as being due to the severe damage to the liver. Yet even in cases of acute poisoning by ingestion the actual damage is apparently more a disturbance of metabolism than destruction of tissue, and can be prevented from becoming fatal if suitable measures are taken in time. This view is supported by the investigation by Beattie et al. (1944) of a case of acute carbon tetrachloride poisoning following the accidental ingestion of 30 to 40 ml. Although the liver showed rapid palpable enlargement, the only significant abnormality, as shown by the liver function tests, was a rise in serum bilirubin during the acute phase. The successful treatment by methionine as an intravenous transfusion with a papain-trypsin digest of casein gave further opportunity of investigating the nature of the liver damage. Estimations of the nitrogen and sulphur balances of the patient, and of the partition of the urinary sulphate between the oxidized (total sulphate) and the unoxidized (neutral) fractions, showed that the nitrogen balance was negative during the first 2 days. The suggestion is made, therefore, that during the first 2 days the synthesis of protein by the liver was disturbed, but that this disturbance was in the nature of a metabolic derangement rather than one of actual destruction of liver substance. This view is also supported by the results obtained by Stewart and Witts (1944).

Rapid Fatal Cases

The cases of death following acute narcosis by carbon tetrachloride inhalation include the case reported in 1909 arising from the use of a hair shampoo (Veley, 1909; Waller, 1909); Pagniez, Plichet and Koang (1932) reported the case of a workman who died after 2 hours' painting in a dyeing works saturated with carbon tetrachloride; one case resulted from the use of a fire extinguisher (Leoncini, 1934), and three similar cases were reported by André and Feillard, 1946; one, also from the use of a fire extinguisher, was reported by the Home Office in 1934. The post-mortem examination of the last case showed the presence of haemorrhage into the base of the brain and pulmonary emphysema. In all these cases death was preceded by unconsciousness.

Fatal Cases with delayed Symptoms

Among the cases in which death has occurred from 2 to 12 days after the initial exposure are those in the dry-cleaning industry recorded by Poindexter and Greene (1934), from a spray used as an insecticide (Hausmann and Helly, 1929), from the use of a fire extinguisher (Lecornu and Pecker, 1932; André and Feillard, 1946), from opening a vessel containing carbon tetrachloride used in a leather factory (von Scheurlen and Witzky, 1935; Carozzi, 1936), from painting with a lacquer containing carbon tetrachloride (Koelsch, 1916, and reported in the Swiss Factory Inspection Report, 1910).

The case recorded by von Scheurlen and Witzky (1935) may be taken as typical, and the post-mortem appearances are characteristic of the severe toxic lesions produced by concentrated inhalation of carbon tetrachloride. The patient opened a vessel containing carbon tetrachloride and inhaled a large amount. Two days later he complained of headache and nausea and had a raised temperature. On the next day he had a severe attack of tonic and clonic convulsions followed by unconsciousness. Later he became cyanosed, dyspnoeic and covered with cold perspiration; his pulse was weak and rapid, his urine scanty and contained 5·5 per cent of albumin with red and white corpuscles. Finally he passed into a state of uraemia and died 9 days after the initial exposure.

Post mortem, the brain and both lungs were oedematous, the liver very soft and fatty, the kidneys enlarged, fatty and oedematous, the spleen hyperaemic, and the cardiac valves inflamed and thickened. Histologically the liver showed very severe injury, necrosis of cells, fatty degeneration and haemorrhages, the kidneys, marked oedema and fatty degeneration of the tubules, while the glomeruli were anaemic.

Von Scheurlen remarks that since reporting this case he has observed another fatal case in a tannery, also of "acute yellow atrophy". He does not mention the occurrence of jaundice, but in the case reported by Lecornu and Pecker

(1932) there was jaundice and an enlarged liver 2 days after the initial exposure, and also epistaxis, purpuric spots and ecchymoses, oedema of the face and both lungs, and a blood urea of 288 mg. per 100 ml. Fatal cases with very similar symptoms and post-mortem appearances are described by Duvoir, Guibert and Desoille (1933), Kehrer and Oudendal (1926), and Martin, Dyke, Coddington and Snell (1946).

In Smetana's (1939) cases there were oliguria, anuria, nitrogen retention and subsequent hypertension, and one of Eddy's patients (1945) died from acute diffuse pyelo-nephritis 2 weeks after exposure.

Oedema, both pulmonary and peripheral, is described in several of these cases of nephritis and hepato-nephritis. Definite anasarca was present in one of the cases reported by Pagniez and co-workers (1932), and Duvoir and co-workers (1933) specially emphasize the fact that this form of poisoning with oedema secondary to hepatic or renal injury is to be differentiated from a purely "pulmonary" form, which, they state, is characterized by a crisis of acute oedema of the lungs. In Blanc and Carrière's (1935) case, acute oedema of the lungs, as well as of the face and lower limbs, was present. Pulmonary lesions are also mentioned by Hagen, Alexander and Peppard (1940), and Thompson (1946) describes radiological changes (increase of hilar shadows and prominence of lung markings) in four out of twenty men poisoned by inhalation of carbon tetrachloride used for cleaning purposes. One of these men had haemorrhagic pneumonia, and one of Eddy's delayed fatal cases had a terminal pulmonary haemorrhage.

In the fatal cases in which oedema of the lungs has been found after the use of fire extinguishers, it is suggested that this condition is caused by the decomposition product, phosgene, but it should be noted that in the fatal case of von Scheurlen and Witzky (1935), where oedema of the lungs and brain were found post mortem, there had been no contact of the carbon tetrachloride with any agent likely to bring about decomposition. An investigation by Elkins and Levine in 1939 indicates that the extent of phosgene production from contact of carbon tetrachloride with burning tobacco, at any rate, does not constitute a serious health hazard.

Cases with Temporary Narcosis but no apparent Liver or Kidney Injury

One of the earliest reported cases of carbon tetrachloride intoxication was recorded by Lehmann in 1911, when a man, after entering a reservoir, showed excitement, delirium and commencing anaesthesia, but recovered in 8 days. Another early case of intoxication from the use of the newly introduced carbon tetrachloride hair shampoo was described by Colman in 1907; temporary unconsciousness was followed by prolonged vomiting and severe headache, but complete recovery took place in 1 or 2 days. At the inquest on the fatal case from a similar cause in 1909, it was established that a condition of narcosis was not infrequent in women who had their hair shampooed in this way. Three such cases were reported by Sandilands in 1909 and three more by Møller (1933).

In one of Møller's cases there was complete unconsciousness, in the others early anaesthesia with absence of inhibition, followed by dizziness for some hours. Møller also reports the case of a woman using carbon tetrachloride as a dry-cleaning fluid in a room where the concentration of the toxic inhalation was apparently increased by the precautions taken to provide a through draught. The user of the fluid became unconscious, and a man who went to her assistance had giddiness, staggering gait and vomiting. A case of deep narcosis following the bursting of a fire extinguisher was reported by Dingley in 1926. Recovery took place after artificial respiration. Four such cases of temporary unconsciousness were reported to the Home Office between 1931 and 1933. In all these, as in the remaining cases reported, the recovery period was attended by symptoms of gastric disturbance, nausea, vomiting, abdominal pain, but not by any delayed symptoms of "hepato-nephritis."

Non-fatal Cases with Liver and/or Kidney Injury

The majority of cases showing severe symptoms of what is described by some authorities as toxic nephritis (Dudley, 1935), hepato-nephritis (Lecornu and Pecker, 1932; Hébert and Phélébon, 1931, etc.), or nephrosis (Smetana, 1939; Clinton, 1947) have occurred in connection with the use of fire extinguishers, but others have followed exposure in the dry-cleaning industry (McGuire, 1932; Blanc and Carrière, 1935), in degreasing processes (Boveri, 1929) and from the use of insecticide sprays (Lande and Dervillée, 1935). Duvoir, Guibert and Desoille (1933) differentiate the group of cases showing renal symptoms predominantly from the group with liver symptoms also, and describe them as two separate forms of carbon tetrachloride poisoning—the hepato-renal and the pure renal. The nature of renal lesions in such cases has been described by Woods (1946) as like those of the "crush syndrome." In a case reported by Clinton (1947) of a man employed for only 2 days in degreasing gears, the symptoms included dizziness, nausea, vertigo, chills and severe headache; there was no evidence of liver damage, but hypertension was present, and the urinary findings were suggestive of nephritis. Recovery took place after 6 weeks. In a case reported by Butsch (1932) the symptoms were apparently those of liver injury only; though there was no clinical jaundice, ascites was present and Butsch states that it was probably a case of cirrhosis of the liver.

The majority of the more severe and acute cases reported, however, appear to have suffered injury to both liver and kidneys, though in some cases, notably those reported by Dudley (1935), Smetana (1939), and Clinton (1947), the symptoms of nephritis seem to have been more conspicuous. In two of Dudley's four cases the symptoms of kidney involvement were very severe—almost complete anuria, a very high blood urea, and in one case acute uraemia; in this case there was also slight jaundice. In both cases the initial symptoms were delayed for 15 and 24 hours respectively after exposure, and in both, recovery took place with the sudden onset of polyuria 9 and 11 days later respectively. In neither of these cases, nor in two slighter ones also reported by Dudley (1935), were casts, bile or blood found in the urine, though albumin was present. In a very similar case of toxic nephritis reported by Gautier, Chatron and Seidmann (1933), however, albumin and casts were present in the urine, and there was also haematemesis and melaena, while in Boveri's case (1929) besides albuminuria, casts and blood in the urine, there was tenderness and enlargement of the liver.

Metabolic Disturbances

Chatron (1934) considers that the essential disturbance produced by the injury inflicted on the liver and kidneys is that of an alteration of the renal ammonia-genetic function, which is the cause of a progressive state of acidosis, complicated by low blood chloride and high blood urea. He notes that the chief sign of persisting renal injury is the low urinary urea, the ammonia-genetic function being still insufficient.

Treatment

Alkalis and chlorides. In view of the above findings, Chatron (1934) administered alkalis to suppress the acidosis and chloride to raise the lowered blood chloride, together with glucose, as follows:—30 g. sodium bicarbonate by mouth, 1·2 g. sodium chloride intravenously and 1·1 g. by mouth, 500 ml. glucose serum by rectum. Further amounts of these substances were given at intervals so that finally, during 3 days, the patient had received 50 g. sodium bicarbonate, and 61·8 g. of sodium chloride.

The clinical results of this treatment were satisfactory; the haemorrhage, somnolence, incessant vomiting, and headache showed distinct improvement within 24 hours of the first administration.

Concentrated blood plasma. In cases of acute nephrosis the administration of concentrated blood plasma has been recommended and found successful by Melick (1946). He bases the rationale of this therapy on the fact that the injury to the kidney in acute nephrosis is primarily epithelial, not vascular, and that the depletion of plasma protein is chiefly the loss of albumin. In his case, where there was a greatly decreased output of urine and generalized oedema, the albumin/globulin ratio was reduced from the normal 1·51 to 0·71.

Methionine. Although the favourable effect of methionine in human beings reported by Beattie et al. (1944) and Eddy (1945) appears to have no parallel in animal experiments (p. 138), Eddy believes that the good results which he obtained in six cases showing varying degrees of liver damage were due largely to the administration of 2 g. of methionine every 4 hours. These men received also intravenous glucose and a high protein, high carbohydrate, low fat diet.

Peritoneal lavage. In cases where symptoms of renal injury predominate, peritoneal lavage has been tried (Pearson, 1947) with a view to maintaining kidney function, until the reparative process can be established. In the case described by Clinton (1947), Pearson (1947), Baggenstoss (1947) and Snell (1947), where oliguria had been present for 9 days before peritoneal lavage was instituted, it was thought that if life could have been maintained for a few days longer, complete restoration of renal function might have been achieved, since anatomically there was evidence that new cells had already formed in the lining tubular epithelium. These, however, were not sufficiently mature to be capable of full function.

Persistent Effect of Carbon Tetrachloride Poisoning

A case in which symptoms and signs of hepato-renal injury persisted after recovery from acute intoxication had taken place is recorded by Ceresa (1945). The acute symptoms, including jaundice and anuria, the latter persisting for 6 days, had followed the use of a fire extinguisher by a cinema operator. Two months later the man complained of asthenia, dyspepsia and lumbar pain, and showed a sub-icteric conjunctiva, slight liver enlargement and urobilin and casts in the urine. Ceresa suggests that the hepato-renal lesions caused by the original acute poisoning may, in some degree, remain permanently.

Chronic Poisoning

Cases of chronic, slight intoxication have seldom been recorded. Amongst the symptoms in such cases are headache, nausea, loss of appetite, vomiting, loss of weight, nervousness, mental confusion, secondary anaemia, slight jaundice, disturbances of vision and dermatitis. The last condition is mentioned by Davies (1934) and by Flandin (1932), and is regarded by them as being due to the solvent action of carbon tetrachloride on the fatty layers of the skin. Most workers, however, including Lehmann (1930), Smyth (1935) and Smyth and Smyth (1936), regard the danger of chronic industrial poisoning from carbon tetrachloride as comparatively slight.

While certain authors have reported cases in factory workers with symptoms which have been diagnosed as chronic carbon tetrachloride poisoning, none of them, with the exception of Smyth and Smyth (1936), has estimated the actual content of carbon tetrachloride vapour in the factory air, and it would appear that these cases do not represent the average conditions in which carbon tetrachloride is used in industry. For example, in Brandt's (1932) series of 20 cases, of whom 15 showed tenderness and 5 enlargement of the liver and others complained of nausea, vomiting and loss of appetite, the shoe cement used contained 75 to 80 per cent of carbon tetrachloride, 5 per cent of methylene chloride and 15 to 20 per cent ethylene chloride. Similarly Loewy (1935), who examined 20 mechanics using carbon tetrachloride as a cleaning agent in an automatic telephone exchange, states that on account of the sensitivity of the apparatus the ventilation was not good. An epidemic of poisoning in a parachute plant reported by Doyle and Baker in 1944 was similarly attributed to lack of ventilation. The case reported by Lyon (1935) as possibly one of carbon tetrachloride poisoning, though not industrial, is specially interesting from the point of view of the metabolic findings.

In a large number of manufacturing plants and factories Smyth and Smyth (1936) found that the average concentration of carbon tetrachloride was low; the incidence of symptoms and evidence of injury were correspondingly slight, and they concluded that their tests showed that no one could be considered seriously, or even unmistakably, injured by the exposure to carbon tetrachloride. This view, especially with regard to the absence of hepatic or renal damage even when symptoms causing temporary disablement are present, was confirmed by an investigation by Stewart and Witts in 1944. The characteristic symptoms in a group of workers in a chemical factory examined by them were gastro-intestinal disturbance and mental hebetude, but there was little evidence of hepatic or renal injury and the symptoms usually cleared up in a few days or weeks after removal from exposure.

Symptoms

Gastro-intestinal disturbance

Nausea, vomiting, loss of appetite (McGuire, 1932; Brandt, 1932), colic (Wurm, 1931) and constipation (Loewy, 1935) are amongst the commonest symptoms recorded. In Wurm's series of cases, where carbon tetrachloride was used as the solvent for rubber, and in Brandt's, these gastro-intestinal disturbances were associated in nearly all the cases with weakness, fatigue, headache, and in some, with burning of the eyes. McGuire reported the occurrence of haematemesis in 3 out of 7 workers using carbon tetrachloride for the removal of spots from felt, and diarrhoea in 6, while Schutz (1937–38) and Smetana (1939) have mentioned a haematemesis and melaena among the characteristic symptoms. In the examination of 96 workers carried out by Smyth and Smyth (1936) the following symptoms were looked for and one or more of them were found in the majority, though in very slight degree:—nausea, vomiting, distress, constipation, loss of weight, burning abdomen, dyspnoea, head colds, nose and throat irritation, eye irritation, dizziness, headache, sleepiness during the day and insomnia at night.

In some cases the gastro-intestinal symptoms may be so acute as to lead to an early diagnosis of food poisoning, or even of "acute abdomen" as in two cases recorded by Graham (1938). Smith (1947) describes two cases of duodenal ulcer following within a short time of a first exposure to carbon tetrachloride. She believes that the development of peptic ulcer may not be an uncommon complication, and quotes the experimental evidence (Bollmann and Mann, 1931) of extensive gastro-intestinal haemorrhage in animals in support of this view. In the series of cases recorded by Stewart and Witts (1944) where the exposure was intermittent and irregular in degree, the severity of gastro-intestinal symptoms varied from mild queasiness to violent and repeated attacks of nausea, vomiting, colic and diarrhoea. Residual symptoms often persisted from one episodic exacerbation to the next, but were rapidly relieved by a brief period of non-exposure. Gastric analyses and gastroscopic and radiographic examinations showed that these alimentary symptoms were associated with hypersecretion, hypertonicity, irregular peristalsis and spasmodic contraction of the stomach and intestines. Since the blood urea, serum-bilirubin, phosphatase and plasma proteins were all well within normal limits, and the Takata Ara reaction was negative throughout, it was considered that these symptoms could not be attributable to lesions of the liver at an early stage as suggested by Elkins (1942). Stewart and Witts (1944) incline to the view that they are due to the action of carbon tetrachloride on the central nervous system, by way of stimulation of the parasympathetic or inhibition of the sympathetic centres in the hypothalamus. They emphasize, however, that the difference between the concentration of carbon tetrachloride which affects the central nervous system and that which affects the liver and kidneys does not seem to be large, and that it would therefore be unwise to overlook the significance of this gastro-intestinal disturbance.

Liver damage

It appears that the occurrence of liver damage after repeated exposure to carbon tetrachloride depends upon the size of the dose and the interval between successive doses. In some of the recorded cases of intoxication, liver damage has been clinically recognizable, e.g., enlargement of the liver with jaundice in two of McGuire's cases (1932), tenderness and enlargement of the liver in some of Brandt's (1932), a marked and noteworthy yellowish-green tinge of the skin in Wurm's (1931), and a preliminary enlargement followed by a decrease to less than normal size in Lyon's case (1935). On the other hand, Smyth and Smyth (1936) found evidence of liver damage in only one of 96 workmen, and in this case the probable cirrhosis was associated with chronic alcoholism. Stewart and Witts (1944) also found neither clinical nor laboratory data which would indicate liver damage, even though exposure must have been fairly heavy on some occasions. They explain this by comparison with the animal experiments of Cameron and Karunaratne (1936), which have shown the capacity of the liver for regeneration when exposure to carbon tetrachloride has ceased. In the rat, repair of the essential acute lesion, central necrosis, is complete in 14 days, and if the intervals between dosage with carbon tetrachloride are spaced beyond this, it can be administered indefinitely without producing any alteration in the liver. If each dose is too small to produce an initial tissue lesion, no matter how frequently it is repeated, there will be no hepatic damage. In rats, even when the liver change has progressed as far as cirrhotic bands around the liver lobules, the process may still be reversible with non-exposure, but after a certain time the cirrhosis may become permanent. To bring about permanent damage, therefore, there must be exposure sufficient to produce an immediate toxic effect and the dose must be repeated at short intervals.

Alterations in metabolism. In a personal communication Bollmann and Mann, quoted by Lyon (1935), emphasize the fact that in the hepatic cirrhosis produced by chronic carbon tetrachloride injury, as distinct from the acute necrotic effect, metabolic and functional alterations are rarely detected but depend upon the extent of damage to the liver. They state that bilirubinaemia, dye retention and decreased galactose tolerance are usually present, but the changes in blood and urine nitrogen are less. There is a slight increase in the uric acid content of the blood, but no change in the total non-protein nitrogen, urea, amino-acids, ammonia or creatinine, is usually found, though at times a slight elevation of amino-acids and low urea values were noted. Stewart and Witts (1944) found the blood urea, serum bilirubin, phosphatase and plasma protein within normal limits.

From analogy with animal studies in which intraperitoneal injection of carbon tetrachloride caused cellular changes in the liver characteristic of protein inanition, Rosin and Doljanski (1946) believe that disturbed protein metabolism is one of the first effects of carbon tetrachloride poisoning.

Icteric index and indirect van den Bergh test. An icteric index of 9 or over and a van den Bergh of 0·2 mg. per 100 ml. or over were taken by Smyth and Smyth as possibly significant. None of the 8 workers exposed to high concentrations showed any abnormality in these respects, while of the 88 not exposed to concentrations above 117 p.p.m. 21 had high icteric indices and 6 high van den Bergh values.

Urinary changes

Urobilinuria. An increase in urobilin or urobilinogen in the urine is regarded by some workers as a criterion of damage to the liver by carbon tetrachloride. Loewy (1935), in particular, regards it as the initial symptom of functional liver disturbance and found urobilinogen present in 13 out of his 20 cases. In Smyth and Smyth's (1936) workmen, however, no change in the urine urobilin was detected.

Uric acid and amino-acids. Bollmann and Mann (1931) state that the urine may show an increase in uric acid and amino-acids.

Total nitrogen and total sulphate. Following the statement by Takahashi (1929) that in animals both total nitrogen and total sulphate of the urine are increased in carbon tetrachloride injury, Smyth and Smyth (1936) examined these constituents in the urine of their workmen. The maximum figures taken were 35 g. for total nitrogen and 2·25 g. total sulphate per 24 hours. Only one of the 8 "high concentration" workers had a high urinary nitrogen (58·8 g.), while none of the 88 "low concentration" workers had high values for either nitrogen or sulphate.

Ratio of inorganic to total sulphate. Since an increased ratio of inorganic to total sulphate is considered by the U.S. Bureau of Mines to be a significant criterion of exposure to benzol, this ratio was estimated by Smyth and Smyth (1936) in a number of their workers exposed to carbon tetrachoride. While a ratio of over 90 per cent (85 to 90 per cent is considered the normal ratio) was found in 5 out of 8 workers examined, this ratio showed very poor correlation with other findings, such as monocytosis, visual field restriction and high icteric index, and also with the known exposure of the individual workers. Smyth and Smyth did not therefore regard it as a significant test for carbon tetrachloride exposure or poisoning.

Blood changes

There appears to be no pathognomonic blood change associated with carbon tetrachloride intoxication. Smyth and Smyth (1936), Stewart and Witts (1944) and Browning (unpublished observations) have found no severe anaemia in any of the workers examined; variation in the total number of white cells, and in the polymorph/lymphocyte/monocyte ratio in some cases (Smyth and Smyth, 1936) were not uniform or capable of being correlated with the degree of exposure to carbon tetrachloride. Only one case of severe fatal anaemia, that of a man who a year earlier had been exposed to carbon tetrachloride, was reported to the Factory Department in 1940. The cause of death was acute dilatation of the stomach with heart failure, a secondary cause being "anaemia haemolytic and dyshaemopoietic". The bone-marrow, post mortem, was found to be hyperplastic.

Blood calcium. Although, as has already been seen, there is a difference of opinion as to whether calcium deficiency can be considered a factor in carbon tetrachloride poisoning, Smyth and Smyth (1936) considered a level below 7 mg. per 100 ml. as possibly significant. Only one of the "high concentration" workers showed a low blood calcium (6·8 mg. per 100 ml.) while 14 of the 88 "low concentration" workers showed levels ranging from 4·6 to 6·9 mg. per 100 ml.

Toxic amblyopia

Toxic amblyopia has been described by several observers, including Wirtschafter (1933), Gocher (1944), Gray (1947) and Smith (1950).

Blurring of vision, pallor of the optic discs and restriction of the visual fields were the chief findings, and in the three cases described by Smith (1950) the final diagnosis was optic atrophy. In two of these cases visual acuity was recovered only to a limited extent.

Restriction of the visual fields is regarded by Wirtschafter and by Smyth, Smyth and Carpenter (1936) as a diagnostic sign of early eye injury due to carbon tetrachloride, though Smyth and Smyth did not find the restriction to be closely correlated with the degree of exposure. In only one instance was a high "peak" exposure (up to 1,000 p.p.m.) associated with a restricted visual field, while of 5 men whose average exposure was 117 p.p.m., 2 had definitely restricted fields and a third had moderate restriction; no restriction was present in 2 men exposed to an average of 111 p.p.m. These findings were not confirmed by Stewart and Witts (1944) in their investigation, which included tests of visual acuity, field of vision and colour vision and of scotomata; they found no evidence of toxic amblyopia or restriction of visual fields.

There appears to be little correlation between eye injury and other symptoms of carbon tetrachloride intoxication. In Wirtschafter's cases, visual effects were present in only 3 out of 5 men who all complained of gastro-intestinal disturbance, while in the 3 cases of optic atrophy described by Smith (1950) only 1 had any sign of associated systemic poisoning.

Skin manifestations

Skin manifestations in the form of dermatitis are mentioned by Davies (1934) and Flandin (1932); the latter relates these specially to the use of dry shampoos; he states that in severe cases necrosis may finally result.

Summary of Symptoms

Table 25 is taken from Smyth and Smyth (1936) and summarizes the symptoms found by them in 96 workmen.

TABLE 25
Summary of symptoms of carbon tetrachloride poisoning

	Men examined	Men with no abnormalities	Men with 1 abnormal test	Men with 2 abnormal tests	Men with 3 abnormal tests	Visual fields	Lymphocytes	Monocytes	Icteric index	Blood calcium	van den Bergh	Urine nitrogen	Urine sulphate	Urine bilirubin
Limit adopted for normal range	—	—	—	—	—	−30°	6,000	600 10%	9	7·0	0·2	35	2·25	0
Exposures averaging above 100 p.p.m. maximum 117 p.p.m.	8	4	4	0	0	2	0	0	0	1	0	1	0	0
Exposures averaging below 100 p.p.m. (5 to 90 p.p.m.)	78	38	28	9	3	7	3	10	18	12	5	0	0	0
Exposures not measured but known not to be extreme	10	5	2	3	0	1	0	1	3	2	1	0	0	0
Totals	96	47	34	12	3	10	3	11	21	15	6	1	0	0

Treatment

Measures of treatment are directed towards combating the various manifestations of poisoning (Drill, Loomis and Belford, 1947).

Hepatic insufficiency. Daily intravenous injections of glucose, protein hydrolysates and vitamin supplements are given (Pearson, 1947).

Acidosis. Sodium bicarbonate is administered in 5 per cent solution (Pearson, 1947).

Disturbed renal function.

(a) 0·25 g. of aminophyllin is injected intravenously twice daily (Pearson, 1947).

(b) Hot packs are applied to the flanks (Pearson, 1947).

(c) Dried blood plasma, equivalent to 500 ml. of whole blood, is dissolved in 300 ml. of sterile water and given intravenously (Melick, 1946).

(d) A modified Tyrode's solution is used for peritoneal lavage with the addition of the sodium salt of heparin and penicillin. This was, however, unsuccessful in the case reported by Snell (1947).

Hypocalcaemia. Calcium gluconate is administered intravenously (Pearson, 1947).

Methionine therapy has proved of debatable value in the treatment of poisoning of human beings by carbon tetrachloride (Snell, 1947), and Drill and Loomis (1946, 1947) have found that animals on a normal or low protein diet show no beneficial effects.

Physical Examinations as Precautions against Poisoning

The tests which Smyth and Smyth (1936) consider most reliable as a source of information, and which would presumably detect injury at the earliest possible date are: determination of icteric index, blood calcium, visual fields, van den Bergh reaction. They recommend that these tests should be carried out at least twice a year on workmen exposed to carbon tetrachloride, and state that the development of one significant finding should call for watchfulness and an early re-examination, preferably one month after the finding. If two significant findings should develop in a worker his exposure should be lessened and he should be examined again in one month. If three significant findings are developed he should be removed from exposure entirely for a month. If at the end of that time he has returned to normal, his exposure may be resumed, under supervision, unless he has been found to be especially susceptible, in which case he should be given a permanent job without exposure. "On the other hand, it is more likely that he will have developed an increased tolerance, the free period having sufficed for the development of regenerated and more resistant parenchymatous cells in the liver."

While emphasizing the need for better design of plant, improved ventilation and better use of masks, Stewart and Witts (1944) suggest that alternation of work with successive weeks in and out of exposure to carbon tetrachloride may increase efficiency if conditions of work cannot be made more ideal.

4. *sym.*-Dichloroethane
(*Ethylene Dichloride*)

$C_2H_4Cl_2$, i.e. $CH_2Cl \cdot CH_2Cl$

sym.-Dichloroethane is to be distinguished from dichloroethylene, with which in the literature it is sometimes confused, and from α-dichloroethane (or ethylidene chloride), its isomer, which is a by-product of the manufacture of chloral hydrate. According to some workers (Kiessling, 1921; Steindorff, 1922), α-dichloroethane is more toxic, but according to Müller (1925), less toxic than *sym.*-dichloroethane.

PROPERTIES

A colourless mobile liquid with an odour like chloroform, an intermediate product in the manufacture of ethylene glycol. Soluble in water to the extent of 0·1 per cent at 20° C.; dissolves 0·17 per cent of water at 20° C. (Durrans, 1950). Slowly hydrolysed by water, but causes only slight corrosion of metals.

B.R. of the commercial product 80–85° C. Sp.Gr. 1·250–1·257. Fl.P. 17° C., but burns only with difficulty.

Solvent for rubber, some resins, benzyl cellulose, benzyl abietate, turpentine, paraffin wax, etc., and partly for soft copals, sandarac, mastic, japan wax, etc.; will not dissolve cellulose nitrate or acetate except after admixture with ethyl alcohol, methyl alcohol, ethyl acetate, etc.

USES

(1) In the separation of mineral oil and paraffin wax. Its stability makes it specially suitable for the extraction of edible oils and medicinal products (Fife and Reid, 1930).

(2) In the chemical industry, in the synthesis of new compounds (Fife and Reid).

(3) As a cleaning fluid often in combination with carbon tetrachloride, and also with trichloroethylene.

(4) As an insecticide and fumigant, especially for cleaning and disinfecting furs (Russ, 1930), also in combination with carbon tetrachloride (Hoyt, 1928 a, b, c). One of the advantages claimed by Hoyt for this mixture, which consists of 3 vols. of dichloroethane and 1 vol. of carbon tetrachloride, is that in an ordinary 8,000 cubic foot room the removal of the fumigant, owing to diffusion or absorption, is so easy that it is possible to enter and remain in the room 24 hours afterwards without discomfort, the odour of the gas being scarcely perceptible.

(5) As a filling liquid for fire extinguishers (Boye, 1935).

(6) As a stimulant for sprouting potatoes (Boye, 1935).

(7) As a constituent of a plasticizing bath.

(8) For coating components of wireless sets in combination with lanoline.

(9) As an intermediate in the manufacture of synthetic rubber.

TOXICITY

Dichloroethane is a powerful narcotic. It was in fact used as an anaesthetic by Simpson, Snow and Nunnely (1848–1849) and is also listed in the Merck's Index (1930) as an anaesthetic, rubefacient and antispasmodic. It is an irritant, and has a toxic action on the heart (von Oettingen, 1935). Severe effects upon the liver like that of tetrachloroethane were recorded by Müller (1925) but are unconfirmed by more recent observers.

The narcosis produced by dichloroethane in animals is about equal to that of chloroform; its haemolytic action, however, is only about half as strong (Kiessling, 1921) and its action on isolated heart muscle is less. It has also a special capacity for producing corneal opacity in a much greater degree than dichloroethylene, and also some degeneration of the liver and kidneys, and oedema of the lungs. Heppel, Neal, Endicott and Porterfield (1944a) consider dichloroethane as one of the more toxic of the halogenated hydrocarbons; to animals it is more acutely toxic than trichloroethylene.

Maximum Permissible Concentration

According to Cook (1945), who based his value on industrial experience, the results of animal experiments and the sensory response of persons exposed to the vapour, the maximum allowable concentration is 100 p.p.m.

TOXIC EFFECTS IN ANIMALS

Lethal and Narcotic Concentrations

Lethal dose

By oral administration. According to Kistler and Luckhardt (1929) it is difficult to produce death by oral administration in dogs, since the animals vomit back quantities larger than 0·5 g. per kg. body weight.

By intravenous injection. The results obtained by Barsoum and Saad (1934) in dogs show a slightly greater toxicity than those of Kistler and Luckhardt. The former give 175 mg. per kg. body weight as the minimum lethal dose (as compared with 90 mg. for chloroform and 60 mg. for tetrachloroethane) while Kistler and Luckhardt found 0·25 ml. per kg. lethal after 24 hours.

By inhalation. The figures given by Lazarew (1929), Pantelitsch (1933) and Sayers, Yant, Waite and Patty (1930) are in close agreement. For mice the two former give the minimum lethal dose as 35 mg. per l. (8,700 p.p.m.) for 2 hours, 100 mg. per l. (24,800 p.p.m.) for ¼ hour, while for guinea-pigs Sayers and co-workers found that 40 mg. per l. (10,000 p.p.m.) for 25 minutes produced death on the following day. Death occurred in a few minutes with 400 to 800 mg. per l. and in ½ hour with 240 mg. per l. Later figures given by Heppel, Neal *et al.* (1944) appear to indicate a higher toxicity than these. They give 3,000 p.p.m. as the fatal single dose, and 1,000 and 400 p.p.m. for repeated exposures.

Narcotic dose

Lazarew (1929) induced light narcosis in mice with a concentration of 15 to 20 mg. per l. (3,700 to 5,000 p.p.m.) and deep narcosis in mice and guinea-pigs with a concentration of 20 mg. per l. (5,000 p.p.m.) and 40 mg. per l. (10,000 p.p.m.) respectively.

Symptoms

Great irritation of the eyes and air passages, vertigo, retching, rapid respiration, ataxia and inco-ordination of the extremities preceded unconsciousness in the guinea-pigs examined by Sayers and co-workers (1930). In dogs given intravenous injections by Kistler and Luckhardt (1929), death was sometimes delayed and was preceded by a period of depression, emaciation and general sick condition. Anaesthesia was often preceded by salivation and excitement, and Steindorff (1922) also described excitement, fall in blood pressure, restlessness and clonic movements in dogs.

The narcotic concentrations and symptoms observed by Sayers and co-workers were summarized by Zernik (1933), and are shown in Table 26.

TABLE 26
Effects from different concentrations of dichloroethane

| Concentration | | | Duration | Effect |
(per cent)	(mg./l.)	(p.p.m.)		
10–20	400–800	100,000–200,000	Few min.	Death
6	240	60,000	30 min.	After 10 min. eye and nose irritation, giddiness, ataxia, convulsions, loss of consciousness; death in 30 min.
1	40	10,000	25 min.	Same symptoms. Death during following day after several inhalations
0·6	24	6,000	30–60 min.	Dangerous
0·35	14	3,500	60 min.	No severe disturbance
0·1	4	1,000	Several hr.	Very slight symptoms

Sayers and co-workers (1930) state the results as follows:—inhalation of 1 per cent in the air for 20 minutes will cause death after a day or two with congestion and oedema of the lungs and degenerative changes in the kidneys. The maximum amount which can be breathed for an hour without producing serious symptoms is 0·35 per cent, and 0·1 per cent will cause slight symptoms after several hours. The narcotic effect is about equal to that of chloroform and carbon tetrachloride, but dichloroethane is less toxic when inhaled for periods of less than an hour.

Effect of Diet

The effect of dietary factors on the toxicity of dichloroethane to animals has been studied by Heppel, Neal, Perrin, Endicott and Porterfield (1945). They found that deficiency of choline and methionine greatly increased susceptibility. Deficiency of protein alone had no effect if adequate provision for methionine and choline were made.

Corneal Opacity

Many of the earlier investigators, including Dubois and Roux (1887), Faranelli (1892), Erdmann (1912), etc., had observed that narcosis with dichloroethane was followed within 24 hours by corneal opacity. This effect was confirmed by Steindorff (1922), by Kistler and Luckhardt (1929) and by Heppel et al. (1944a). Steindorff showed that it was a specific toxic action, not due to a local irritation, by producing it by subcutaneous injection without inhalation. Endothelial necrosis and infiltration of the cornea by lymphocytes and connective tissue cells took place and generally did not disappear for several months. Heppel et al. (1944a) found symmetrical turbidity of the cornea after exposure to 1,000 p.p.m. Repeated exposures, separated by rest periods, produced tolerance to the vapour so that eventually no cloudiness developed.

Post-mortem Appearances

Congestion and oedema of the lungs and congestion and degenerative changes of the kidneys were the chief findings in the guinea-pigs of Sayers and co-workers (1930), while Kistler and Luckhardt (1929) also observed oedema of the lungs, together with engorgement of the blood vessels, liver, mesentery and intestinal lumen, and albumin and bile in the urine. Müller (1925) stated that, besides hyperaemia, the liver and kidneys showed fatty degeneration. No changes in the liver other than severe congestion and cloudy degeneration have been recorded by any other workers. In the animals investigated by Heppel et al. (1945), death was not due to toxic hepatitis; its mechanism was obscure. They found that dichloroethane did not produce the liver lesions characteristic of poisoning by chloroform or by carbon tetrachloride.

Effect on the Isolated Heart

The results obtained by Kiessling (1921) and by Barsoum and Saad (1934) show that, estimated by its effect on the isolated heart, dichloroethane is slightly less toxic than chloroform. Barsoum and Saad found that the lowest concentration which would stop the action of the isolated toad's heart was 1 : 1,250, and the highest which produced no effect 1 : 20,000. Comparable figures for chloroform were 1 : 2,000 and 1 : 100,000, and for tetrachloroethane 1 : 5,000 and 1 : 200,000. Kiessling gave the toxicity relative to that of chloroform as 0·8 : 1.

TOXIC EFFECTS IN MAN

The cases of industrial poisoning reported would appear to show that while both kidney and liver damage may occur after sufficient exposure to dichloroethane vapour, liver necrosis is much less likely to occur than with carbon tetrachloride. The fatal cases reported by Hueper and Smith (1935), Hulst, Steenhauer and Kledden (1946) and by Bloch (1946) were due to the drinking of dichloroethane. The five cases of poisoning reported to the

Factory Department since 1932 and three recorded by Wirtschafter and Schwartz (1939), following exposure to the vapour, were of only moderate severity and of short duration.

Acute Poisoning

By Ingestion

When ingested, as exemplified in a fatal case described by Hueper and Smith (1935), dichloroethane produces many of the clinical and pathological manifestations observed in experimental animals: mydriasis, dizziness, increasing stupor and gradually progressive circulatory failure, shown by cyanosis and rapid pulse, and not yielding to treatment by glucose, adrenaline, oxygen or atropine. The patient died 22 hours after the ingestion of 2 ounces of dichloroethane, and though there was no jaundice, the liver showed fatty degeneration and the kidneys extensive tubular nephrosis resembling that seen in mercuric chloride poisoning. The urine contained large amounts of albumin and sugar and there was an extensive haemorrhagic colitis. Hueper and Smith (1935) suggest that these findings indicate excretion by the kidney of the substance itself or of a decomposition product, possibly oxalic acid. Hulst and co-workers (1946) found pericardial bleeding, haematin in the gastric contents and haemorrhages in the gastro-intestinal mucosa, kidneys and brain. The post-mortem findings in Bloch's case also included diffuse fatty degeneration of the liver.

By Inhalation

The most severe of the five cases reported to the Factory Department in 1945 was that of a man who was exposed to a sudden outflow of dichloroethane from a faulty valve. He suffered from dizziness but had no loss of consciousness at the time; 3 hours later he had severe nausea and vomiting, followed by violent epigastric pain which persisted for 3 days. The other four patients had only slight nausea and headache.

Chronic Poisoning

Two cases reported by McNally and Fostvedt (1941) may be regarded as examples of chronic intoxication by dichloroethane. These men suffered from anorexia, vomiting, epigastric distress, drowsiness, tremors, nervousness and nystagmus. Carozzi (1936) states that when trichloroethylene is mixed with dichloroethane, as was the case in a dry-cleaning establishment mentioned by him, workmen may present symptoms (presumably not exclusively attributable to trichloroethylene) of giddiness, trembling of the hands, disturbances of cardiac activity, somnolence and headache, only improved after changing the solvent. Three cases with a similar, though somewhat more severe symptomatology, were reported by Wirtschafter and Schwartz in 1939. The men were employed in cleaning yarn immersed in an open vat of dichloroethane. The symptoms, arising 4 hours from the beginning of the exposure, included dizziness, nausea, vomiting, weakness, tremors, headache and abdominal cramps, the last being promptly relieved by the intravenous administration of 10 per cent calcium gluconate solution. Hypoglycaemia was present 3 days after exposure and Wirtschafter and Schwartz regarded this as evidence of liver damage, though the injury was apparently insufficient to produce jaundice. There was no clinical evidence of kidney damage or of haemolysis in these cases, but all three showed leucocytosis. Severe dermatitis, probably largely due to a defatting action of the dichloroethane, was present.

5. Tetrachloroethane
(Acetylene Tetrachloride)
$C_2H_2Cl_4$, i.e. $CHCl_2 \cdot CHCl_2$

"Westron", "Alanol", "Emaillet", "Quittnerlack" (Germany), "Novania" (France). This substance was also formerly known as "Tetraline" but the name is no longer used owing to its confusion with "Tetralin" (tetrahydronaphthalene).

PROPERTIES

A colourless mobile liquid with an odour resembling that of carbon tetrachloride and chloroform; non-inflammable. Attacks iron, copper, lead and nickel slightly and aluminium vigorously. In presence of moisture slowly liberates hydrochloric acid.

B.R. of the commercial product 140–150° C. Sp.Gr. 1·600–1·602.

Solvent for cellulose acetate and nitrate, bitumen, waxes, resins, pitch, tar, sulphur, rubber and oils.

USES

Although the use of tetrachloroethane in the aviation industry has become much restricted, in fact forbidden in many countries, because of the toxic effects which it produced, it has a very wide field of application in other industries, chiefly on account of its high solvent power. The principal uses are as follows:

(1) In the rubber industry (apparently to a much smaller degree than the benzol and benzine group or carbon tetrachloride). Würm (1931) does not mention tetrachloroethane among the solvents described as "the most used in the rubber industry".

(2) As a dry-cleaning agent usually mixed with benzol, sometimes under the name "benzinoform".

(3) As a lacquer solvent. Hamilton (1934) states that she knows of no instance where tetrachloroethane has been used for paint coatings, but its use as a lacquer solvent is described by Hofmann and Reid (1929), although it does not appear to be so used now in this country; its special use as a constituent, together with amyl and butyl acetate and benzol, of "zapon-lacquer,"* which is used in the straw hat industry, is described by Ohnesorge (1930).

(4) In the film industry as a celluloid solvent.

(5) In the artificial silk industry particularly in the "lustron" process.

(6) In the artificial pearl industry; in Germany and France this now appears to be the chief application of tetrachloroethane.

(7) In the leather industry especially in the manufacture of shoes when tetrachloroethane is a constituent of the shoe cement; Zollinger (1931) gives the composition of such a cement called "resistin" as follows: alcohol 6 parts, benzol 4, tetrachloroethane 14, acetylcellulose 3, and soda in variable quantities.

(8) As a parasiticide, especially as a hairwash for dogs, with the addition of perfumed substances.

(9) In the estimation of the water content of tobacco and many drugs; Zangger states that this is a very dangerous manipulation and that other solvents should replace tetrachloroethane for this purpose.

---

* "Zapon-lacquer" is a term of German origin meaning a nitrocellulose lacquer. It is used occasionally in this report as a translation of the word "Zaponlack" employed by the German author cited.

(10) In the manufacture of tube cements, adhesives and floor waxes.
(11) In the manufacture of gas masks as a constituent of the adhesive.
(12) For the impregnation of furs and skins; chromium chloride dissolved in tetrachloroethane is used for this purpose (Zollinger, 1931).
(13) In fire extinguishers; according to Flury and Zernik (1931) and Glaser and Frisch (1928), decomposition of tetrachloroethane gives rise to very little phosgene in contrast to carbon tetrachloride.
(14) In the manufacture of safety glass.

TOXICITY

Tetrachloroethane is generally considered on the basis of Lehmann's figures (1911) for monophasic toxicity to be the most toxic of the chlorinated hydrocarbons, exerting not only a narcotic effect, but producing profound metabolic injury, especially atrophy of the liver.

Toxicity in Industrial Processes

It seems unnecessary to recapitulate in detail the accounts of the cases of tetrachloroethane poisoning observed during the 1914–18 war from its use in aeroplane "dope". Between 1914 and 1916, at least seventy cases of jaundice with twelve deaths came to the notice of the Factory Department, and it was established, chiefly from the investigations of Willcox, Spilsbury and Legge (1915), that tetrachloroethane was a most powerful liver poison. Similar cases during those years had occurred in France, Holland and Germany, and towards the end of the war the use of tetrachloroethane in dope was forbidden in those countries as well as in England.

This restriction applied only to the aviation industry, and although tetrachloroethane apparently is not much used now in Great Britain, German and French records indicate that it still has a fairly wide application, especially in the manufacture of artificial pearls. Two cases (one fatal) were described by Fiessenger, Brodin and Wolf in 1923, and one fatal case of acute yellow atrophy of the liver by Boidin, Rouquès and Albot (1930), from its use in this industry. Tetrachloroethane is used as a solvent for a mixture of acetylcellulose and *essence d'orient* in which the pearls are dipped, and the operation involves in some cases actual manipulation of the solvent. Two cases of polyneuritis so caused were recorded by Léri and Breitel (1922).

Twenty-one cases of slight chronic intoxication in the artificial silk industry were described by Parmenter in 1923.

Zollinger (1931) states that the household use of tetrachloroethane as a drycleaning agent is probably much wider than is generally believed, and quotes the case of a servant who suffered from gastric symptoms after using it for domestic clothes cleaning. Describing the use of tetrachloroethane as a constituent of soap preparations, Zangger (1931) observes that such is the ignorance of its properties among manufacturers that they sometimes do not even first make the solutions neutral, but add alkalis direct to the tetrachloroethane.

Zollinger (1931) reported three fatal cases and three of less severity from the use of tetrachloroethane in the manufacture of shoes.

The dangerous properties were emphasized by Kohn-Abrest in 1924, when he stated that concentrations of 2 to 3 mg. per l. (290 to 435 p.p.m.) should be avoided, and that 1 mg. per l. (145 p.p.m.) was detectable by odour. Lehmann and Schmidt-Kehl (1936) found that 0·02 mg. per l. could be detected by odour, 0·09 mg. per l. could be tolerated for 10 minutes with no effect, and that 1 mg.

per l. for 30 minutes or 2·3 mg. per l. for 10 minutes caused irritation of mucous membranes, pressure in the head, vertigo and fatigue.

Relative Toxicity

The relative narcotic toxicity of tetrachloroethane is usually considered as four times that of chloroform, but it will soon be seen from Table 27 of Lehmann's values that this holds only for the "monophasic" toxicity (i.e. in the form of vapour, representing the pure narcotic potency) but not for the "diphasic" (i.e. in the form of condensation in the room, depending on temperature and moisture, and representing the secondary and cumulative effects). The values of Lazarew (1929) also show a difference in the relative toxicity when estimated by narcosis and toxicity respectively produced in white mice after 2 hours' inhalation. Lazarew's values were drawn up in a table by Flury and Zernik (1931) and are shown here in Table 28. It is interesting to note in this connection that animal experiments by Meerssemann, Perrot and Franque (1934), which emphasize the functional effect of chloroform and tetrachloroethane on the liver, as contrasted with the severe lesions inflicted by a poison such as phosphorus, have shown that the congestion and mitochondrial lesions observed in the early stages of chloroform and tetrachloroethane poisoning are less with the latter than with the former.

TABLE 27

Monophasic and diphasic toxicities of chlorinated hydrocarbon solvents

Solvent	Relative toxicity	
	Monophasic	Diphasic
Tetrachloromethane (carbon tetrachloride)	1·0	4·1
Perchloroethylene	1·6	1·0
Trichloroethylene	1·7	3·8
Dichloroethylene	1·7	12·5
Chloroform	2·2	10·8
Pentachloroethane	6·2	1·0
Tetrachloroethane	9·1	1·9

TABLE 28

Chlorinated hydrocarbon solvents arranged in order of increasing toxicity

Solvent	Order producing light narcosis	Solvent	Order of lethal toxicity
Ethyl chloride	1	Ethyl chloride	1
Carbon tetrachloride	2	Ethylidene chloride	2
1 : 1 : 1–Trichloroethane	3	Carbon tetrachloride	3
Ethylidene chloride	} 4	1 : 1 : 1–Trichloroethane	4
Methylene dichloride		1 : 1 : 2–Trichloroethane	5
Trichloroethylene	5	Methylene dichloride	6
Chloroform	6	Trichloroethylene	7
Ethylene dichloride	7	Perchloroethylene	} 8
Perchloroethylene	8	Tetrachloroethane	
1 : 1 : 2–Trichloroethane	9	Chloroform	9
Tetrachloroethane	10	Ethylene dichloride	} 10
Pentachloroethane	11	Pentachloroethane	

Intravenous injection. Barsoum and Saad (1934) judge tetrachloroethane to be the most toxic of the chloroethanes. The minimum lethal dose was 60 mg. per kg. as compared with 90 mg. per kg. for chloroform.

Effect on isolated frog's heart and muscle. From experiments on perfusion of the isolated frog's heart, Barsoum and Saad again find tetrachloroethane considerably more toxic than chloroform, the lowest effective concentration being 1 : 5,000 (chloroform 1 : 2,000) and the highest concentration producing no decided action, 1 : 200,000 (chloroform 1 : 100,000).

Experiments by Veley (1910) on the isolated sartorius muscle of the frog showed that tetrachlorethane was four times more toxic than chloroform, molecule for molecule, but that recovery from anaesthesia or paralysis was more regular in the case of the former than in the latter.

Maximum Permissible Concentration

In the United States, for satisfactory working conditions 10 p.p.m. (0·068 mg. per l. at 25° C.) is recommended as the maximum allowable concentration by Bowditch, Drinker, Drinker, Haggard and Hamilton (1940).

Method of Estimation

A method for rapid estimation of low concentrations of tetrachloroethane in the air of industrial plants has been described by Goldenson and Thomas (1947). The apparatus involves the use of a quartz combustion tube packed with platinum foil. The chlorides liberated by combustion are absorbed in alkaline arsenite solution, and determined volumetrically with 0·01 N hydrochloric acid solutions.

TOXIC EFFECTS IN ANIMALS

Acute Poisoning

Lethal and Narcotic Concentrations

Lethal dose

According to Zernik (1933), the researches of Pantelitsch (1933) and Kummeth (1933) have shown that the lethal and narcotic doses of tetrachloroethane are considerably less than appeared from the earlier work of Lehmann (1911). Lehmann's cats, for example, did not die even with 57 mg. per l., while Pantelitsch gives 30 mg. per l. (4,200 p.p.m.), Lazarew (1929) 40 mg. per l. (5,800 p.p.m.), and Müller (1931) 34 mg. per l. (5,000 p.p.m.) as the fatal dose for mice.

Narcotic dose

10 to 15 mg. per l. (1,450 to 1,800 p.p.m.) after about 2 hours is given as the dose for producing deep narcosis in mice by both Lazarew and Pantelitsch. The dose for light narcosis for mice is stated to be 7·5 to 10 mg. per l. after 20 minutes (Lazarew and Pantelitsch) and for cats, 5·7 mg. per l. after 5¼ hours (Lehmann).

Symptoms

The acute effects of tetrachloroethane inhalations resemble those of the other chlorinated hydrocarbons, preliminary symptoms of irritation of mucous membranes and nerve centres followed by central nervous paralysis.

Narcosis is preceded by lachrymation, salivation and irritation of the nose (Lehmann, 1911). Loss of weight, vomiting and diarrhoea, with blood in the faeces, bilirubinuria and albuminuria were also observed by Grimm, Heffter and Joachimoglu (1914), and their findings have been in all essentials confirmed by Müller (1931). In Müller's animals a characteristic feature was the delay in the onset of fatal symptoms after apparent recovery from inhalations. With parenteral administration the delay did not take place, the animals dying with convulsions a short time after injection of 0·2 ml.

Lesions of the Internal Organs

Fatty infiltration and degeneration of the liver have been the outstanding lesions observed by practically all workers, including Lehmann, Grimm, Heffter and Joachimoglu, and Müller. Müller points out that these changes only occur when the animal has survived acute poisoning and has been re-subjected to it, and that they do not include the reactive inflammation characteristic of the acute yellow atrophy observed in tetrachloroethane poisoning in man. He suggests that this difference may be due to the fact that owing to their different alimentary metabolism, death occurs in animals before the reactive inflammation of acute yellow atrophy has time to develop.

In the experiments of Meerssemann (1934) and of Meerssemann, Perrot and Franque (1934) the liver lesions were examined, both by special staining for mitochondrial lesions and by urinary tests for changes of urea and ammonia excretion. They concluded that the mitochondrial lesions, which were less severe than those produced by chloroform, appeared to run parallel with the severity and permanence of the damage to the ureogenous function. The urinary tests, taken as signs of liver insufficiency, showed that this was not of the same order as that produced by a severe liver poison such as phosphorus. These workers consider that the liver disturbance produced by chloroform and tetrachloroethane is more definitely functional, and more susceptible to regression, when exposure to the poison has ceased. That the hepatic injury is not due to the release of hydrochloric acid has been shown by the animal experiments of Barrett, Cunningham and Johnston (1939). These workers have shown that trichloroethylene is partially converted in the body to trichloroacetic acid. No such conversion, or at least only about one thousandth of the amount, takes place with tetrachloroethane.

Chronic Poisoning

In Lehmann's experiments, inhalations of 1 to 3 mg. per l. (160 to 480 p.p.m.) for 6 to 7 hours daily repeated over a period of 4 weeks were well tolerated, the only sign of injury being loss of weight.

Müller's inhalation experiments cannot be regarded as providing examples of chronic intoxication, since he used concentrations of 80 mg. per l., and the animals did not survive a second exposure. He attempted to imitate the progress of chronic poisoning by giving injections of tetrachloroethane dissolved in paraffin which would only be slowly absorbed. The typical appearance of fatty infiltration and degeneration of the liver was observed, but no symptoms, other than loss of weight, were found. Benzi (1925) also found fatty degeneration of liver, kidneys and spleen as a result of chronic intoxication, and states that susceptibility to poisoning increases with repeated inhalation.

Effect on Blood System

Grimm, Heffter and Joachimoglu (1914) recorded that in a few of their animals after subcutaneous injection there was some anaemia and basophil granulation, and they stated that the haemolytic effect of tetrachloroethane was 7·6 times that of chloroform. In Müller's (1931) and Benzi's (1925) chronic experiments no changes in the red blood cells were observed; Benzi found a diminution of leucocytes which he attributed to their concentration in the deeper capillaries.

Absorption through the Skin

Tetrachloroethane, according to Schwander (1936), can exert its typical toxic effects, both narcotic and metabolic, after absorption through the skin. He demonstrated the vapour in the air expired by a rabbit to whose shaved abdomen tetrachloroethane had been applied. Typical narcosis was produced, and fatty degeneration of the liver and slight degeneration of the kidneys were found post mortem. Tetrachloroethane was the most active of the group of halogens tested, which included trichloroethylene, pentachloroethane and perchloroethylene.

TOXIC EFFECTS IN MAN
Acute Poisoning

Acute intoxication in the form of narcosis leading rapidly to death has apparently not been reported from tetrachloroethane inhalation. An example of its acute narcotic effect when taken internally, with death occurring before liver injury could be produced, is shown in the cases reported by Hepple (1927) and Elliott (1933) when the "silk cleansing fluid" which had been drunk proved to have been tetrachloroethane. Both men became completely unconscious and cyanosed with loss of corneal reflexes and died, one 8 hours later and one 12. Post-mortem examinations showed hyperaemia of the gastric mucosa, slight congestion of the kidneys and acute congestion of the lungs, but no lesion of the liver, except in Hepple's case where there were congestion and cloudy swelling.

One case of severe giddiness, causing the patient to fall twice, from the use of "Aviatol" as a lacquer, was mentioned by Grimm in 1914.

Chronic Poisoning

In many of the fatal cases reported, the course of the disease is so rapid after the first onset of acute symptoms that the term "chronic" seems almost a misnomer. In practically all cases these acute symptoms have been preceded by a period of initial disturbance of an indefinite nature, such as headache, giddiness, palpitation, loss of appetite and nausea. Zollinger (1931) points out that up to a certain point, which he calls the end of the first stage, the fatal and non-fatal cases run a similar course, and that this period may last from 3 days to 3 months. About that time, in the non-fatal cases, the symptoms gradually decrease, while the fatal ones enter a second stage of exacerbation of symptoms, and finally pass into the third stage of coma and death. A fatal case reported to the Home Office in 1929 was that of a female worker making safety-glass where the exposure, described as "relatively short", was of 9 weeks. The composition of the solution used was: tetrachloroethane 30 parts, methylated spirit 15, *iso*propyl alcohol 5, diacetone alcohol in variable quantities.

Symptoms

Most authorities appear to recognize two more or less distinct groups of cases of tetrachloroethane poisoning in which the predominating symptoms are related either to the nervous or to the gastro-intestinal system. Parmenter (1923) considers that both nervous and gastric symptoms might appear together, the gastric representing a more severe manifestation and a more advanced stage of the intoxication. Zollinger (1931), who has made a very extensive study of the subject, inclines to the belief that patients in whom nervous symptoms predominate usually show no liver disturbance, and particularly no icterus. This was not so in a case reported by Grimm *et al.* (1914), or in a case of acute liver atrophy reported by Schibler (1929) which showed also a Babinski reflex, but on the whole Zollinger's opinion does seem to be borne out by most of the accounts in the literature.

Gastro-intestinal or Hepatic Form

First stage (*early pre-jaundice*, Willcox, 1914). This form of intoxication, which is much more frequently observed than the nervous, begins with slight symptoms—loss of appetite, fatigue, headache, usually followed by vomiting, a feeling of pressure in the region of the stomach and liver, and irregular bowels. According to Zollinger, stomach pain may be so prominent as to suggest lead poisoning. Parmenter (1923) shows that the loss of weight during this stage may be as much as 5 to 15 lb. during a period of 2 weeks to 2 months. The difficulty of diagnosis of tetrachloroethane poisoning at this stage is emphasized by Lejeune (1934). He quotes four cases, two of which were eventually fatal and two with symptoms of an indefinite nature, which were at first diagnosed as those of gastro-enteritis and gastric neurosis complicated by jaundice; only the occurrence of other cases of jaundice in the same factory led to the diagnosis of tetrachloroethane poisoning. The most practical and efficient test to determine early toxic effects and progress of the intoxication, according to Wilson and Brimley (1944), is the icterus index estimation. In an examination of 1,000 workers exposed to tetrachloroethane they found 25 with a raised icterus index.

Second stage (*jaundice without toxaemia*). An increase in the severity of the symptoms is usually accompanied by slight jaundice. Zollinger (1931) notes the early appearance of yellow finger nails and constipation with clay coloured stools. In his series of cases, most of the workers at this stage had to cease work through exhaustion. Slight rise in temperature, vomiting, albuminuria and swelling of the legs were also present. In 1933, the Home Office records mentioned two cases at this stage, both showing jaundice, and in one there were pale stools and bile in the urine.

Third stage (*jaundice with toxaemia*). The jaundice increases, the liver becomes enlarged and tender, and toxic symptoms, somnolence, delirium, convulsions and coma appear, leading usually to death.

Fourth stage. Willcox (1914) associates this with disturbances of the portal system (ascites).

In the Home Office (1929) fatal case, the immediate cause of death was haematemesis and epistaxis. Coyer (1944) reported seven cases, one of which was fatal. These cases were all of the gastro-intestinal type, showing severe jaundice, bleeding into the gastro-intestinal tract and ascites.

Lesions of the internal organs

The outstanding lesion found post mortem in fatal cases is that of acute necrosis and atrophy of the liver, but Zollinger also found lesions of the kidney, cholaemic nephrosis and fat phanerosis, haemorrhage into the lungs and serous membranes, and oedema of the intestine and brain. In the Home Office (1929) case the kidneys were swollen, pale, fatty, the cortex jaundiced and haemorrhages were present in the heart muscle, endocardium and sub-pleural regions. In the majority of fatal cases the liver is found to be much reduced in size; that of the Home Office case weighed only 26 oz.

According to the report of Spilsbury in 1914, the jaundice is haemohepatogenous, obstructive in nature due to inflammation of the small intrahepatic bile ducts by the direct action of tetrachloroethane on the liver cells. A toxic fatty degeneration thus develops in the central zones and spreads through the lobules and leads to necrosis. If the patient survives long enough, this process is followed by one of replacement, and death may occur later from the formation of contracting scar tissue. Koelsch (1915) described fatty degeneration of the liver as a secondary development on the primary lesion of the liver cell protoplasm. A later detailed account of the pathology of the liver is that given by Boidin, Rouquès and Albot (1930), in which they record a case of fatal acute yellow atrophy in a female worker engaged in making artificial pearls.

Cutaneous manifestations

Boidin, Rouquès and Albot (1930) describe a vesiculo-papular eruption, apparently due to a superimposed infection, as an outstanding feature in a case where the only other early sign of tetrachloroethane poisoning, with rapidly ensuing death, was jaundice. The occurrence of pruriginous erythematous plaques and of purpura is also mentioned by Desoille and Mélissinos (1933).

Changes in the blood picture

The evidence of changes in the blood picture, as typical particularly of early tetrachloroethane poisoning, rests chiefly upon the findings of Minot and Smith (1921), confirmed later by Parmenter (1923). Willcox (1914, 1916) merely stated that "the jaundice is hepatogenous, not haemolytic", and that "no appreciable anaemia or changes in the red corpuscles occur", while Zollinger (1931) states that no other evidence is forthcoming to confirm Minot and Smith's results. It should be noted, however, that tetrachloroethane has been shown to be 7·6 times as haemolytic as chloroform (Grimm, Heffter and Joachimoglu, 1914), and that in one of their cases anaemia and poikilocytosis were present.

Minot and Smith (1921) and Parmenter (1923), however, insist that the changes which they regard as characteristic—especially the increase in mononuclear cells—are of valuable diagnostic significance in cases of early and slight intoxication. Nearly all of 25 persons examined by Minot and Smith, whose blood showed definite abnormalities, had, or soon developed, some symptoms attributable to poisoning, especially if there was a progressive increase of mononuclear cells, while none of 24 persons showing symptoms failed to show increases of the large mononuclear cells, which often formed over 30 per cent of the leucocytes. The blood changes in a typical case of poisoning were as follows: (1) progressive increase of large mononuclears (up to 40 per cent); (2) progressive increase of young large mononuclears, some formed and some broken; Parmenter (1923) regards the broken forms as specially diagnostic; (3) a somewhat elevated white cell count (up to 15,600); (4) a slight but progressive anaemia; (5) a slight increase of platelets.

Minot and Smith (1921) suggested that routine blood examinations should be carried out on all workers exposed to tetrachloroethane and that an arbitrary figure of more than 12 per cent monocytes should be regarded as indicative of possible poisoning.

A blood picture typical of early poisoning is described by Parmenter (1923) as follows: polymorphs 32 per cent, lymphocytes 5 per cent, large mononuclears 47 per cent, mast cells 1 per cent, broken cells 15 per cent. This picture, in Parmenter's experience, was accompanied by mild symptoms—nervousness, lack of sleep, distaste for food, nausea and malaise. He states that while sudden heavy doses of tetrachloroethane produce symptoms far in advance of the blood changes and far more severe than are justified by the blood picture, the latter ultimately reaches the same degree of severity. He believes that acquired tolerance may be manifested by the absence of severe symptoms and of jaundice combined with the presence of 34 to 40 per cent of large mononuclear cells.

Nervous Form

Léri and Breitel (1922) describe a special polyneuritic syndrome characterized by paralysis of the interossei muscles of the hands and feet, and areas of hypoaesthesia. Feil and Heim de Balsac (1924) also observed nervous symptoms without definite polyneuritis in 8 out of 11 workers in the artificial pearl industry, the chief signs being paralysis of the palate and dilated pupils; these signs were also observed by Léri and Breitel.

Zollinger (1931) states that the nervous syndrome of tetrachloroethane poisoning usually begins with numbness and tingling of the fingers and toes, tremor of the hands and prickings in the soles of the feet, and in some instances, abnormal perspiration and twitchings of the facial muscles. Disturbances of the motor and sensory reflexes of the upper, and especially of the lower, extremities, were recorded in a case by Koelsch in 1915. These symptoms have been specially noted among workers in the artificial pearl industry, where they are exposed to tetrachloroethane, not only by inhaling the vapour but also by manipulating it. Léri and Breitel have tentatively suggested that the symptoms may be due to cutaneous absorption since in their series of cases many of the patients who were exposed to the vapour suffered only from giddiness, whereas the two who showed polyneuritis developed it a month after manipulating the tetrachloroethane. Zollinger, however, suggests that these symptoms may not have been due to tetrachloroethane alone, but to its action in combination with some other substance, probably arsenic.

An organic injury of the central nervous system, ascribed by Schultze (1920) to the use of tetrachloroethane in a lacquer solution, is also regarded by Zollinger as an example of a combined toxic action, probably that of trichloroethylene. Schultze's patient had paralysis due to a lesion of the pyramidal and cerebellar tracts, loss of taste and melaenic diarrhoea. He believed this to be a case of encephalomyelomalacia. Another example of a mixed intoxication in which tetrachloroethane is believed to have played the chief part is Lutz's (1931) description of a worker in a chemical laboratory who was accustomed to measure various solvents by means of a pipette; the chief symptoms were weakness, loss of weight, thirst and a pain in the tongue; post-mortem examination showed atrophy of all the internal organs and severe liver injury, characteristic of tetrachloroethane poisoning; sections of the tongue showed degenerative nervous lesions.

Prophylactic Measures and Treatment

The measures to be taken at the various stages of poisoning were outlined by Willcox in 1916. Prophylactic measures include removal from exposure on the earliest appearance of slight symptoms of anorexia, general malaise, headache, constipation, drowsiness and occasional vomiting. Zollinger also emphasizes the desirability of closed apparatus for cleaning purposes, adequate ventilation, warnings to the workers of the toxicity of the substance, and against eating food in the rooms where cleaning processes are carried out, periodical medical examinations, and protective masks when the work is of long duration. Young people, women, and all persons with constitutional weaknesses, including the feeble-minded, the anaemic, the obese and alcoholics, should be forbidden to work with tetrachoroethane. Special precautions in the artificial pearl industry in the form of glass cages with openings for the arms of the worker manipulating the pearls are suggested by Desoille and Mélissinos (1933). The wearing of a commercial type of organic vapour canister containing activated charcoal is recommended by Goldenson and Thomas (1947). Tests made by them have shown that a good gas mask canister should provide almost permanent protection against the concentrations usually found in and around plants where tetrachoroethane is used.

Injections of liver extract in daily doses corresponding to 60 g. or more of fresh liver are recommended by Desoille and Mélissinos, and were considered of value in the six non-fatal cases recorded by Coyer (1944).

6. Pentachloroethane
(Pentaline)
C_2HCl_5, i.e. $CHCl_2 \cdot CCl_3$

PROPERTIES

A colourless, non-inflammable liquid with a camphor-like smell. B.P. 162° C. Sp.Gr. of the commercial product 1·685–1·709. A good solvent for cellulose acetate and some resins.

USES

Pentachloroethane is used, though to a much smaller extent, in industries similar to those employing tetrachloroethane (p. 154). Its use as a dry-cleaning agent, especially in France and Switzerland, in small apparatuses allowing the cleaning of 6 to 8 kg. of clothes, is mentioned by Rochaix (1936), and its danger emphasized.

TOXICITY

The exact degree of acute toxicity of pentachloroethane is much disputed; some workers regard it as more and some as less toxic than tetrachloroethane. It must at any rate be regarded as a narcotic and metabolic poison of considerable potency, exerting toxic effects similar in nature to those of tetrachloroethane.

TOXIC EFFECTS IN ANIMALS
Acute Poisoning

Lethal and Narcotic Concentrations

Lethal dose

By inhalation. Lazarew (1929) found pentachloroethane slightly more toxic than either tetrachloroethane or chloroform, the minimum lethal dose for mice being 35 mg. per l. (4,260 p.p.m.) as compared with 40 mg. (5,800 p.p.m.) for tetrachloroethane and 30 to 40 mg. (6,150 to 8,200 p.p.m.) for chloroform.

By intravenous, subcutaneous or oral routes. Pentachloroethane appears to be the least toxic of the three solvents according to the results of Barsoum and Saad (1934). The minimum lethal dose for dogs by intravenous injection, for example, was 100 mg. per kg. body weight, while for tetrachloroethane and chloroform it was 60 and 90 mg. respectively.

Narcotic dose

The actual concentrations of pentachloroethane necessary to produce narcosis agree fairly well as reported by Lehmann (1911) and by Lazarew (1929) but Lehmann's results as to the comparative potency of pentachloroethane are complicated by his criteria of monophasic or diphasic toxicity. For monophasic toxicity, i.e. inhalation of the pure vapour in a small closed space, he found the ratio between chloroform, pentachloroethane and tetrachloroethane to be 2·2 : 6·2 : 9·1; for diphasic toxicity, i.e. inhalation of the free mixture of the vapour with air in a room, and depending upon volatility and temperature, pentachloroethane was the least toxic of the three (pentachloroethane 1, tetrachloroethane 1·9, chloroform 10·8).

Lazarew, on the other hand, finds pentachlorethane the most toxic for the onset of light narcosis, and slightly less toxic for deep narcosis (Table 29).

TABLE 29

Toxic concentrations of pentachloroethane

Solvent	Concentration producing light narcosis		Concentration producing deep narcosis	
	(mg./l.)	(p.p.m.)	(mg./l.)	(p.p.m.)
Pentachloroethane	7·5	900	25	3,000
Tetrachloroethane	7·5–10	1,100–1,450	10–15	1,450–2,000
Chloroform	20	4,100	20	4,100

From Lehmann's point of view, however, the danger of pentachloroethane, represented by its narcotic potency when used in well-ventilated rooms, is small, about half as great as that of tetrachloroethane and one tenth to one twelfth as great as that of chloroform. This relative lack of toxicity is explained by the relatively low volatility of pentachloroethane at a constant temperature as shown by the following figures for the evaporation in 20 minutes at 25° C.: pentachloroethane, 3·1 g.; tetrachloroethane, 3·8 g.; chloroform, 146 g.; or, expressed as a ratio, 1: 1·2: 47.

Joachimoglu (1921a) found that in aqueous solution pentachloroethane had greater narcotic potency for fishes than either tetrachloroethane or chloroform, the ratio being: chloroform : tetrachloroethane : pentachloroethane, 1 : 13·1 : 20.

Effect on the Mucous Membranes

Lehmann (1911) found the vapours of pentachloroethane strongly irritant to the mucous membranes of cats and Joachimoglu (1921a) to those of dogs. Joachimoglu observed a purulent secretion from the nose during life, and purulent secretion in the trachea and bronchi with acute congestion of the lungs after death.

Effect on the Internal Organs

In Joachimoglu's dogs, the inhalations were repeated daily over periods of from 10 to 20 days, but the effects must be regarded as acute since quantities sufficient to produce narcosis were given. Injury to the kidneys was evidenced by the presence of albuminuria and haematuria after four or five inhalations, while the livers were enlarged and yellowish to the naked eye and showed marked fatty degeneration.

Effect on the Isolated Frog's Heart

While the results of both Kiessling (1921) and Barsoum and Saad (1934) show that pentachloroethane has a powerful toxic action on the isolated frog's heart, these workers differ as to its relative toxicity when compared with tetrachloroethane and chloroform. Kiessling found it considerably more toxic than either, the heart's action being brought to a standstill by a concentration of 0·1 ml. per l., as compared with 1·25 ml. for tetrachloroethane and 5·0 ml. for chloroform, while Barsoum and Saad found it about the same as chloroform (1 in 2,000) and less than tetrachloroethane (1 in 5,000).

Chronic Poisoning

The investigation of Lehmann (1911) showed that pentachloroethane given in small doses over a long period, while producing practically no symptoms, yet acts both as an irritant to respiratory mucous membranes and as a liver poison. Cats tolerated well exposures to an average of 1 mg. per l. (120 p.p.m.) for 8 to 9 hours daily for 20 days and no symptoms other than drowsiness and fatigue were produced with no loss of weight. Post-mortem examination, however, showed chronic bronchitis or purulent pneumonia and fatty degeneration of the liver.

Estimation in Animal Tissues

Pentachloroethane is amongst the group of substances, including chloroform and trichloroethylene, stated by Brüning and Schnetka (1933) to give a positive reaction with pyridine and caustic soda, as described by Fujiwara (1914), Cole (1926) and others. Brüning and Schnetka describe special modifications of the method with regard to temperature and strength of the caustic soda solution, and claim great sensitivity and accuracy for their technique.

TOXIC EFFECTS IN MAN

No toxic effects of pentachloroethane in man have been reported, partly perhaps because of its low volatility.

7. Dichloroethylene
(Acetylene Dichloride, Dielene, Dioform)
$C_2H_2Cl_2$, i.e. CHCl:CHCl

PROPERTIES

A colourless liquid with a slight chloroform-like odour, consisting of a mixture of two stereo-isomers, *cis*- and *trans*-dichloroethylene. One of these (*trans*-, constituting about 40 per cent of the mixture) boils at 48° C., the other (*cis*-) at 60° C. Sp.Gr. of the mixture is 1·25–1·28 at 15° C. Fl.P. 2° C.

The hot vapour can be ignited, but burns with a cold flame which will extinguish itself; in practice there is no danger from fire.

The physical and chemical properties (surface tension, stability, volatility, lipoid-solvent capacity, etc.) of the two isomers are very similar (Zangger, 1930). From the results of animal experiments, the *trans*-isomer would appear to be the more toxic and irritative.

It is a solvent for cellulose acetate, rubber, oils, shellac, waxes and resins. It is insoluble in water, but slowly liberates hydrochloric acid when in contact with water. It has no action on metals when anhydrous.

USES

(1) As a dry-cleaning agent.

(2) In the rubber industry, particularly as an adhesive in the shoe industry, because it is a solvent for crêpe rubber.

(3) In the perfume industry, as an extractive for delicate perfumes.

(4) As a surgical cleansing agent, as in iodine solutions for the skin.

(5) As an insecticide; for this purpose according to Beck and Susstrunk (1931) it is much less toxic than ethylene oxide.

(6) The *trans*-form is used as a refrigerating agent, according to Flury and Zernik (1931).

(7) In the artificial silk industry, as a solvent for cellulose acetate.

(8) In the dyeing industry.

TOXICITY

It is generally agreed that dichloroethylene is a narcotic, but opinions are divided as to whether it resembles other members of the chlorinated aliphatic hydrocarbons in exerting a toxic metabolic effect. It apparently has a special irritative effect on the cornea of animals. Except for one fatal case quoted by Hamilton (1934) no instance of acute poisoning has been reported from its use in industry. Two complaints of digestive disorder were investigated by the Home Office in 1930 and 1934, but neither was considered definitely due to dichloroethylene.

TOXIC EFFECTS IN ANIMALS

In general, from the investigations of Lehmann (1911) and others, the toxicity of dichloroethylene as a narcotic is less than that of chloroform, while its toxicity compared with that of other solvents depends upon whether it is estimated from the "monophasic" or "diphasic" standpoint. Lehmann regards the "diphasic" toxicity, dependent on volatility and temperature, as expressing its toxicity in ordinary use, while the "monophasic" represents the action of the pure vapour. He gives the following figures of relative toxicity: trichloroethylene: dichloroethylene : chloroform, $1.7 : 1.7 : 2.2$ (monophasic) and $3.8 : 12.5 : 10.8$ (diphasic). He states further that for practical use dichloroethylene and chloroform are ten to twelve times as toxic as pentachloroethane and perchloroethylene and three times as toxic as trichloroethylene and carbon tetrachloride. This estimate apparently does not hold good if dichloroethylene is viewed as a metabolic poison. Lewin (1920) and Joachimoglu (1921) reported toxic effects on the liver and kidneys from repeated administration, but these effects were apparently relatively mild. Other observers, notably Wittgenstein (1918) and Beck and Susstrunk (1931) failed to find any lesions of the internal organs, apart from congestion, which were severe enough to cause death in acute intoxication.

Acute Poisoning

Lethal and Narcotic Concentrations

Lethal dose

By inhalation. Lehmann (1911) obtained inconstant results in cats, one animal dying after exposure to 50 mg. per l. (12,500 p.p.m.), another recovering after 218 mg. per l. (55,000 p.p.m.). For mice and guinea-pigs, Wittgenstein (1918) found the lethal doses to be 76 and 155 mg. per l. (19,000 and 39,000 p.p.m.) respectively. Schoenes (1931) was unable to kill cats even with concentrations up to 230 mg. per l.

By subcutaneous injection. Barsoum and Saad (1934) found the lethal dose for dogs to be 225 mg. per kg. body weight as compared with 150 mg. for trichloroethylene.

Narcotic dose

Authors are in closer agreement, both as to the actual concentration of dichloroethylene necessary to produce narcosis, and as to its action relative to other narcotics. Thus about 40 mg. per l. (10,000 p.p.m.) is given by Wittgenstein (1918), Müller (1925) and Meyer and Gottlieb-Billroth (1921) as the lowest concentration which will produce light narcosis in $\frac{1}{2}$ to $\frac{3}{4}$ hour, while Lehmann's figures for cats are 72 mg. per l. (18,000 p.p.m.) for $1\frac{1}{2}$ hours for light narcosis and $2\frac{3}{4}$ hours for deep narcosis. A distinction between the narcotic potency of *cis*- and *trans*-dichloroethylene was made by Beck and Susstrunk (1931), who found the *trans*-isomer to be the stronger. The figures in Table 30 are those of the later investigator (Schoenes, 1931, quoted by Zernik, 1933) for cats.

TABLE 30

Narcotic concentrations of dichloroethylene

Concentration		Duration	Time before effect		
(mg./l.)	(p.p.m.)		side position (min.)	light narcosis (min.)	deep narcosis (min.)
51·2	13,000	4 hr. 10 min.	246	—	—
76·8	19,500	6 ,, 20 ,,	40	185	380
102·4	26,000	2 ,, 55 ,,	10–13	80–100	175
128·0	32,000	1 ,, 15 ,,	10	30–47	68
204·8	52,000	45 ,,	3–4	20	45
230·4	58,500	15 ,,	7	14	15

Symptoms

Dichloroethylene apparently produces central nervous irritation as well as narcosis. Restlessness, twitching and clonic convulsions were observed by Wittgenstein (1918), Steindorff (1922), Schoenes (1931) and Beck and Susstrunk (1931), the latter finding them much more marked with the *trans*- than with the *cis*- isomer.

Rapid recovery on removal from exposure is characteristic of dichloroethylene intoxication. Local irritation of mucous membranes of the nose and the eyes has also been a characteristic finding. Corneal injury is stated by Steindorff (1922) to be a special feature, though not so severe as that produced by ethylene dichloride. Corneal opacity preceded by lachrymation was an invariable sequel of narcosis, but recovery began 24 hours after removal from exposure and was complete after a further 24 hours.

Effect on the Internal Organs

No lesions of the liver and kidneys, in the sense of fatty degeneration, have been observed in acute poisoning by dichloroethylene. Beck and Susstrunk (1931) observed acute congestion of all the internal organs and punctate haemorrhages in the lungs which they were doubtful whether to attribute to the direct toxic action of dichloroethylene or to obstruction. Wittgenstein found no change in the heart or blood vessels; the blood pressure was not influenced and no albuminuria was present after narcosis lasting for 2 hours.

Effect on the Isolated Frog's Heart

According to Barsoum and Saad (1934), 1 : 500 is the lowest concentration of dichloroethylene which will arrest the action of the isolated heart and 1 : 16,000 the highest which has no effect. So judged, dichloroethylene is less toxic than trichloroethylene for which the corresponding figures are 1 : 2,000 and 1 : 32,000.

Chronic Poisoning

The only experiments with repeated inhalation recorded are those of Lewin (1920) and of Joachimoglu (1921a, c). The latter observed loss of weight, corneal inflammation and the presence of bile pigment in the urine in dogs which inhaled 10 ml. on successive days. Post mortem he found fatty degeneration of the liver and congestion of the intestinal mucous membrane. Joachimoglu believes that dichloroethylene is a stronger metabolic poison than perchloroethylene. Lewin (1920), quoting experiments carried out by him in 1912, states that animals subjected to the inhalations of the vapour of the chemically pure substance repeated on consecutive days showed much fat in the liver and also in the kidneys.

TOXIC EFFECTS IN MAN

Acute Poisoning

The only fatal case from the industrial use of dichloroethylene is that quoted by Hamilton (1934) of a man who entered a vat containing rubber dissolved in dichloroethylene and who was found dead the next morning. A slight case of "gassing" from the use of a paint remover which contained both dichloroethylene and methylene dichloride was reported to the Factory Department in 1944. A female worker complained of faintness and a choking feeling, and later of continued retching which lasted for 4 weeks.

Two fatal cases from the attempted use of dichloroethylene as an anaesthetic have been quoted by Fraenkel (1919), but an extensive use of a mixture of "dichloren" (the *cis*-isomer of dichloroethylene) and ether by Albrecht (1927) seems to show that dichloroethylene is a much less dangerous narcotic than chloroform. Albrecht records 2,000 cases of "dichloren"-ether anaesthesia, only two cases being fatal, and in both of these the patients suffered from such severe pathological conditions that death could not be attributed to the "dichloren" itself. Albrecht found that "dichloren" produced an initial stage of muscle twitching, but states that this is transient and harmless. It had no effect on blood pressure or upon the internal organs, as judged from function tests of the liver and kidneys.

Chronic Poisoning

No cases of chronic poisoning definitely attributable to dichloroethylene have been recorded, though two complaints of workers using it were investigated by the Home Office in 1930 and 1934. The first patient was the manager of a factory making an adhesive crêpe rubber in which dichloroethylene was the solvent, and the symptoms complained of were "dyspeptic". As a result of the investigation these were stated to be "partly due to dichloroethylene, partly to worry and lack of exercise". In the second case, where dichloroethylene was being used as a solvent for cellulose acetate in the plasticising of celluloid, and where the smell was perceptible in the room, a woman complained of sickness. Later, her own doctor modified his original view that the sickness was due to her work, and no other cases occurred in the same room.

8. Trichloroethylene
(Chlorylene, Trilene)

C_2HCl_3, i.e. $CHCl:CCl_2$

CONSTITUTION AND PROPERTIES

Trichloroethylene, an unsaturated derivative of ethylene, belongs to the group of chlorinated hydrocarbons of the aliphatic series of which chloroform is the best known representative.

It is a colourless liquid with a not unpleasant smell. Small quantities of stabilizers are sometimes added to prevent the slight decomposition which may take place from long storage or exposure to light. Two qualities are accordingly available commercially: viz. technical and stabilized.

British Standard Specification, No. 580 (1934). B.R. (technical) 86·2–87·2° C. 95 per cent minimum; (stabilized) 86–88° C. 95 per cent minimum. Sp.Gr. 1·469–1·475. Residue 0·01 per cent maximum. Acidity as HCl 0·002 per cent maximum. Free from free chlorine.

It is insoluble in water, but mixes freely with alcohol, benzene, acetone and other solvents. Addition of water causes it to separate from alcohol or acetone mixtures, and therefore its use in place of benzene has been advocated for the preparation of absolute alcohol by distillation. It vaporizes easily at ordinary temperatures, is relatively stable in air and non-inflammable, but under certain physical and chemical conditions, which include exposure to strong light and the presence of effective catalysts (e.g. aluminium dust), it may decompose with formation of hydrochloric acid. In the presence of alkalis it may give rise to dichloroacetylene and in contact with a naked flame, phosgene.

USES

(1) In the metal industry; the degreasing of metal parts in metalware factories, motor, machine and railway works, electrical works, etc. constitutes the chief industrial application.

(2) In painting and enamelling; trichloroethylene is a good solvent for tar, pitch, etc., but less so for acetyl- and nitro-cellulose. It is, therefore, not much used as a solvent for lacquer, but occasionally for coloured paints, housepainting, the painting of casks and vats, furniture stains, etc.

(3) In dyeing and dry-cleaning in place of petrol and benzene since it is noninflammable, readily recoverable and there is no excessive loss in use. In the hat industry the solvent property of trichloroethylene is applied to the removal of marks from sheep wool and felt. It is also a constituent of many cleansing soap solutions.

(4) Sometimes in place of carbon disulphide for the extraction of oil from corn, grape seeds, olives, etc., since it is more readily volatilized and has less smell and taste; also in extraction from wool, leather, hide, bones, fish (fish-meal factories); in the recovery of fat-free glue from residues of glue-boilers in tanneries; in the recovery of wax and paraffin from refuse; in the extraction of fat from feeding-stuffs.

(5) In the boot and shoe industry mixed with rubber cement and used as an adhesive for crêpe soles.

(6) In textile manufactures and the printing industry (cylinders and type); a limited use.

(7) In the disinfection of hides and skins from anthrax; as a hairwash for animals, but rarely for human beings; as an egg preservative; for surgical

sterilization of the hands (Winkelbauer and Musger, 1933). It is pointed out by Rimpau (1931) that pure trichloroethylene is not a disinfecting agent, but that in watery emulsion and with phenol derivatives it becomes strongly bactericidal.

(8) For impregnating and dressing purposes in the artificial silk industry, impregnation of leather and cardboard, etc.

(9) Cleaning of films and photographic plates.

(10) Cleaning of optical lenses polished with fat-containing grinding pastes.

(11) In the chemical industry for the extraction of phenol from waste water, extraction of resin from wood, production of pulverized sulphur, etc. A large number of patents exist for its use in chemical processes.

(12) In gas purification as a solvent of sulphur and phosphorus.

(13) As a rubber solvent; a limited use.

(14) As an anaesthetic. Trichloroethylene has been widely used as a general anaesthetic during recent years; it is generally considered safe and convenient, especially as a substitute for nitrous oxide and air, in midwifery, dentistry, etc. (Hewer, 1948); most of the undesirable features such as convulsions (Culbert, 1942; Garland, 1942; Condor, 1948) and cranial palsies (Humphrey and McClelland, 1944; McAuley, 1943) are believed to occur only when trichloroethylene is used in a closed circuit, and to be due to toxic oxidation products formed by the interaction of soda lime and trichloroethylene (Carden, 1944).

In industrial processes trichloroethylene may be used pure or as an addition to other solutions under various proprietary names, e.g. "Westrosol", "Trielin (e)", "Benzinol", "Tetraline" (for dry-cleaning), "Triklone", "Triol", "Tripur", "Vestrol" (soaps containing it), "Petazinol-Pfannhauser" (in galvanising), "Lanadin-Lap" (in felt hat industry), "Lithuin E" (waterproof material for brickwork), "Dukeron" (for painting tanks, etc.), "Crawshawpol" (for degreasing). In America trade names for trichloroethylene are many and varied, in many cases giving no hint of their actual connection with it, e.g. "Blanco-solv"; "Cocolene"; "Cinco-solv"; "Flock-flip"; "Lith Lithurin"; "Pern-a-chlor".

TOXICITY

Trichloroethylene is an acute narcotic and causes death from respiratory failure if exposure is severe and prolonged. There appears to be little evidence of a chronic or cumulative effect, but a few cases have occurred where there has been a latent period between exposure to trichloroethylene and sudden death after exertion.

Comparison with other Solvents

As a narcotic trichloroethylene lies between chloroform and carbon tetrachloride, their relative efficacies being shown by the following ratio, carbon tetrachloride : trichloroethylene : chloroform 1 : 1·7 : 2·2 (Lehmann, 1911). Lazarew (1929) places their toxicity in the same order. An interesting comparison between dichloroethylene, trichloroethylene and perchloroethylene has been made by Génévois (1936), who based his conclusions not only on the amounts required to induce narcosis, but also on the two factors, speed of narcosis and interval between the narcotic and lethal dose. From this point of view trichloroethylene is regarded as being slightly more toxic than dichloroethylene but considerably less toxic than perchloroethylene, in contrast to the values given by earlier workers (Lehmann, 1911; Joachimoglu, 1925). Lehmann gave the ratio for "diphasic" toxicity as perchloroethylene : trichloroethylene : dichloroethylene

1 : 3·8 : 12·5. Génévois points out that narcosis takes place more rapidly as the vapour tension decreases (Table 31).

TABLE 31

Relation between vapour tension and narcotic concentration

	Vapour tension at 20° C. (mm. Hg)	Time necessary for narcosis (min.)	Concentration (mg./l.)
Dichloroethylene ..	200	73	40
Trichloroethylene ..	56·5	32	40
Perchloroethylene ..	17	11	40

He shows that with a concentration double the narcotic dose perchloroethylene may be fatal, while trichloroethylene is fatal only with 2½ times, and dichloroethylene 3 times the narcotic dose. He considers that trichloroethylene is the most suitable of these three substances as a solvent and a degreaser, dichloroethylene being too volatile and perchloroethylene too rapidly narcotic and toxic. From the point of view of chronic exposure, carbon tetrachloride is considerably more toxic than trichloroethylene (Barrett, Maclean and Cunningham, 1938).

Toxicity in Industrial Processes

One of the chief applications of trichloroethylene lies in the degreasing of metal parts, and it is from such processes that the largest number of cases of poisoning have been reported. The greatest danger appears to occur from the cleaning of degreasing vats, danger from exposure to the fumes during the actual process of immersion of articles in the fluid having been to some extent overcome in apparatus of modern design. In Germany, apparently, such danger is more probable than in England; for, according to Stüber (1931), "open apparatus" is in more general use than the "half-closed" or "closed". Although in the open apparatus a cooling device is provided which condenses the vapour and diminishes the amount of it entering the room, the workers are exposed in other ways, e.g. bending over the vessels to make sure that the articles are dry, taking them out while they are still moist, etc. A type of container widely used in England—the new "A" type of I.C.I.—obviates some of these dangers.

Witheridge and Walworth (1940) compared various methods of ventilating trichloroethylene degreasers. Of the four systems investigated, the vertical slot, the horizontal slot, round holes and elongated holes, it was found that in the operatives' breathing zone, either the horizontal slots or the round holes reduced the average concentrations of trichloroethylene of 400 p.p.m. without ventilation to 100 p.p.m. with ventilation. The vertical slot system required a ventilation rate of 1,400 cubic feet per minute to achieve the same result.

The use of a paint called "Dukeron", containing trichloroethylene, was responsible for several of the cases reported in the Annual Reports of the Chief Inspector of Factories and Workshops from 1921 onwards.

In connection with the extraction of fats it should be mentioned that "Dürener's disease", occurring when soya bean extracted with trichloroethylene was used as a food-stuff, and thought to be directly due to its toxic action, has been shown by Bleyer and Mayer (1927) to be an avitaminosis due to the loss of fat-soluble vitamins.

It is from the metal industry that the majority of cases of poisoning are now reported. Of the 284 cases reported by Stüber (1931), for example, 104 were listed as due to "cleaning in Tri apparatus", while of the 39 cases reported to the Home Office between 1921 and 1935, 16 were in connection with the degreasing of machinery and cleaning of degreasing tanks.

ESTIMATION

In Air

According to Sappington (1935) the amount of trichloroethylene in air may be estimated by a modification of the method suggested by Patty, Schrenk and Yant (1932) for methyl chloride, the principle of which is to mix the organic halide in the air with natural gas and burn it in a micro-burner. By contact with ammonium carbonate, chloride is formed and is determined by the Volhard method. For determining the total weight per litre, Sappington states that an activated charcoal absorption apparatus devised by Cook and Coleman is very satisfactory. Another method is described by Barrett and Johnston (1939).

In the Blood

A method for the estimation of trichloroethylene in blood, by extraction with toluene and treatment of an aliquot of the extract with pyridine and alkali, has been described by Habgood and Powell (1945). They found that a substance, probably trichloroacetic acid, developed in the blood 24 hours later, when the blood concentration of trichloroethylene itself had fallen from 6·5–12·5 mg. per ml. to 0·1 mg. after half an hour. The estimation of trichloroacetic acid in the urine has been described by Frank and Westendorp (1950).

TOXIC EFFECTS IN ANIMALS

The outstanding feature of the results of animal experiments with trichloroethylene is the absence of any severe tissue damage or lesion of the liver and kidneys, such as is found with other halogenated hydrocarbons, like tetrachloroethane. Only Castellino (1932) appears to have recorded slight fatty degeneration of the liver and slight granular degeneration of the kidneys in chronic inhalation experiments, while Barsoum and Saad (1934) found fatty degeneration of the liver in the dogs and rabbits receiving lethal doses of trichloroethylene by intravenous and subcutaneous injection. Disturbance of liver function, following repeated inhalation of trichloroethylene, has been postulated by Seifter (1944), together with evidence of depletion of glycogen and hydropic parenchymatous degeneration.

Acute Poisoning

Lethal and Narcotic Concentrations

Lethal dose

By subcutaneous and intravenous injection. According to Barsoum and Saad (1934) trichloroethylene is considerably less toxic than chloroform, the minimum lethal dose by intravenous injection into dogs being 150 mg. per kg. compared with 90 mg. for chloroform. Those animals which died 2 days or more after receiving the drug are stated by these workers to have shown fatty degeneration of the liver to practically the same degree as animals dying from the effects of the other halogenated compounds tested.

By inhalation. The lethal concentration varies with different animals from 40 to 45 mg. per l. (7,400 to 8,300 p.p.m.) after 2 hours for the mouse (Lazarew, 1929) to 146 mg. per l. (27,000 p.p.m.) for the guinea-pig (Carrieu, 1929).

Narcotic dose

The narcotic concentration is approximately 20 to 25 mg. per l. (3,700 to 4,600 p.p.m.) (Lazarew, 1929; Flury and Zernik, 1931; Lehmann, 1911, etc.). Taylor's (1936) values agree closely with these; he found that 0·2 per cent produced only slight drowsiness, and 0·4 and 0·5 per cent deep narcosis. McCord (1932) found that the threshold of danger for rabbits was 500 p.p.m.

Symptoms

Trichloroethylene apparently produces narcosis in animals without any profound effect on the respiratory or vascular system (Taylor, 1936) and recovery from a sublethal dose is rapid and complete (Lehmann, 1911, etc.).

The interval between the narcotic and the lethal dose is comparatively wide— $2\frac{1}{2}$ times the narcotic dose to produce death (Génévois, 1936). It is pointed out by Jackson (1934) that trichloroethylene is not very volatile, and that a little goes a long way, so that an overdose may easily be given, and that animals in which respiratory failure occurs are difficult to revive. He observed a striking development of tolerance to trichloroethylene in animals anaesthetized by it for several hours.

Lesions of the Internal Organs

Practically all workers, including Herzberg (1934), Taylor (1936), Castellino (1932), etc., but with the exception of Meyer (1929), are agreed that acute poisoning by trichloroethylene has no specific effect upon the liver or kidneys. They have usually found only acute congestion of all internal organs in animals. Herzberg (1934) did observe fatty infiltration of the liver in two of his dogs subjected to prolonged anaesthesia and vacuolation of the liver cells in two others, but in view of the fact that similar changes, though not so great, have been found in the livers of animals not exposed to trichloroethylene, he does not consider that they can be regarded as due to trichloroethylene poisoning. Meyer (1929), however, has recorded severe fatty infiltration of the liver in dogs repeatedly narcotized and subcutaneously injected; some of the animals also showed fatty infiltration of the tubular epithelium of the kidneys. Carrieu (1929) and Nebuloni (1930) also noted inflammation of the tracheal mucosa.

Effect on Smooth Muscle and Isolated Frog's Heart

Barsoum and Saad (1934) found trichloroethylene slightly less toxic to the isolated toad's heart than chloroform; though both arrested the heart's action in concentrations of 1 in 2,000, the strongest concentration that had no decided action was 1 in 32,000 for trichloroethylene as compared with 1 in 100,000 for chloroform. The experiments of Kiessling (1921) and of Fühner (1921) appear to indicate an irritative effect on the isolated frog's heart, as do those of Joachimoglu (1925) on the smooth muscle of the leech.

Haemolytic Effect

Plötz (1920) found no haemolytic effect.

Chronic Poisoning

Symptoms

In Lehmann's (1911) original experiments on cats the only effect of inhalation of 4·1 mg. per l. (760 p.p.m.) for 6 hours a day over a period of 10 to 17 days was drowsiness, fatigue and loss of weight, leading in some cases to death. Similar results have been obtained by Taylor (1936) using concentrations of 0·05, 0·1, 0·2 and 0·3 per cent for 6 hours a day, the exposures totalling up to 122 in the animals which survived. Only the animals exposed to the highest concentration died; no effect on the growth or reproduction of the others was observed. An investigation by Barrett, Maclean and Cunningham (1938) also showed no significant effects on guinea-pigs or rabbits following the inhalation of 1200 p.p.m. for 473 hours. In the experiments of Seifter (1944) symptoms of **chronic** intoxication in dogs supervened 3 to 8 weeks after inhalation of 500 to 750 p.p.m. for 4 to 8 hours daily, 5 to 6 days per week. The symptoms consisted of lethargy, anorexia, nausea, vomiting and loss of weight.

Liver dysfunction was also shown by the bromsulphalein test and by the progressive failure of the intoxicated animals to conjugate chloral; i.e. in these dogs excess free chloral was excreted in the urine.

Changes in the Internal Organs

Castellino (1932) recorded the appearance of fatty degeneration of the liver and granular degeneration of the kidneys in some of his animals exposed to chronic inhalation, but he gives no definite values for the concentrations used, merely stating that the animals inhaled 8 to 10 drops every day up to the time of death 20 to 60 days later. Lehmann (1911) found no liver lesions in his cats, nor did Taylor (1936) in his rats inhaling concentrations below 2 and 3 per cent. In the higher concentrations some congestion of liver, kidneys and lungs was found; in the lower only a tendency to collapse of the lung, which by comparison with normal animals was not regarded as significant. The only demonstrable change in guinea-pigs, subjected by Barrett *et al.* (1938) to 1,200 p.p.m. of commercial trichloroethylene vapour for over 1,100 hours of exposure, was a very slight degenerative change in the liver of some animals, the significance of which, according to these observers, is doubtful. In the dogs subjected by Seifter (1944) to repeated inhalations of 500 to 750 p.p.m., the liver showed depletion of glycogen and hydropic parenchymatous degeneration. The intestines, heart, adrenals and kidneys showed no pathological changes.

Effect on the Blood

Seifter (1944) observed a gradual decrease, up to 10 per cent, in the total number of erythrocytes.

Absorption by the Skin

McCord (1932) states that trichloroethylene may exert its toxic effect through absorption by the skin, though the threshold of danger is not easily determined owing to the local skin damage. In rabbits application to the skin produced the following results:

7·7 ml. per kg. body weight three times daily	: dead after 4 days
6·6 ml. ,, ,, ,, ,,	: ,, 5 ,,
3·8 ml. ,, ,, ,, ,,	: ,, 5 ,,
3·6 ml. ,, ,, ,, ,,	: alive after 14 days
1·2 ml. ,, ,, ,, ,,	: ,, 7 ,,

The two surviving animals were made ill.

Goldberg (1924) and Bieder (1923) have suggested, on the basis of experiments with chlorylene on dogs, that substances such as trichloroethylene may be more freely absorbed by the skin and mucous membranes than other narcotics, and may form irreversible combinations with the peripheral nerve endings. This suggestion is considered by Stüber (1931) to be applicable to the nervous lesions described by her as due to trichloroethylene intoxication—an affinity for the roots of the cranial nerves.

TOXIC EFFECTS IN MAN
Acute Poisoning

There can be no doubt that the chief danger of trichloroethylene in industry is that of acute narcosis following prolonged exposure to high concentrations. Like chloroform, it invariably causes death if given for long enough periods and in high enough concentrations, but there appears to be a relatively wide interval between the narcotic and the lethal dose. According to Génévois (1936), $2\frac{1}{2}$ times the narcotic dose is necessary to cause death. According to McCord (1932) the narcotic action appears at a concentration of about 10,000 p.p.m., and after the threshold of danger is reached the damage depends upon the concentration and length of exposure. He points out that 1 ml. of liquid trichloroethylene yields on evaporation at room temperature 271 ml. of vapour, so that dangerous degrees of concentration are liable to occur readily.

Fatal Cases

Most of the fatal cases recorded in the literature have been preceded by unconsciousness. Among the 284 cases collected by Stüber (1931), 25 were fatal, and of the 19 patients concerning whom details were known, 12 were acutely narcotized and unconscious before death. In one of these 12, death was due to suffocation from aspirated vomit. Three fatal cases reported in 1935 in a German leather factory were due, not only to inhalation of the vapour, but to actual immersion of the men's heads. Between 1938 and 1948, 224 cases of gassing from trichloroethylene were reported to the Chief Inspector of Factories in Great Britain; of these 10 were fatal. In 1946 there were, in addition, two fatalities in workers using trichloroethylene, though at the inquest it was decided that the evidence was insufficient to account definitely for the cause of death. Both these cases presented the interesting feature, observed on several earlier occasions, of a latent period between the exposure and the sudden death following exertion. The only lesion found post mortem in both cases was a slight congestion of the lungs.

Thuresson (1942) reported a fatal case, a soldier using a trichloroethylene fire extinguisher; symptoms and post-mortem appearance were related to pulmonary oedema, which was caused, according to Thuresson, by phosgene produced by the decomposition of trichloroethylene.

Of Stüber's remaining fatal cases, some were due to apoplectic seizures, which may or may not have been attributable to the action of trichloroethylene. Stüber (1931) regards them as analogous to similar sequelae of carbon monoxide poisoning—i.e. due to a capillary injury such as thrombosis or capillary degeneration. In the cases reported by Plessner (1916) and Kalinowski (1927) 3 out of 5 individuals, who had suffered from acute trichloroethylene poisoning, developed apoplectiform symptoms after 1 to 2 years, and another had a definite apoplectic seizure. No such occurrence seems to have been reported in England.

Temporary Unconsciousness

Temporary unconsciousness, described as "overcome by fumes" or "gassing", is by far the commonest manifestation of poisoning by trichloroethylene. Of the 284 cases collected by Stüber (1931), 117 were rendered unconscious, while of the 87 actual cases reported to the Home Office between 1939 and 1948, the majority were acute attacks of greater or less severity. It is noteworthy that recovery from acute poisoning by trichloroethylene is usually complete; Striker, Goldblatt, Warn and Jackson (1935) found no harmful consequences from its use as an anaesthetic in 304 cases investigated.

The results of Stüber's (1931) investigation apparently revealed the occurrence of a number of residual symptoms after recovery, as described below. It should be noted that such disturbances are now regarded by many authorities as being associated only with commercial preparations of trichloroethylene; the pure substance probably produces no such pathological changes.

Symptoms

General nervous disorders. Among the acute cases recorded by Stüber, 11 patients had residual symptoms of headache, giddiness, loss of appetite, disturbance of cardiac action. The corresponding figures for deferred symptoms in the Factory Department's cases are headache 27, giddiness 10, drowsiness 9, loss of appetite, pain in the back and headache 6 (Cotter, 1950). Stüber states that it is difficult to decide whether these general nervous disorders are to be attributed to slight organic lesions due directly to the action of trichloroethylene or are simply functional. She appears to regard the symptoms as having an organic rather than a functional basis.

A case with epileptiform symptoms was described by Starkenstein, Rost and Pohl in 1929, while one with apoplectic symptoms accompanied by speech paralysis was described by Lewin in 1920. Stüber's view that such cases are due to actual injury of the cerebral vessels has already been discussed.

Trigeminal paralysis. Isolated trigeminal paralysis was found in 10 of Stüber's 284 cases, and van Themsche (1934) has also described instances. In England this apparently has never been observed. Stüber states that it is not clear from the neurological observations made by Plessner (1916) and Kalinowski (1927) whether the paralysis is to be regarded as a neuritis or as an injury to the medullary centre, analogous to the trigeminal paralysis of tabes. Jackson (1934), however, suggests that since no such local paralysis has been known to occur in cases of trigeminal neuralgia treated by trichloroethylene, it is probable that the paralysis was due to some other poison. Other workers, including Gerbis (1928), Glaser (1931), Barrett *et al.* (1938) and Powell (1945) incline to the view that the effects originally observed by Plessner were probably due to impurities in the trichloroethylene, chiefly dichloroacetylene. This view is confirmed by the experience of Humphrey and McClelland (1944), Carden (1944) and others using trichloroethylene as an anaesthetic in a closed circuit.

Optic nerve lesions. Nine cases showing optical disturbances are reported by Stüber (1931); these disturbances include blindness from optic atrophy (2 cases), pallor of the papilla with marked reduction of visual acuity, retrobulbar neuritis and disturbances of colour vision. Of the last, only 1 of 3 showed pathological appearances on ophthalmoscopic examination. Stuber regards this as evidence that trichloroethylene, like methyl alcohol, has an affinity for the roots of the cranial nerves, but she considers the specific affinity of any individual poison for individual nerves unexplainable. Zangger (1929) ascribes them to impurities in the trichloroethylene, while Goldberg (1924) and Bieder (1923), as already mentioned, suggest that this substance, being more freely absorbed by the skin and mucous membranes than some other narcotics, may form irreversible combinations with the peripheral nerve-endings. Jackson (1934) appears to think that it may have some specific action on the eye; he states that both dogs and men develop nystagmus during trichloroethylene anaesthesia and that dogs appear to show some slight visual disturbance during recovery.

Meyer (1929) on the basis of animal experiments considers that trichloroethylene has no specific action on the optic nerve. He made careful preparations, using Weigert's stain, of the optic nerve of dogs subjected to repeated narcosis by inhalation and to subcutaneous injection of large amounts, and observed no pathological changes.

Addiction to trichloroethylene. Baader (1927) described a case in which the addiction led to anaemia, general weakness and cachexia. In the report of the Chief Inspector of Factories and Workshops for 1934, it was stated that at least two cases had come to his knowledge where workers had voluntarily inhaled trichloroethylene.

Gastric disorders. Nausea, vomiting and loss of appetite after or during an acute attack of trichloroethylene poisoning are fairly frequent complaints. Cotter (1950) described cases in whom abdominal cramps and diarrhoea lasted sometimes for many weeks. Stüber considers that the vomiting and loss of appetite of acute poisoning are due to the swallowing of the substance condensed in the mouth, and may be brought about by direct irritation of the smooth muscle, as suggested by Joachimoglu's (1921b) experiments on the leech. Trichloroethylene appears to have very little corrosive action, unless decomposed by physical or chemical agencies, though, as will be seen on page 178, cases showing varying degrees of respiratory irritation have been described.

Nephritis. Most authorities appear to consider that nephritis as a sequel is unknown. Stüber (1931) states that the kidneys are never affected. Roholm (1933), however, mentions acute nephritis as a manifestation of severe poisoning.

Lesions of the liver. Hepatitis and cirrhosis of the liver are caused by trichloroethylene poisoning much more rarely than by tetrachloroethane, carbon tetrachloride and chloroform; Stüber (1931) goes so far as to state that the liver is never affected. Such affections have been mentioned in the literature, but the evidence for their attribution to a specific effect of trichloroethylene is not definite. Roholm (1933), for example, states that acute hepatitis may occur as late as 60 hours after the initial exposure, and may produce acute degeneration of the liver like that due to delayed chloroform poisoning; Willcox (1931) claims to have seen a case of jaundice from the use of trichloroethylene; Perrault (1944) also refers to a similar case following ingestion of trichloroethylene which was reported by Lechelle in 1944.

In the fatal case reported by Vallée and Leclercq (1935), the enlarged "yellowish" liver already described was not attributed to the trichloroethylene, which had produced an acute intoxication, but it was suggested that pre-existing liver disease might have been a predisposing factor to the fatal effect. In a case reported to the Chief Inspector of Factories and Workshops in 1932, a boy developed jaundice after 6 months' exposure to trichloroethylene in a degreasing process. The liver was palpable and smooth, the left lobe being larger than the right. The possibility of tetrachloroethane having been substituted by mistake for trichloroethylene during the period of exposure was inquired into, but was apparently denied by the firm supplying the fluid.

Only one case of liver damage following the use of trichloroethylene as a general anaesthetic has been reported (Dodds, 1945), and this was regarded as most probably not due to trichloroethylene since less than $\frac{1}{2}$ drachm was used.

Cardiac disturbances. Disturbances of the heart action are stated by Stüber (1931) to occur sometimes after acute poisoning, but are attributed by her to nervous stimulation.

Menstrual disturbances. Stüber (1931) states that menstrual disturbances have been observed, but these are not listed in actual numbers in her table of symptoms.

Skin lesions. Reddening and burning of the skin may, according to Stüber, occur after comparatively short contact (she gives 13 cases in her table), but real inflammation with blistering is rare and occurs only after prolonged immersion and after a delay of 1 or 2 days. Neither of these varieties of skin affection is considered to be due to a true corrosive effect; the slighter degrees are regarded as a chemical stimulation of pain and temperature nerve endings, as demonstrated by Goldscheider and Joachimoglu (1924), or a reflex vaso-motor stimulation; the more severe grades, occurring after long immersion, are attributed to cell injury following absorption of trichloroethylene by the protective fatty layers of the skin. Only one case of dermatitis due to inhalation of trichloroethylene has been reported by Baker and White (1946). In this case severe erythema of the whole body appeared whenever the worker failed to wear a mask, but did not appear when a mask was worn.

Mucous membrane lesions. Stüber (1931) states that symptoms of irritation of the air passages have never been recorded. This is not in accordance with the findings of McCord (1932), or with those in the fatal case recorded by Vallée and Leclercq (1935) where acute oedema and congestion of the lungs were observed at autopsy. Bronchitis occurred in two cases in the Home Office Reports, and congestion of the lungs was found in one fatal case. McCord (1932) states that the mildest degree of trichloroethylene poisoning may be represented by moderate respiratory irritation, and that more severe involvement may be characterized by pronounced bronchitis, diffuse haemorrhage into pulmonary tissues, pneumonia and enteritis. A case is recorded by Christiansen (1933) in a woman, aged thirty, originally diagnosed as miliary tuberculosis of the lungs, with a 2 months' history, the symptoms beginning after trichloroethylene had been used for cleaning shoes near a hot stove. A skiagram led to the diagnosis of poisoning by this substance and the lung shadows cleared up after 81 days in hospital.

Stüber (1931) regards any symptoms of acute irritation of mucous membranes as due, not to trichloroethylene itself, but to the products of its decomposition (hydrochloric acid or phosgene) by the action of light, catalysts or a naked flame. She states that after exposure to trichloroethylene in contact with a naked flame, owing to the production of phosgene, the symptoms may then be those of inflammation of mucous membranes of intensity varying from catarrhal inflammation of the upper air passages to fatal lung oedema. Two such cases, both fatal, are included in her table. This argument may possibly apply to Christiansen's (1933) case, but in that of Vallée and Leclercq (1935) no such decomposition with either hydrochloric acid or phosgene production appears likely, since the subject was employed in a vinegar factory, painting the inside of a vat. Concerning the possible liberation of phosgene by contact of trichloroethylene with burning tobacco, regarded by some authorities as a hazard, Elkins and Levine (1939) have stated that this is not so and that decomposition is of a low order.

Stüber (1931) regards the action of metals, resins, fatty acids, light, and aluminium dusts as potent in decomposing trichloroethylene to form hydrochloric acid. She states that with the ordinary use of trichloroethylene for metal cleaning or for extraction of fats and mineral oils, any hydrochloric acid which may be given off is insufficient to produce a corrosive action. When trichloroethylene is heated to 120° C. with resin solutions or fatty acids, however, 0·06 to 1·2 per cent of hydrochloric acid may be formed. She also states that when the vapour is mixed with air in diffuse light it is easily decomposed with formation of hydrochloric acid and that high concentrations, leading to complaints of respiratory catarrh, may be produced when trichloroethylene is kept in open vessels in small, light, but badly ventilated rooms.

She describes the formation of hydrochloric acid with an explosive effect under the catalytic action of aluminium. This effect apparently takes place only in the presence of aluminium chips and dust, not with smooth aluminium parts. She also quotes instances where the room, in which trichloroethylene had been left contaminated with aluminium dust in closed iron containers, became filled with bluish, sharply corrosive vapour, producing cough and inflammation of mucous membranes in the workers.

Decomposition by alkalis and oxidizing agents with a similar production of a corrosive vapour is exemplified by a case reported by Zangger (1930) in which a woman added a trichloroethylene solution to hot soapy water, with resultant scalding of the face. These observations on the decomposition of trichloroethylene by various agencies, especially the production of dichloroethylene, have been confirmed by van Themsche (1934), McNally (1937), von Oettingen (1937) and Converse (1938).

Chronic Poisoning

It should be mentioned that in the United Kingdom at any rate, there has been little confirmation of Stüber's (1931) experience with regard to the incidence of slight or chronic cases of trichloroethylene poisoning. In Stüber's report 82 cases out of 284 are listed as chronic. One case out of the 39 reported to the Chief Inspector of Factories and Workshops between 1921 and 1935, another reported in 1946, as well as the unusual case with jaundice, might be described as chronic. One of these patients, a female worker in a dyeing and cleaning plant, complained of pain in the chest, dyspnoea, loss of appetite and slight anaemia, all of which had persisted for several weeks.

The other was one of the two fatal cases referred to on page 175, but there was not sufficient evidence to ascribe death as being directly due to trichloroethylene poisoning. This patient, a girl employed in cleaning woollen knitted wear, which included occasional "spotting" with trichloroethylene, had shown gradual deterioration of health, especially extreme drowsiness, for about 3 months before death.

There appears to be little or no evidence of the cumulative action of trichloroethylene in industry, unless these two instances of death with insufficient evidence to account for it can be regarded as such.

Three workers, who had been exposed for 15 months, 1 year and 2 years respectively to trichloroethylene under unfavourable conditions, were examined by the Factory Department in 1935 and 1936. Two of these patients, sisters, were employed in dry-cleaning gloves with trichloroethylene. Both were somewhat hysterical and complained of sleepiness, headaches, a lessened tolerance of fatty foods, and one of "unappeased hunger". The other had fallen down a flight of stairs after an attack of giddiness. On examination, which included tests for liver and kidney efficiency, no abnormality was discovered. The third case was that of a man engaged in making spirit soap, which consisted of 25 per cent trichloroethylene and 75 per cent potassium oleate. He had been exposed for 2 years to large quantities of trichloroethylene fumes under conditions of bad ventilation; he had the appearance of being anaemic, and complained of dizziness, fatigue and indigestion. A blood examination did not confirm the presence of anaemia; on the contrary, the red cell count was surprisingly high, 6·9 millions, and the haemoglobin, 100 per cent. The white cell count was 8,800 and the film showed nothing of note. Van den Bergh and laevulose tolerance tests revealed no indication of liver damage.

Effect on the Blood

The general opinion is that trichloroethylene is not a chronic haematopoietic poison. A few investigations by Meyer (1929) gave normal values for both absolute and differential counts. Dérobert (1944–5) also has found no significant variation in the blood picture of trichloroethylene workers. Browning (unpublished) examined 9 girls employed in applying a gum dissolved in trichloroethylene to paper stencils by hand; the chief complaint was headache but a few reported drowsiness and lassitude; abnormality in the blood picture was very slight; one girl showed a rather low haemoglobin value and two others, slight relative lymphocytosis. Tara (1944–5) examined 36 workers and mentions that 27 reported headache, fatigue, vertigo, and a feeling of intoxication due to exposure to trichloroethylene; there was slight leucopenia. Baader (1927) and Stüber (1931) and a few other observers have noted anaemia.

Hyperglobulinaemia. Cotter (1950) observed hyperglobulinaemia in 4 out of 10 workmen who had suffered from acute trichloroethylene intoxication; this persisted for some weeks. Cotter considers that this change in globulin level suggests a factor of liver damage, though there was no clinical evidence of jaundice, or of bilirubin or phosphatase retention.

Symptoms

Gastro-intestinal disturbance. These symptoms were specially prominent in a case described by Glibert (1935) in a man employed in degreasing electrical machinery. The man had pain and vomiting, and later stomatitis, with symptoms of respiratory irritation—cough, pre-sternal pain, dyspnoea, bronchitic râles and expectoration.

Nervous and psychic symptoms. A case of retrobulbar neuritis described by Isenschmid and Kunz (1935) and two cases of a "neuromyelitic" complex by Persson (1934) are regarded as evidence of the affinity of trichloroethylene for nervous tissue. In the case of Isenschmid and Kunz, a man aged fifty-six, who had been exposed for a year to the vapour of trichloroethylene in the process of cleaning steel cylinders, retrobulbar neuritis was accompanied by left-sided paralysis of the hypoglossal nerve, polyneuritis of all four limbs, with disappearance of reflexes and increased sensibility of the peripheral nerve endings. In Persson's (1934) two cases, men employed in cleaning machine parts, the first symptoms complained of were giddiness, abnormal fatigue during the day, loss of appetite, nausea, headache, tremor and intolerance of alcohol. Later much diminution of sensation about the second lumbar segment was found, with spastic ataxia, a strongly positive Romberg sign and disturbance of sensation in the feet, hands and forearms. Improvement took place after work was discontinued, but some paraesthesia and stiffness of the legs remained.

Somewhat similar cases have been described by Antonioli and Rigola (1946). Three workers employed in making a shoe cement complained of "drunkenness", pains in the arms, paraesthesia and loss of sensation in the feet; in addition one of them had loss of heat sensation in the left forearm and hand, colic, vomiting, nausea and loss of weight. Symptoms in all three patients disappeared within a few weeks or months of cessation of exposure.

It has already been mentioned that some authorities consider that nerve lesions following exposure to trichloroethylene are due probably more to impurities or decomposition products than to trichloroethylene itself. Cases reported by McAuley (1943), Carden (1944) and Humphrey and McClelland (1944), in which nerve lesions have appeared after trichloroethylene anaesthesia, seem to have occurred under conditions where dichloroethylene might have been formed during the passage of the trichloroethylene through soda lime used for absorption of carbon dioxide.

Skin lesions (*chronic eczema*). In Stüber's table only 18 cases of this kind were listed but she states that it occurs more frequently, and that in 3 years she saw 4 cases in one establishment.

Corneal inflammation. Stüber (1931) describes corneal inflammation as being fairly frequent and suggests that it may be due to condensation of the vapour on entering the eye, with a resorptive, not corrosive, effect.

Intolerance. The only case of intolerance which appears to have been described was reported by Gilbert-Dreyfus, Zarachovitch and Herrault (1947). The patient was a man who developed violent attacks of asthma every time he was exposed to trichloroethylene. Desensitization was not effective, though skin and subcutaneous reactions were highly positive.

TABULAR ANALYSIS OF CASES OF TRICHLOROETHYLENE POISONING

Table 32 illustrates the differences in the incidence of acute and chronic toxic effects of trichloroethylene, and of various symptoms as shown by a summary of the cases collected by Stüber.

TABLE 32

Summary of Stüber's table of reported cases

Nature of work	No. of cases (including deaths)			No. of fatal cases		Symptoms										
	Acute	Chronic	Total	No.	Per cent	Unconsciousness	Excitation	Nervous excitability	Nervous sequelae	Gastric symptoms	Eye symptoms	Acute skin effects	Chronic skin effects	Mucous membrane effects	Trigeminal paralysis	Addiction
Cleaning in Tri*: apparatus	51	53	104	7	6·75	23	26	36	2	8	8	3	15	8	7	12
Handling Tri:	17	9	26	—	0·0	—	11	10	—	—	—	—	3	3	—	—
Leaky apparatus	11	—	11	2	18·1	8	1	—	—	—	—	—	—	—	1	—
Cleaning of Tri: containers	37	—	37	3	8·1	29	3	2	7	1	—	6	—	3	—	—
Internal painting of vessels	33	—	33	4	12·1	23	6	—	1	—	—	—	—	—	—	—
Tri: as solvent	19	14	33	1	3·3	16	—	15	—	3	—	2	—	3	—	—
Extraction by means of Tri:	20	5	25	1	4·0	16	4	1	1	—	—	2	—	7	—	—
Other causes	14	1	15	7	46·6	2	4	1	—	—	1	—	—	—	2	—
Total	202	82	284	25	8·8	117	55	65	11	12	9	13	18	24	10	12

* Trichloroethylene

METABOLISM IN ANIMALS AND MAN

Animal experiments carried out by Barrett and Johnston (1939) and Barrett, Cunningham and Johnston (1939) indicate that trichloroacetic acid may be an end-product of trichloroethylene decomposition in the body, either as a conjugated product or as a decomposition product of some more complex unstable compound actually excreted in the urine. Estimation of the amount of trichloroethylene excreted as trichloroacetic acid in the urine of dogs showed it to be from 5 to 8 per cent. Only minute amounts of trichloroacetic acid or of trichloroethylene could be detected in the organs of dogs 66 hours after the termination of repeated exposures to trichloroethylene vapour. A single experiment on a human being indicated that trichloroethylene was similarly metabolized in the human organism. From an investigation on the absorption, elimination and metabolism of trichloroethylene in human beings carried out by Powell (1945), it appears probable that trichloroethylene is not actually metabolized in the blood stream, but that the slow appearance of trichloroacetic acid in the circulation may be due to diffusion from some organ where fixation and metabolism of trichloroethylene have taken place. Although tetrachloroethane, carbon tetrachloride and chloroform—all liver poisons—undergo no such conversion, Barrett and co-workers (1939) do not consider it safe to assume that the absence of similar severe toxic effects on the liver in trichloroethylene intoxication is directly related to its partial conversion to trichloroacetic acid. Frank and Westendorp (1950) found little correlation between the symptoms of trichloroethylene intoxication and the amount of trichloroacetic acid in the urine, but suggest that prolonged exposure to high concentrations may cause acidosis.

9. Perchloroethylene
(*Tetrachloroethylene*)

C_2Cl_4, i.e. $CCl_2:CCl_2$

PROPERTIES

A colourless, non-inflammable liquid. B.R. 119–121° C. Sp.Gr. 1·624–1·632. Stable to moisture. Good solvent for oils, fats, and resins and for cellulose acetate. Solubility in water at 20° C. 0·0014 per cent; dissolves 0·01 per cent water at 25° C. (Durrans, 1950).

USES

(1) As a dry-cleaning agent under the names of "Perawin", "Tetralix", etc.
(2) In the printing industry as a fat solvent for duplicator operators.
(3) As a constituent of solvent soaps.
(4) As a degreaser.
(5) As resin solvent for fabric and fibre impregnation.

TOXICITY

Perchloroethylene is to some extent a narcotic, and in high concentration it also has a slight toxic action upon the liver and kidneys, but both these effects are less than those produced by chloroform or carbon tetrachloride. It must be emphasized, however, that its smaller toxicity in comparison with the latter has been estimated chiefly from its effects when given orally as an anthelmintic;

administered by subcutaneous injection (Barsoum and Saad, 1934) perchloroethylene proved more toxic than carbon tetrachloride. By inhalation its narcotic dose is slightly less.

Maximum Permissible Concentration

Most authorities recommend 200 p.p.m., the same as that for trichloroethylene, as the maximum allowable concentration for continuous exposure to perchloroethylene, though Morse and Goldberg (1943) suggest that it should be less than this. Carpenter (1937), on the basis of his results from experiments on human beings, suggests 100 to 500 p.p.m.; he found that 2,000 p.p.m. produced light narcosis in a few minutes, 1,000 p.p.m. produced inebriation in 45 minutes, 500 p.p.m. slight discomfort in 2 hours. At 50 p.p.m. there was a definite odour.

An investigation by Crowley, Ford and Stern (1945) of the actual concentrations obtaining in plants where perchloroethylene was used in degreasing tanks of the non-condensing type showed that even the 200 p.p.m. level was exceeded during some stages of the process. While the work was being put into the tanks, the air concentration averaged 180 p.p.m., and while it was being removed 484 p.p.m. It is recommended, therefore, that such tanks should be equipped with condenser and local exhaust ventilation, though of course this reduces their portability.

TOXIC EFFECTS IN ANIMALS

Lethal and Narcotic Concentrations

Lethal dose

By mouth. There is some disagreement as to the actual amount lethal to cats and dogs, especially in the experiments of Lamson, Robbins and Ward (1929), who found that some animals survived doses as large as 25 ml. per kg. body weight while others died with 4 ml. Barsoum and Saad (1934) also found that it did not kill dogs when given in an emulsion in doses up to 6 g. per kg. Maplestone and Chopra (1933), however, found 5 ml. per kg. invariably fatal to cats.

By subcutaneous injection. The minimum lethal dose for dogs was found by Barsoum and Saad (1934) to be 85 mg. per kg. as compared with 125 mg. for carbon tetrachloride. They suggest that the greater toxicity by injection may indicate that when given by mouth the substance is either destroyed in the alimentary tract or rapidly excreted or absorbed with difficulty.

By inhalation. The lethal potency of perchloroethylene vapour is very close to that of trichloroethylene and considerably less than that of chloroform. Lehmann (1911) found that it did not kill cats even in concentrations of 112 mg. per l. (16,500 p.p.m.) for $2\frac{1}{2}$ hours. According to Lazarew (1929), however, the minimum lethal dose for mice is 40 mg. per l. (6,000 p.p.m.) as compared with 40 to 45 mg. (7,400 to 8,300 p.p.m.) for trichloroethylene. Taylor (unpublished experiments) found concentrations of 0·5 per cent (5,000 p.p.m.) fatal to rats after 6 hours. Carpenter (1937) states that air saturated with perchloroethylene (3 per cent) is lethal in a few minutes, and that death occurs in animals anaesthetized at 19,000 p.p.m.

Narcotic dose

The narcotic potency by inhalation is, according to Lazarew, slightly greater than that of trichloroethylene. Lehmann's figures for the relative monophasic toxicity (i.e. of the pure vapour uninfluenced by volatility and temperature) are as follows, perchloroethylene : trichloroethylene : chloroform, 1·6 : 1·7 : 2·2. Later results of Lehmann and Schmidt-Kehl (1936) are given in Table 33.

TABLE 33
Narcotic concentrations of perchloroethylene

Concentration (mg./l.)	(per cent)	Time before effect (min.)	Narcosis
61	0·9	21–39	Deep
46	0·68	36–51	Deep
29	0·43	182	Deep
29	0·43	16	Slight

These results agree closely with those of Lamson, Robbins and Ward (1929) who give 62 mg. per l. (9,132 p.p.m.) as the narcotic dose.

Effect on the isolated frog's heart. Barsoum and Saad found the lowest concentration which would arrest the frog's heart to be 1 : 3,000 and the highest concentration producing no action 1 : 100,000. (The corresponding figures for trichloroethylene are 1 : 2,000 and 1 : 32,000.)

Acute Poisoning
Symptoms

Although it appears from experiments with oral administration that perchloroethylene is a metabolic poison of the nature of tetrachloroethane, pentachloroethane and carbon tetrachloride, i.e. with a toxic action on the liver and kidneys, this action is comparatively mild. Lamson, Robbins and Ward (1929) in fact reported no symptoms and no pathological changes (with the exception of a slight fatty metamorphosis of the liver, which they regarded as a more or less normal condition) in animals which had been given doses ranging from 0·3 to 25 ml. per kg. body weight. Maplestone and Chopra (1933), however, found that cats began to exhibit symptoms (giddiness, restlessness, drowsiness and unsteadiness in the hind limbs) when the dosage reached 1 ml. per kg., and in those receiving doses from 1 to 5 ml. per kg., the livers showed congestion and fine fatty change with a small amount of necrosis in one case, and the kidneys congestion, albuminous exudation and granular and fatty changes of the tubular epithelium. These changes, however, were not so great as those produced by as little as 0·5 ml. of carbon tetrachloride per kg. In dogs, however, Schlingmann and Gruhzit (1927) produced hepatic degeneration with as little as 0·2 ml. per kg. By inhalation the effects are much less. Lehmann observed no secondary effects of any kind in his animals, nor did Lamson, Robbins and Ward (1929) with high concentrations.

Chronic Poisoning

With long-continued exposure the effects of perchloroethylene are slight. Joachimoglu (1921b) observed no pathological changes in dogs which inhaled 7 ml. daily from 19 to 24 days. Taylor (1936) found that concentrations of 0·2 per cent (2,000 p.p.m.) produced very slight fatty change, but without cirrhosis, in the liver of rats after long-continued exposure. Carpenter (1937) observed no deaths in animals exposed to concentrations up to 7,000 p.p.m. for 8 hours a day, 5 days a week for 7 months. Slight non-progressive changes in the liver, kidneys and spleen were produced by exposure to concentrations of 2,300 p.p.m. and upwards.

TOXIC EFFECTS IN MAN

Perchloroethylene is usually placed practically lowest in toxicity in the ascending series of chlorinated hydrocarbons when used as a dry-cleaner (Kohn-Abrest 1924, Rochaix, 1936) and, except for one case reported by Beyer and Gerbis (1931) and quoted by Zernik (1933), no toxic effects from its use have been recorded. One man examined by the Home Office in 1936 stated that he never felt any ill-effects after using "Perawin" for 18 months, though he had previously felt ill when using trichloroethylene. The case reported by Beyer and Gerbis, a fatal one, from the chronic effects of a dry-cleaning substance called "Tetralix", is apparently somewhat doubtful, since perchloroethylene was stated only to be the chief constituent of the solvent, and it is not known how far the other constituents were responsible. The symptoms were those of gastro-intestinal disturbance, and liver degeneration was found to be the cause of death.

Zangger (1930) points out the special danger of using solvent soaps hot; he states that the rapid evaporation of perchloroethylene so produced may cause intoxication. Bird (1934) claims perchloroethylene is free from many of the disadvantages of trichloroethylene and carbon tetrachloride for this purpose, but its price is too high for it to be widely used. Rochaix (1936) includes it in the list of materials widely used in France and Switzerland for small hotel and shop apparatus for dry-cleaning.

10. Propylene Dichloride
(1:2-*Dichloropropane*)

$C_3H_6Cl_2$, i.e. $CH_2Cl \cdot CHCl \cdot CH_3$

PROPERTIES

A colourless liquid with an unpleasant chloroform-like odour. B.P. 96° C. Sp.Gr. 1·56 at 20° C. Insoluble in water; miscible with most common solvents. Solvent for fats, oils, waxes and resins (Durrans, 1944).

USES

(1) As an insecticide when mixed with dichloropropylene; it is applied to the roots of pineapple plants and other crops to kill nematode worms.
(2) As a solvent, especially for Buna-*N* synthetic rubber.
(3) As a dry-cleaning agent mixed with other solvents.
(4) As a metal degreaser (Sterner, 1949).

TOXICITY

Though no systemic toxic effects from the industrial use of propylene dichloride have been reported, animal experiments indicate that in high concentrations it is a narcotic and may produce lesions in the liver and kidneys. Dermatitis among workmen using a mixture of propylene dichloride and dichloropropylene has been reported.

Maximum Permissible Concentration

100 p.p.m. is recommended as the maximum allowable concentration for prolonged exposure (Greenburg and Moskowitz, 1945).

TOXIC EFFECTS IN ANIMALS
Acute Poisoning

Inhalation experiments by Heppel, Neal, Highman and Porterfield (1946) indicate that propylene dichloride is more lethal than dichloromethane, trichloroethylene or carbon tetrachloride, but less so than dichloroethane.

Symptoms

High concentrations over a short period produced lachrymation, conjunctival infection, loss of weight, excitement and inco-ordination. Continued exposure to concentrations of 1,500 and 1,000 p.p.m. caused lack of appetite, impaired growth and early narcosis to which tolerance developed.

Lesions of the Internal Organs

The organs chiefly affected were the liver and kidneys; fatty degeneration was observed in animals that died after short exposure (Highman and Heppel, 1946). Those animals that survived thirty-five or more exposures showed no significant lesions, and the results of liver function tests did not indicate important damage. In some animals, fatty lesions of the adrenal cortex and a marked splenic haemosiderosis were revealed.

Influence of Diet on Organic Lesions

Heppel, Highman and Porterfield (1946) found that the fatty infiltration of the liver produced by the higher concentrations (5,000 and 10,000 p.p.m.) of propylene dichloride was significantly decreased by a low protein, choline-deficient diet. These observers believe that the deficiency of sulphur-containing amino-acids is a more important factor in the fatty infiltration of the liver than actual deficiency of choline.

Chronic Poisoning

According to Heppel, Highman and Peake (1948) repeated inhalation of relatively low concentrations (400 p.p.m.) is innocuous to most species of animals except mice. Other than a lower rate of gain in weight in rats, no ill-effects were found in rats, guinea-pigs or dogs, and no significant lesions in the internal organs were observed. Concentrations of 400 p.p.m. for mice caused a heavy mortality, and, in one susceptible strain, hepatomas of the liver were produced.

TOXIC EFFECTS IN MAN

No toxic effects other than dermatitis in man have been recorded, but Greenburg and Moskowitz (1945) state that "it is likely that propylene dichloride acts in a manner similar to ethylene dichloride".

11. Amyl Chloride
$C_5H_{11}Cl$
PROPERTIES

The technical product consists of a mixture of *n*- and *iso*- amyl chlorides, 3- and 2-chloropentanes, 1-chloro-2-methyl butane and *tertiary* amyl chloride.

B.R. of the technical product, 95 per cent between 85 and 109° C. Sp.Gr. 0·88 at 20° C.

Solvent for waxes, oils, tars and resins. Non-miscible with water.

CHLORINATED HYDROCARBONS

USES

Chiefly in the textile industry as a scouring agent (Clayton and Clark, 1931).

12. Amylene Dichloride

$C_5H_{10}Cl_2$, i.e. $C_2H_5 \cdot CH(Cl) \cdot CH(Cl) \cdot CH_3$

Recently produced from natural pentane. Consists mainly of 2: 3-dichloropentane.

PROPERTIES

B.P. 138° C. Sp.Gr. 1·085 at 20° C. *Technical product*—B.R. 95 per cent between 130° and 200° C. Fl.P. 43° C.
Solvent for tars, fats and waxes; not for cellulose esters.

USES

Chiefly in the textile industry as a scouring agent.

TOXICITY

No animal experiments have been carried out on either amyl chloride or amylene dichloride and no ill-effects in human beings recorded.

13. Monochlorobenzene
(*Chlorobenzene, Chlorobenzol*)

C_6H_5Cl, i.e.

PROPERTIES

A colourless liquid with faint, not unpleasant odour. B.P. 131·7° C. Sp.Gr. 1·113.
Practically insoluble in water (0·0488 g. per 100 g. water). Miscible with most organic solvents. Does not develop acidity.
Solvent for ethyl cellulose, oils, grease, fat, soft copal resins, bakelite, gums, etc.

USES

(1) In the dyeing and chemical industries.
(2) As a lacquer constituent.
(3) As a cleansing agent, especially for household use, usually mixed with ethers and esters.
(4) In colour-printing, as a substitute for toluol.

TOXICITY

Chlorobenzene is a narcotic of considerable potency, producing effects, according to Flury and Zernik (1931), like those of benzene but rather stronger. Hamilton (1934) remarks, however, that "the entrance of chlorine into an aromatic compound does not increase its toxicity as it does in a fatty compound.

In fact chlorobenzene is less toxic than benzene." Evidence for its chronic or cumulative effect on human beings is not very definite, the cases described by Möhr in 1902 being attributed to chlorobenzol with no reference to its actual constitution, while in some of the cases the subjects were apparently exposed also to nitro-compounds of benzene.

Maximum Permissible Concentration

The maximum allowable concentration for daily exposure recommended by the U.S. Public Health Service (cited by Cook, 1945) is 75 p.p.m.

TOXIC EFFECTS IN ANIMALS

Table 34 shows Götzmann's (1904) results for the lethal and narcotic concentrations for cats (quoted in Flury and Zernik, 1931).

TABLE 34

Lethal and narcotic concentrations of monochlorobenzene for cats

Concentration (mg/l.)	(p.p.m.)	Effect
1–3	220–660	Tolerated for 1 hr.
5·5	1,200	Definite narcotic symptoms
11–13	2,400–2,900	Unsteadiness after about 1 hr.; tremor; twitching; if removed within 7 hr. no severe injury
17	3,700	Death after 7 hr. (P.M. lung haemorrhage)
37	8,000	Severe narcosis after ½ hr.; death 2 hr. after removal from exposure

TOXIC EFFECTS IN MAN

The effect of chlorobenzene as a central nervous poison has been best demonstrated in a case of poisoning by accidental swallowing reported by Reich in 1934. The most striking feature of this case was the delay in the onset of symptoms. The subject, a child, who had swallowed only a small amount (5 to 10 ml.) of "Puran" (a household cleansing agent apparently consisting of chlorobenzene) showed no immediate symptoms, but 2 hours later suddenly became unconscious, pale, with cyanosed lips, with loss of reflexes and of response to skin irritation, and fibrillary twitchings of the facial muscles. The unconsciousness lasted for 3 hours then gradually disappeared. There was an aromatic smell of "Puran" in the breath and in the urine, which contained traces of albumin. Wybert (1934) points out that apart from accidents of this kind due to carelessness of parents, "Puran" is a much safer cleaning agent for household use than benzol. Reich states that two similar cases were observed by Francorii, which were treated by giving saline water and which showed less severe symptoms.

In connection with its use in colour-printing, Zangger (1930) on the other hand remarks that it is not well-known that, while the use of chlorobenzene and dichlorobenzene mixtures diminish the danger of explosion, they do not diminish the danger of intoxication.

Toxic symptoms from the inhalation of chlorobenzene vapour, used in the chemical industry, were reported by Möhr in 1902, and also, according to Flury and Zernik (1931), by Navrozkij (1925). Möhr's four cases are apparently the same as those quoted later by Lewin (1929). They were brought to his notice by the occurrence of acute symptoms following the ingestion of alcohol, which

CHLORINATED HYDROCARBONS 189

he apparently regards as the precipitating factor, but questioning elicited a history of previous symptoms which are regarded as chronic toxic effects. It should be emphasized again, however, that only "chlorobenzol" is mentioned, with no distinction between monochlorobenzene, *ortho-* or *para-*dichlorobenzene, and that in one case the subject had also been exposed to "binitro".

Acute Poisoning

Symptoms

In the cases exposed to chlorobenzene the symptoms described ranged from severe somnolence to loss of consciousness, lasting in one case for 15 hours. The breathing was deep and rapid, and the pulse small, frequent and irregular. In one case there was tremor of the hands and in another, twitchings of the arms and legs. One case showed a yellow coloration of the skin, while two others were deeply cyanosed.

The blood in two cases was chocolate brown and showed the presence of methaemoglobin.

The red cells in the two cases in which a microscopical examination was made showed both degenerative (fragmentary cells of different size) and regenerative (nucleated cells) changes, which Möhr compares with those due to poisoning by nitrobenzene derivatives. No leucopenia was present, as in benzene poisoning; on the contrary there was a slight leucocytosis in two of the cases.

The urine was "burgundy red" but contained no bile pigment or haematoporphyrin, albumin or sugar. According to Lewin (1929) the urine contained o-chlorophenylmercapturic acid.

Chronic Poisoning

Symptoms

Two of the patients complained of previous headache, and one of giddiness and some difficulty in micturition.

14. *ortho*-Dichlorobenzene

$C_6H_4Cl_2$, i.e.

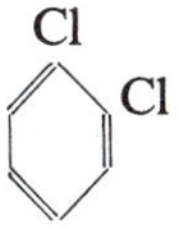

CONSTITUTION AND PROPERTIES

A liquid at ordinary temperature, F.P. − 17·6° C.

Commercial *ortho*-dichlorobenzene is practically always a complex mixture containing both monochlorobenzene and *para*-dichlorobenzene, trichlorobenzene etc. According to Carswell (1928), the isomers are so closely related in chemical and physical properties that it is very difficult to prepare pure *ortho*-dichlorobenzene from this mixture.

The following properties are given by him for the pure substance: density 1·3112–1·3088, B.R. 179·5–180·2° C. (180·3° C. at 760 mm.).

USES

(1) As a fumigant, disinfectant (e.g. for garbage cans), and general insecticide; also for wood preservation, when it is usually mixed with cedar wood and Turkey red oils.

(2) As an intermediate in dyestuffs.
(3) In metal cleaning and degreasing.
(4) In the manufacture of shoe dyes and waxes.
(5) In the removal of sulphur from illuminating gas.
(6) In the synthesis of organic chemical compounds.
(7) In the artificial silk and wool industries.
(8) In oil extraction.

TOXICITY

A certain amount of confusion as to the actual toxicity of *ortho*-dichlorobenzene has inevitably arisen from the fact that most authorities who mention it do so on the assumption that its effects must be similar to those of either monochlorobenzene or *para*-dichlorobenzene. Thus the Answer to a Question as to the toxicity of chlorobenzenes in the *Industrial Chemist* (1932) states "A mixture of *ortho*- and *para*-dichlorbenzene is more toxic than the monochlorbenzene just described, the action being similar, and since the commercial varieties of each of these compounds would contain the other, it appears definite that *ortho*-dichlorbenzene is a poison." Jordan (1932) finds that *ortho*-dichlorobenzene acts like monochlorobenzene, while according to Fraenkel (1912) the *para*-isomer is more toxic than the *ortho*-. The toxicity of *ortho*-dichlorobenzene compared with that of halogenated hydrocarbons such as carbon tetrachloride and trichloroethylene must apparently be considered as depending on volatility as well as molecular constitution. Thus the Special Development Bulletin of the Dow Chemical Co. (1932) reports that the lethal concentration in air relative to that of carbon tetrachloride is as 2·5:1, and Zangger (1930) states that *ortho*-dichlorobenzene should from its molecular constitution be more toxic than trichloroethylene, but it is much less volatile. The results of experiments carried out at Porton by the Ministry of Supply in 1928 (unpublished report) showed it to be less toxic than tetrachloroethane, but according to Cameron, Thomas, Ashmore, Buchan, Warren and McKenny-Hughes (1937) it is more acutely toxic to the blood of animals than benzene.

ESTIMATION IN AIR OF ROOMS

The amount of *ortho*-dichlorobenzene used in fumigation is 1 gallon per 1,500 to 2,000 c.ft. The concentration is estimated by the amount of silver chloride formed when air containing *ortho*-dichlorobenzene is passed over a spiral of heated silver gauze (Cameron *et al.*, 1937). These workers found that the air of a room sprayed with 1 gallon per 1,500 c.ft., ventilated after being closed for 9 hours, gives a concentration of 0·026 per cent 3 hours later, while with free ventilation, 0·001 per cent could be detected a fortnight later.

TOXIC EFFECTS IN ANIMALS
Acute Poisoning

The results of the Porton experiments in 1928 and the investigation by Cameron *et al.* in 1937 show that the acute effects of *ortho*-dichlorobenzene are those of injury to the liver and kidneys with fairly strong narcosis.

Lethal and Narcotic Doses

For guinea-pigs, the Porton workers found 1 in 1,000 concentration (0·1 per cent) lethal after 20 hours.

Cameron et al. found that inhalation of concentrations from 0·005 to 0·08 per cent caused severe irritation of eyes and nose, drowsiness, coma and sometimes death, in rats, mice and guinea-pigs.

Lesions of Internal Organs
  Liver. The lesions varied from patchy hydropic degeneration and slight fatty change, to massive necrosis, similar to that produced by chloroform and carbon tetrachloride. Such necrosis could follow exposure for ½ hour to concentrations of 0·039 per cent (Cameron et al.).
  Kidneys. In the Porton experiments guinea-pigs showed catarrhal degeneration of the convoluted tubules of the kidney. The effects on rabbits were much less severe (0·1 per cent concentration for 20 hours). Cameron et al. found only slight kidney injury in their animals.

Effect on the Blood
  According to Cameron et al., subcutaneous injection of 3 doses of 1 ml. (0·5 ml. per kg. body weight) to rabbits was followed by a fall in the white cell count, in some animals leucopenia, in others complete agranulocytosis. There was no marked change in the total number of red blood corpuscles or in the haemoglobin level.

Chronic Poisoning

According to the Porton experiments, the effects of prolonged exposure to *ortho*-dichlorobenzene are not severe, even when the air contains vapour equivalent to one sixth of the saturated concentration.

TOXIC EFFECTS IN MAN

In spite of the rather vague statements which have been published regarding the toxic effect of *ortho*-dichlorobenzene based on the assumption of its similarity to monochlorobenzene and *para*-dichlorobenzene, there is little or no evidence that its use in industry has ever produced either acute or chronic intoxication in human beings. Thus, after a statement to the Home Office in 1932 that the use of a timber fluid containing *ortho*-dichlorobenzene mixed with cedar wood oil and Turkey red oil made the workmen sick, two workers were examined. One of these had worked with the substance for 15 years, the other 1½ to 2 years. Neither showed any ill-effects. A manufacturer of *ortho*-dichlorobenzene reported to the Chlorine Institute, New York, in 1924 that "Our workmen can work in *ortho*- fumes without any special protection at all, and as far as we have been able to observe without any harmful effects whatsoever." It must be noted, however, that the Porton investigators on the basis of their animal experiments considered it too toxic for a paint solvent and Cameron et al. (1937) state that exposures to concentrations of 100 mg. per l. (17,000 p.p.m.) may have a detrimental effect on human beings. *ortho*-Dichlorobenzene was included in a list of solvents suitable for shoe dyes investigated by the Chicago Department of Health in 1927 and was said to be satisfactory.

BIBLIOGRAPHY

ALBRECHT, P. (1927). Dichloren als Ersatz des Chloroforms in Äthermischung. *Arch. klin. Chir.*, **146**, 273.
ALTHAUSEN, T. L. and THOENES, E. (1932). Influence on carbohydrate metabolism of experimentally induced hepatic changes. *Arch. intern. Med.*, **50**, 257.
ANDRÉ, L. and FEILLARD, R. (1946). Nouveaux cas d'intoxication grave par le tétrachlorure de carbone. *Bull. Soc. méd. Hôp.*, Paris, 4me sér., **62**, 418.

ANSWER TO QUESTION, Industrial Chemist (1932). Poisoning by chlorbenzenes. *Industr. Chem. chem. Mfr.*, **8**, 33.
ANTONIOLI, E. and RIGOLA, A. (1946). Tre casi di intossicaziona cronica da tricloroetilene. *Med. d. Lavoro*, **37**, 119.
BAADER, E. W. (1927). Tätigkeitsbericht der Abteilung für Gewerbekrankheiten des Kaiserin Auguste-Viktoria-Krankenhaus in Berlin-Lichtenberg. *Zbl. GewHyg.*, **14**, 385.
BAGGENSTOSS, A. H. (1947). Carbon tetrachloride intoxication treated by peritoneal lavage; pathologic aspects. *Proc. Mayo Clin.*, **22**, 321.
BAKER, K. C. and WHITE, C. J. (1946). Occupational dermatitis due to inhalation of trichlorethylene gas. *Industr. Med.*, **15**, 389.
BARRETT, H. M., CUNNINGHAM, J. G. and JOHNSTON, J. H. (1939). Study of the fate in the organism of some chlorinated hydrocarbons. *J. industr. Hyg.*, **21**, 479.
BARRETT, H. M. and JOHNSTON, J. H. (1939). The fate of trichloroethylene in the organism. *J. biol. Chem.*, **127**, 765.
BARRETT, H. M., MACLEAN, D. L. and CUNNINGHAM, J. G. (1938). Comparison of the toxicity of carbon tetrachloride and trichlorethylene. *J. industr. Hyg.*, **20**, 360.
BARSOUM, G. S. and SAAD, K. (1934). Relative toxicity of certain chlorine derivatives of the aliphatic series. *Quart. J. Pharm.*, **7**, 205.
BEATTIE, J., HERBERT, P. H., WETCHEL, C. and STEELE, C. W. (1944). Studies on hepatic dysfunction. I. Carbon tetrachloride poisoning treated with casein digest and methionine. *Brit. med. J.*, i, 209.
BECK, G. and SÜSSTRUNK, M. (1931). Versuche über akute Vergiftungen mit cis- und trans-Dichloräthylen und Äthylenoxyd. *Arch. Gewerbepath. Gewerbehyg.*, **2**, 81.
BENZI, T. (1925). Indagini ed osservazioni sperimentali sull'intossicazione professionale da tetrachloretano $C_2H_2Cl_4$ simm. *Boll. Soc. med. chir. Pavia*, **37**, 537.
BEYER, A. and GERBIS, H. (1931). Jahresbericht über die Tätigkeit der preussischen Gewerbemedizinalräte. *Veröff. Med. verw.*, **39**. (Cited in Lehmann and Flury, Toxicology and hygiene of industrial solvents, Williams and Wilkins, Baltimore, 1943, p.190).
BIANCALANI, A. (1934). Ricerche sperimentali sulle alterazioni del sistema nervoso centrale nella intossicazione da tetracloruro di carbonio. *Riv. Patol. nerv. ment.*, **44**, 352.
BIEDER, E. (1923). Dissertation, Breslau.
BIESALSKI, E. (1924). Pyrogene Phosgenbildung. *Z. angew. Chem.*, **37**, 314 (abstr. in *Chem. Abstr.*, **18**, 2480).
BIRD, C. L. (1934). Solvents for dry cleaning. White spirit or chlorohydrocarbons? *Chem. Tr. J.*, **94**, 192.
BLANC, F. and CARRIÈRE, M. (1935). Réflexions sur un nouveau cas d'intoxication par le tétrachlorure de carbone. *Marseille méd.*, **72**, 1, 719.
BLEYER, B. and MAYER, K. (1927). Zur Frage der Ursachen der Dürener Rinderkrankheit. *Fortschritte der Landwirtschaft*, Jahrgang 2, Heft 6, 173.
BLOCH, W. (1946). Zwei Intoxikationen durch Dichloraethan bei Verwendung als Berauschungsmittel. *Schweiz. med. Wschr.*, **76**, 1078.
BOIDIN, L., ROUQUÈS, L. and ALBOT, G. (1930). Ictère grave toxique par le tétrachloréthane chez une ouvrière perlière. *Bull. Soc. méd. Hôp., Paris*, **54**, 1305.
BOLLMANN, J. L. and MANN, F. G. (1931). Experimentally produced lesions of the liver. *Ann. intern. Med.*, **5**, 699.
BORDAS (1935). L'emploi des liquides inflammables ou toxiques dans les salons de coiffure. *Ann. Hyg. publ., Paris, N.S.*, **13**, 167.
BOURNE, W. and STEHLE, R. L. (1923). Methylene chloride in anaesthesia. *Canad. med. Ass. J.*, **13**, 432.
BOVERI, P. (1929). Empoisonnement d'origine professionnelle par le tétrachlorure de carbone. *Progr. méd., Paris*, **56**, 1198.
BOWDITCH, M., DRINKER, C. K., DRINKER, P., HAGGARD, H. H. and HAMILTON, A. (1940). Code for safe concentrations of certain common toxic substances used in industry. *J. industr. Hyg.*, **22**, 251.
BOYE, E. (1935). Die chemischen Feuerlöschmethoden. *Chemiker Ztg.*, **59**, 155, 175.
BRANDT, A. (1932). Angebliche Trichloräthylenerkrankungen in Schuhausbesserungswerkstätten, hervorgerufen durch Tetrachlorkohlenstoff. *Arch. Gewerbepath. Gewerbehyg.*, **3**, 335.
BRANDT, A. (1932). Zur Vergiftung durch Lösungsmittel in Tiefdruckereien. *Arch. Gewerbepath. Gewerbehyg.*, **3**, 527.
BRÜNING, A. and SCHNETKA, M. (1933). Über den Nachweis von Trichloräthylen und andern halogenhältigen organischen Lösungsmitteln. *Arch. Gewerbepath. Gewerbehyg.*, **4**, 740.
BUTSCH, W. L. (1932). Cirrhosis of the liver caused by carbon tetrachloride. *J. Amer. med. Ass.*, **99**, 728.
CAMERON, G. R. and KARUNARATNE, W. A. E. (1936). Carbon tetrachloride cirrhosis in relation to liver regeneration. *J. Path. Bact.*, **42**, 1.
CAMERON, G. R., THOMAS, J. C., ASHMORE, S. A., BUCHAN, J. L., WARREN, E. H. and McKENNY-HUGHES A. W. (1937). The toxicity of certain chlorine derivatives of benzene, with special reference to o-dichlorobenzene. *J. Path. Bact.*, **44**, 281.

CARDEN, S. (1944). Hazards in the use of the closed-circuit technique for trilene anaesthesia. *Brit. med. J.*, **i**, 319.
CAROZZI, L. (1936). Les solvants, leur physiologie et leur pathologie. *Rev. Path. Physiol. Trav.*, **12**, 130.
CARPENTER, C. P. (1937). Chronic toxicity of tetrachlorethylene. *J. industr. Hyg.*, **19**, 323.
CARRIEU, M. F. (1929). Contribution à l'étude de l'intoxication par le trichlorure d'éthylène. Sa prophylaxie par le masque à l'huile. *Rev. Hyg. Police sanit.*, **51**, 338.
CARSWELL, T. S. (1928). Physical properties of ortho-dichlorobenzene. *Industr. Engng. Chem.*, **20**, 728.
CASTELLINO, N. (1932). Il trichloroetilene. *Folia med.*, Napoli, **18**, 415. (Abstr. *Bericht. Physiol.*, **68**, 584.)
CERESA, C. (1945). Un cas d'intoxication aiguë par le tétrachlorure de carbone. *Med. d. Lavoro*, **36**, 159.
CHATRON, M. (1934). De l'interprétation des résultats de laboratoire dans un cas d'hyperazotemie avec chloropénie et abaissement de la réserve alcaline. *Bull. Soc. Chim. biol.*, **16**, 405.
CHICAGO DEPARTMENT OF HEALTH (1927). Toxic shoe dyes. Report for the years 1926–30 inclusive. pp. 340, 643. Chicago, 1931.
CHIEF INSPECTOR OF FACTORIES AND WORKSHOPS (1932). Annual Report, p. 107. H.M. Stationery Office, 1933.
CHIEF INSPECTOR OF FACTORIES AND WORKSHOPS (1934). Annual Report, p. 70. H.M. Stationery Office, 1935.
CHRISTIANSEN, T. (1933). Et bidrag til trikloraethylenforgiftningens klinik. *Ugeskr. Laeg.*, **95**, 1187.
CLAYTON, E. and CLARK, C. O. (1931). Modern organic solvents. Part I. Development of solvents industry. A new conception of the constitution of cellulose nitrate. Solvent action in dyeing and allied processes. *J. Soc. Dy. Col.*, Bradford, **47**, 183.
CLAYTON, E. and CLARK, C. O. (1931). Modern organic solvents. Part II. Classification. Applications in the textile and allied trades. *J. Soc. Dy. Col.*, Bradford, **47**, 247.
CLINTON, M. (1947). Renal injury following exposure to CCl_4. *New Engl. J. Med.*, **237**, 183.
COLE, W. H. (1926). Pyridine test as a qualitative method for estimation of minute amounts of chloroform. *J. biol. Chem.*, **71**, 173.
COLE, W. H. (1927). Measurements of fatal doses of chloroform in brains of white rats. *Proc. Soc. exp. Biol.*, N.Y., **24**, 340.
COLLIER, H. (1936). Methylene dichloride intoxication in industry. *Lancet*, **i**, 594.
COLMAN, H. (1907). A dangerous dry shampoo. *Lancet*, **i**, 1709.
CONDOR, H. A. (1948). Convulsions under trilene anaesthesia. *Brit. med. J.*, **ii**, 340.
CONVERSE, J. D. (1938). Chlorinated solvents. *Canad. Chem. Process. Indust.*, **22**, 361.
COOK, W. A. (1945). Maximum allowable concentrations of industrial atmosphere contaminants. *Industr. Med.*, **14**, 936.
COTTER, L. H. (1950). Trichloroethylene poisoning. *Arch. indust. Hyg. and occup. Med.*, **1**, 319.
COYER, H. A. (1944). Tetrachloroethane poisoning; 7 cases; review of several treated. *Industr. Med.*, **13**, 230.
CROWLEY, R. C., FORD, C. B. and STERN, A. C. (1945). A study of perchlorethylene degreasers. *J. industr. Hyg.*, **27**, 140.
CULBERT, T. D. (1942). Convulsions under trilene anaesthesia. *Brit. med. J.*, **ii**, 679.
CUTLER, J. T. (1932). Influence of diet on carbon tetrachloride intoxication in dogs. *J. Pharmacol.*, **45**, 209.
DAROGA, R. P. and POLLARD, A. G. (1941). Colorimetric method for the determination of minute quantities of carbon tetrachloride or of chloroform in air and soil. *J. Soc. chem. Ind.*, **60**, 218.
DAVIES, P. A. (1934). Carbon tetrachloride as an industrial hazard. *J. Amer. med. Ass.*, **103**, 962.
DAVIS, N. C. (1924). Influence of diet upon liver injury produced by carbon tetrachloride. *J. med. Res.*, **44**, 601.
DAVIS, N. C. and WHIPPLE, G. H. (1919). Influence of fasting and various diets on liver injury effected by chloroform anaesthesia. *Arch. intern. Med.*, **23**, 612.
DÉROBERT, M. (1944–5). Le sang dans l'intoxication par le trichloréthylène. *Arch. Mal. prof.*, **6**, 321.
DERVILLÉE, P. M. and CASTAGNOU, R. (1934). Étude sur les variations du taux de la glycémie chez le lapin soumis à l'intoxication par le tétrachlorure de carbone en ingestion. *C. R. Soc. Biol.*, Paris, **117**, 365.
DESOILLE, H. and MÉLISSINOS, J. C. (1933). Les lésions hépatiques dues au tétrachloréthane. *Médecine*, **14**, 533.
DINGLEY, L. A. (1926). A case of CCl_4 poisoning due to the bursting of a patent fire extinguisher. *Lancet*, **i**, 1037.
DOCHERTY, J. F. and BURGESS, E. (1922). The action of carbon tetrachloride on the liver. *Brit. med. J.*, **ii**, 907.
DOCHERTY, J. F. and NICHOLLS, L. (1923). Report of three autopsies following carbon tetrachloride treatment. *Brit. med. J.*, **ii**, 753.
DODDS, G. H. (1945). Necrosis of liver and bilateral massive suprarenal haemorrhage in puerperium. *Brit. med. J.*, **i**, 769.

Dow Chemical Co. (1932). Improved methods of industrial metal degreasing. *Spec. Development Bull.*, No. 18. Michigan.
Doyle, W. E. and Baker, C. (1944). Carbon tetrachloride poisoning; epidemic in a parachute plant. *Industr. Med.*, **13**, 184.
Drescher (1920). Dissertation, Würzburg.
Drill, V. A. and Loomis, T. A. (1946). Effect of methionine supplements on hepatic injury produced by carbon tetrachloride. *Science*, **103**, 199.
Drill, V. A. and Loomis, T. A. (1947). Methionine therapy in experimental injury produced by carbon tetrachloride. *J. Pharmacol.*, **90**, 138.
Drill, V. A., Loomis, T. A. and Belford, J. (1947). Effect of protein and carbohydrate intake on liver injury produced in dogs by carbon tetrachloride. *J. industr. Hyg.*, **29**, 180.
Dubois, R. and Roux, L. (1887). Action du chlorure d'éthylène sur la cornée. *C. R. Acad. Sci., Paris*, **104**, 584.
Dudley, S. F. (1935). Some atmospheric hazards encountered in naval life. *Proc. R. Soc. Med.*, **28**, 1291.
Dudley, S. F. (1935). Toxic nephritis following exposure to carbon tetrachloride and smoke fumes. *J. industr. Hyg.*, **17**, 93.
Durrans, T. H. (1944). Solvents. 5th ed. Chapman & Hall, London.
Durrans, T. H. (1950). Solvents. 6th ed. Chapman & Hall, London.
Duvoir, M., Guibert and Desoille, H. (1933). Les intoxications par le tétrachlorure de carbone. *Ann. Méd. lég.*, **13**, 533.
Eddy, J. H. (1945). Carbon tetrachloride poisoning in industry. *Industr. Med.*, **14**, 283.
Eddy, J. H. (1945). Carbon tetrachloride poisoning; a preliminary report on the use of methionine in hepatitis. *J. Amer. med. Ass.*, **128**, 994.
Edwards, J. D. (1917). Gas interferometer calibration. Technologic Papers of the U.S. Bureau of Standards, No. 316. Government Printing Office, Washington.
Elkins, H. B. (1942). Maximal allowable concentrations. I. Carbon tetrachloride. *J. industr. Hyg.*, **24**, 233.
Elkins, H. B. and Levine, L. (1939). Decomposition of halogenated hydrocarbon vapors by smoking. *J. industr. Hyg.*, **21**, 221.
Elliott, J. M. (1933). Report of a fatal case of poisoning by tetrachlorethane. *J. R. Army med. Cps.*, **60**, 373.
Emara (1935). Toxicity of carbon tetrachloride. *J. Egypt. med. Ass.*, **18**, 3.
Erdmann (1912). Über Augenveränderungen durch Äthylenchlorid. *Klin. Mbl. Augenheilk.*, **14**, 370.
Eulenberg, H. (1876). Handbuch der Gewerbehygiene. Berlin, Hirschwald.
Faranelli, E. H. (1892). A proposito dell' azione delle inhalazioni di biclorurio di etilene sulla cornea. *Arch. Sci. med.*, **16**, 79.
Feil, A. and Heim de Balsac, F. (1924). L'intoxication professionelle par le tétrachloréthane. *Progr. méd., Paris*, **51**, 306.
Fieldner, A. C. and Katz, S. H. (1921). Gases produced in the use of carbon tetrachloride and foamite fire extinguishers in mines. *U.S. Bur. Mines Rep. Invest. Serv.*, No. 2262.
Fieldner, A. C., Katz, S. H. and Kinney, S. P. (1921). Gas masks for gases met in fighting fires. *U.S. Bur. Mines Technical Paper*, **248**.
Fieldner, A. C., Katz, S. H., Kinney, S. P. and Longfellow, E. S. (1920). Poisonous gases from carbon tetrachloride fire extinguishers. *J. Franklin Inst.*, **190**, 543.
Fiessinger, N., Brodin, P. and Wolf, M. (1923). Les ictères des perlières et les hépatites dues au tétrachloréthane. *Ann. Méd. lég.*, **3**, 76.
Fife, H. R. and Reid, E. W. (1930). New industrial solvents: ethylene dichloride, dichlorethyl ether and isopropyl ether. *Industr. Engng. Chem.*, **22**, 513.
Flandin (1932). Un cas d'intoxication par le tétrachlorure de carbone. *Bull. Soc. méd. Hôp., Paris*, 3me sér., **56**, 1246.
Flury, F. (1928). Moderne gewerbliche Vergiftungen in pharmakologisch toxikologischer Hinsicht. *Arch. exp. Path. Pharmak.*, **138**, 65.
Flury, F. and Zernik, F. (1931). Schädliche Gase, Dämpfe, Nebel, Rauch- und Staubarten. Springer, Berlin.
Fohlen, J. (1922). Carbon tetrachloride as a fire extinguisher. *Technique Mod.*, **14**, 593. (Abstr. in *Chem. Abstr.*, **17**, 1111.)
Fohlen, J. (1924). Dangers incident to the use of carbon tetrachloride fire extinguishers. *Amer. J. Pharm.*, **96**, 66.
Fraenkel, S. (1912). Die Arzneimittel-Synthese auf Grundlage der Beziehungen zwischen chemischen Aufbau und Wirkung. 3 Aufl. Springer, Berlin.
Fraenkel, S. (1919). Die Arzneimittel-Synthese auf Grundlage der Beziehungen zwischen chemischen Aufbau und Wirkung. 4 Aufl. Springer, Berlin.
Frank, R. and Westendorp, J. (1950). Medical control on exposure of industrial workers to trichlorethylene. *Arch. industr. Hyg. and occup. Med.*, **1**, 308.
Führer, H. (1921). Die Wirkungsstärke der Narkotica. *Biochem. Z.*, **120**, 143.
Führer, H. (1923). Die Wirkungsstärke von Chloroform und Tetrachlorkohlenstoff. *Arch. exp. Path. Pharmak.*, **97**, 86.
Führer, H. (1929). Versuche zur Entgiftung des Chloroforms. *Dtsch. med. Wschr.*, **55**, 1331.

FUJIWARA, K. (1914). (Cited by Brüning and Schnetka, 1933, in *Arch. Gewerbepath. Gewerbehyg.*, **4**, 740.)
GARDNER, G. H., GROVE, R. C., GUSTAFSON, R. K., MAIRE, E. D., THOMPSON, M. J., WELLS, H. S. and LAMSON, P. D. (1925). Studies on pathological history of experimental carbon tetrachloride poisoning. *Bull. Johns Hopk. Hosp.*, **36**, 107.
GARLAND, Y. (1942). Convulsions under trilene anaesthesia. *Brit. med. J.*, **ii**, 607.
GAUTIER, C., CHATRON, M. and SEIDMANN, P. (1933). Intoxication par le tétrachlorure de carbone. *Bull. Soc. méd. Hôp., Paris*, 1638.
GÉNÉVOIS, L. (1936). Sur les propriétés physiologiques des solvants chlorés. *Ann. Hyg. publ., Paris*, **14**, 139.
GERBIS, H. (1928). Irreparable Gesichtsnervenlähmung durch gewerbliche Vergiftung. *Zbl. Gewhyg.*, **15**, 97.
GILBERT-DREYFUS, ZARACHOVITCH, M. and HERRAULT, A. (1947). Asthme professionel par intolérance au trichloréthylène. *Bull. Soc. méd. Hôp., Paris*, **63**, 19.
GLASER, M. A. (1931). Treatment of trigeminal neuralgia with trichlorethylene. *J. Amer. med. Ass.*, **96**, 916.
GLASER, E. and FRISCH, S. (1928). Zum Phosgennachweis in chemischen Feuerlöschern. *Z. angew. Chem.*, **41**, 263.
GLIBERT, D. (1935). À propos d'un cas d'intoxication professionnelle par le trichloréthylène. *Scalpel, Liège*, **88**, **ii**, 1446.
GOCHER, T. E. P. (1944). Carbon tetrachloride poisoning. *Northwest Med.*, **43**, 228.
GOLDBERG, E. (1924). Über die Wirkungsweise des Trichloräthylens (Chlorylens) und die Indikationen für seine therapeutische Anwendung. *Dtsch. Z. Nervenheilk.*, **82**, 10.
GOLDENSON, J. and THOMAS, J. W. (1947). Determination of acetylene tetrachloride in air. *J. industr. Hyg.*, **29**, 14.
GOLDSCHEIDER, A. and JOACHIMOGLU, G. (1924). Untersuchungen über den Temperatursinn. III. Über die Wirkung von Chlorderivaten des Methans, Äthans und Äthylens sowie einiger anderer Stoffe auf die Hautnerven. *Pflüg. Arch. ges. Physiol.*, **206**, 325.
GÖTZMANN (1904). Dissertation, Würzburg. Quoted by Flury and Zernik, 1931, in Schädliche Gase.
GRAHAM, W. H. (1938). Simulation of the "acute abdomen" in carbon tetrachloride poisoning. *Lancet*, **i**, 1159.
GRAY, I. (1947). Carbon tetrachloride poisoning. *N.Y. St. J. Med.*, **47**, 2311.
GREENBURG, L. M. and MOSKOWITZ, S. (1945). Safe use of solvents for synthetic rubber. *Industr. Med.*, **14**, 359.
GRIMM, V., HEFFTER, A. and JOACHIMOGLU, G. (1914). Gewerbliche Vergiftungen in Flugzeugfabriken. *Vjschr. gerichtl. Med.*, **48**, Suppl. 2, 161.
VON GRONOW, W. E. (1927). Die Anwendung chemischer Sondernassfeuerlöscher in den gewerblichen Betrieben unter dem Gesichtspunkte ihrer Einwirkung auf die Benutzer. *Zbl. Gewhyg.*, **14**, 161.
GYÖRGY, P., EIFTER, J., TOMARELLI, R. M. and GOLDBLATT, H. (1946). Influence of dietary factors and sex on the toxicity of carbon tetrachloride in rats. *J. exp. Med.*, **83**, 449.
HABGOOD, S. and POWELL, J. F. (1945). Estimation of chloroform, carbon tetrachloride and trichlorethylene in blood. *Brit. J. industr. Med.*, **2**, 39.
HAGEN, J. (1939). Ein Fall von Tetrachlorkohlenstoff-Vergiftung mit symptomatisch toxischer Epilepsie. *Samml. Vergiftungsf.*, **10**, 169, A184.
HAGEN, W. S., ALEXANDER, H. A. and PEPPARD, T. A. (1940). Toxic effects of carbon tetrachloride; report of a case. *Minn. Med.*, **23**, 715.
HALL, M. C. (1921). Use of carbon tetrachloride for removal of hookworms. *J. Amer. med. Ass.*, **77**, 1641.
HAMILTON, A. (1933). Formation of phosgene in the thermal decomposition of carbon tetrachloride. *Industr. Engng. Chem.*, **25**, 539.
HAMILTON, A. (1934). Industrial toxicology. Harper, New York.
HAMMES, E. M., jr. (1941). Carbon tetrachloride as an industrial hazard. Report of two cases. *J. industr. Hyg.*, **23**, 112.
HASSAN, A. and SALAH, M. (1935). Investigation on carbon tetrachloride intoxication. *J. Egypt. med. Ass.*, **18**, 207.
HAUSMANN and HELLY (1929). Ueber ein Fall von Tetrachlorkohlenstoffvergiftung bei einem Desinfektor. *Schweiz. Ztschr. f. Unfallmed. u. Berufskrankh.* **2–3**, 50.
HÉBERT, P. and PHÉLÉBON (1931). L'intoxication aiguë (hepato-néphrite grave) par inhalation de tétrachlorure de carbone. *J. Prat., Paris*, **45**, 327.
HEIMANN, H. and FORD, C. A. (1941). Low concentrations of carbon tetrachloride capable of causing mild narcosis. *Industr. Bull.*, **20**, 209, issued by Industrial Commissioner, New York State.
HELLWIG, A. (1922). Klinische Narkoseversuche mit Solaesthin. *Klin. Wschr.*, **1**, 215.
HEPPEL, L. A., HIGHMAN, B. and PEAKE, E. G. (1948). Toxicology of 1, 2-dichloropropane (propylene dichloride). IV. Effects of repeated exposures to a low concentration of the vapor. *J. industr. Hyg.*, **30**, 189.
HEPPEL, L. A., HIGHMAN, B. and PORTERFIELD, V. T. (1946). Toxicology of 1–2 dichloropropane (propylene dichloride); influence of dietary factors. *J. Pharmacol.*, **87**, 11.

HEPPEL, L. A., NEAL, P. A., ENDICOTT, K. M. and PORTERFIELD, V. T. (1944a). Toxicology of dichloroethane. *Arch. Ophthal.*, **32**, 391.

HEPPEL, L. A., NEAL, P. A., HIGHMAN, B. and PORTERFIELD, V. T. (1946). Toxicology of 1–2 dichloropropane (propylene dichloride). Studies on effects of daily inhalation. *J. industr. Hyg.*, **28**, 1.

HEPPEL, L. A., NEAL, P. A., PERRIN, T. L., ENDICOTT, K. M. and PORTERFIELD, V. T., (1945). Toxicology of 1–2 dichloroethane (ethylene dichloride). III. Its acute toxicity and the effect of protective agents. *J. Pharmacol.*, **84**, 53.

HEPPEL, L. A., NEAL, P. A., PERRIN, T. L., ENDICOTT, K. M. and PORTERFIELD, V. T. (1946). Toxicology of 1–2 dichloroethane (ethylene dichloride). V. Effects of daily inhalations. *J. industr. Hyg.*, **28**, 113.

HEPPEL, L. A., NEAL, P. A., PERRIN, T. L., ORR, M. L. and PORTERFIELD, V. T. (1944b). Toxicology of dichloromethane (methylene chloride). I. Studies on effects of daily inhalation. *J. industr. Hyg.*, **26**, 8.

HEPPLE, R. A. (1927). An unusual case of poisoning. *J. R. Army med. Cps.*, **49**, 442.

HERZBERG, M. (1934). Histology of tissues taken from animals killed by prolonged administration of concentrated vapors of trichlorethylene. *Curr. Res. Anaesth.*, **13**, 203.

HEUPER, W. C. and SMITH, C. (1935). Fatal ethylene dichlorid poisoning. *Amer. J. med. Sci.*, **189**, 778.

HEWER, LANGTON (1948). Recent advances in anaesthesia. J. and A. Churchill, London.

HIGHMAN, B. and HEPPEL, L. A. (1946). Toxicology of 1–2 dichloropropane (propylene dichloride). III. Pathologic changes produced by a short series of daily exposures. *Arch. Path.*, **42**, 525.

HOFBAUER, (quoted in Occupation and Health, 1929). Chloroform. Encyclopaedia of Hygiene, Pathology and Social Welfare, p.436, issued by International Labour Office, Geneva, 1930.

HOFMANN, H. E. and REID, E. W. (1929). Cellulose acetate lacquers. *Industr. Engng. Chem.*, **21**, 955.

HOYT, L. F. (1928a). Some fumigation tests with ethylene dichloride-carbon tetrachloride mixture. *Industr. Engng. Chem.*, **20**, 460.

HOYT, L. F. (1928b). Comparative tests with certain fumigants. *Industr. Engng. Chem.*, **20**, 835.

HOYT, L. F. (1928c). Further fumigation tests with ethylene dichloride-carbon tetrachloride mixture. *Industr. Engng. Chem.*, **20**, 931.

HULST, A. J., STEENHAUER, A. J. and KLEDDE, D. L. (1946). Fatal poisoning with dichloroethane. *Ned. Tydschrift. Geneesk.*, **90**, 406.

HUMPHREY, J. H. and McCLELLAND, M. (1944). Cranial-nerve palsies with herpes following general anaesthesia. *Brit. med. J.*, i, 315.

INMAN, C. (1915). Lipase studies. II. On experimental chloroform necrosis of the liver. *J. med. Res.*, **32**, 73.

ISENSCHMID, R. and KUNZ, Z. (1935). Gefahren moderner gewerblicher Gifte. Polyneuritis mit Retrobulbär-neuritis nach Arbeit mit "Tri". *Schweiz. med. Wschr.*, **65**, 530, 612.

JACKSON, D. E. (1934). Study of analgesia and anesthesia, with special reference to such substances as trichloroethylene and vinesthene (divinyl ether), together with apparatus for their administration. *Curr. Res. Anesth.*, **13**, 198.

JOACHIMOGLU, G. (1921a). Vergleichende Untersuchungen über die antiseptische Wirkung einiger Chlorderivate des Methans, Äthans und Äthylens. *Biochem. Z.*, **124**, 130.

JOACHIMOGLU, G. (1921b). Die Pharmakologie des Trichloräthylens (Chlorylen Kahlbaum). *Klin. Wschr.*, **58**, 147.

JOACHIMOGLU, G. (1921c).Die Wirkung einiger Verwandten des Chloroforms mit besonderer Berücksichtigung der Traubeschen Theorie über die Wirkung der Narkotica der Fettreihe. *Biochem. Z.*, **120**, 203.

JOACHIMOGLU, G. (1925). Über die Wirkung einiger Narkotica der Fettreihe auf die glatte Muskulatur des Blutegels. *Biochem. Z.*, **156**, 224.

JORDAN, O. (1932). Chemische Technologie der Lösungsmittel. Springer, Berlin.

KALINOWSKI, L. (1927). Gewerbliche Sensibilitätslähmungen des Trigeminus. *Z. ges. Neurol. Psychiat.*, **110**, 245.

KEHRER, J. K. W. and OUDENDAL, A. J. F. (1926). Poisoning with carbon tetrachloride. *Ned. Tijdschr. Geneesk.* **70**, I, 1170. (Abstr. in *Chem. Abs.*, **21**, 610.)

KIESSLING, W. (1921). Vergleichende Untersuchungen über die Wirkung einiger Chlorderivate des Methans, Äthans and Äthylens am isolierten Froschherzen. *Biochem., Z.*, **114**, 292.

KIONKA, H. (1931). Vergiftungsgefahr bei der Verwendung von Tetrachlorkohlenstoff zerstäubenden Feuerlöschapparaten. *Münch. med. Wschr.*, **78**, 2107.

KISTLER, G. H. and LUCKHARDT, A. B. (1929). Pharmacology of some ethylene halogen compounds. *Curr. Res. Anaesth.*, **8**, 65.

KOELSCH, F. (1915). Gewerbliche Vergiftungen durch Zelluloidlacke in der Flugzeugindustrie. *Münch. med. Wschr.*, **62**, 1567.

KOELSCH, F. (1916). Zur Toxikologie des Tetrachlormethans und Tetrachloräthans. *Zbl. GewHyg.*, **4**, 69.

KOELSCH, F. (1916). Vergiftungen; aliphatische Verbindungen. *Handbuch soz. Hyg.*, **2**, 352.

KOELSCH, F. (1916). Vergiftungen; cyclische Verbindungen. *Handbuch soz. Hyg.*, **2**, 390.

KOHN-ABREST, E. (1924). Recherche des poisons gazeux dans le sang. *C. R. Acad. Sci., Paris*, **179**, 903.
KOHN-ABREST, E. (1924). Toxicity of industrial solvents. *Chem. Age.*, *N.Y.*, **32**, 199.
KUMMETH, J. (1933). Neue Versuche über die Giftwirkung des Tetra bei Katzen. Dissertation, Würzburg.
LAMSON, P. D., GARDNER, G. H., GUSTAFSON, R. K., MAIRE, E. D., MCLEAN, A. J. and WELLS, H. S. (1924). Pharmacology and toxicology of carbon tetrachloride. *J. Pharmacol.*, **22**, 215.
LAMSON, P. D. ROBBINS, B. H. and WARD, C. B. (1929). Pharmacology and toxicology of tetrachlorethylene. *Amer. J. Hyg.*, **9**, 430.
LANDE, P. and DERVILLÉE, P. (1934). Recherches expérimentales sur l'action toxique, chez le lapin, du tétrachlorure de carbone commercial, en ingestion. *C.R. Soc. Biol.*, **116**, 225.
LANDE, P. and DERVILLÉE, P. (1935). À propos d'un cas d'intoxication par les vapeurs de tétrachlorure de carbone. *Ann. Méd. lég.*, **15**, 21.
LANDE, P. and DERVILLÉE, P. (1935). Recherches expérimentales sur l'action toxique de tétrachlorure de carbone. *Ann. Méd. lég.*, **15**, 25.
LATTES, L. (1934). Tetrakohlenstoffvergiftung, akut tödliche, medizinale. *Samml. Vergiftungsf.*, **5**, 103, A435.
LAZAREW, N. W. (1929). Über die narkotische Wirkungskraft der Dämpfe der Chlorderivaten, des Methans, des Äthans und des Äthylens. *Arch. exp. Path. Pharmak.*, **141**, 19.
LECHELLE (1944). Quoted by Perrault, 1944, *Arch. Mal. prof.*, **6**, *Discussions*, 397.
LECORNU and PECKER (1932). Intoxication par le tétrachlorure de carbone. *Brux-méd.*, **12**, 480.
LEHMANN, K. B. (1911). Experimentelle Studien über den Einfluss technisch und hygienisch wichtiger Gase und Dämpfe auf den Organismus. Die gechlorten Kohlenwasserstoffe der Fettreihe. *Arch. Hyg., Berl.*, **74**, 1.
LEHMANN, K. B. (1930). Führt die technische Verwendung von Tetrachlorkohlenstoff zu hygienischen Gefahren? *Zbl. GewHyg.*, **17**, 123.
LEHMANN, K. B. and FLURY, F. (1943). Toxicology and hygiene of industrial solvents. (Tr. by E. King and H. F. Smith, jr.) Williams and Wilkins, Baltimore.
LEHMANN, K. B. and HASEGEWA (1910). Studien über die Absorption chlorierter Kohlenwasserstoffe aus der Luft durch Tier und Mensch. *Arch. Hyg., Berl.*, **72**, 327.
LEHMANN, K. B. and SCHMIDT-KEHL, L. (1936). Die 13 wichtigsten Chlorkohlenwasserstoffe der Fettreihe von Standpunkt der Gewerbehygiene. *Arch. Hyg., Berl.*, **116**, 131.
LEHNHERR, E. R. (1935). Acute carbon tetrachloride poisoning. *Arch. intern. Med.*, **56**, 98.
LEJEUNE, E. (1934). Schwierigkeiten der Diagnose beruflicher Vergiftungen für den praktischen Ärzt. Tetrachloräthanvergiftungen. *Arch. Gewerbepath. Gewerbehyg.*, **5**, 274.
LEONCINI, F. (1934). Sopra un caso di avvelenamento acuto mortale da tetracloruro di carbonio. *Rass. med. Lav. industr.*, **5**, 6.
LÉRI, A. and BREITEL (1922). La polynévrite chlorique (polynévrites par tétrachloréthane) chez des perlières. *Bull. Soc. méd. Hôp., Paris*, **46**, 1406.
LEWIN, L. (1920). Über giftige Extraktionsmittel für Fette, Waschse, Harze, und andereähnliche wasserunlösliche Stoffe. *Z. Dtsch. Ol- u. Fettindustr.*, **40**, 421.
LEWIN, L. (1929). Gifte und Vergiftungen. 4th Aufl. Springer, Berlin.
LOEWY, J. (1935). Die chronische Vergiftung mit Tetrachlorkohlenstoff. *Arch. Gewerbepath. Gewerbehyg.*, **6**, 157.
LUTZ, G. (1931). Nervendegeneration durch chronische Lösungsmittelvergiftung. *Arch. Gewerbepath. Gewerbehyg.*, **1**, 740.
LYON, B. B. V. (1935). Instance of possible cirrhosis of the liver induced by a hair tonic containing carbon tetrachloride. *Ann. intern. Med.*, **9**, 470.
MCAULEY, J. (1943). Trichlorethylene and trigeminal anaesthesia. *Brit. med. J.*, **ii**, 713.
MCCORD, C. P. (1932). Toxicity of trichlorethylene. *J. Amer. med. Ass.*, **99**, 409.
MCCORD, C. P., STERNER, J. H., KLINE, L. L. and WILLIAMS, P. E. (1946). Thymol-barbitol test in experimental carbon tetrachloride poisoning. *Occup. Med.*, **1**, 160.
MCGILL, C. M. (1946). Death and illness from the use of carbon tetrachloride. *Northw. Med.*, **45**, 169.
MCGUIRE, L. W. (1932). Carbon tetrachloride poisoning. *J. Amer. med. Ass.*, **99**, 988.
MACLAGAN, N. F., (1944). Thymol turbidity test; a new indicator of liver dysfunction. *Nature, Lond.*, **154**, 670.
MCMAHON, H. E. and WEISS, S. (1929). Carbon tetrachloride poisoning with macroscopic fat in the pulmonary artery. *Amer. J. Path.*, **5**, 623.
MCNALLY, W. D. (1937). Case of phosgene poisoning. *Industr. Med.*, **6**, 539.
MCNALLY, W. D. and FOSTVEDT, G. (1941). Ethylene dichloride poisoning. *Industr. Med.*, **10**, 373.
MAPLESTONE, P. A. and CHOPRA, R.N. (1933). Toxicity of tetrachlorethylene to cats. *Indian med. Gaz.*, **68**, 554.
MARTIN, W. B., DYKE, L. H. jr., CODDINGTON, F. L. and SNELL, A. M. (1946). Carbon tetrachloride poisoning. *Ann. intern. Med.*, **25**, 488.
MEERSSEMAN, F. (1934). Recherches sur l'insuffisance hépatique expérimentale. Les lésions hépatiques chez le cobaye au cours de l'intoxication aigüe par le phosphore, le chloroforme et le tétrachloréthane. *C. R. Soc. Biol., Paris*, **117**, 931.

MEERSSEMAN, F., PERROT, L. and FRANQUE, E. (1934). Recherches sur l'insuffisance hépatique expérimentale. Le coefficient de Maillard chez le cobaye au cours de l'intoxication aigüe par le tétrachloréthane et le chloroforme. *C. R. Soc. Biol., Paris*, **117**, 934.

MELICK, W. F. (1946). Acute toxic nephrosis due to poisoning by carbon tetrachloride. *J. Urol.*, **55**, 342.

MERCK'S INDEX (1930). 4th ed. American Series, Merck, N. J.

MEYER, H. (1929). Untersuchungen über die Giftwirkung des Trichloräthylens besonders auf das Auges. *Klin. Mbl. Augenheilk.*, **82**, 309.

MEYER, K. H. and GOTTLIEB-BILLROTH, H. (1921). Theorie der Narkose durch Inhalationsanästhetika. *Hoppe-Seyl. Z.*, **112**, 55.

MEZEY, K. and STAUB, H. (1935). Giftwirkungen am isolierten Herzkammerstreifen des Frosches. I. Wirkung der Narcotica und Hypnotica. *Arch. exp. Path. Pharmak.*, **180**, 12.

MINOT, A. S. (1927). Relation of calcium to the toxicity of carbon tetrachloride in dogs. *Proc. Soc. exp. Biol., N.Y.*, **24**, 617.

MINOT, A. S. and CUTLER, J. T. (1928). Guanidine retention and calcium reserve as antagonistic factors in carbon tetrachloride and chloroform poisoning. *J. Soc. exp. Biol.*, **26**, 138.

MINOT, G. R. and SMITH, L. W. (1921). The blood in tetrachlorethane poisoning. *Arch. intern. Med.*, **28**, 687.

MÖHR, L. (1902). Über Blutveränderungen bei Vergiftungen mit Benzolkörpern. *Dtsch. med. Wschr.*, **28**, 73.

MØLLER, K. O. (1933). Some cases of carbon tetrachloride poisoning in connection with dry shampooing and dry cleaning, with a survey of the use and action of the substance. *J. industr. Hyg.*, **15**, 418.

MORSE, K. M. and GOLDBERG, L. (1943). Chlorinated solvent exposures at degreasing operations. *Industr. Med.*, **12**, 706.

MÜLLER, J. (1925). Vergleichende Untersuchungen über die narkotische und toxische Wirkung einiger Halogen-Kohlenwasserstoffe. *Arch. exp. Path. Pharmak.*, **109**, 276.

MÜLLER, L. (1931). Experimenteller Beitrag zur Tetrachloräthanvergiftung. *Arch. Gewerbepath. Gewerbehyg.*, **2**, 326.

NAVROZIJ (1925). Urcic. Delo (Russ.) **8**, 1637. (Cited by Flury and Zernik, 1931, in Schädliche Gase.)

NEBULONI, A. (1930). Apparecchio per lo studio delle intossicazioni per via inalatoria. *Med. d. Lavoro*, **21**, 399.

NELSON, R. L. (1934). Methylene blue in treatment of carbon tetrachloride poisoning. *Minn. Med.*, **17**, 344.

VON OETTINGEN, W. F. (1935). Toxic gases and vapors. *Industr. Med.*, **4**, 245.

VON OETTINGEN, W. F. (1937). Halogenated hydrocarbons; their toxicity and potential dangers. *J. industr. Hyg.*, **19**, 349.

OHNESORGE, G. (1930). Über Zaponlackvergiftung. *Dtsch. med. Wschr.*, **56**, 961.

OLSEN, J. C. (1933). Discussion of "Formation of phosgene in thermal decomposition of carbon tetrachloride": a paper by A. Hamilton. *Industr. Engng. Chem.*, **25**, 541.

OPIE, E. L. and ALFORD, L. B. (1915). Diet and the hepatic lesions of chloroform, phosphorus, or alcohol. *J. exp. Med.*, **21**, 1.

PAGNIEZ, P., PLICHET, A. and KOANG, N. K. (1932). Un cas d'intoxication par le tétrachlorure de carbone. *Progr. méd., Paris*, **59**, 1328.

PANTELITSCH, M. (1933). Versuche über die Wirkung gechlorter Methane und Äthane auf Mäuse. Dissertation, Würzburg.

PARMENTER, D. C. (1923). Further observations on the control and prevention of tetrachlorethane poisoning. *J. industr. Hyg.*, **5**, 159.

PATTY, F. A., SCHRENK, H. H. and YANT, W. P. (1932). Determination of small amounts of methyl chloride in air. *Industr. Engng. Chem.*, **4**, 259.

PEARSON, C. C. (1947). Carbon tetrachloride intoxication with acute hepatic and renal failure treated with peritoneal lavage: report of a case. *Proc. Mayo Clin.*, **22**, 314.

PERNELL, C. (1944). The collection and analysis of halogenated hydrocarbon vapors employing silica gel as an absorbing agent. *J. industr. Hyg.*, **26**, 331.

PERRAULT, M. (1944). À propos d'une ingestion de trichloréthylène. *Arch. Mal. prof.*, **6**, *Discussions*, 397.

PERSSON, H. (1934). Über Trichloräthylenvergiftung. *Acta med. scand.*, **59**, *Suppl.*, 410.

PLESSNER, W. (1916). Die Erkrankung des Trigeminus durch Trichloräthylenvergiftung. *Mschr. Psychiat. Neurol.*, **39**, 129.

PLÖTZ, W. (1920). Vergleichende Untersuchungen über die hämolytische Wirkung einiger Chlorderivate des Methans, Äthans und Äthylens. *Biochem. Z.*, **103**, 243.

POINDEXTER, C. A. and GREENE, C. H. (1934). Toxic cirrhosis of the liver. *J. Amer. med. Ass.*, **102**, 2015.

POWELL, J. F. (1945). Trichlorethylene: absorption, elimination and metabolism. *Brit. J. industr. Med.*, **2**, 142.

RAMBOUSEK, J. (1911). Gewerbliche Vergiftungen deren Vorkommen, Erscheinungen, Behandlung Verhütung. Veit, Leipzig.

REGNAULD, J. and VILLEJEAN (1884). Caractères differentiels du chloroforms et du chlorure de méthylène. *C. R. Soc. Biol., Paris*, **8**, 158.

REICH, H. (1934). Puran (Monochlorbenzol) Vergiftung bei einem zweijährigen Kinde. *Samml. Vergiftungsf.*, 5, 193, A463.
RETAIL CREDIT CO. OF ATLANTA (1931). Carbon tetrachloride as a health hazard in dry cleaning. *J. Amer. med. Ass.*, 97, 48.
REUSS, A. (1931). Neuere Versuche über die Giftigkeit des Tetrachlorkohlenstoffes. Dissertation, Würzburg.
RICHARDSON, B. W. (1867). On bichloride of methylene as a general anaesthetic. *Med. Times and Gazette*, pp. 423, 479.
RIMPAU, W. (1931). Die desinfizierende Wirkung des zur chemischen Reinigung benutzten Trichloräthylens. *Z. Hyg. InfektKr.*, 112, 202.
ROCHAIX, A. (1936). Sur les dangers inhérents à l'emploi de certains appareils à solvants chlorés pour le nettoyage des vêtements. *Ann. Hyg. publ., Paris*, 14, 61.
ROHOLM, K. (1933). Trikloraethylenforgiftning i industrien. *Ugeskr. Laeg.*, 95, 1183.
ROSENTHAL, S. M. and LILLIE, R. D. (1931). Functional and histological studies of the effect of fat ingestion upon the normal and damaged liver. *Amer. J. Physiol.*, 97, 131.
ROSIN, A. and DOLJANSKI, L. (1946). Pyroninophilic structures of liver cells in carbon tetrachloride poisoning. *Proc. Soc. exp. Biol., N.Y.*, 62, 62.
RUSS, J. M. (1930). Ethylene oxide and ethylene dichloride; two new fumigants. *Industr. Engng. Chem.*, 22, 844.
SANDILANDS, I. E. (1909). Dangers of carbon tetrachloride. *Lancet*, ii, 570.
SAPPINGTON, C. O. (1935). Control of occupational diseases by laboratory methods. *J. industr. Hyg.*, 17, 17.
SARTORIOUS, F. and BOEDECKER, W. (1931). Zur Frage der Gesundheitsschädigungen durch Bohnermassen. *Zbl. GewHyg.*, 18, 103.
SAYERS, R. T., YANT, W. P., WAITE, C.P. and PATTY, F.A. (1930). Acute response of guineapigs to vapours of some new commercial organic compounds. I. Ethylene dichloride. *Publ. Hlth. Rep., Wash.*, 45, 225.
VON SCHEURLEN and WITZKY, H. (1935). Ein Todesfall durch gewerbliche Tetrachlorkohlenstoffvergiftung. *Zbl. GewHyg.*, 22, 60.
SCHIBLER, W. (1929). Akute gelbe Leberatrophie durch Acetylentetrachlorid. *Schweiz. med. Wschr.*, 10, 1079.
SCHLINGMAN, A. S. and GRUHZIT, O. M. (1927). Studies on the toxicity of tetrachlorethylene, a new anthelmintic. *J. Amer. vet. med. Ass.*, 71, 189.
SCHOENES (1931). Dissertation, Würzburg, (cited by Zernik, 1933, *Ergebn. Hyg.*, 14, 209).
SCHULTE, H. F. (1945). Report on the quartz crystal industry. *Industr. Med.*, 14, 68.
SCHULTZE, E. (1920). Encephalo-myelomalazie als Unfallfolge nach gewerblicher Vergiftung (Tetrachloräthan?). *Berl. klin. Wschr.*, 57, 941.
SCHUTZ (1937–38). Über ein hepatorenales Syndrom bei Tetrachlorkohlenstoff-Vergiftung. *Arch. Gewerbepath. Gewerbehyg.*, 8, 469.
SCHWANDER, P. (1936). Über die Diffusion halogenisierter Kohlenwasserstoffe durch die Haut. *Arch. Gewerbepath. Gewerbehyg.*, 7, 109.
SEIFTER, J. (1944). Liver injury in dogs exposed to trichloroethylene. *J. industr. Hyg.*, 26, 250.
SHAFFER, C. B. and CRITCHFIELD, F. H. (1945). Nitrogen and sulfur content of liver tissue in relation to carbon tetrachloride exposure. *Proc. Soc. exp. Biol., N.Y.*, 59, 210.
SHERMAN, S. R. and BINDER, C. F. (1944). Hazards of carbon tetrachloride in present-day use. *Nav. med. Bull., Wash.*, 43, 590.
SIMPSON, J. Y. (1847). Cited in Chloroform and other anaesthetics by John Snow, 1858. Churchill, London.
SIMPSON, N., SNOW, J. and NUNNELY, T. (1848–1849). New anaesthetics. *Providence med. and surg. J.*
SMETANA, H. (1939). Nephrosis due to carbon tetrachloride. *Arch. intern. Med.*, 63, 760.
SMILLIE, W. G. and PESSÔA, S. B. (1923). Treatment of hookworm disease with carbon tetrachloride. *Amer. J. Hyg.*, 3, 35.
SMITH, A. R. (1947). Duodenal ulcer following exposure to carbon tetrachloride. *J. industr. Hyg.*, 29, 134.
SMITH, A. R. (1950). Optic atrophy following inhalation of carbon tetrachloride. *Arch. industr. Hyg. and occup. Med.*, 1, 348.
SMYTH, H. F. (1935). Carbon tetrachloride in industry. *Industr. Med.*, 4, 12.
SMYTH, H. F. and SMYTH, H. F. jr. (1936). Safe practices in the industrial use of carbon tetrachloride. *J. Amer. med. Ass.*, 107, 1683.
SMYTH, H. F., SMYTH, H. F., jr. and CARPENTER, C. P. (1936). Chronic toxicity of carbon tetrachloride: animal exposures and field studies. *J. industr. Hyg.*, 18, 277.
SNELL, A. M. (1947). Carbon tetrachloride intoxication treated by peritoneal lavage: clinical aspects. *Proc. Mayo Clin.*, 22, 327.
STARKENSTEIN, E., ROST, E. and PÖHL, J. (1929). Toxikologie, ein Lehrbuch für Aertze, Medizinalbeamte und Medizin-studierende. Urban and Schwarzenberg, Berlin.
STEINDORFF, K. (1922). Über die Wirkung einiger Chlorderivate des Methans, Äthans und Äthylens auf die Hornhaut des Tierauges. *v. Graefes Arch. Ophthal.*, 109, 252.
STERNER, J. H. (1949). Halogenated hydrocarbons, in Patty, F.A., Industrial hygiene and toxicology, II, 819. Interscience Publishers, New York.

STEWART, A. and WITTS, L. J. (1944). Chronic carbon tetrachloride intoxication. *Brit. J. industr. Med.*, **1**, 11.
STRIKER, C., GOLDBLATT, S., WARN, I. S. and JACKSON, D. E. (1935). Clinical experience with the use of trichlorethylene in the production of over 300 analgesias and anesthesias. *Curr. Res. Anesth.*, **14**, 68.
STÜBER, K. (1931). Gesundheitsschädigungen bei der gewerblichen Verwendung des Trichloräthylens und die Möglichkeiten ihrer Verhütung. *Arch. Gewerbepath. Gewerbehyg.*, **2**, 398.
TAKADA, Y. (1932). Effect of carbon tetrachloride on the blood and its age variations. *J. pediat. Tokyo*, No. **389**, 1988.
TAKAHASHI, M. (1929). Über das Stoffwechselverhaltnis bei der Kohlenstofftetrachloridvergiftung. *Jap. J. exp. Med.*, **7**, 417.
TARA, S. (1944–5). Le sang dans l'intoxication chronique par le trichloréthylène. *Arch. Mal. prof.*, **6**, 319.
TAYLOR, H. (1936). Experiments on the physiological properties of trichlorethylene. *J. industr. Hyg.*, **18**, 175.
TAYLOR, H. (1936). Unpublished experiments, by courtesy of Imperial Chemical Industries Ltd.
VAN THEMSCHE, M. E. (1934). Industrial toxicology; trichlorethylene. *Chem. Abstr.*, **28**, 7360.
THOMPSON, C. M. (1946). Pulmonary changes in carbon tetrachloride poisoning. *Amer. J. Roentgenol.*, **55**, 16.
THURESSON, F. (1942). Todesfall bei Anwendung eines Feuerlöschapparates mit Trichloräthylen. *Samml. Vergiftungsf.*, **12**, 81, A 911.
TIETZE, A. (1933). Klinische Beobachtungen zur Methylbromid- und Tetrachlorkohlenstoffvergiftung. *Arch. Gewerbepath. Gewerbehyg.*, **4**, 733.
TOMB, J. W. and HELMY, D. M. (1933). Toxicity of carbon tetrachloride and its compounds. *J. trop. Med. (Hyg.)*, **36**, 265, 334.
VALLÉE, C. and LECLERCQ, J. (1935). L'intoxication par le trichloréthylène. *Ann. Méd. lég.*, **15**, 10.
VALLERY-RADOT, P., MAURIC, G., DOMART, A. and GAUTHIER-VILLARS (1938). Lésions des reins et du foie au cours de l'intoxication par inhalation de tétrachlorure de carbone chez le lapin. *C.R. Soc. Biol., Paris*, **128**, 482.
VELEY, V. H. (1909). Dangers of the dry shampoo. *Lancet*, **ii**, 370, 1162.
VELEY, V. H. (1910). Examination of the physical and physiological properties of tetrachlorethane and trichlorethylene. *Proc. roy. Soc., B*, **82**, 217.
WALLER, A. D. (1909). Relative toxicity of chloroform ($CHCl_3$) and of carbon tetrachloride (CCl_4). *Lancet*, **ii**, 369.
WELLS, H. G. (1908). Chloroform necrosis of the liver. *Arch. intern. Med.*, **1**, 589.
WHIPPLE, G. H. (1912). Insusceptibility of pups to chloroform during the first three weeks of life. *J. exp. Med.*, **15**, 259.
WHIPPLE, G. H. and SPERRY, J. A. (1909). Chloroform poisoning: liver necrosis and repair. *Johns Hopk. Hosp. Bull.*, **20**, 278.
WILLCOX, W. H. (1914). Fatal case of poisoning by tetrachloride of ethane. *Lancet*, **ii**, 1489.
WILLCOX, W. H. (1916). Treatment of toxic jaundice due to tetrachloride poisoning. *Brit. med. J.*, **i**, 300.
WILLCOX, W. H. (1931). Toxic jaundice. *Lancet*, **ii**, 1, 57, 111.
WILLCOX, W. H. (1934). Toxic effects of substances of the carbon tetrachloride type. *Proc. R. Soc. med.*, **27**, 455.
WILLCOX, W. H., SPILSBURY, B. H. and LEGGE, T. M. (1915). Outbreak of toxic jaundice of a new type amongst aeroplane workers; its clinical and toxicological aspects. *Trans. med. Soc., Lond.*, **38**, 129.
WILSON, R. H. and BRIMLEY, D. R. (1944). Health hazards in the use of tetrachlorethane. *Industr. Med.*, **13**, 233.
WINKELBAUER, A. and MUSGER, A. (1933). Über die Verwendung von Trichloräthylen zur Händesterilisation. *Chirurg.*, **5**, 1.
WIRTSCHAFTER, Z. T. (1933). Toxic amblyopia and accompanying physiological disturbances in carbon tetrachloride intoxication. *Amer. J. publ. Hlth.*, **23**, 1035.
WIRTSCHAFTER, Z. T. and SCHWARTZ, E. D. (1939). Acute ethylene dichloride poisoning. *J. industr. Hyg.*, **21**, 126.
WITHERIDGE, W. N. and WALWORTH, H. T. (1940). Ventilation of a trichlorethylene degreaser. *J. industr. Hyg.*, **22**, 175.
WITTGENSTEIN, H. (1918). Pharmakologische Untersuchungen über Dichloräthylen als Narkotikum. *Arch. exp. Path. Pharmak.*, **83**, 235.
WOODS, W. W. (1946). Changes in the kidneys in carbon tetrachloride poisoning. *J. Path. Bact.*, **58**, 767.
WÜRM, E. (1931). Die gebräuchlichsten Lösungsmittel in der Gummiindustrie: ihre Gefähren und der Verhütung. *Arch. Gewerbepath. Gewerbehyg.*, **2**, 766.
WYBERT, E. (1934). Puranvergiftung bei einem zweijährigen Kinde. *Schweiz. med. Wschr.*, **15**, 561.
ZANGGER, H. (1929). Über die Bedeutung der fluchtige Gifte. *Schweiz. med. Wschr.*, **10**, 9, 325, 469.

ZANGGER, H. (1930). Über die modernen organischen Lösungsmittel. *Arch. Gewerbepath. Gewerbehyg.*, **1**, 77, 109, 145.
ZANGGER, H. (1931). Toxikologie neuerer Lösungsmittel. *Schweiz. Z. Unfallmed. Berufskrank*, No. 1.
ZANGGER, H. (1931). Vergällung (Denaturierung) als gewerbepathologisch-hygienisches Problem. *Arch. Gewerbepath. Gewerbehyg.*, **2**, 205.
ZERNIK, F. (1933). Neuere Erkenntnisse auf dem Gebiete der schädlichen Gase und Dämpfe. *Ergebn. Hyg. Bakt.*, **14**, 139.
ZOLLINGER, F. (1930). Über sechs Fälle von Tetrachloräthanvergiftung. *Schweiz. Z. Unfallmed. Berufskrank.*, **45**, 92.
ZOLLINGER, F. (1931). Ein Beitrag zur gewerbepathologischen Bedeutung des Tetrachloräthans. *Arch. Gewerbepath. Gewerbehyg.*, **2**, 298.

CHAPTER III

ALCOHOLS

1. Methyl Alcohol
(*Wood Spirit, Methanol, Columbia Spirit*)

$CH_3 \cdot OH$

PROPERTIES

A COLOURLESS, permanently neutral liquid of faint odour, scarcely distinguishable from ethyl alcohol. Prepared both by distillation of wood and synthetically from carbon monoxide and hydrogen. The wood alcohol contains considerable quantities of acetone and also other impurities such as allyl alcohol. The synthetic product, methanol, is nearly pure. An analysis by Agnoli (quoted by Benedicenti, 1935) of synthetic methanol manufactured in Turin showed this variety of alcohol to be free from allyl alcohol, allyl compounds, arsenic and arsenic compounds. Its chief impurities were as follows: aldehyde (in terms of formaldehyde) 0·001 per cent, acetone 0·02–0·08 per cent, amines 0·012 per cent, chlorine less than 1 p.p.m. Allyl alcohol is usually present in wood alcohol to the extent of about 0·5 per cent and, according to McCord (1932), is considerably more toxic than methyl alcohol by inhalation; by ingestion it is 150 times more toxic according to Sollmann (1920–21), Reid Hunt (1902), Reif (1923), Sayers and Yant (1930), etc.

British Standard Specification, No. 506 (1933). Sp.Gr. 0·799 maximum. B.R. 64·5–65·5° C., 95 per cent. Residue 0·01 per cent maximum. Aldehydes and ketones 0·06 per cent maximum. Acidity 0·004 per cent as acetic acid maximum. Total sulphur 0·001 per cent maximum.

Evaporation time, 6·3 (compared with ethyl ether, 1).

Miscible in all proportions with water and most organic solvents especially with the aromatic hydrocarbons, but not with linseed oil or petroleum hydrocarbons.

An attempt has been made to explain some of the discrepancies in the reports of the effects of comparable exposures to, or ingestion of, methyl alcohol, by its varying content of impurities, but it has been shown (Keeser, 1931) that highly purified methyl alcohol when ingested is toxic *per se*, and that its toxic action is specific, not merely characteristic of its nature as a primary alcohol but due probably to its oxidation within the body to formaldehyde.

Solvent for most resins, ethyl cellulose, and for "½ sec" cellulose nitrate, but not for cellulose acetate or high viscosity nitrate, except when a small proportion of acetone is added.

USES

(1) In the paint and varnish industry. Methyl alcohol has been used in the manufacture of nitrocellulose lacquers, which dry more quickly when made with this solvent than with ethyl alcohol; wood alcohol is also largely used as a solvent for shellac coverings; varnish removers also frequently contain methyl alcohol.

(2) In the dye industry.

(3) As a cleaning agent (*a*) for furniture and bricks, (*b*) for dry cleaning.

(4) In the hat industry as a stiffening agent for the frames.

(5) In the manufacture of artificial flowers.

(6) In boot and shoe manufacture as a solvent of the shellac cement used.

(7) In the manufacture of incandescent mantles as the solvent for collodion lacquer in which they are dipped (International Labour Office (1926), *Brochure No. 55*).
(8) In the chemical industry as a solvent for recrystallizing substances.
(9) In printing and lithography.
(10) In hairdressing as a constituent of hair lotions.
(11) As an antifreeze mixture in radiators; here it is rarely designated "wood alcohol" but usually "completely denatured" or "188–proof"
(12) As a motor-fuel, blended with petrol.
(13) In the manufacture of safety glass as a solvent for cellulose acetate.
(14) In the manufacture of photographic films.
(15) In the denaturing of alcohol (Zangger, 1931).
(16) In the manufacture of artificial silk (Kober and Hayhurst, 1924).
(17) In the manufacture of patent leather.
(18) In removing spray residues from apples (Neller, 1931).
(19) In the manufacture of xylonite.

TOXICITY

Methyl alcohol is a powerful narcotic with an additional severe cumulative metabolic effect owing to the toxic decomposition products formed in the body. It is also an irritant to mucous membranes. In both animals and human beings, but particularly in the latter, it has a special tendency to produce blindness which may become permanent. These toxic effects may be caused by inhalation of the vapour of methyl alcohol, as well as by ingestion, but apparently individual susceptibility plays a considerable role in the development of toxic symptoms during its industrial use.

Relative Toxicity

Compared with ethyl or other alcohols. Methyl alcohol given by injection to animals is less poisonous than ethyl alcohol, and still less than propyl, butyl or amyl alcohols, following Richardson's (1869) law that the toxicity of the alcohols belonging to the fatty acid series increases with their molecular weight. By oral administration and inhalation methyl alcohol is more toxic than ethyl alcohol. Thus, most workers who have studied both the comparative lethal doses for animals (Baer, 1898; Macht, 1920) and the effects of injection on special organs (Mario, 1929; Efron, 1885; Macht, 1920) have confirmed Richardson's law. The figures given differ somewhat according to the method of administration, but for intravenous injection Baer's results appear to be fairly representative. He gave the relative toxicities as follows:

Methyl alcohol	0·8
Ethyl "	1·0
Propyl "	2·0
Butyl "	3·0
Amyl "	4·0

Macht (1920), Kuno (1913, 1914), Fühner (1904) and others have arrived at the same conclusion as to the relative toxicity of the alcohols when injected, with respect to their effects on isolated organs and their inhibition of neuromuscular activity. By injection, when acting acutely on the heart and brain, ethyl alcohol has constantly proved more toxic than methyl. Thus, Munch and

Schwartze (1925) found the lethal dose for the rabbit to be 18 ml. per kg. body weight for methyl alcohol and 12·5 ml. for ethyl alcohol; Fühner (1921) in his experiments on the isolated frog's heart found the amount producing arrest to be 3·7 mols per litre for methyl alcohol as compared with 1·2 mols per litre for ethyl alcohol. Similarly, Gradinesco and Degan (1934) state that, as judged by the determination of the chronaxie on the frog nerve-muscle preparation, ethyl alcohol is more toxic than methyl in paralysing the nerve.

Macht (1920) and others have explained the difference in the secondary action of ethyl and methyl alcohols by stating that when methyl alcohol is introduced by mouth, it exerts its most deleterious effects through its decomposition products, especially formaldehyde and formic acid (Keeser, 1931). Henderson and Haggard (1927) and Colebatch (1946) also point out that while ethyl alcohol is more toxic for equal concentrations maintained in the body, its destruction in the body is more rapid, while Brückner (1924) emphasizes the fact that since methyl alcohol remains in the body five to ten times as long as the same quantity of ethyl alcohol, repeated administration leads to a cumulative effect of a highly toxic nature.

The remote or chronic effects of methyl alcohol, according to most observers, are considerably more toxic than those of ethyl alcohol. Rost and Braun (1926) found that the repeated oesophageal introduction of 5 to 10 ml. of 10 per cent solution of methanol to rabbits produced severe loss of weight and finally death, while according to Rost and Wolf (1925) ethyl alcohol in these amounts was tolerated for almost any length of time without causing any ill-effects.

It appears from the researches of Weese (1928) that Richardson's law of increasing toxicity holds for the inhalation of saturated but not of unsaturated concentrations of air-alcohol mixtures (corresponding more or less to acute and chronic effects respectively). Mice were exposed to the vapour of known amounts of ·methyl, ethyl, *n*-propyl, *iso*propyl, *n*-butyl, *iso*butyl, *sec*.- and *tert*.- butyl alcohols respectively in a 5 l. flask. The narcotic potency was estimated by the length of time necessary for equimolar amounts of the alcohols to produce narcosis, and the toxicity by the length of time for a known concentration to produce death. With saturated mixtures methyl alcohol showed the weakest narcotic and toxic effect. With unsaturated mixtures, however, methyl alcohol, though still very weakly narcotic, proved to be by far the most toxic. Daily repeated slow narcosis with methyl alcohol caused death after nine to eighteen repetitions, while the other alcohols in the same dosage were never lethal. This effect is explained by Weese as a summation effect of methyl alcohol.

Compared with acetone. That methyl alcohol is a comparatively weak poison if judged by the criterion of acute narcosis, but a strong one if judged by its secondary or cumulative effect, is supported by some experiments of Sklianskaya, Urieva and Mashbitz (1936) and Mashbitz, Sklianskaya and Urieva (1936). By examining its action on the isolated frog's heart the former showed that while, during the entrance phase of action on the heart, methyl alcohol produced a slighter depression of cardiac activity than acetone, it allowed much poorer recovery during the elimination phase.

In its narcotic action on white mice, methyl alcohol was in the experiments of Mashbitz *et al*. (1936) less powerful than acetone, but it caused a higher mortality rate, a result which they, like Pohl (1893), Keeser (1931) and others, explain by its slow elimination and oxidation to formic acid.

Dangerous Concentrations

Exact estimation of the minimal concentration of methyl alcohol in the air which might cause toxic symptoms is difficult, since the cumulative capacity of methyl alcohol on the body must be taken into consideration. According to Greenburg, Mayers, Goldwater and Burke (1938), 22 to 25 p.p.m. (0·32 mg. per l.) can be tolerated without hazard; Humperdinck (1941), on the basis of actual estimations made in the workshop where the case described on p. 214 occurred, suggests that efforts should be made to keep the methyl alcohol concentration under 1 mg. per l. (about 750 p.p.m.).

In the Answer to a Query in the *Journal of the American Medical Association* (1931) the International Labour Office regards a concentration of 100 p.p.m. as the threshold of toxicity, though Rogers (1945) states that "the presence of 200 p.p.m. has become widely accepted as the maximum permissible concentration for continuous exposure during an eight hour day". McCord (1932), arguing from his findings that 1,000 p.p.m. are dangerous to monkeys, states that the vapours from 1 oz. of methyl alcohol entering the human body constitute a threat to life when the exposure is distributed over 2 or 3 days. He states that "a practical hazard may readily arise from any open container, such as a pan or tub, in any small closed workroom, particularly if the methyl alcohol is agitated in processes of manipulation".

Yant, Schrenk and Sayers (1931) state that "there are conditions of exposure to methanol vapour in the air which cause no apparent harm; also that there are conditions which will produce serious poisoning". Cases of severe poisoning are usually reported from its use as a solvent for paints and varnishes in relatively small, confined and unventilated places.

Estimation in Air

The most useful method of estimation is the oxidation of the methanol with potassium permanganate, followed by the production of a colour with Schiff's reagent (a solution of basic fuchsin containing sodium bisulphite and sulphurous acid). The efficiency of this method has been investigated by Rogers (1945), using various sampling absorbers. He found either a midget impinger or a fritted glass bubbler more satisfactory in performance than a tower packed with glass beads, which gave an efficiency of only 37 to 87 per cent as compared with 91 to 96 per cent with the two other methods.

TOXIC EFFECTS IN ANIMALS

Metabolism

From experiments on animals it appears that the toxic action of methyl alcohol must be regarded as specific, and not characteristic only of its action as a primary alcohol.

Many of the earlier writers (Pohl, 1893, 1918; Kròl, 1913; Harnack, 1912) related its specific action to its decomposition within the body to formic acid, while later workers (Flury, 1928; Keeser, 1931) state that formaldehyde is the decomposition product formed. McCord (1932), on the other hand, obtained results which were not in accord with the theory that the secondary, as distinct from the narcotic, injury from methyl alcohol is due to formaldehyde. He failed to find formaldehyde in distillates of the tissues of animals poisoned with methyl alcohol either by inhalation or absorption; on the other hand he demonstrated the presence of methyl alcohol itself, and only when the distillates were subjected to oxidation under conditions which would transform methyl alcohol into formaldehyde did they give regularly positive tests for formaldehyde.

Flury (1928), however, states that formaldehyde has been found in the aqueous humour, cerebrospinal and abdominal fluids of rabbits poisoned with methyl alcohol, and Keeser (1931) claims to have demonstrated *in vitro* the capacity for transformation of methyl alcohol into formaldehyde in the tissues. He showed that when the brain substance and aqueous humour were incubated with both methyl alcohol and ammonium carbonate, no formaldehyde was found owing to the reaction between them. These substances together form hexamethylenetetramine, and Keeser demonstrated its presence in crystalline form. He also found that the administration of ammonium carbonate prevented organic lesions in animals poisoned with methyl alcohol. Leo (1925) obtained this result in dogs but not in mice, rabbits or rats, and he was unable to explain his results on the basis of a true acidosis. Keeser believes that failure by some workers to demonstrate formaldehyde in the tissues of animals poisoned with methyl alcohol is due to the fact that this reaction only takes place over a short period, and that such observations did not take place during that time. The formaldehyde is believed to act by interfering with the oxidizing processes of the cells, and formic acid by withdrawing oxygen from them. Taking this hypothesis as the basis of investigation into suitable methods of treatment for methyl alcohol poisoning, Uri and Jeney (1946) state that pectins, which form an important part of the cell membranes of green vegetables and fruits, act as a detoxicating agent by binding the methyl alcohol molecule and preventing its oxidation.

Another theory of the specific toxic action of methyl as contrasted with ethyl alcohol is that its unchanged molecule acts directly upon the blood. Egg (1927) and Weese (1928) relate this supposed action to the formation of a complex between methyl alcohol and the iron of the haemoglobin, while Simon (1933) suggests that it combines with the proteins of the blood and transforms them into a precipitable form. A hypothesis based on the investigation of sixteen cases of poisoning by ingestion (Røe, 1943) embodies most of the conflicting elements of the long controversy on formic acid versus formaldehyde as the specifically harmful decomposition product of methyl alcohol. This hypothesis suggests that the factor primarily responsible for the acidosis is formic acid, which forms a complex compound with the iron of the respiratory enzyme, thus inhibiting oxidation and causing an increase of lactic acid.

It appears to be universally agreed that the greater toxicity of methyl alcohol as compared with that of ethyl alcohol (at any rate by oral administration or inhalation) is due to its much slower oxidation in the body. Experiments with alcohol dehydrogenase have shown that methyl alcohol is oxidized at only one ninth the rate of ethyl alcohol and this process is inhibited by giving ethyl alcohol simultaneously (Zatman, 1946).

Excretion by the urine is slow, part being excreted as methanol and part as formic acid (Rost and Braun, 1926; Pohl, 1922). According to Rost and Braun, methanol administered orally to rabbits could be detected in the urine for 4 days, the maximum amount occurring usually on the second day, that of formic acid on the second and third days. Pohl gives these maximum excretions as occurring on the third and fourth days respectively. Excretion by human subjects has been shown to be increased by the simultaneous ingestion of ethyl alcohol, and Zatman (1946) suggests that ethyl alcohol may diminish the toxicity of methyl alcohol by permitting the excretion of a larger fraction of the ingested dose in an unchanged condition.

Absorption by inhalation of fairly high concentrations (10,000 p.p.m.) during repeated brief exposures, is apparently on about the same level as that for lower concentrations (450 to 500 p.p.m.) during continuous daily exposure (Sayers, Yant, Schrenk, Chornyak, Pearce, Patty and Linn, 1944). These workers found the blood concentration in both series of dogs investigated to be from 6·5 to 14 mg. per 100 ml.

Acute Poisoning

It has already been emphasized that methyl alcohol is a comparatively weak narcotic for animals, but a strong cumulative poison with lethal metabolic effects, which are related to its slow elimination and to its decomposition within the tissues to formaldehyde and formic acid.

It has been shown by Scott, Helz and McCord (1933) that the characteristic lesions of methyl alcohol poisoning in animals are degenerative, affecting only the parenchymal tissues and neurones with no injury to connective tissue, while Tyson and Schoenberg (1915) believed it to act by depriving the tissues of nutrition by interference with the circulation. Bracq (1944) states that fatty degeneration of the liver and haemorrhages of the kidney, bladder and central nervous system are constant findings in rats.

The eye lesions so characteristic of methyl alcohol poisoning in man are apparently not so constant in animals, but they have been observed (Scott, Helz and McCord, 1933; Mocard, 1933) following both inhalation of the vapour and absorption through the skin, and some experiments by Marinesco, Lissievici-Draganesco, Draganesco and Grigoresco (1929), in which the methyl alcohol was given by intra-oesophageal introduction, appear to indicate a special affinity of the eye tissues.

Lethal and Narcotic Concentrations

Lethal dose

By inhalation. The immediate lethal effect of methyl alcohol inhalations pushed to narcosis is not great. Recovery usually takes place on removal from exposure (Rost and Braun, 1926; Flury and Wirth, 1934; Mashbitz *et al.*, 1936), but death takes place at some later period, usually 1 to 3 days. This occurred with concentrations of 86 to 170 mg. per l. (66,000 p.p.m. to approximately saturation point) in Flury and Wirth's investigations, and with 40 to 200 mg. per l. (30,000 p.p.m. upwards) according to the duration of narcosis in those of Mashbitz *et al.* (1936). In Scott, Helz and McCord's (1933) animals (monkeys) short exposures of 1 hour daily to 40,000 p.p.m. (52 mg. per l.) produced little immediate effect but longer exposures were eventually lethal.

Narcotic dose

The concentration necessary for narcosis varies with the species of animals as well as with the length of exposure. Rats appear to be more sensitive than cats and dogs (Loewy and van der Heide, 1914; Flury and Wirth, 1934), while rabbits are the most resistant (Scott, Helz and McCord, 1933; Rost and Braun, 1926). Most of the investigations on acute narcosis have not been quantitative. Flury and Wirth (1934) found a saturated air mixture (which they state is under 170 mg. per l. (130,000 p.p.m.) at 20° C.) slightly narcotic to cats in 6 hours. With higher concentrations, when the alcohol was in vapour form and its exact concentration not determinable, deep narcosis was produced after 6 hours. More exact quantitative estimations of the narcotic dose have been made on rats and mice, however, as they are more sensitive than cats. In rats, slight narcosis is produced in 8 hours by a concentration of 29·5 mg. per l. (23,000 p.p.m.), and deep narcosis in 2½ hours by 67 mg. per l. (52,000 p.p.m.) (Loewy and van der Heide).

For mice the figures for deep narcosis according to concentration and length of exposure are shown in Table 35 and are taken from Mashbitz et al. (1936):

TABLE 35
Concentrations of methyl alcohol producing deep narcosis

Concentration		Time before deep narcosis (min.)
(mg./l.)	(p.p.m.) (approx.)	
200	160,000	94
133	100,000	91
120	90,000	95
100	76,000	89
80	60,000	134
60	46,000	153
40	30,000	190

Symptoms

The immediate symptoms are depression amounting to drowsiness, then excitation, with inco-ordination of movement and paralysis of the hind legs (Mashbitz et al.), tremor, stupor and unconsciousness (Flury and Wirth, 1934). With concentrations from 10 mg. per l. (7,500 p.p.m.) up to about 90 mg. per l. (69,000 p.p.m.), there is evidence of slight irritation of mucous membranes—salivation and lachrymation. The irritation becomes intolerable at higher concentrations (Flury and Wirth).

The narcosis produced by higher concentrations disappeared, in Mashbitz's experiments, usually in 1 to 2½ hours, but sometimes lasted 24 hours.

The effects of repeated administration of high concentrations are, as already stated, often lethal within a few days. In dogs exposed to concentrations of 1 per cent (10,000 p.p.m.) of methanol vapour for a total of 800 brief exposures of 3 minutes eight times a day, Sayers et al. (1944) observed no significant symptoms, abnormalities of the blood count or ophthalmoscopic findings.

Symptoms of involvement of the nervous system, apart from the depressant central effect evidenced by narcosis have been shown by failure of sight and partial deafness. Holden (1899) noted that "the hearing was not completely abolished in a dog given large doses of methyl alcohol by mouth" and evidence of peripheral neuritis was reported by McCord (1932).

Effect on the Eyes

Many observers from Holden (1899) and Birch-Hirschfeld (1901) to McCord (1932) have recorded the occurrence of blindness in experimental animals poisoned by methyl alcohol, though not all the animals received the alcohol by inhalation. In McCord's experiments, however, inhalation produced visual disturbance in various species of animals. In some small animals, the cornea became milk-white, the condition sometimes occurring early, and being preceded only by dilation of the pupil. Monkeys sometimes became apparently blind; often they recovered, but occasionally recurrences took place.

There has been much discussion as to whether the blindness produced by methyl alcohol is due primarily to lesions of the optic nerve itself or to degenerative changes in the ganglion cells of the retina.

Most of the early workers, including Birch-Hirschfeld and Holden, believed the primary change to be a degeneration of the retinal ganglion cells, preceding optic nerve degeneration; with this opinion Scott, Helz and McCord agree, since they found few animals in their series with optic nerve degeneration, the changes in the retina, oedema and patchy degeneration of ganglion cells being more constant. Other workers (de Schweinitz, 1931) have failed to show changes in the retinal cells of animals acutely or chronically poisoned with methyl alcohol. The fact that in some animals the blindness has been only transitory has also led to the suggestion (Kasass, 1913) that the retinal ganglion cell degeneration is due partly to oedema of the sheath of the optic nerve associated with circulatory disturbances in the choroid. An acid reaction of the aqueous humour in animals poisoned with methyl alcohol has been observed by Tyson and Schoenberg (1915) and by Grignolo (1913). A special affinity of the eye tissues for methyl alcohol appears to be shown by the results of Marinesco *et al.* (1929). When methyl and ethyl alcohols respectively were administered to rabbits by the oesophagus, the largest amounts of methyl alcohol were found to be fixed by the tissues of the ocular globe, while ethyl alcohol was found chiefly in the brain. According to Røe (1943) the retina contains more carbohydrate and consumes it more rapidly than other tissues of the eye. Methyl alcohol reduces the respiration of these tissues while causing general inhibition of the organic oxidation processes. Though there is general agreement as to the high non-immediate toxicity of methyl alcohol to animals, the results of Elhardt (1931) provide an exception. He found that methyl alcohol repeatedly injected into the crops of chickens was less injurious than propyl alcohol. The growth and vigour, feathering, combs and disposition of the chicks were unfavourably affected, but there was no evidence of blindness, deafness, or paralysis after 4 months of daily injections.

Effect on the Respiratory Centre

According to Gradinesco (1934) methyl alcohol acts primarily upon the respiratory centre, large doses producing typical respiratory failure. With small doses (intravenous injections of 1 to 2 ml. in the dog) he observed an increase of the respiratory amplitude and at first a lowering of blood pressure, but he states that the heart does not cease to contract until long after the respiration has ceased.

Effect on the Internal Organs

The widespread toxic manifestations of methyl alcohol poisoning were first observed by Poincaré in 1878, who recorded lesions in the liver, heart, kidneys and lungs of his experimental animals.

According to McCord the lesions found in the liver, kidneys and heart are essentially a parenchymatous degeneration. In the liver he observed that this sometimes went on to focal necrosis. Fatty degeneration of the liver was also observed by Poincaré (1878) and Weese (1928) (also in inhalation experiments). In the kidneys parenchymatous degeneration of the epithelium of the convoluted tubules was observed (McCord, 1932; Weese, 1928; Poincaré, 1878). In the lungs, these workers found various inflammatory lesions—desquamation of alveolar epithelium, severe broncho-pneumonia, oedema, congestion, etc.

Flury and Wirth (1934) also observed slight emphysema. In the heart, there was oedema, granular degeneration, and sometimes necrosis of muscle fibres. In Flury and Wirth's animals the heart blood remained fluid.

Effect on the Blood

Tyson and Schoenberg (1914) observed changes in the blood of animals inhaling methyl alcohol, which led them to believe that it had a direct action on the blood. They found an increase in erythrocytes, haemoglobin and total leucocytes; in the latter it was due to polymorphonuclears, the lymphocytes being decreased. Scott, Helz and McCord (1933) found evidence of a stimulative effect on the blood-forming organs, especially the spleen; the lymph nodes also were hyperplastic.

The question of a true acidosis, demonstrable by changes in the blood, has been much discussed. Kròl (1913) assumed that acidosis was present on the evidence of an increased ammonia excretion in the urine, but Loewy and Münzer (1923) found no essential disturbance of carbon dioxide binding in the blood and did not consider methyl alcohol poisoning a true acidosis, an opinion shared by Leo (1925). Tyson and Schoenberg (1915), however, state that an acid reaction of the serum, an increase of electro-conductivity and increased viscosity occurred in animals poisoned by methyl alcohol. The question apparently remains unsettled.

Effect on the Central Nervous System

The lesions found by Ruhle (1912) and Holden (1899) in the central nervous system of animals receiving methyl alcohol by subcutaneous injection or orally were mainly congestive and haemorrhagic in nature. Scott, Helz and McCord (1933) in their inhalation experiments did not find these conditions so clearly developed. They observed capillary congestion, oedema and patchy degeneration of neurones more prominent in the spinal cord than in the brain. They also found evidence of peripheral nerve involvement on staining. Flury and Wirth (1934) observed only congestion of the vessels of the pia.

Chronic Poisoning

The effect of repeated daily exposures to comparatively low concentrations of methyl alcohol vapour is apparently influenced by their duration. Scott, Helz and McCord (1933) found that 1,000 p.p.m. were tolerated for 7 hours for 6 days a week, but that the animals died when the daily exposure was prolonged to 16 hours. In Weese's (1928) experiments no quantitative data are given, but the alcohol was placed in a 15 l. flask and an unsaturated mixture, insufficient to produce narcosis, inhaled. Under these conditions there was no lethal effect and very little apparent injury during the life of the animal. Post mortem, the liver and kidney parenchyma showed slight fatty degeneration.

According to Loewy and van der Heide (1914) 2,000 p.p.m. will produce toxic effects if inhaled for long enough.

Poisoning by Skin Absorption

While Rost and Braun (1926) and Bertarelli (1934) have been unable to demonstrate any ill-effects of cutaneous absorption of methyl alcohol by animals, McCord and Cox (1931), Mocard (1933) and Scott, Helz and McCord (1933) state definitely that toxic effects are readily produced by absorption. That cutaneous absorption does occur was proved by Rost and Braun (1926), who detected the alcohol quantitatively in the urine of animals to whose skin it had been applied. According to both McCord and Mocard the pathological effects are practically the same by skin absorption as by inhalation. Mocard observed atrophy of the optic nerve, modification of the body lipoids and death after the repeated application of 30 ml. to monkeys, while Scott, Helz and McCord state that the toxic dose by skin absorption is 0·5 ml. per kg. body weight applied four times daily.

TOXIC EFFECTS IN MAN

Methyl alcohol has occupied a peculiar place among solvents on account of the temptation to use it as a beverage. Many of the fatal cases reported and also those of severe blindness have been due to drinking brandy adulterated with methyl alcohol, as in the outbreak of poisoning in Berlin in 1911 when approximately 70 people died and about 85 others were ill, and in the Ruhr in 1920 when 15 died and 3 were totally blinded (Monier Williams, 1929); similar effects are recorded from fruit juices diluted with methyl alcohol in Japan (Kaplan and Levreault, 1945), or from drinking methyl alcohol itself as in the cases in Hamburg described by Reif in 1923. Of the 275 cases recorded by Wood and Buller in 1904, including 122 deaths and 153 cases of blindness, only 6 were due strictly to industrial exposure; and of the 725 recorded by Baskerville in 1913, including 390 deaths, 90 cases of total and 85 of partial blindness were due to the drinking of methyl alcohol. There were other outbreaks of severe poisoning of a similar origin in Germany in 1922 and 1926.

For this reason the earlier literature on the subject tends to give the impression that methyl alcohol is one of the most dangerous of solvents. This impression is not supported by the cases in which definite proof exists that the toxic effects reported were entirely due to exposure to the vapour or to skin absorption. Certainly cases of severe poisoning have occurred where there has been severe exposure to the fumes of methyl alcohol in confined spaces, but the accounts of some of these cases (Gerbis, 1931) are accompanied by the suggestion that the patient had probably drunk some of the alcohol. Wood and Buller (1904) remark that "taking a drink from the supply of alcohol kept for dissolving gums in making varnishes is a very common habit among varnishers".

Bertarelli (1934) has called attention to the fact that there have been many instances of prolonged exposure without symptoms, as in the investigation of Yant, Schrenk and Sayers mentioned later (p. 213).

Individual susceptibility. It is believed by some authorities that individual sensitivity plays some part in the unequal incidence of toxic symptoms under conditions of equal exposure. Loewy (1912), in a questionnaire to forty doctors closely connected with chemical works producing or using methyl alcohol in large quantities, found no specific ill-effects reported, except two cases of dermatitis, and Koelsch (1921) and Yant, Schrenk and Sayers (1931) had similarly favourable experiences with men employed in the production of wood alcohol. Koelsch believes that the greatly varying accounts of the effects of industrial exposure to methyl alcohol are only explainable on the grounds of individual susceptibility. He quotes a typical case which he himself observed—that of a woman using a shoe cement called "Kernol" in which methyl alcohol appears to have been the chief though not the only constituent. This woman, who was debilitated and anaemic, complained of throat irritation, headache and burning of the eyes though no signs of eye irritation were visible, and suffered from gradual failure of sight leading to blindness. No other workers in the same room were affected, except for slight irritation of mucous membranes, and it was decided that this was a true case of methyl alcohol poisoning in an individual with a special idiosyncrasy. Schwarzmann (1934) describes a case in whom nervous symptoms predominated, who was regarded as probably one of special susceptibility acquired by repeated exposure.

Industrial Poisoning

In the earlier literature, cases of industrial poisoning by methyl alcohol were regarded as questionable, especially when the symptoms were severe, although in 1855 MacFarlan referred to eye affections among cabinet workers, hatters, etc. and its industrial dangers were stressed by Poincaré in 1878. The toxicity of methyl alcohol from ingestion was of course recognized comparatively early, and the descriptions of the symptoms of such cases are very like those of severe poisoning from inhalation.

Among the first cases of industrial poisoning by absorption through the lungs and skin were those recorded by Patillo and Wood in 1899 of men working in beer vats, with blindness as the outstanding symptom. In 1906, when a measure was passed in the U.S.A. for providing revenue-free denatured grain alcohol, the National Society for the Prevention of Blindness presented a number of histories of death or blindness in men using wood alcohol varnish on the interior of beer vats (Hamilton, 1925). Five fatal cases, in whom the exposure was only 2 to 4 days, were recorded by the New York State Department of Labor in 1917. A rather unusual effect was described by Goltdammer in 1878, when a 26-year-old worker accidentally inhaled hot methyl alcohol vapour and died from purulent bronchitis.

Two cases of poisoning in women, in whom the symptoms included dimmed or blurred vision, were described in the Special Bulletin of the New York State Department of Labor in 1917 from the use of a shellac covering in methyl alcohol solution for lead pencils. A case of peripheral neuritis with reduction in the field of vision and a central scotoma was reported to the Home Office in 1924, the patient being employed in varnishing metal with a solution of shellac in wood spirit. Poisoning in lacquer workers was also described by Wood and Buller in 1904.

A case of blindness was reported by Robinson (1918) from the use of a black dye in which the solvent contained only 4 per cent of methyl alcohol, while Hamilton (1925) states that there was "a good deal of illness" among men in a dye-works in which the solvent was denatured alcohol containing 10 per cent of methyl alcohol.

Up to 1912, about 100 cases of amblyopia and death from inhalation had been reported; in 1913, Baskerville added a further 64 cases from inhalation in America, occurring in painters, lacquerers and chimney sweeps using methyl alcohol to clean the bricks; the alcohol used was either undiluted or 50 per cent strength in badly ventilated rooms. Koelsch (1921) also quotes numerous cases of blindness during the war period in painters, lacquerers and polishers. Most of these cases are cited by McCord (1932), who also states that "the practice of freely sponging women's dresses in dry-cleaning establishments, transferring methyl alcohol on a sponge from a basin of that substance, is to be regarded as a practical hazard".

Danbury hatters presented 75 affidavits from men who had suffered impairment of sight or injury to health from using shellac containing wood alcohol for stiffening felt hats. Twenty cases with slight symptoms, chiefly dermatitis, conjunctivitis and anaemia were reported in the New York Special Bulletin in

1917 in a factory making artificial flowers, where the concentration was about 2 vols. per 10,000 of air at a distance of 6 ft., while the odour of the vapour was noticeable 75 ft. from the place of dipping and drying. Five cases of poisoning in the boot and shoe industry were mentioned in Hayhurst's occupational survey in Ohio in 1915.

There has been much controversy as to the danger from the use of methanol as an antifreeze mixture. In 1931 the Legislature of Arkansas passed an Act regulating the sale of antifreeze mixtures containing more than 10 per cent methyl alcohol. It is considered by Reid Hunt (1930) that the chief danger to be apprehended from its use is the temptation to drink it, while Yant, Schrenk and Sayers (1931) found no evidence of ill-health in drivers of trucks where methyl alcohol antifreeze had been used exclusively for $2\frac{1}{2}$ months. On the other hand Trumper (1931) points out that the synthetic methyl alcohol used as an antifreeze has no odour, and that therefore the automobilist has no warning when it is evaporating and he is breathing it. In discussing this statement de Schweinitz (1931) stated that with proper ventilation there should be no risk of visual disturbances from methyl alcohol intoxication. A fatal case, with two other cases of temporary unconsciousness, was reported to the Home Office in 1923, following leakage of motor spirit containing methanol from a storage tank.

Six men employed in the patent-leather industry where methyl alcohol was used as a solvent for the nitrocellulose were examined by the Factory Department in 1929 but no ill-effects were observed.

Apples for storage are dipped in methyl alcohol before being washed with hydrochloric acid; in connection with the possible danger of operators becoming poisoned by the fumes of the liquid, Neller (1931) remarks that the incoming fruit is generally cool or cold (10 to 15° C.) and vaporization of the methyl alcohol is therefore slight.

Poisoning by Skin Absorption

McCord (1932) and Mocard (1933) have shown that poisoning by methyl alcohol can be produced in animals by skin absorption (p. 210). It appears that in man also, toxic effects can be produced in this way. Wood and Buller (1904) recorded a case of poisoning following the application of methyl alcohol for rheumatism, while Brown (quoted by Ziegler, 1921) described toxic symptoms in a painter who spilled it over his feet, soaking his shoes. In two cases of amblyopia observed by Tyson (1912) in females polishing lead pencils, there was a suggestion that the methyl alcohol had been used for washing the hands and bathing the head and might have been absorbed.

Poisoning by Inhalation

Symptoms

Most of the cases recorded must be regarded as examples of acute intoxication. It would appear that exposure to concentrations high enough to produce any symptoms at all may be enough to cause the rapid development of eye lesions, beginning with blurring or dimming of the vision and photophobia.

Koelsch (1921) remarks, however, that acute signs of eye involvement may suddenly be superimposed upon slighter, more chronic (resorptive) symptoms and lead to complete blindness. The case reported by Schwarzmann (1934) in which the symptoms were chiefly related to the central nervous system, and those of ocular injury due to the use of a shoe cement recorded by Vayntsvayg, Kleybes and Pasternak (1933), must be regarded as chronic rather than acute poisoning by methyl alcohol; so also must a case recorded by Humperdinck

(194[1]), in which amaurosis supervened on previous abnormalities of vision. The patient, a man employed in handling material soaked in methyl alcohol (30 to 40 per cent content), showed also symptoms of liver disturbance. No other case of poisoning had been observed in workers in the same process for 10 years, so that individual susceptibility probably played some part, but conditions of black-out and bad weather had raised the concentration of methyl alcohol in the atmosphere to 5 to 10 mg. per l., and there was also opportunity for absorption through the skin.

Severe exposure may be followed in a very short time by giddiness, unconsciousness, sighing respiration, depressed cardiac action, cold sweats, coma and death, sometimes preceded by violent delirium, as in the Home Office case of 1923. In Gerbis' (1931) case, death was preceded by several attacks of vomiting and dyspnoea.

Less severe exposure may cause acute gastro-intestinal disturbance, headache, vertigo, feeling of intoxication, dilated pupils, accompanied by fogginess of vision and even sudden blindness.

Local Irritation

The fatal case of purulent bronchitis recorded by Goltdammer in 1878 has already been referred to. Koelsch (1921) also describes cases showing respiratory inflammation, progressing to broncho-pneumonia, in workers using wood oil as a constituent of "Zapon" lacquer, and Dublin and Leiboff (1922) mentioned inflammation of the throat and respiratory mucous membranes following inhalation of methyl alcohol fumes. Local eye irritation in the form of conjunctivitis was also observed by Koelsch, and he quotes Grunow (1912) as having reported severe conjunctivitis with ulceration in workers employed in dipping collars in a mixture of methyl alcohol and collodion.

Eye Lesions

Loss of vision, most frequently permanent, is the most important consequence of methyl alcohol poisoning.

Amblyopia, which is bilateral, and which occurs sometimes within a few hours of exposure, sometimes after several days, is often transient at first but recurs and may be followed by complete and permanent loss of sight. In the case recorded by Hawes in 1905, for example, total blindness supervened within 5 days after transitory improvement. Before actual blindness occurs there may be pain and tenderness on pressure of the eyes, dilatation of the pupils, fogginess and blurring of the vision, diplopia or ptosis. Ophthalmological examination shows an increasing contraction of the visual fields for form and colour, congestion of the sclera, pallor of the disc, central scotoma and much reduction of visual acuity. In a typical case investigated by Tyson in 1912, for example, colours were not recognized, the field for white was contracted about 10°, and vision was $R\frac{1}{200}$ and $L\frac{3}{200}$.

Pathology of eye lesions. It is generally believed that atrophy of the optic nerve is the basis of the failure of vision (Hale, 1901). The exact aetiology of the optic atrophy and especially whether it is preceded by retinal changes is still in question. De Schweinitz (1931) explains it as an optic neuritis, temporarily subsiding, but followed by atrophy.

Ziegler (1921) considers that the original lesion is a slow retrobulbar neuritis, which is followed by shrinking of the nerve head or by sudden sclerosis, occurring rapidly and with no appearance of shrinkage. This theory is supported by the results of a recent investigation by Joiris (1935) of a case of acute methyl alcohol poisoning by ingestion. He states that the retrobulbar neuritis observed in methyl alcohol poisoning is distinguishable from retrobulbar neuritis due to other causes (e.g. syphilis, diabetes) by the presence of a positive transitory central scotoma. From the results on experimental animals, however, and from the fact that a few observers have found retinal changes similar to those found in animals, McCord (1932) and others are inclined to think that the retinal ganglion cells are affected first, and that the changes in the optic nerve may be due not only to the direct action of methyl alcohol but also to accompanying oedema.

The association between restoration of the collateral circulation, disappearance of oedema, and partial restoration of vision pointed out by Kasass in 1913 may possibly be the explanation of actual cases of transitory blindness due to methyl alcohol, e.g. some of those recorded by the New York State Department of Labor in 1917, where several chronic cases showed recurrent transitory blindness with gradual recovery. From a study of cases of blindness due to ingestion of methyl alcohol and especially from the results of lumbar puncture in these cases (Mathewson and Alexander, 1932) it appears that this supposition of acute oedema as the primary change is correct. Mathewson and Alexander, and Hämäläinen and Teräskeli (1928) have reported remarkably rapid improvement of vision following lumbar puncture and the former workers state that "it is clear . . . that the action of the formaldehyde" (following Keeser's (1931) work on the decomposition of methyl alcohol in the body into nascent formaldehyde and formic acid) "on the cells of the optic nerve and retina is at first not so great as to cause the death of the cells but is strong enough to cause marked dysfunction. The actual change in the cells is probably oedema . . . It is also clear that while this poison will eventually kill these cells, as has resulted in hundreds of cases, still some length of time must elapse before this happens, or putting it in another way, the formaldehyde keeps on exerting its deleterious action for a long time". They believe that the secondary later deterioration of vision characteristic of methyl alcohol poisoning is due to subsequent atrophy of severely injured retinal vessels. The hypothesis of Røe (1943) that methyl alcohol acts on the retinal tissues by inhibiting the cellular respiration has already been referred to on pp. 206, 209. He suggests treatment by administration of ethyl alcohol because it may displace methyl alcohol from the cell surface and interrupt its oxidation into formic acid.

Other Effects on the Nervous System

Deafness. A case of amblyopia with increasing deafness was reported by Tyson in 1912, and he suggested that the auditory as well as the optic nerve might be affected.

Peripheral neuritis. Neuritis of the upper extremities was reported by Jelliffe in 1905 in two painters using wood alcohol varnish, and peripheral neuritis was also present in a case, showing reduction in the field of vision and a central scotoma, reported to the Home Office in 1924. Ptosis, due to paresis of the ocular muscles, has also been recorded (Tyson, 1912).

Cerebrospinal lesions. A case presenting unusual features has been reported by Schwarzmann (1934), who regards it as one of acquired susceptibility by long exposure. The symptoms in this case were chiefly located in the nervous system and simulated those of multiple sclerosis, disseminated encephalomyelitis or cerebrospinal syphilis. The subject, who was a worker in a straw-hat factory, had used "technical formaldehyde", containing 12 to 16 per cent of methyl alcohol, and Schwarzmann suggests that in a very similar case described by Weger in 1927 and attributed by him to formaldehyde poisoning the symptoms may also have been due to methyl alcohol. These symptoms included nystagmus, ataxia, a strongly positive Romberg sign, hyperaesthesia of the left thigh and soles of both feet, paraesthesia of the left leg and paresis of the facial and abducens nerves. There was also giddiness, vomiting, loss of appetite, alternating constipation and diarrhoea. There was practically no diminution of vision, but there was a double central colour scotoma. The condition showed improvement during the intervals of non-exposure.

2. "Wood Spirit"
(*Wood Naphtha, Wood Alcohol*)

CONSTITUTION AND PROPERTIES

Crude wood spirit, obtained by distilling pyroligneous acid, is a liquid of variable constitution, containing usually about 80 per cent of methyl alcohol, with varying quantities of acetone, methyl acetate, acetal, allyl alcohol, acetaldehyde and methylamine. It is a greenish-yellow liquid with a disagreeable odour. No standard specification of its properties exists.

USES

In the paint and varnish industry, especially as a constituent of varnishing solutions for bakelite, usually in combination with other substances, such as formaldehyde and phenol.

TOXICITY

While no animal experiments on "wood spirit" itself have been carried out, and no cases of industrial poisoning from its exclusive use reported, it would be expected that its effects would be at least as toxic as those of either methyl alcohol or acetone, its chief constituents. If allyl alcohol is present in considerable proportion, wood spirit would be more toxic than pure methyl alcohol, since according to McCord (1932) allyl alcohol is a highly toxic substance (by feeding experiments to animals 150 times as toxic as methyl alcohol) leading, when inhaled, to pulmonary oedema, haemorrhage, gastro-enteritis, vomiting, diarrhoea, nephritis and haematuria (p. 241).

The toxic effects of mixtures of methyl alcohol and acetone on animals have been investigated by Mashbitz, Sklianskaya and Urieva (1936), who concluded that in the combined action of these substances, acetone predominates in high concentrations and methyl alcohol in low concentrations.

TOXIC EFFECTS IN MAN

Twelve workers using a bakelite varnishing solution containing wood spirit were examined by the Home Office in 1929. Three of these complained of irritation of the eyes, but this was attributed to the action of either the phenol or the formaldehyde which were also constituents of the varnish. A fourth man had had an attack of "intoxication", not due to drinking alcohol, but which his doctors attributed to fumes inhaled during his work. This effect was not definitely attributed to the varnish solvent, but there was no recurrence of the condition after exhaust ventilation was installed.

ALCOHOLS

3. Ethyl Alcohol
(*Ethanol, Spirits of Wine, Industrial Spirit, Methylated Spirit*)
$C_2H_5 \cdot OH$, i.e. $CH_3 \cdot CH_2 \cdot OH$

PROPERTIES

A colourless, mobile liquid with a characteristic spirituous odour. Burns with an almost non-luminous flame. Exceedingly hygroscopic and miscible with water and ether in all proportions.

Prepared chiefly from molasses, also by fermentation of wheat, barley, maize, rice, potatoes, etc.; and synthetically by the hydration of acetylene or ethylene (International Labour Office, *Brochure No.* 50).

Pure anhydrous ethyl alcohol (*absolute alcohol*). B.P. 78·3° C. Sp.Gr. 0·7937. In the United States the most readily available form of anhydrous ethyl alcohol is "Formild 12A", which contains 5 per cent benzene.

British Standard Specification, No. 507 (1933). *Not denatured.* Sp.Gr. 0·8171 (66° O.P.; 92 per cent by weight; 94·7 per cent by volume). Aldehydes 0·1 per cent maximum. Residue 0·01 per cent maximum.

Denatured alcohol (*methylated spirit*). Usually consists of British Standard alcohol to which 5 per cent of methyl alcohol has been added. Sometimes other denaturants and aniline dyes are present which have been said to give rise to skin irritation.

Excellent solvent for many resins and oils, also sulphur and phosphorus to some extent. Anhydrous ethyl alcohol dissolves ethyl cellulose, colophony and sandarac, etc. The addition of 20 to 30 per cent benzene makes it a solvent for benzyl cellulose, ester gum and cumarone.

USES

(1) In the manufacture of ether.
(2) In the manufacture of collodion.
(3) In the manufacture of artificial vinegar (acetic acid).
(4) In the manufacture of explosives.
(5) In the varnish and lacquer industries.
(6) In the photographic industry.
(7) In the preparation of tannin, chloral, etc. and pharmaceutical products in the chemical industry.
(8) In the artificial silk industry.
(9) In motor spirit.

TOXICITY

While it is known that inhalation of the vapour of ethyl alcohol to the point of intoxication produces the same effects as a toxic dose taken by the mouth, actual reports of intoxication by this route are rare. Hamilton (1934) remarks that "in industry . . . experience shows that the lowest member of the series, methyl alcohol, is the only one to be feared" while in the International Labour Office's *Brochure No.* 50, it is stated that the effects recorded by different authorities refer to different products used in denaturing methyl alcohol, pyridine and many others (Zangger, 1931), and that in any case such effects must be held to be of rare occurrence to-day. It has already been stated that ethyl alcohol is more acutely toxic than methyl alcohol, but that methyl alcohol is a much more dangerous chronic intoxicant. One of the chief reasons for this difference has already been discussed, viz. the rapid and ready oxidation of ethyl alcohol to carbon dioxide and water, as opposed to the slow oxidation of methyl alcohol to formic acid and formaldehyde with a cumulative effect. Vollmer (1931) has also shown that the toxic effects of ethyl alcohol are diminished if substances which increase tissue oxidation are given beforehand. No difference in toxicity between grain and synthetic ethyl alcohol has been found by oral, intraperitoneal or intravenous injection in animals (Barlow, 1936) or in man by oral ingestion (Muehlberger, 1935).

Estimation in Air

A method of estimating the amount of ethyl alcohol in air is described by Haggard and Greenberg (1934); it consists of the passage of a measured sample of air into a tube containing hot iodine pentoxide. The alcohol is thus decomposed with the liberation of iodine and hydriodic acid from the pentoxide. The volatile products are collected by absorption and are estimated directly by titration.

TOXIC EFFECTS IN ANIMALS
Acute Poisoning

Lethal and Narcotic Concentrations

Lethal dose

By oral administration. For rabbits 9·5 g. per kg. (Barlow, 1936) or 10 g. per kg. (Heffter, 1920–23).

By intraperitoneal injection. For rabbits 3·5 g. per kg. (Barlow, 1936).

By subcutaneous injection. For mice 4·73 g. per kg. (Vollmer, 1931).

By inhalation. For mice, 55 mg. per l. (30,000 p.p.m.), time not given (Bachem, 1927). For rats, 19 to 24 mg. per l. (10,000 to 12,500 p.p.m.) after 21 hours; 85 mg. per l. (45,000 p.p.m.) after 4½ hours. For guinea-pigs, 85 mg. per l. (45,000 p.p.m.) after 10¾ hours.

Narcotic dose

By inhalation. For rats, 12 mg. per l. (6,400 p.p.m.) for light narcosis; 19 to 24 mg. per l. (10,000 to 12,500 p.p.m.) for deep narcosis. For guinea-pigs, 12 mg. per l. (6,400 p.p.m.) for light narcosis; 85 mg. per l. (45,000 p.p.m.) for deep narcosis (Loewy and van der Heide, 1918).

Symptoms

The effects of inhalation of high concentrations are apparently identical with those of administration by mouth or intravenously. According to Gradinesco and Degan (1934) ethyl alcohol acts primarily upon the respiratory centre, producing in small doses a slowing of respiration and in large doses typical asphyxia from respiratory syncope. They state that the hearts of dogs, injected intravenously with ethyl alcohol, continued to contract long after respiration had ceased. According to Reid Hunt (1902) the acute effects of ethyl and methyl alcohols do not differ much, except that methyl alcohol intoxication is more prolonged, the coma lasting 2 to 4 days while that due to ethyl alcohol lasts only a few hours, but death is more rapid with ethyl alcohol given in the same dosage.

Animals subjected to inhalation, including fowls (Carpenter, 1929), become somnolent, react at first to stimuli, but later become completely narcotized and non-reactive.

ALCOHOLS   219

Table 36 shows the behaviour of the animals (rats and guinea-pigs) observed by Loewy and van der Heide (1918), whose results were summarized by Flury and Zernik (1931).

TABLE 36
Toxic effects of ethyl alcohol in rats and guinea-pigs

Concentration			Exposure (hr.)	Effect	
(vol. per cent)	(mg./l.)	(p.p.m.)		Rats	Guinea-pigs
0·3–0·45	5·6–8·5	3,000–4,500	6 8 24	No visible effect Sleepy Slight stupor	
0·64	12	6,400	8	Light narcosis	Light narcosis
0·7–1·15	18–22	7,000–11,500	Up to 24		Only light narcosis
1–1·25	19–24	10,000–12,500	4 8 21	Stupor Deep narcosis Death	
2–2·3	38–43	20,000–23,000	4 6½ 10 24	Light narcosis Side position Death	No special effect Stupor Stupor Deep narcosis
4·5	85	45,000	4 6½ 8¾	Deep narcosis Death	Stupor, light narcosis Deep narcosis, corneal reflex retained Death after 2 more hours

Changes in the Internal Organs

In the experiments of Weese (1928) repeated exposure to narcotic concentrations (4·5 to 6·7 vols. per cent) produced only moderate fatty infiltration of the liver and kidneys.

Chronic Poisoning

According to Weese (1928), repeated administration of non-narcotic doses of ethyl alcohol produces very different results from similar administration of methyl alcohol to animals. None of the animals died as the result of 36 inhalations and all recovered rapidly when removed from the anaesthetizing chamber. Examination of the internal organs revealed only slight and reversible fatty infiltration of the liver, kidneys and heart muscle. Similar results were obtained by Smyth and Smyth (1928). They used ethyl alcohol denatured for industrial use and exposed guinea-pigs to concentrations of 3,000 p.p.m. for 64 exposures. No injurious effects were observed except slight albuminuria in one animal.

Absorption by Inhalation

An examination of the amount of ethyl alcohol absorbed by animals exposed to different amounts of vapour revealed the interesting fact that the highest fatal amount of ethyl alcohol absorbed was less than the corresponding fatal amount for methyl alcohol (Loewy and van der Heide, 1918). Thus, for ethyl alcohol the highest amount absorbed by a rat from inhalation of a concentration of 12,700 p.p.m. for 21¾ hours was 5·78 g. per kg., while for methyl alcohol the corresponding amount was 8·7 to 12·8 g. per kg. The lowest amount of ethyl alcohol absorbed was 0·14 g. per kg. from exposure to a concentration of 18,200 p.p.m. for 2 hours. (These amounts were determined by estimating the alcohol content of the whole animal after exposure.) In Carpenter's (1929) experiments on fowls he found the lethal amount absorbed to be 3·7 to 5·6 g. per kg. body weight, and that signs of abnormal behaviour appeared when the amount absorbed was 1·6 g. per kg. in the whole body and 2·5 g. per kg. in the blood.

Relative concentrations in the internal organs. The greatest amount of ethyl alcohol is absorbed by the brain and the blood, the liver and fat taking up the smallest amounts (Carpenter, 1929).

Absorption through the Skin

According to Schwenkenbecher (1904), white mice were completely narcotized by immersion in a 16 per cent solution of ethyl alcohol. The action began soon after immersion, but loss of reflexes only occurred after about 9 hours and many of the animals died, presumably from respiratory paralysis, after removal from the immersion chamber.

TOXIC EFFECTS IN MAN

Acute Poisoning

As already observed, acute intoxication from inhalation of ethyl alcohol, either by accident or during industrial exposure, has rarely been reported. One of the very few instances is that of a child who had inhaled it during the night from an application to the chest (recorded by Leschke, 1932, and others). It is stated in International Labour Office, *Brochure No.* 50 that "the clinical picture from inhalation of alcohol vapour is identical with that seen in drinkers, and chronic alcoholics. As soon however as a workman combines with alcohol absorbed from vapour the swallowing of it in liquid form, he shows symptoms which do not yield to ordinary treatment". No actual recorded instances are quoted. In this connection it may be noted that the symptoms of acute ethyl alcohol poisoning from ingestion differ considerably from those of methyl alcohol poisoning in that they are chiefly psychic with disturbances of speech and gait, and of heat regulation, and that they are not related to eye injury, which is characteristic of methyl alcohol poisoning (Neiding, Goldenberg and Blank, 1932).

Effects of Inhalation

From their investigations on the effects on human beings of various concentrations of ethyl alcohol vapour, Loewy and van der Heide concluded that industrial intoxication from ethyl alcohol was not likely to occur unless the air contained more than 1,000 p.p.m. of ethyl alcohol. They found that this concentration produced very slight symptoms of intoxication; 5,000 p.p.m. caused stupor and drowsiness after some time but 5,000 to 10,000 p.p.m. within 1 hour. Persons who were unaccustomed to taking alcohol were more susceptible. The findings of Loewy and van der Heide were summarized by Flury and Zernik (1931) and are shown in Table 37.

Absorption by the Skin

Most of the earlier workers (Fleischer, 1880; Winternitz, 1891; Buchner, Fuchs and Megele, 1901) considered that the amount of alcohol absorbable by the human skin was so small that its toxic effects need not be considered. Rohrig

(1876) produced indirect evidence of its absorbability from the fact that certain substances, such as sodium iodide, were absorbed in alcoholic but not in aqueous solution. The clearest evidence for this view is a case (James, 1931) of acute intoxication following the repeated application of towels wrung out in surgical spirit to the skin of both legs. The subject, a boy aged eight, presented typical symptoms of acute alcohol poisoning and alcohol was found in both the vomit and the urine.

TABLE 37

Effect of inhalation of ethyl alcohol on human beings accustomed and unaccustomed to alcohol

Habit	Concentration (vol. per cent)	(mg./l.)	(p.p.m.)	Exposure (min.)	Effect
Unaccustomed to alcohol	0·138	2·6	1,380	39	No visible effect up to 28 min.; after 33 min., headache; after-effect, slight stupor
Unaccustomed to alcohol	0·23	4·3	2,300	50	At first, feeling of warmth in head and trunk, then in lower limbs, hands and feet cold, nasal irritation; after-effect, drowsiness and feeling of pressure in forehead
Accustomed to alcohol	0·5	9·4	5,000	120	After 20 min., increasing headache
Accustomed to alcohol	0·6	11·3	6,000	120	Odour unpleasant; sometimes slight tendency to sleep
Accustomed to alcohol	0·7	13·2	7,000	110	Odour as above; after 30 min., pressure above the eyes, feeling of heat; after 90 min., fatigue and desire to sleep
Unaccustomed to alcohol	0·88	16·7	8,800	30	Odour at first intolerable, soon tolerated, but burning of eyes; increasing feeling of heat in forehead and ears; after 30 min., fatigue and desire to sleep

Chronic Poisoning

Symptoms

The only accounts of any symptoms specifically related to slight repeated exposure to ethyl alcohol vapour are those of Loewy (1914), who noted irritation of mucous membranes (eyes, larynx, bronchi), headache, tremor, drowsiness and a tendency to nausea in workers using denatured alcohol. Loewy concluded that these symptoms must be due to ethyl and not methyl alcohol, since the concentration in the air produced by the evaporation of the amount of methanol used for denaturing was about 0·1 per cent lower than that which had been shown to produce no symptoms in animals (Loewy and van der Heide, 1918). Koelsch (1921) examined a number of workers using denatured ethyl alcohol for lacquering and polishing and found no severe toxic effects; the minor symptoms complained of related chiefly to intolerance of the odour and irritation of the eyes and upper respiratory passages, with slight headache, drowsiness, disturbance of appetite and sense of oppression. He considered these to be more probably due to the pyridine used as a denaturing agent than to the ethyl alcohol itself.

Flury and Zernik quote Kochmann (1923) as stating that repeated inhalation of alcohol vapour may under certain conditions give rise, like chronic alcohol drinking, to cirrhosis of the liver. No evidence of such an occurrence is given Brezina (1929) mentions that in Germany cardiac disturbances were observed from the effects of inhaling warm alcohol, presumably ethyl alcohol, during the manufacture of alcoholic liquors. He gives no reference or detailed description.

4. n-Propyl Alcohol
(Propanol)
$C_3H_7 \cdot OH$, i.e. $CH_3 \cdot CH_2 \cdot CH_2 \cdot OH$

PROPERTIES

A colourless fluid with strong alcoholic odour and miscible with water. It is a constituent of fusel oil, from which it can be isolated in a state of high purity. Its properties are similar to those of *iso*propyl alcohol, but it has a higher boiling point and more rapid evaporation rate.

B.P.—pure, 97·3° C., technical, 95–102° C. Sp.Gr. 0·805. Fl.P. 22° C Evaporation time, 11·1 (compared with ethyl ether, 1 and *iso*propyl alcohol, 21).

Solvent for many gums and resins, castor, linseed and other oils; does not dissolve cellulose esters, but dissolves some cellulose ethers.

USES

(1) In the lacquer industry (Weber and Koch, 1933).
(2) In laboratory work as a dehydrating agent for tissues (Sheridan, 1929).

TOXICITY

n-Propyl alcohol appears to possess narcotic and toxic properties very similar to those of *iso*propyl alcohol (see p. 224) but is more toxic if judged by injection into animals (Macht, 1920; Weese, 1928; Rost and Braun, 1926), experiments on isolated frog's heart (Fühner, 1921), and on contractions of isolated nerve and muscle (Efron, 1885; Macht, 1920). Its bactericidal action is greater than that of ethyl or methyl alcohols (Buchner, Fuchs and Megele, 1901).

On inhalation, however, according to Weese's results, it is less narcotic in saturated concentrations than *iso*propyl, and it is interesting to note that the toxicity of its vapour to non-incubated eggs is less than that of ethyl or methyl, but greater than that of butyl alcohol (Lallemand, 1929).

n-Propyl alcohol seems to carry little risk for human beings; no actual experiments on its inhalation by man have been performed, but no ill-effects have been recorded from its industrial use.

TOXIC EFFECTS IN ANIMALS
Acute Poisoning

Lethal and Narcotic Concentrations

Lethal dose

By intraperitoneal injection. Macht found the dose (in terms of pure alcohol) to be 2 ml. per kg. body weight as compared with 2·5 ml. for *iso*propyl. Lendle (1928) gives 4 ml. (in terms of 100 per cent solution) as compared with 8 ml. for ethyl, 1·2 ml. for *n*-butyl and 0·6 ml. for *n*-amyl alcohol. According to Rost and Braun (1926) *n*-propyl alcohol is about 1·5 times more toxic than ethyl alcohol by this route, *iso*propyl being 1·3 times greater. These workers also state that *n*-propyl alcohol seems to possess a powerful excitant action on mucous membranes, since administration to dogs was impossible on account of severe vomiting.

Narcotic dose

By intraperitoneal injection. The dose is 1·5 ml. per kg. body weight as compared with 4·5 ml. for ethyl, 0·76 ml. for butyl and 0·36 ml. for amyl alcohol (Lendle, 1928).

According to Lendle, *n*-propyl alcohol occupies a peculiar position among the normal alcohols with regard to the tolerance of animals to immediate lethal effect. This he judges from the breadth of narcosis, expressed as the quotient of the lethal divided by the narcotic dose. For all the other normal alcohols this figure is about 1·8, whereas for *n*-propyl alcohol it is over 2 (Table 38). Lendle believes this distinction to arise from the fact that *n*-propyl alcohol is more slowly eliminated.

TABLE 38
Lethal and narcotic doses of some alcohols

Alcohol	Narcotic dose (ml.)	Lethal dose (ml.)	Breadth of narcosis	Average time of recovery (min.)
Ethyl	4·5	8·0	1·8	210–360
Propyl	1·5	4·0	2·66	90–200
Butyl	0·76	1·2	1·6	54–106
Amyl	0·36	0·6	1·7	35–50

By inhalation. None of the animals examined by Weese (1928) died under exposure to saturated or unsaturated air mixtures.

Symptoms

In the only experiments on inhalation of *n*-propyl alcohol (Weese) no symptoms other than those of narcosis were observed.

Lesions of the Internal Organs

Only slight reversible fatty infiltration of the liver, kidneys and heart were observed by Weese in animals killed after narcosis.

Chronic Poisoning

No ill-effects were observed by Weese after repeated inhalation. It is interesting to note, however, that its effect on the growth and general health of chickens, when repeatedly injected into the crop over a period of several months, was greater than that of either methyl or butyl alcohols (Elhardt, 1931).

TOXIC EFFECTS IN MAN

No toxic effects from prolonged local application of *n*-propyl alcohol to the skin were observed by Buchner, Fuchs and Megele (1901).

5. isoPropyl Alcohol

(isoPropanol, Secondary Propyl Alcohol, Dimethyl Carbinol. Commercially, "Perspirit", "Petrohol", "Avantine", etc.)

$$C_3H_7 \cdot OH, \text{ i.e. } \genfrac{}{}{0pt}{}{CH_3}{CH_3}{>}CH \cdot OH$$

PROPERTIES

A stable, colourless liquid, miscible with water in all proportions, and with ethyl alcohol, ether, glycerin and glycol.

B.P. 82·4° C. Sp.Gr. 0·790. Fl.P. 12° C. Evaporation time, 21 (compared with ethyl ether, 1).

Not a usual constituent of fusel oils, but is manufactured either from acetone by hydrogenation in presence of catalysts, or from olefine gases arising in the cracking of petroleum.

Solvent for colophony, mastic, shellac, sandarac and most copals; not a solvent for cellulose esters, but the presence of a relatively small quantity of ester makes it a solvent for cellulose nitrate; it is a better solvent than ethyl alcohol for fats, oils, lipoids and camphor.

USES

(1) Largely as a substitute or denaturant for ethyl alcohol in the manufacture of toilet remedies, perfumes, cosmetics, gargles, culinary essences (Zangger, 1931). Its solvent power at 40 per cent strength is equal to that of ethyl alcohol at 80 per cent and it is cheaper (Answer to Question, *Industrial Chemist*, 1930).

(2) In the manufacture of safety glass.

(3) As a dehydrating agent, for sugars, starches, gelatine, animal and vegetable tissues, histological specimens, etc.

(4) For cleaning and drying photographic films and plates.

(5) In the lacquer and varnish trade as a "latent" solvent, as a substitute for ethyl alcohol in nitrocellulose lacquers, and in spirit varnishes for use on food containers (Garlick, 1927).

(6) As an emulsifier (Ruemele, 1948).

TOXICITY

While *iso*propyl alcohol shares with the other alcohols the properties of a narcotic of the central nervous system, the danger from inhalation in the course of industrial processes would seem to be very slight owing to its low volatility and high boiling point. No actual cases of such injury have ever been reported nor any ill-effects from its use undiluted as a disinfectant for the skin, etc. (Garlick, 1927). When used as a constituent of hair lotion it is non-irritant to the scalp (Ruemele, 1948). Detailed discussions of its suitability from a physiological point of view as compared with ethyl alcohol have been made by Heffter and Juckenack (1919) and Boruttau (1921). The latter decided, as the result of his investigations on men and animals, that there was no contra-indication to its use for the care and cleansing of the skin, hair and nails, or as a gargle or mouthwash. It has also been recommended as a germicide for skin sterilization (Grant, 1923; Bernhardt, 1922), being effective in lower concentrations than ethyl alcohol.

ALCOHOLS

TOXIC EFFECTS IN ANIMALS

The effects of *iso*propyl alcohol are similar to those of ethyl alcohol, high dosage causing depression, narcosis and paralysis. Salivation, retching and vomiting are more frequent than with ethyl alcohol (Lehman and Chase, 1944). The majority of investigations have shown it to be 1·5 to 2 times as toxic as ethyl alcohol, but it is pointed out by Morris and Lightbody (1938) that on account of pharmacological differences the toxicity of ethyl alcohol and *iso*propyl alcohol cannot strictly be compared except on the basis of narcosis after single large doses. Since the rate of disposal, detoxification, or destruction of *iso*propyl alcohol in the body is slow, there is a possibility of a cumulative effect even in a small, rather widely separated dosage, whilst the acetone produced by metabolic processes may have a toxic action distinct from the pharmacological action of the alcohol itself.

Comparison with other Alcohols

By ingestion

*iso*Propyl alcohol has been found less toxic than *n*-propyl and *iso*butyl alcohols (Weese, 1928). Lehman, Schwerma and Rickards (1944) state that it is twice as narcotic as ethyl alcohol when given by mouth or by intravenous injection. The investigation by Morris and Lightbody (1938), however, gives a somewhat different result with regard to ethyl alcohol; they found that animals given ethyl alcohol (4·3 ml. per kg. daily) were at first more deeply narcotized and for a longer time than those given 2·5 ml. per kg. of *iso*propyl alcohol, but in later periods the effect was reversed, and the animals given *iso*propyl alcohol showed a marked decrease in the rate of recovery and a greater average loss of weight. In the experiments of Lehman and Chase (1944) no delayed toxic effect after sub-lethal doses was found.

By subcutaneous or intravenous injection

According to Richardson's law, which states that the toxicity of the alcohols of the fatty acid series increases with their molecular weight, *iso*propyl alcohol is more toxic than ethyl and less toxic than butyl or amyl alcohols. It is, according to Macht (1920) and Vollmer (1931), slightly less toxic than *n*-propyl alcohol.

By inhalation

According to Weese (1928) and Nelson, Ege, Ross, Woodman and Silverman (1943), in unsaturated concentrations it holds the same place in the series of alcohols enunciated by Richardson, that is, it is more narcotic than ethyl and less than butyl alcohol.

In saturated concentrations, however, Macht (1920) and Weese (1928) agree that it holds the same relative toxicity as that by ingestion, i.e. it is less toxic than ethyl, methyl or secondary butyl alcohols, and more toxic than *n*- or *iso*butyl alcohols. Its relative toxicity to nerve and muscle preparations (Efron, 1885), isolated frog's heart (Führer, 1921) and mammalian heart (Kuno, 1913) is similar to that by subcutaneous and intravenous injection.

Acute Poisoning

Lethal and Narcotic Concentrations
  Lethal dose

By ingestion. For rabbits, rats and dogs, about 166 ml., according to Lehman and Chase (1944), but it was found difficult to administer so much on account of its emetic and depressant effect.

By intravenous injection. 2·5 g. per kg. body weight compared with 5 g. per kg. for ethyl, 0·3 g. for butyl and 1·5 g. for amyl alcohol (Macht, 1920). Vollmer's (1931) figures are 3·2 g. for *iso*propyl and 1·6 g. for *n*-propyl alcohol.

By inhalation. Exposure for a week to air saturated with the vapour was not fatal to rats (Macht) nor did they show any sign of disturbance of health.

Narcotic dose

By inhalation. Weese used the amount of vapour produced by 0·1 to 0·17 ml. (20 to 34 p.p.m.) of *iso*propyl alcohol in 5 l. of air to produce narcosis, but does not give the actual concentrations.

Symptoms

In Macht's experiments practically no symptoms were produced. Weese apparently produced narcosis, but states that a few hours after removal from exposure the animals were normal. According to Nelson *et al.* (1943), mild irritation of the eyes, ears and throat was produced by inhalation of concentrations of 400 p.p.m. and above, but this was not severe, even at 800 p.p.m.

Chronic Poisoning

Symptoms

Long continued administration of fairly small doses by ingestion, in the experiments of Lehman and Chase (1944), had only slight effects on growth and body weight, but large doses caused depression and paralysis.

Weese found no symptoms or after-effects in mice exposed daily to "non-narcotic" concentrations for 8 to 12 hours a day.

Lesions of the Internal Organs

Weese found very slight reversible fatty infiltration of the liver, kidneys and heart muscle, which he did not regard as seriously injurious to the health of the animal. Macht remarks that no blindness or defects in vision were noted after exposure of the rats for a week to the fumes of *iso*propyl alcohol.

Absorption and Excretion

The earlier findings of Pohl (1922) that ingestion of *iso*propyl alcohol by animals was followed by excretion of acetone through the lungs has been confirmed by Willcox (1929), Cook and Smith (1929), and Morris and Lightbody (1938) who found it also in the urine. It is for this reason that Morris and Lightbody emphasize the difference and the difficulty in comparing the toxicity of *iso*propyl with that of ethyl alcohol. The rate of disposal, detoxification or destruction of *iso*propyl alcohol is also relatively slow, so that a cumulative action may result from repeated dosage. Its rate of disappearance from the blood, according to Lehman, Schwerma and Rickards (1944), depends on the functional efficiency of the excretory organs. Vollmer (1931) considers that the toxic action of *iso*propyl alcohol is not so readily influenced by substances which increase oxidative process as that of ethyl alcohol.

Absorption through the Skin

*iso*Propyl alcohol is apparently not readily absorbed through the skin of animals; Boruttau (1921) applied dressings soaked in a 60 per cent solution and renewed daily to the shaved skin of the abdomen of rabbits, but could find no ill-effects.

TOXIC EFFECTS IN MAN

No cases of industrial poisoning have been reported. Boruttau (1921) observed no ill-effects from local application, nor from subcutaneous injections into himself. Ingestion by six men of single doses of 22·5 ml. diluted with water was followed by immediate acute symptoms of dizziness and flushing of the face and by a fall in the blood pressure (Thompson, cited by Morris and Lightbody, 1938). Secondary symptoms, headache, mental depression, nausea and vomiting, which all developed 2 to 3 hours later, were attributed to the formation of acetone. The excretion of acetone by human beings after ingestion of propyl alcohol has been confirmed by Fuller and Hunter (1927) and by Kemal (1927).

6. n-Butyl Alcohol
(Butanol)

$C_4H_9 \cdot OH$, i.e. $CH_3 \cdot CH_2 \cdot CH_2 \cdot CH_2 \cdot OH$

PRODUCTION AND PROPERTIES

n-Butyl alcohol has assumed great industrial importance since 1915 when, owing to a shortage of acetone obtained by older methods, a fermentation process was introduced giving this material with n-butyl alcohol as a by-product. The latter serves as the raw material for the manufacture of n-butyl acetate, an ideal solvent for reducing the viscosity of nitrocellulose without limiting its resistance to weathering (Gabriel, 1928). It is manufactured chiefly by the fermentation of maize-flour by a special strain of bacillus; also from acetaldehyde, which is converted to crotonic aldehyde, and thence by hydrogenation to n-butyl alcohol. It is not a normal constituent of fusel oil but may be obtained from the fusel oil formed when certain species of yeast (*Sacch. ellipsoideus*) are used in alcoholic fermentation. It is a colourless liquid with a pungent but not unpleasant odour.

British Standard Specification, No. 508 (1933). Sp.Gr. 0·810–0·816. B.R. 115–118° C. 95 per cent. Residue 0·01 per cent maximum. Acidity 0·01 per cent maximum. Aldehyde 0·5 per cent maximum (butyraldehyde).

A.S.T.M. Specification, D 304–33. Sp.Gr. 0·810–0·815 at 20° C. B.R. up to 100° C. none; 105° C. 2 per cent maximum; 115° C. 10 per cent maximum; above 118° C. none. Residue 0·005 per cent maximum. Acidity 0·03 per cent maximum.

Evaporation time, 33 (compared with ethyl ether, 1).

Solvent for many gums and resins and especially for hard copals; not a solvent for cellulose esters or ethers, but when mixed with solvents increases their solvent power and dilution ratios (I.G. Farbenindustrie Aktiengesellschaft, 1930).

USES

(1) In the lacquer industry as a diluent and blender (Holden and Doolittle, 1935; Gardner, 1925; Cunningham, 1934). According to Bogin (1935) the replacement of small amounts of naphtha by butyl alcohol considerably reduces the viscosity of lacquers. As a solvent for metal and bronze lacquers it is usually mixed with butyl acetate.

(2) In the dye industry as a dehydrating agent.

(3) In the separation of oils and waxes (Poole, 1929).

(4) As a constituent of shoe cements.
(5) In the manufacture of safety glass (Holden and Doolittle).
(6) In the hat industry as a constituent of the lacquer (Krüger, 1937).
(7) In the manufacture of artificial silk (Sisley, 1934).
(8) In the manufacture of raincoats, etc. as a solvent for the synthetic resin polyvinyl butyral (Tabershaw, Fahy and Skinner, 1944).

TOXICITY

Injection into animals shows butyl alcohol to be more toxic than methyl, ethyl and propyl alcohols and less toxic than amyl alcohol in accordance with Richardson's law; on account of its slight solubility and volatility, its vapour, at any rate in saturated solution, is less narcotic than that of propyl alcohol and is only slightly toxic for animals (Weese, 1928; Flury and Zernik, 1931). Its toxicity compared with *secondary* butyl alcohol is also different according to the method of administration. By injection it is more toxic (Macht, 1920), by inhalation less so (Weese).

It has been suggested (Smyth and Smyth, 1928) that the vapour of butyl alcohol may have an injurious effect on the blood-forming organs of animals, though Gardner (1925) was able to trace only a very mild degree of anaemia to its effect. He, however, observed irritation of mucous membranes, particularly conjunctival and respiratory, an effect which may be correlated with the observations of Krüger (1932) in hat workers, and Tabershaw *et al.* (1944) in the raincoat industry.

Metabolic effects as evidenced by liver and kidney injury appear uncertain both in animals and human beings.

Butyl alcohol, at any rate in the concentrations likely to be present in the air of factories where it is used, appears to have little or no general systemic effect. Krüger (1932) observed that potentially it is a considerable hazard with regard to injury of the eyes, and this has been confirmed by later investigations.

Maximum Permissible Concentrations

By experiments at the Harvard School of Public Health it has been shown that above 25 p.p.m. the odour of butyl alcohol becomes disagreeable to many people. According to Tabershaw *et al.* (1944), eye irritation occurs above 50 p.p.m., but they found no evidence of systemic effects at about this level. Cook (1945) also recommends 50 p.p.m. as the maximum allowable concentration.

TOXIC EFFECTS IN ANIMALS
Acute Poisoning

Lethal and Narcotic Concentrations

Lethal dose

By intraperitoneal injection. 0·3 ml. (in terms of pure alcohol) per kg. body weight, compared with 5·0 ml. for ethyl, 2·0 ml. for *n*-propyl and 0·9 ml. for *iso*butyl alcohol (Macht, 1920). Lendle (1928) gives 1·2 ml., compared with 4 ml. for propyl and 8 ml. for ethyl alcohol.

Narcotic dose

By intraperitoneal injection. 0·76 ml. per kg. body weight compared with 1·5 ml. for propyl and 4·5 ml. for ethyl alcohol (Lendle, 1928).

By oral administration. Rost and Braun (1926) produced deep and rapid narcosis in rabbits by 5 ml. per kg. body weight in 25 per cent mixture.

By inhalation. None of the animals examined by Weese died under exposure to saturated or unsaturated vapour. Starrek (1938, cited by Lehman and Flury, 1943) records that some mice died after deep narcosis. He gives the narcotic dose as 20 mg. per l. (0·66 per cent).

Symptoms and Changes in the Internal Organs

In Weese's inhalation experiments deep narcosis was produced, but the animals showed no other symptoms and only slight reversible fatty infiltration of the liver and kidneys, and they recovered quickly.

Chronic Poisoning

The toxicity of *n*-butyl alcohol when administered repeatedly to animals either orally or by inhalation is apparently not great, but there is some difference of opinion as to its remote effects. Elhardt (1931) found it less injurious than either methyl or propyl alcohols as estimated by the growth and general health of chickens after it had been injected repeatedly into their crops for several months. Weese (1928) found very little ill-effect from repeated exposure to inhalation of non-narcotic concentrations. Similarly, Gardner (1925) found that repeated inhalations produced no definite poisonous effect except mucous membrane irritation. He observed no effect on the blood picture except a mild degree of anaemia, and stated that the slight congestions of the kidneys and intestine detected were definitely traceable to the action of the butyl alcohol. Smyth and Smyth (1928), however, found that air containing 100 p.p.m. of butyl alcohol vapour produced a red cell decrease with a relative and absolute lymphocytosis and also signs of early liver and kidney degeneration. Post mortem there were slight changes in the liver and kidneys in Weese's animals, but he did not regard them as of any significance.

TOXIC EFFECTS IN MAN

Symptoms

Ocular injury. Krüger (1932) noted severe conjunctival and corneal irritation among workers in a hat factory where butyl alcohol was only one of the lacquer constituents used; the others included butyl acetate, amyl acetate, acetone, etc., but Krüger believed that butyl alcohol was the essential toxic agent. Tabershaw *et al.* (1944) confirmed this view by investigations on a series of workers in the waterproof industry where butyl alcohol, while not the sole constituent used in some of the processes, was used alone in others. It was in the plant where the highest concentrations of butanol were found that the ocular irritation was at its greatest.

A later investigation of these more serious cases was made by Cogan and Grant (1945), from which it appears that in addition to conjunctivitis, butyl alcohol may produce definite and characteristic changes in the corneal epithelium leading to keratitis.

Irritation of nose and throat. This was only slight in the workers investigated by Tabershaw *et al.*

Dermatitis. Slight fissured eczema around the finger nails and along the sides of the fingers can be prevented (Tabershaw *et al.*, 1944) by the use of protective ointment and emollients.

Slight headache and dizziness. These were also reported by Tabershaw *et al.*

Metabolic Effects

The relation of the supposed irritation of the liver caused by inhalation of small amounts of the solvents used in spraying lacquers (Burger and Stockmann, 1932) rests on very slight evidence. These authors found a higher incidence of urobilinuria in lacquer workers than in controls (39 per cent compared with 10 per cent in wallpaper workers), but butyl alcohol was only one of the constituents; others were amyl alcohol, amyl and butyl acetates and acetone.

7. Secondary Butyl Alcohol
(*Methyl Ethyl Carbinol, Butanol*-2)

$$C_4H_9 \cdot OH, \text{ i.e. } \genfrac{}{}{0pt}{}{CH_3}{CH_3 \cdot CH_2}{>}CH \cdot OH$$

PROPERTIES

A colourless liquid with a somewhat choking odour, but less pungent than *n*-butyl alcohol. Produced by the hydrogenation of methyl ethyl ketone, also during the "crack" distillation of petroleum.

B.R. of the technical product 98–102°C., pure 99·5°C. Sp.Gr. 0·810–0·815 at 15°C. Solvent for ester gum, shellac, kauri, mastic, sandarac, etc. but not for cellulose esters.

USES

(1) In the lacquer industry, but not to any great extent.
(2) In the separation of wax from oil (Smith, 1927).

TOXICITY

In agreement with Overton's (1901) statement that the normal alcohols are always more toxic than their secondary isomers, *secondary* butyl alcohol should be less toxic by injection to animals than *n*-butyl alcohol. According to Hufferd (1932), this was the case for intraperitoneal injection into guinea-pigs in contrast to *n*-amyl alcohol and its isomers, while in Rost and Braun's (1926) experiments with oral administration, the results were inconclusive owing to very slight solubility.

TOXIC EFFECTS IN ANIMALS

Weese (1928) found *secondary* butyl alcohol (saturated air-vapour mixture) more narcotic on inhalation than either *n*-butyl, *iso*butyl or the propyl alcohols. This he explains by stating that its higher vapour tension allows it to enter more freely into the blood and central nervous system.

The symptoms and changes in the internal organs were practically identical with those produced by *n*-butyl alcohol (p. 229). No difference between "optically active" and "optically inactive" isomers of *secondary* butyl alcohol was found by Viditz (1933) with regard to their anaesthetic property on tadpoles or their toxicity for isolated frog's heart and nerve-muscle preparations.

TOXIC EFFECTS IN MAN

No ill-effects have been recorded.

ALCOHOLS

8. isoButyl Alcohol

$$C_4H_9 \cdot OH, \text{ i.e. } \genfrac{}{}{0pt}{}{CH_3}{CH_3}{>}CH \cdot CH_2 \cdot OH$$

PROPERTIES

A colourless liquid with a rather suffocating odour of fusel oil. Obtained by fractional distillation of fusel oil, or synthetically.

B.P. 108° C. Sp.Gr. 0·806. Fl.P. 22° C. Evaporation time, 24 (compared with ethyl ether, 1). Solubility in water, up to 12·5 per cent (Rost and Braun, 1926).

USES

Chiefly in the lacquer industry (Weber and Koch, 1933).

TOXIC EFFECTS IN ANIMALS

Acute Poisoning

Although by injections into animals and by experiments on isolated organs *iso*butyl alcohol is more toxic than ethyl and propyl alcohols (Richardson's law) and is less toxic than *n*-butyl alcohol (Macht, 1920), these relationships do not hold for inhalation of saturated air-vapour mixtures (Weese, 1928) or for oral administration (Rost and Braun).

Lethal and Narcotic Concentrations

Lethal dose

By intraperitoneal injection. For rats, 0·9 ml. (in terms of pure alcohol) per kg. body weight, compared with 0·3 ml. for *n*-butyl alcohol (Macht).

Narcotic dose

By oral administration. Rost and Braun found the narcotic potency of *iso*butyl alcohol high; rapid and deep narcosis was produced in dogs by 2·5 ml. per kg. body weight compared with 5 ml. per kg. for *n*-butyl alcohol.

By inhalation. Weese found the narcotic potency of saturated air-vapour mixtures less than that of propyl alcohol and about the same as that of *n*-butyl alcohol.

Symptoms and Changes in the Internal Organs

Weese produced narcosis without other symptoms and the animals recovered rapidly. He found very slight changes, similar to those caused by the other alcohols except methyl alcohol, i.e. slight reversible fatty infiltration of liver and kidneys. Repeated inhalations of non-narcotic concentrations produced no injury to health and no significant changes in the internal organs.

TOXIC EFFECTS IN MAN

No ill-effects have been recorded.

9. Tertiary Butyl Alcohol

(*Trimethylcarbinol, 2-Methyl-2-propanol*)

$C_4H_9 \cdot OH$, i.e. $(CH_3)_3C \cdot OH$

PROPERTIES

A colourless, inflammable liquid. B.P. 82·8° C. V.P. 42·0 mm. Hg at 25° C. Saturated air contains 5·53 per cent *tertiary* butyl alcohol vapour at 25° C.

(Treon, 1949). Fl.P. 52° F. (11° C.). Eight times more volatile than *n*-butyl alcohol (Lehmann and Flury, 1943).

USES

(1) In fruit essences.
(2) As an intermediate.
(3) In the plastic lacquer industry (Schwartz and Tulipan, 1939).
(4) In the perfume industry to a limited extent.

TOXICITY

tertiary Butyl alcohol is essentially a narcotic of a potency greater than that of *n*- or *iso*butyl alcohols, and with a toxic effect between that of ethyl and propyl alcohols (Weese, 1928).

TOXIC EFFECTS IN ANIMALS

According to Treon (1949), the symptoms of intoxication in animals following inhalation of *tertiary* butyl alcohol are similar to those induced by the other butyl alcohols. Weese (1928) states that two thirds of the amount inhaled is exhaled within 6 hours. Repeated (up to eighteen) narcotic doses were not fatal and there were no injurious effects from a long-continued slight dosage.

TOXIC EFFECTS IN MAN

No injurious effects from its industrial use have been recorded. Schwartz and Tulipan (1939) state that it may be a skin irritant, but according to Oettel (1936) application to the skin produced only slight hyperaemia and erythema.

10. Amyl Alcohol
$C_5H_{11} \cdot OH$

CONSTITUTION AND PROPERTIES

There are eight isomeric amyl alcohols, three existing each in two modifications.

Fermentation Amyl Alcohol

The amyl alcohol of commerce, obtained from fusel oil, is a mixture of two of the eight isomers:—

(i) *primary iso*amyl alcohol (*iso*butyl carbinol): $\begin{matrix} CH_3 \\ CH_3 \end{matrix} > CH \cdot CH_2 \cdot CH_2 \cdot OH$.

(ii) "active" (i.e. optically active) amyl alcohol (*secondary* butyl carbinol): $\begin{matrix} CH_3 \cdot CH_2 \\ CH_3 \end{matrix} > CH \cdot CH_2 \cdot OH$.

The fusel oil from potatoes or cereals contains chiefly *iso*butyl carbinol, with only 13 to 22 per cent of *secondary* butyl carbinol; that from beet molasses contains 48 to 58 per cent of the latter. Fermentation, or fusel oil, amyl alcohol is available in several grades:

(i) the pure quality of commerce. ..R. 128–132° C. Sp.Gr. 0·813–0·817. Water dissolves 3 per cent at 20° C. Miscible with benzene in all proportions.

(ii) less pure anhydrous fusel oil. B.R. 105–132° C. Sp.Gr. 0·813–0·817. Contains small quantities of *n*-propyl, *iso*butyl and *n*-hexyl alcohols with traces of pyridine, furfural and esters. The technical product is a yellowish, clear fluid with a penetrating alcoholic odour, irritating to the throat. Good solvent for soft copals, ester gum, shellac, sandarac, etc.; moderately good for ethyl cellulose, copal ester, mastic and coumarone; renders glyceryl phthalate resin soluble in other solvents; does not dissolve cellulose esters.

A.S.T.M. Specification, D 319–33. Sp.Gr. 0·812–0·820 at 20° C. B.R. up to 118° C. none; 120° C. 5 per cent maximum; 125° C. 50 per cent maximum; 130° C. 85 per cent maximum; above 140° C. none. Residue 0·005 per cent maximum. Acidity nil.

Synthetic Amyl Alcohol

Synthetic amyl alcohol is also available, known as "Pentasol". It is manufactured from petroleum (Aschan, 1927) and is a mixture of five of the eight isomeric alcohols, those obtained depending on the concentration of acid used in hydrolysis (Pipik and Mezhebovskaya, 1933). According to Wilson and Worster (1929) synthetic amyl alcohol is practically a duplicate of the high test fusel oil used in the past. Solvent for some gums and resins, but not for cellulose esters or coumarone.

USES

(1) In the lacquer industry chiefly as a diluent (Yarsley, 1933) and as a constituent of "Zapon" lacquer (Baader, 1933, etc.).

(2) In the textile industry, but its suffocating odour limits its use (Sisley, 1934).

(3) In the chemical industry in the preparation of amyl nitrite, valeric acid, etc. (Flury and Zernik, 1931).

(4) In the manufacture of smokeless powder (Eyquem, 1905).

(5) In spirit refining.

(6) In shoe cements e.g. "Stabilin" (Vorob'eva, 1932).

(7) In laboratories as a dehydrating and clearing agent for microscopical preparations (Hartridge, 1920).

TOXICITY

Amyl alcohol is the most toxic of all the commonly used alcohols, as shown by its effect when injected into animals (Macht, 1920; Hufferd, 1932; Salant, 1909), and by its action on smooth muscle, isolated frog's heart, etc., but it is not known whether it would also be the most toxic when inhaled, since inhalation experiments have not apparently been carried out. It is interesting to note, however, that the toxicity of its vapours to the striated muscle of the frog, as estimated by the diminution of phosphagen, according to Martino (1928), is greater than that of either methyl or ethyl alcohols, though less than that of chloroform.

Its effect on human beings from an industrial point of view does not seem to be agreed upon. Hamilton (1934) states that while amyl alcohol is a well-known poison when taken through the mouth, causing headache, dizziness, nausea, vomiting, diarrhoea, delirium, coma and death, there is no record of poisoning from its industrial use. Flury and Zernik (1931), on the other hand, give a list of toxic symptoms resulting from its use, which are almost identical with those given by Eyquem (1905) and by Ebert (1925). One record of fatal results possibly due to amyl alcohol exists, that of Zangger (1933), although the information obtained from the factory was vague and unsatisfactory. In most of the cases of chronic effects possibly attributable to the use of amyl alcohol, e.g. the occurrence of urobilinuria in lacquer workers described by Burger and

Stockmann (1932), a case described by Baader (1933), and a special case in a neurasthenic patient quoted by Flury and Zernik (1931), there also appears to be some doubt as to whether the symptoms were due actually to amyl alcohol or to some other constituent.

Estimation of Vapour in Air

A method for estimating colorimetrically the amount of amyl alcohol vapour in air is described by Korenman (1932). The method is based on the reaction of Komarowsky, in which the addition of amyl alcohol to an alcoholic solution of furfural and concentrated sulphuric acid gives a colour varying from pink to violet red.

Maximum Permissible Concentration

Cook (1945) recommends 100 p.p.m. as the maximum allowable concentration for industrial purposes.

TOXIC EFFECTS IN ANIMALS

Metabolism

A study of the rate of metabolism of the amyl alcohols after intraperitoneal injection into animals was made by Haggard, Miller and Greenberg (1945). This investigation showed that due to the difference in their rate of metabolism, there was a marked diversity in the blood concentration reached by the various isomers.

With regard to the primary alcohols the blood concentration 1 hour after injection of 1 g. per kg. ranged from 14 to 55 mg. per 100 ml., for the secondary alcohols 51 to 65 mg. per 100 ml., and for the tertiary 125 mg. per 100 ml.; the length of time for disappearance similarly varied from $3\frac{1}{2}$ to 9 hours for the primary, 13 to 16 for the secondary, and 50 for the tertiary alcohols.

Excretion through the lungs and kidneys showed a corresponding variation. The primary alcohols were completely metabolized except for 1·2 to 7·6 per cent being converted, as in the case of ethyl alcohol, to the corresponding aldehydes which are rapidly metabolized. The liver plays a considerable part in this conversion. The metabolism of the secondary alcohols involves conversion of a considerable amount to ketones, 38 to 53·5 per cent of the ketone being excreted in the air and urine. The liver plays a less striking role here than in the conversion of the primary alcohols to aldehydes. 65 per cent of the tertiary amyl alcohol is metabolized, and 35 per cent appears in the expired air and urine as such, no volatile metabolites being found.

Lethal Concentrations

Most investigators used the amount necessary to cause death as the minimum dose for amyl alcohol, but the results have not been entirely consistent. Macht (1920) gives 0·15 ml. per kg. body weight, Lendle (1928) 0·48, and Munch and Schwartze (1925) 2·0, the values varying with the method of administration and the species of animal used. Basing the comparative toxicities of the primary and tertiary alcohols on the amount in g. per kg., which when present in the body will yield the concentration in the blood causing respiratory failure, Haggard, Miller and Greenberg (1945) find that the primary alcohols have a basic toxicity 12 times as great as that of ethyl alcohol, the tertiary alcohols 8·4 and 4·9 times as great while the secondary alcohols lie between the primary and the tertiary.

According to Oswald (1924) amyl alcohol in narcotic dosage has a strong influence on the respiratory centre and is rapidly lethal.

Effect on Nerve and Muscle

In agreement with earlier workers, Macht found amyl alcohol the most toxic of the alcohols to nerve and muscle. Saviano (1929), noting the effect on the gastrocnemius of the frog when amyl alcohol was injected into the dorsal lymph sac, made the observation that its action differed from that of methyl, ethyl, propyl and butyl alcohols, in that it affected the nerve differently from the muscle. The toxic action was greatest on the muscle and slight on the nerve.

Absorption through the Skin

According to Schwenkenbecher (1904) narcosis can be produced in animals by immersion in amyl alcohol. A difference in the nature of the toxic effect of the different isomers was noted by Haggard, Miller and Greenberg (1945). The primary alcohols differed in two respects from the secondary and tertiary ones, first, in a sleepiness or sedation of the animals long after all alcohol and aldehyde had disappeared from the blood, and second, in marked irritation of the peritoneum and redness of the lung after injection. They suggest that these effects may be explained by the conversion of the primary alcohols first into aldehyde, thus producing irritation, and then into valeric acid, producing sedation.

TOXIC EFFECTS IN MAN
Acute Poisoning

Two fatal cases have been reported by Zangger (1933) in men painting the inside of a benzine container with "Zapon" lacquer containing amyl alcohol. Two other men who went to their rescue became unconscious. He states, however, that there was a possibility that tetrachloroethane was also used.

Chronic Poisoning

Symptoms

Flury and Zernik (1931) have described the following symptoms as due to amyl alcohol: irritation of mucous membranes of eyes and air passages; rush of blood to the head, headache, giddiness, nausea, vomiting, diarrhoea (faeces and perspiration smelling of amyl alcohol), shallow respiration; later, double vision, stupor, delirium, in some cases death with severe nervous symptoms. Eyquem (1905) found these symptoms in workers in a smokeless powder factory. Other alcohols were used, but Eyquem states that a recrudescence of symptoms occurred whenever the use of amyl alcohol was increased and that all the workers themselves dreaded amyl alcohol.

Nervous symptoms. Eyquem describes headache, giddiness, and disturbances of vision, smell and taste, trembling of the limbs, insomnia or somnolence, agitation and finally a state of cachexia. Although the faeces smelt of alcohol, examinations of the urine revealed no traces of it. Nervous or psychic disturbances were also prominent in the case described by Hilbert (1925) (the "special case" quoted by Flury and Zernik). The subject, a foreman in a brewery, was a neurasthenic. After inhalation of the vapour from a fermentation cask he showed psychic irritability and sleeplessness, and complained that objects appeared alternately scarlet and blue. There were no pathological findings in the eyes, and Flury and Zernik remark that it is questionable whether these symptoms were due to amyl alcohol or to carbon dioxide.

Digestive symptoms. In a case described by Baader (1933), digestive symptoms predominated, but he does not state whether they were to be attributed to amyl alcohol or amyl acetate, both being present in the lacquer used. The patient suffered from loss of appetite, a feeling of fullness, and eructations. The dyspepsia was followed by loss of weight and a secondary anaemia which had led at first to a presumptive diagnosis of lead poisoning or pernicious anaemia. It may be mentioned that amyl alcohol was one of the constituents of the solvents stated by Burger and Stockmann (1932) to be associated with urobilinuria in the workmen using them.

11. Methyl*iso*butylcarbinol
(*Methyl Amyl Alcohol*)

$$C_6H_{13} \cdot OH, \text{ i.e. } \begin{matrix} CH_3 \\ CH_3 \end{matrix} {>} CH \cdot CH_2 \cdot CH {<} \begin{matrix} CH_3 \\ OH \end{matrix}$$

PROPERTIES

Prepared by reduction of mesityl oxide obtained either from ketone oils or by condensation of acetone (Durrans, 1935).

B.P. 125–131° C. Sp.Gr. 0·807. Solubility in water at 20° C. 1·7 per cent (Durrans, 1950). Fl.P. 46° C. Vapour pressure 4·6 mm. Hg at 30° C.

USES

(1) In the lacquer industry to a small extent as a blending agent in cellulose lacquers and varnishes and as a substitute for amyl alcohol.

(2) In the perfumery industry as an intermediate in synthetic perfumery (Yarsley, 1934).

(3) As an extraction solvent for concentrating fatty acids.

TOXIC EFFECTS

No animal experiments have been carried out and no ill-effects from its industrial use recorded.

In tests on human subjects for sensory response to various concentrations, Silverman, Schulte and First (1946) have shown that eye irritation is produced at 50 p.p.m., though the odour is not objectionable at this level.

12. *cyclo*Hexanol
(*Hexahydrophenol, Sextol, Hexalin, Adronol, Anol*)

$$C_6H_{11} \cdot OH, \text{ i.e. } CH_2 {<} \begin{matrix} CH_2 \cdot CH_2 \\ CH_2 \cdot CH_2 \end{matrix} {>} CH \cdot OH$$

PROPERTIES

An oily, neutral, water-white liquid, with an odour like camphor and amyl alcohol.

B.P. 160° C. Sp.Gr. 0·947. Fl.P. 68° C. Very low volatility; evaporation period, 400 (compared with ethyl ether, 1); evaporation rate 0·4 mg./min./sq. cm. at 45° C.

Solubility in water at 20° C. 6 per cent; dissolves 12 per cent water at 20° C. (Durrans, 1950).

Solvent for fats, oils, waxes, rubber, cellulose esters. When added to mineral oils has a powerful emulsifying action. As a solvent for alkyl resins, it is a little stronger than xylene; poor solvent for nitrocellulose.

USES

(1) In soap manufacture. It is used to a small extent to form a clear solution with aqueous soap solutions and increase their detergent properties.

(2) In textile and artificial silk industry as an addition to fulling and scouring soap emulsions (Clayton and Clark, 1931) and wetting-out agents; also to produce a better penetration of the bleach in the peroxide bleaching of cotton by dissolving the natural wax, and for removal of paint and tar-print marks from raw wool and oil spots from cloth (Reid, 1934, etc.).

(3) In the straw hat industry as a penetrant in impregnation of the straw before bleaching (Reid, 1934).

(4) In the cleansing of skins and fabrics.

(5) In the lacquer industry as a blending agent to prevent blushing (Sanderson, 1933).

(6) In the leather industry as a spray in combination with butyl acetate, etc.

(7) In the printing industry (Sato, 1928).

(8) In liquid wax emulsions and furniture, automobile and metal polishes (Sanderson, 1933).

TOXICITY

*cyclo*Hexanol in high concentration is a narcotic with a powerful paralytic effect, but, except when inhaled in very high dosage, with no convulsive action. In this respect it is more toxic than either benzene or *cyclo*hexane, but does not produce the injury to the blood system characteristic of benzene. Its chronic effects in animals are slightly greater than those of *cyclo*hexane; it produces some slight degenerative lesions in the liver and kidneys.

On account of its low volatility the danger from industrial use would appear to be small, and in fact the only record of any symptoms possibly associated with its use is that of one case investigated by the Home Office in 1932 (p. 239). When used, as it generally is, as an addition of not more than 10 per cent to soap solutions, it should not be dangerous.

Maximum Permissible Concentration

The maximum allowable concentration suggested by Cook (1945) is 100 p.p.m. (400 mg. per c.m.).

TOXIC EFFECTS IN ANIMALS

From the results of animal experiments by Treon, Crutchfield and Kitzmiller (1943), using the oral, cutaneous or inhalation route, it appears that the toxicity of *cyclo*hexanol is less than that of the methylated oxygenated derivatives, but is greater than that of the non-oxygenated compounds, *cyclo*hexane and methyl-*cyclo*hexane.

Lethal and Narcotic Concentrations

Lethal dose

By oral administration. For rabbits, 2·2 to 2·6 g. per kg. (Treon *et al.*, 1943).

By cutaneous application. For rabbits, 12·4 to 22·7 g. per kg.

By intraperitoneal injection. For rabbits, 1·42 g. per kg. (Sato, 1928); for mice, 0·5 ml. (Filippi, 1914).

By inhalation. For rabbits subjected to repeated inhalations 6 hours per day for 5 to 11 weeks, above 4 mg. per l. (977 p.p.m.) (Treon *et al.*, 1943). Owing to the low vapour pressure and the low rate of evaporation, the range of experimental concentrations is limited. Pohl (1925) observed no effects in a dog exposed to air saturated with *cyclo*hexanol for 10 minutes a day on 7 successive days.

Narcotic dose

By oral administration. For rabbits, 0·8 to 2·2 g. per kg. (Treon *et al.*, 1943); 1·3 g. per kg. (Pohl, 1925), i.e. about the same as for methyl*cyclo*hexanol.

By cutaneous application. For rabbits, application of 10 ml. daily caused local skin injury, narcosis and some evidence of irritation of the central nervous system after the second application; there was also marked hypothermia. Cutaneous application of soap containing 5, 10 or 15 per cent *cyclo*hexanol in a potassium oleate base produced only a slight temporary effect; the skin lesions produced by the higher concentrations became normal a short time after the applications were discontinued and no symptoms of hypothermia were observed (Treon *et al.*).

By intraperitoneal injection. For rabbits, 0·002 ml. per kg. (Sato, 1928).

By inhalation. For rabbits subjected to repeated inhalations 6 hours per day for 5 to 11 weeks, 4 mg. per l. No symptoms or signs of discomfort were present during exposure to 0·58 mg. per l. (145 p.p.m.), which is therefore regarded as near the maximum safe level for rabbits, but not entirely harmless, since after 56 6-hour periods of exposure there was evidence of slight toxic effects on the liver and kidneys.

Acute Poisoning

Symptoms

The narcosis produced by intraperitoneal and oral administration is not accompanied by convulsions (Sato, 1928; Treon *et al.*, 1943; Pohl, 1925). With inhalation a few convulsive movements were observed by Treon and his co-workers at the highest concentrations only.

Effect on the Internal Organs

The evidence of toxic effect, like that of *cyclo*hexane and its other derivatives, was not specific, but was expressed as a general vascular injury and the inflammatory reaction to it.

Chronic Poisoning

Symptoms

*cyclo*Hexanol in sub-lethal repeated dosage causes irritation to the eyes, salivation and lethargy.

Lesions in the Internal Organs

Even in the lowest concentrations employed (145 p.p.m.) exposure to *cyclo*hexanol produced some degenerative lesions in the liver and kidney, but these were slight.

Effect on the Blood

Although intraperitoneal injection of *cyclo*hexanol was stated by Sato (1928) to produce some hyperfunction of the bone marrow, as indicated by a slight polymorphonuclear leucocytosis, it is evident from his observations and those of Treon *et al.*, using the oral, cutaneous or inhalation routes, that *cyclo*hexanol has no specific effect, like that of benzene, on the haematopoietic system.

Effect on Metabolism

A decrease in the inorganic urinary sulphates and an increase in the excretion of glucuronic acid were found by Treon *et al.* with all methods of administration. The increased excretion of glucuronic acid occurred following inhalation of even the lowest concentrations. With oral administration, 40 to 50 per cent of the *cyclo*hexanol was excreted conjugated with glucuronic acid, compared to 5 per cent with *cyclo*hexane. A diminution of the glutathione content of the blood as opposed to an increase produced by true narcotics is stated by di Prisco (1932) to be produced by *cyclo*hexanol.

TOXIC EFFECTS IN MAN

Owing to the low vapour pressure and the low evaporation rate of *cyclo*hexanol the danger of its industrial use is unlikely to be very great. A case of ill-health in a worker using a spray for leather containing butyl acetate and three parts *cyclo*hexanol was investigated by the Home Office in 1932. The symptoms were chiefly related to the gastro-intestinal system, vomiting, a coated tongue and slight tremor. It was not decided that these symptoms were definitely attributable to the use of *cyclo*hexanol. Beyond this case, no ill-effects in industry have been recorded.

13. Methyl*cyclo*hexanol

(*Hexahydrocresol, Sextol, Hydrolin, Methyl-adronol, Methyl-hexalin,* etc.)

$$CH_3 \cdot C_6H_{10} \cdot OH$$

PROPERTIES

An oily liquid with an odour similar to *cyclo*hexanol and to amyl alcohol, but weaker. Solubility in water 3 per cent (Durrans, 1950). B.R. of the technical product 170–180° C. Sp.Gr. 0·925–0·930. Very low volatility; evaporates at half the rate of *cyclo*hexanol (Flury and Zernik, 1931); evaporation rate 0·2 mg./min./sq. cm.

A mixture of three isomeric secondary alcohols, each of which can exist in two geometric modifications. Prepared by hydrogenation of cresol, but the products vary according to the composition of the original cresol.

Solvent for fats, oils, waxes, rubber, cellulose esters, etc., similar to but not so powerful as *cyclo*hexanol.

USES

(1) In soap manufacture, a specially good auxiliary solvent for dry-cleaning soaps (Bird, 1932).

(2) In textile and artificial silk industry; when added to soap emulsions in the latter industry it is said to affect the lustre of cellulose acetate silk (Clayton and Clark, 1931).

(3) As a degreaser, e.g. in the compound known as "Cykloran M", which consists of methyl*cyclo*hexanol 25 per cent, soap and fatty acids 75 per cent.

TOXICITY

In animals by oral and intraperitoneal injection, methyl*cyclo*hexanol has been shown to be more toxic than *cyclo*hexane or any of its derivatives except methyl*cyclo*hexanone. By inhalation, however, owing to its low volatility, it is not likely to be highly dangerous at ordinary temperatures, nor has it proved injurious to animals when applied cutaneously as an addition to soap in amounts up to 15 per cent. No significant toxic effect from its industrial use has been reported.

Maximum Permissible Concentration

The maximum allowable concentration suggested by Cook (1945) is 100 p.p.m. (500 mg. per c.m.).

TOXIC EFFECTS IN ANIMALS

Lethal and Narcotic Concentrations

Lethal dose

By oral administration. For rabbits, 1·75 to 2·0 g. per kg. (Treon *et al.*, 1943).

By cutaneous application. 6·8 to 9·4 g. per kg. proved fatal when applied to the clipped skin of rabbits.

By inhalation. No concentration has been found lethal (Pohl, 1925; Treon *et al.*). In the experiments of Treon, concentrations of 0·56 and 2·30 mg. per l. were used.

Narcotic dose

By oral administration. For rabbits, any amount greater than 1 g. per kg. (Treon *et al.*). For dogs, 1·4 ml. per kg. (Pohl, 1925).

By cutaneous application. Convulsive movements, local irritation and thickening of the skin were noted, and deep anaesthesia developed after repeated applications for 2 days.

By inhalation. Narcosis was not produced even at 2·30 mg. per l.; only slight lethargy was observed.

Symptoms

By oral administration, twitching and narcosis followed doses greater than 1 g. per kg. body weight. With inhalation of 2·30 mg per l., salivation, conjunctival irritation and slight lethargy were the only signs of intoxication (Treon *et al.*).

Effect on Metabolism

Like *cyclo*hexane and its other derivatives, the administration of methyl-*cyclo*hexanol causes a decrease in the urinary inorganic sulphate excretion and an increase in glucuronic acid. When administered by mouth about 45 to 50 per cent of the dose is excreted conjugated with glucuronic acid, as compared to 4 to 5 per cent for *cyclo*hexane and 90 to 92 per cent for methyl*cyclo*hexanone (Treon *et al.*, 1943).

TOXIC EFFECTS IN MAN

An examination of workers using a cellulose solvent for safety glass containing 5 parts of methyl*cyclo*hexanol was carried out by the Home Office in 1933, 1934 and 1935. In 1933, three were examined and showed no significant symptoms, but one had had slight nose-bleeding and two had slightly low white blood cell counts. In 1934, six workers were examined; two again had very slightly diminished absolute white cell counts (5,600 and 5,100 respectively) and one a slight relative lymphocytosis. In 1935, seven workers, not including the two who had previously shown low white cell counts, were examined. None complained of symptoms of any significance, though it was stated that the fumes were irritating to the eyes and throat; in two, blood examinations showed no abnormality.

14. Allyl Alcohol
(Propenol, Propenyl Alcohol)

$CH_2{:}CH{\cdot}CH_2{\cdot}OH$

PROPERTIES

Allyl alcohol is the lowest member of the unsaturated aliphatic alcohols. It is prepared by heating glycerol and oxalic acid at 200° C. It may be obtained from crude wood alcohol by fractional distillation (Flury and Zernik, 1931).

A colourless liquid. B.P. 96·6° C. Sp.Gr. 0·854 at 20° C. Solidifies at −129° C. Inflammability—upper limit—2·40 per cent (Jones, 1938). It is antiseptic in 0·5 per cent solution.

USE

In organic synthesis.

TOXICITY

It is absorbed through the lungs, the gastro-intestinal tract and through the skin of animals, according to Sander (1933, cited by von Oettingen, 1943). It is rapidly diffused through the system, causing general congestion and swelling of the parenchymal cells. Atkinson (1925) considers it to be 150 times more toxic than methyl alcohol; it is not, however, believed that the severe eye injuries resulting from the use of methyl alcohol are caused by its content of allyl alcohol as an impurity, in spite of an observation by Atkinson of eye effects in one animal.

TOXIC EFFECTS IN ANIMALS
Acute Poisoning

Lethal Dose

By mouth. Dogs receiving 0·05 ml. of a 1 per cent solution died within 7 hours (Atkinson, 1925). Death was preceded by vomiting, irritation of the gastric mucosa, convulsions and coma. For rabbits, Miessner (1891) found 0·2 ml. to be rapidly fatal, but 0·15 to 0·1 ml. to be more slowly so.

By inhalation. McCord (1932) found that one exposure of 1,000 p.p.m. for 3 to 4 hours was fatal to monkeys, rabbits and rats, that 200 p.p.m. caused death after 3 to 18 repeated exposures, and 450 p.p.m. after an average of 30 exposures. Symptoms included vomiting, diarrhoea, dyspnoea, evidence of severe pain, and exudation from the nose and mouth.

Lesions of the Internal Organs

Severe inflammation of the gastro-intestinal tract, spleen and brain (McCord, 1932), kidneys (Miessner, 1891, Piazza, 1915) and generalized petechial haemorrhages (McCord, 1932) indicate the acute systemic injury caused by entry of allyl alcohol into the organism by any mode of administration.

Local Irritation

Intense irritation of the mucous membranes of the nose and mouth shown by erythema followed by discharge has been noted by practically all investigators. Conjunctivitis, lachrymation and disturbance of accommodation are recorded by Lewin and Guillery (1913). Only one of Atkinson's animals, a monkey, showed corneal opacity and blindness, but this proved to be transitory.

TOXIC EFFECTS IN MAN

Irritation of Mucous Membranes

Concentrations even as low as 5 p.p.m. will cause some irritation of the eyes and of the mucous membranes of the nose and mouth, and 50 p.p.m. will produce oedema and excessive secretion with severe conjunctivitis and lachrymation (McCord, 1932).

Effect on the Skin

According to Oettel (1936) 70 minutes' direct contact with the skin caused pain, erythema and hyperaemia.

Systemic Effects

Few symptoms of systemic injury have been recorded, but Flury and Zernik (1931) describe the case of a chemist who, in addition to the local irritation, showed general malaise, dyspnoea, and disturbance of accommodation. Gastrointestinal disturbances, nausea, vomiting, severe headache and very slight haemoptysis occurred in one of two cases observed by Browning in 1949 (unpublished). These men were removing allyl alcohol from a cask when some was spilt on the floor and on to their clothing. Although the accident occurred in the open air some of the vapour was blown into the shed where they were working. Later both men vomited, and the one more seriously affected complained of very severe headache which persisted for 48 hours. In hospital he was treated by nasal irrigation and coughed up some sputum slightly spotted with blood. A fortnight later he had recovered.

15. Benzyl Alcohol
(*Phenylmethanol, Phenylcarbinol*)
$C_6H_5 \cdot CH_2 \cdot OH$

PROPERTIES

A neutral colourless liquid of faint aromatic odour; solubility in water, up to 4 per cent at 17° C.; dissolves about 8 per cent of water at 20° C. (Durrans, 1950).

B.P. 204·7° C. Sp.Gr. 1·0507–1·0628. Evaporation time, 1767 (compared with ethyl ether, 1).

Occurs naturally in oil of jasmine, and in the form of benzoic and cinnamic esters in Peru and Tolu balsams. It is usually prepared by the action of concentrated aqueous potash on benzaldehyde.

USES

(1) In airplane manufacture as a softener of "dope" (Britton, 1927; Chapman, 1920).

(2) In the perfume industry (Blanc, 1922).

TOXICITY

No ill-effects from the industrial use of benzyl alcohol have been reported and the effects of its inhalation have not been investigated in animals. From experiments on injection in animals and on its use as a local anaesthetic in human beings, it would appear to be a narcotic of comparatively low toxicity, but in large doses producing in animals a fall in blood pressure, a depressant effect on nerve and muscle and finally respiratory paralysis.

TOXIC EFFECTS IN ANIMALS

Lethal Dose

By *intravenous injection*. For mice, rats and guinea-pigs respectively 1 ml., 1 to 3 ml., and 1 to 2·5 ml. per kg. body weight (Macht, 1918). Smaller doses were sedative and narcotic.

Symptoms

Fall in blood pressure follows injection; Macht believes this to be due to vaso-dilation through direct action on the muscle fibres, not to depression of the vaso-motor centre. After large doses primary stimulation of respiration is followed by death from paralysis of the respiratory centre (Macht, 1918). Gruber (1923) states that the heart may be paralysed before the respiratory centre. Gruber (1924) mentions a decided action on the kidney function. When injected intraperitoneally into dogs, it produced a sudden fall in blood pressure accompanied by a slowing of the secretion of urine, then a more gradual fall of blood pressure, then a return to normal with a great increase in urine secretion.

Effect on Nerve-Muscle Preparations

According to Supniewski and Macht (1926) benzyl alcohol has a pronounced depressor effect on nerve and muscle, probably owing to the presence of the free hydroxyl group in its molecule.

Effect on Leucocytes

From the effect of benzyl alcohol on the survival of leucocytes in the frog, Paglioni (1927) concludes that in this respect it is more toxic than either ethyl or methyl alcohol.

Insecticidal and Bactericidal Effect

According to Moore (1917) benzyl alcohol is very toxic to insects, more so than carbon disulphide. Its bactericidal effect is also considerable. Macht and Nelson (1918) noted that although when injected it produced irritation and necrosis of tissue the slough remained sterile, while Jakobson (1920) observed that it had a solvent action on the tubercle bacillus.

TOXIC EFFECTS IN MAN

Local anaesthetic effect. The local application of a 1 per cent solution of benzyl alcohol to mucous membranes produces anaesthesia lasting half an hour or longer (Macht, 1918); Sollmann (1919) found it in this respect about equal to β-eucaine, but that its action was weaker than and not so lasting as that of cocaine.

By oral administration. It is less toxic than the other alcohols, probably on account of the fact that it is rapidly oxidized and excreted as hippuric acid (Macht, 1918; Snapper, Grünbaum and Sturkop, 1924; Diack and Lewis, 1928, etc.).

16. Diacetone Alcohol
(*Dimethylacetonylcarbinol, Pyranton A, 4-Hydroxy-4-methylpentanone–2*)

$C_6H_{12}O_2$, i.e. $\genfrac{}{}{0pt}{}{CH_3}{CH_3}{>}C(OH)\cdot CH_2\cdot CO\cdot CH_3$

PROPERTIES

A clear colourless liquid, which may become yellowish on standing. Miscible with water in any proportion, also with alcohol and petroleum ether. Free acid, calculated as acetic, not more than 0·025 per cent. Technical product contains small amounts of acetone and occasionally of mesityl oxide.

B.R. of the technical product 160–170° C. Sp.Gr. at 20° C. 0·935–0·941 (British Standard Specification, 1934). Fl.P. 40° C. Evaporation time, 147 (compared with ethyl ether, 1). Good solvent for nitrocellulose, cellulose acetate, cellulose ethers and many resins; not a solvent for rubber or copal esters.

USES

(1) In the lacquer industry especially for nitrocellulose lacquers as an "anti-blush" agent (Yarsley, 1933).

(2) In the dyeing industry as a special solvent for certain pigments (Clayton and Clark, 1931).

(3) In the textile industry for mercerization (Sisley, 1934).

(4) In the manufacture of quick-drying writing ink (Ottley, 1933).

TOXICITY

The results of ingestion, injection and introduction by stomach tube into animals indicate that its action by these routes is similar to that of acetone, but somewhat more toxic (Walton, Kehr and Loevenhart, 1928). Toxic effects upon the haematopoietic system and the liver have been demonstrated by Keith (1932) in rats given sub-lethal doses by stomach tube. Gross (unpublished, cited by Lehmann and Flury, 1943) states that by inhalation, diacetone alcohol is slightly narcotic and possibly injurious to the liver. From its industrial use, only irritation of mucous membranes has been recorded.

TOXIC EFFECTS IN ANIMALS

Lethal and Narcotic Concentrations

Lethal dose

By intravenous injection. 3·25 ml. per kg. body weight, compared with 6 to 8 ml. for acetone.

By intramuscular injection. For rabbits, 3 to 4 ml. per kg. body weight in 48 to 60 hours; 5·0 ml. acetone is not lethal.

By stomach tube. For rabbits, 5·0 ml. per kg. body weight, compared with 10 ml. acetone.

Narcotic dose

By intravenous injection. 1·0 to 1·15 ml. per kg. body weight; results with acetone vary from 1 to 2 ml.

By intramuscular injection. Drowsiness but not unconsciousness is produced in rabbits by 2·0 ml. per kg. body weight.

By stomach tube. 2·4 to 4·0 ml. per kg. body weight, compared with 7 ml. for acetone.

By inhalation. No narcosis but sleepiness following initial restlessness and excitation was produced in Gross's experiments by inhalation of 10 mg. per l. (0·21 per cent) for 1 to 3 hours.

Symptoms

In rats, the soporific effect of diacetone alcohol was more immediate than that of acetone and the injection itself was less irritating. Death was due to respiratory failure, the heart continuing to beat for a few minutes after cessation of respiration. In rabbits, sub-lethal doses produced drowsiness and relaxation, followed by narcosis and accompanied by slowing of the respiration, greater than with acetone. With inhalation there was restlessness, symptoms of irritation, excitation, then sleepiness (Gross, unpublished).

Effect on the Blood Pressure

A fall in blood pressure was noted in dogs and rabbits, which was not altered by section of the vagi. Walton, Kehr and Loevenhart (1928) therefore conclude that it was probably due to decreased cardiac output, as shown with acetone by Schwartz (1898) and Salant and Kleitman (1922). An increased susceptibility produced by repeated injections was evidenced by a greater fall in blood pressure.

Effect on the Internal Organs

Kidneys. In one rabbit, dying after 48 hours from intramuscular injection of 4 ml. per kg. body weight, Walton and co-workers noted acute nephritis. Gross also noted kidney injuries in one rabbit after inhalation of 10 mg. per l.

Liver. In rats receiving 2 ml. per kg. body weight by stomach tube, Keith (1932) observed changes in the liver, beginning 6 hours after administration and reaching a maximum after 24 hours, but proceeding to complete recovery. The early changes consisted of peri-portal infiltration, going on to cloudy swelling and vacuolization and granulation of the cytoplasm. The swelling of the parenchymal cells was so acute as to obliterate the sinusoids. There was also an early destruction and disappearance of histiocytes, but later a great increase in their number, which Keith suggests may have been due to haemopoietic stimulation from the haemolytic action of diacetone alcohol. Recovery began after 48 hours and was practically complete in 7 to 14 days.

Blood. Destruction of erythrocytes and a reduction of haemoglobin were observed by Keith, lasting for 4 to 5 days. The blood was normal again on the sixth day.

TOXIC EFFECTS IN MAN

In the investigations by Silverman, Schulte and First (1946) on the sensory response of human beings to various concentrations of diacetone alcohol it was observed that irritation of the eyes, nose and throat appeared in the majority at 100 p.p.m., and also that at this level, many found both the odour and the taste unpleasant. Owing to the variety and number of complaints made by subjects exposed to 100 p.p.m., even though the majority indicated that they could work an 8-hour day at this level, Silverman *et al.* recommend 50 p.p.m. as a desirable limit.

17. Ethylene Chlorohydrin
(*Glycol Chlorohydrin, 2-Chloroethyl Alcohol*)

$CH_2Cl \cdot CH_2 \cdot OH$

PROPERTIES

Ethylene chlorohydrin is a clear glycerine-like fluid with an odour like that of alcohol. It is fully miscible with benzine, water, alcohol, etc. When heated with water to 100°C. it decomposes into glycol and acetaldehyde; when heated alone to 184°C., into ethylene chloride and acetaldehyde (Koelsch, 1927). The anhydrous product (96–98 per cent) has a boiling range of 125 to 132°C. Sp.Gr. 1·21. Fl.P. 55°C.

The chief impurity in the commercial product is ethylene dichloride, and according to Goldblatt (1944), dichlorodiethyl ether.

The volatility of ethylene chlorohydrin is as follows (Koelsch, 1927): at 15°C., about 0·8 mg./sq. cm./hr.; at 23°C., about 1·8 mg./sq. cm./hr.; at 37°C., about 3·5 mg./sq. cm./hr.

USES

(1) In the lacquer industry, especially as an addition to plasticisers for paint and varnish, because of its solvent action for cellulose acetate, resin and wax.

(2) In the dyeing and cleaning industry for the removal of tar spots, as a cleaning agent for machines, and as a solvent in fabric-dyeing.

(3) In the oilcloth, paper and pharmaceutical industries.

(4) In the treating of seeds and ripening of fruit and for speeding up the sprouting of potatoes (Denny, 1928). The danger of such applications has been pointed out by Pratt (1930).

TOXICITY

The general character of the toxic effects of ingestion, application to the skin, intraperitoneal injections or inhalation of ethylene chlorohydrin is that of severe nervous and metabolic poisoning, with symptoms appearing at an interval after the exposure, and accompanied by local irritation of surface mucous membranes and deeper lung tissue. These effects have been observed in both animals and human beings. It is toxic at relatively low atmospheric concentrations, and its odour, resembling that of ethyl alcohol, is not particularly pronounced even at near-lethal levels of the vapour.

Maximum Permissible Concentration

The maximum allowable concentration recommended by Cook (1945) for ethylene chlorohydrin is 10 p.p.m., but Goldblatt and Chiesman (1944), in view of the possibility of its cumulative action, suggest an even lower concentration, i.e. 2 p.p.m. The actual concentration was 18 p.p.m. at the time when their nine non-fatal cases occurred. The concentration which proved fatal in the case described by Dierker and Brown (1944) was 305 p.p.m. (1 mg. per l.).

TOXIC EFFECTS IN ANIMALS
Acute Poisoning

Lethal Dose

By oral administration. According to Goldblatt (1944) the L.D.50 is 7·2 mg. per 100 g.

By intraperitoneal injection. L.D. 50 is 5·6 mg. per 100 g. for a single dose.

By skin absorption. Smyth and Carpenter (1945) found that the L.D.50 of ethylene chlorohydrin when applied as a poultice to the skin was 0·07 ml. per kg. body weight; most of the deaths occurred within 24 hours. When contact was limited to 2 hours, the L.D.50 was about 0·3 ml. per kg.; when a 10 per cent solution in water was applied, it was about 1·14 ml. of ethylene chlorohydrin per kg. In Koelsch's experiments the application of a cloth soaked in a solution of 5 g. of ethylene chlorohydrin caused death to the animals on the following day. Goldblatt (1944) found that the application of a single dose of 0·03 to 0·09 ml. to the depilated skin of mice was lethal. There was no evidence of direct skin irritation.

By inhalation. In the experiments carried out by Koelsch (1927) on guinea-pigs and cats, none of the animals, even with inhalations of high concentration of the vapour, died immediately, but even with small concentrations (4 mg. per l. for guinea-pigs), they died at varying intervals after being removed from exposure. Koelsch emphasizes the fact that the symptoms were only slight during the actual exposure, and often only appeared an hour after removal. With small doses and only 1 hour's exposure, it was found possible to keep the animals alive, but with severe nervous and metabolic disturbance, recovery being very slow and often taking weeks or months. The minimum lethal concentration for guinea-pigs was found to be 3·6 mg. per l. Koelsch, quoting from a later report from the Badische Anilin- und Soda-Fabrik, Ludwigshafen, states that in cats the poisoning is more delayed than in mice, guinea-pigs or rabbits.

These results agree closely with those of Goldblatt (1944) whose figures for lethal dosage in rats, mice and guinea-pigs are given in Table 39.

TABLE 39

Lethal doses of ethylene chlorohydrin for some animals

Animal	Concentration (mg./l.)	Time of exposure (min.)
Rat	4 (1,120 p.p.m.)	30
Mouse	3	60
Guinea-pig	3	112

Goldblatt confirms Koelsch's observation that the animals do not die during exposure.

Symptoms

After oral administration, no true narcosis precedes death, which is delayed in onset. The symptoms include general prostration, diminished or absent corneal reflexes and spasmodic movements. In both Koelsch's and Goldblatt's experiments the delay in the onset of lethal symptoms was quite characteristic. Death appeared to be due to respiratory paralysis and was preceded by depression, inco-ordination of movement and convulsions.

Lesions of the Internal Organs

The chief damage observed was in the kidneys, interstitial haemorrhage, oedema and, according to Koelsch, epithelial casts in the tubules with commencing severe degeneration of nuclei. The lungs showed some emphysema, and in Koelsch's animals but not in Goldblatt's, some interstitial haemorrhage and oedema.

Effect on Blood Pressure and Respiration

While inhalation of the vapour of ethylene chlorohydrin failed to cause a manifest fall in blood pressure or significant changes in respiratory depth or rhythm, intravenous injection to anaesthetized cats was followed by an immediate drop in blood pressure and inhibition of respiration. Vagal action and cardiovascular reflexes were not affected (Goldblatt, 1944).

Effect on Nerve Impulses

When a 1 per cent solution of ethylene chlorohydrin was placed in contact with nerve preparations it was followed by complete nerve block. This was reversible (Goldblatt, 1944).

Effect on Smooth Muscle

Ethylene chlorohydrin inhibits both tone and rhythm of smooth muscle (small intestine and uterus of rabbit (Goldblatt, 1944)).

Chronic Poisoning

From the experiments of both Koelsch and Goldblatt, it appears that the kidneys are first attacked, followed by some injury to the liver and the lungs. A dosage of 2·2 to 2·5 mg. per l. in Koelsch's experiments and 3 mg. per l. in Goldblatt's, repeated daily, led eventually to the death of all the animals. The only symptoms were apathy and loss of weight, and in Koelsch's animals, refusal to eat.

Lesions of the Internal Organs

The kidneys showed multiple small haemorrhages and complete disintegration of the cells of the convoluted tubules, but no lesions of the glomeruli. Since the secretion of urine was unimpaired and the urine showed surprising absence of signs of kidney damage, Goldblatt suggests that the effect of ethylene chlorohydrin on the kidneys is that of a vascular poison, perhaps only secondarily affecting the secretory cells. The liver showed intense congestion and fatty degeneration. The lungs showed collapse and areas of haemorrhage.

TOXIC EFFECTS IN MAN

Several cases of fatal poisoning by inhalation of ethylene chlorohydrin vapour have been recorded as well as two fatal cases from ingestion of solvent mixture containing 25 per cent of ethylene chlorohydrin. It also appears from both animal experiment and human experience that this substance can exert its toxic action by absorption through the skin. A number of non-fatal cases, including twenty cases between 1940 and 1947 reported to the Factory Department, have also been recorded, and of these, some may be regarded as due to sub-acute intoxication (Goldblatt and Chiesman, 1944).

Fatal Cases

Following ingestion. In two fatal cases reported by Güthert (1943) death followed the drinking of a mixture containing methylene chloride and toluene as well as ethylene chlorohydrin. Death was, however, attributed to the ethylene chlorohydrin constituent on account of the severe degenerative lesions in the kidneys, liver, brain and heart muscle in the one man who survived 36 hours. Both patients died after a period of improvement under treatment, both showed extreme excitement, motor irritability and convulsions, and in the longer delayed case, the kidneys showed severe nephrotic changes.

Following inhalation. Most of these cases, two reported by Koelsch (1927), one by Cavallazzi (1942), one by Dierker and Brown (1944), and one by Middleton (1930), have occurred in connection with the use of ethylene chlorohydrin as a cleaning agent, a constituent of adhesive fluid or the cleaning of a still which had contained it as in Middleton's case. In the cases recorded by Goldblatt and Chiesman, however, the toxic effects occurred during the actual manufacture of ethylene chlorohydrin. In the case reported by Middleton the toxic effects appear to have been due not only to inhalation of the vapour, but also to absorption through the skin, the man having entered a still in a dye-works in order to mop out water containing ethylene chlorohydrin in solution.

Symptoms

In all these cases except one, the first symptoms, nausea, vomiting and headache, appear to have occurred after $1\frac{1}{2}$ to 2 hours of intensive exposure. In one of the cases of Goldblatt and Chiesman, which may be described as sub-acute rather than acute, there had been no sudden increase in the concentration to which the man had been exposed for 2 months before he developed the acute intoxication which proved fatal. Vomiting is usually followed by drowsiness, dyspnoea and a narcosis suggestive of a toxic action on the central nervous system. In some cases, as in that of Cavallazzi, vomiting is followed by delirium and violent excitement. In only one case, that of Goldblatt and Chiesman, were signs definitely significant of kidney injury observed. This patient had haematuria shortly before death. In all the cases death took place after an interval of some hours after the onset of the symptoms. This appears to indicate a cumulative action of ethylene chlorohydrin.

Post-mortem appearances

In none of the fatal acute cases did the post-mortem appearances point directly to kidney injury as the predominant cause of death. In one of Koelsch's cases the kidneys were unaffected; the other case showed degeneration. In the case of Dierker and Brown, the kidneys showed deep congestion; in that of Goldblatt and Chiesman they appeared normal macroscopically, but sections were not available. In no case were the lung and liver changes severe enough to account for death. From the data available it is difficult to decide the exact cause of death in fatal acute intoxication by ethylene chlorohydrin. Goldblatt and Chiesman suggest that the toxic effect is that of a violent cerebral and vascular poison.

Non-fatal Acute Cases

In addition to describing one fatal case occurring in a paper factory, Koelsch includes six of fairly acute poisoning and one with no symptoms other than slight eye irritation. In a linoleum factory there was one fatal case and three fellow-workmen were less severely affected. Two non-fatal cases were reported to the Factory Department, one being that of a man doing the same kind of work as the one in the fatal case described by Middleton, the other, a workman in a dye-works employed in dissolving cellulose acetate in ethylene chlorohydrin. Goldblatt and Chiesman describe two fatal cases which occurred in a factory during a period in which a fault had developed in the plant, also nine other cases, all relatively mild in nature, from the same factory.

Symptoms

The nature of the symptoms is not specific, but in nearly all the cases, nausea, vomiting, giddiness and drowsiness were the chief features. Among the nine cases described by Goldblatt and Chiesman the vapour of ethylene chlorohydrin was mixed with that of ethylene dichloride, and they suggest that the mild narcosis which was observed was probably due to the latter. Slight albuminuria was present in even the mild cases, but evidence of a toxic action on the liver was not obtained except in one case which showed liver enlargement. Goldblatt and Chiesman suggest that the nausea and vomiting are central in origin and that the toxic action of ethylene chlorohydrin in man is directed towards the nervous system, the cardiovascular system and the kidneys. The secondary points of attack are the lungs and the liver.

Conditions Conducive to Intoxication

Personal Idiosyncrasy

In both the fatal cases reported by Koelsch existing organic lesions seem to have favoured the toxic effect, old pleuritic adhesions and evidence of healed pulmonary tuberculosis in one, and degeneration of the heart and arteries in the other. Cavallazzi also suggests that a constitutional factor must have played an important part in this case, as other workers exposed for much longer periods had shown only very slight disturbance. Poor physical standard is mentioned by Goldblatt and Chiesman as being a factor in the liability to intoxication, and in such cases, symptoms and signs were more severe.

Sex

Goldblatt and Chiesman consider women to be more susceptible than men. Five of their non-fatal cases were women.

Surface Area and Temperature

Large evaporating surfaces and high temperatures causing increased volatilization were conditions present in both factories in which Koelsch's two fatal cases occurred.

Contact of Skin with Solutions

In view of the case reported by Middleton, and the fact that one of Koelsch's fatal cases had a skin contact with the solvent, and the animal experiments of both Koelsch and Goldblatt, it must be assumed that absorption through the skin is a factor of some importance in increasing the effects of inhalation. Goldblatt and Chiesman point out that though very large applications are required to endanger life, and though the main hazard is probably inhalation of the vapour, there are certain operations in which repeated contamination of the clothes and skin readily occur unless adequate precautions are taken.

18. Monochlorohydrin
(*Chloropropylene Glycol*)

$C_3H_5Cl(OH)_2$, i.e. $CH_2Cl \cdot CH(OH) \cdot CH_2 \cdot OH$

PROPERTIES

Commercial monochlorohydrin is a mixture of two isomers, in which the above preponderates. B.R. of the commercial product 213–228° C. Sp.Gr. 1·28–1·35.

It is miscible with water and with most organic solvents, but not with hydrocarbons or vegetable oils. It is hygroscopic and develops acidity on prolonged contact with moisture.

Solvent of limited power for cellulose acetate and glyceryl phthalate resins; to a smaller extent for ester gum, benzyl abietate, shellac and mastic; does not dissolve cellulose nitrate, copals, coumarone or hard resins.

USES

Similar to those of dichlorohydrin (see below) but is not much used in the lacquer industry because it is expensive, hydrolyses easily and darkens on ageing.

TOXIC EFFECTS IN ANIMALS

The experiments of Marshall and Heath (1897) carried out by introduction of a solution of the substance (4 ml. per kg. body weight) into the stomach of rabbits showed that monochlorohydrin has a narcotic effect, and a depressant action on the heart, circulation and respiration, similar to that of dichlorohydrin but weaker. The slighter narcotic action was attributed by them to the smaller number of chlorine atoms in the molecule. The onset of effects on the central nervous system was much later with monochlorohydrin than with dichlorohydrin. During the first 6 hours nothing was noted beyond a fall in the pulse rate and the number of respirations. The next day there was muscular weakness and cerebral depression. During the next few days the temperature fell, heart beats became weaker, the muscular paresis increased and the sphincters became completely paralysed. Greater somnolence developed but the reflexes remained. Albuminuria was present during the last few days and death occurred on the eighth day. Post-mortem examination showed congestion of the stomach and duodenum and cloudy degeneration of the kidneys.

TOXIC EFFECTS IN MAN

Except for the statement by Zernik (1933) that monochlorohydrin was the effective toxic agent in the fatal case reported by Molitoris (1931) as due to dichlorohydrin (p. 252), no toxic effects of monochlorohydrin in human beings have been recorded.

19. Dichlorohydrin

(*Dichloro*iso*propyl Alcohol*, 1:3-*Dichloropropan-2-ol*)

$C_3H_5Cl_2 \cdot OH$, i.e. $CH_2Cl \cdot CH(OH) \cdot CH_2Cl$

(Technical dichlorohydrin is a mixture of 1:3-dichloropropanol-2 and 1:2-dichloropropanol-3; the former preponderates and is represented by the formula.)

PROPERTIES

Solubility in water, about 10 per cent (Durrans, 1950).

B.P. 174° C. Sp.Gr. 1·34–1·38. Solvent for acetyl cellulose, resin, celluloid, etc.

USES

(1) In the manufacture of pharmaceutical products.

(2) In the textile industry as a solvent of acetyl cellulose.

(3) In the lacquer industry; not much used as it has the same defects as monochlorohydrin.

(4) As raw material in the explosive industry.

TOXICITY

Animal experiments have shown that dichlorohydrin is a narcotic of considerable strength, with toxic depressant effects upon the heart and circulation greater than those of monochlorohydrin. While two fatalities occurring in Germany in workmen using solvents containing dichlorohydrin were at first attributed to, and reported as, dichlorohydrin poisoning, there now seems to be some doubt as to whether these were in fact due to dichlorohydrin itself.

TOXIC EFFECTS IN ANIMALS

In the earlier experiments carried out by Romensky (1872) and Marshall and Heath (1897), the dichlorohydrin was injected into the stomach or given by mouth to mice and rabbits. Romensky gave detailed data only for trichlorohydrin which acted as a narcotic, producing a decided fall of blood pressure; he said that dichlorohydrin acted in a similar manner. Marshall and Heath also produced narcosis, fall in temperature, and slowing of pulse and respiration, preceded by a slight rise and finally death after 6 hours, by the administration of 1·5 g to a rabbit weighing 1·5 kg. Smaller doses produced similar results, but the narcosis was slighter and more prolonged, the animal recovering temporarily but dying 24 hours later. The depressant action upon the heart was clearly shown by experiments on the perfusion of the isolated frog's heart, dichlorohydrin producing a distinct effect in 1 in 1,000 concentration.

These results have been confirmed, by intraperitoneal and intravenous injection, by Kaminski and Seelkopf (1933), who have also carried out experiments on the inhalation of dichlorohydrin, both pure and commercial. By intraperitoneal injection of pure dichlorohydrin (1 ml. per kg. body weight) in a rabbit, deep narcosis with loss of corneal reflexes was produced in 5½ minutes and death took place after 4 hours. Intravenous injection of 1 ml. produced a rapid fall of blood pressure, and of 2 ml., irregular heart action and respiratory rate.

Kaminski and Seelkopf reported (1933) that the effect on mice inhaling pure dichlorohydrin vapour by exposure in a flask to the evaporation of 0·1 ml. was less than that of either the commercial product or the distillation residue, although it produced slight convulsions and incomplete narcosis, followed by death the next day. The fraction of the commercial product boiling at 105–107° C. was found to be very much the most toxic. They concluded that the impurities in commercial dichlorohydrin were responsible for the more acute toxic effects, but that pure dichlorohydrin itself produced severe central nervous and narcotic effects when given over a longer period.

Local irritative effects. Acute inflammation of the eyes was produced by the instillation of a drop of either pure or commercial dichlorohydrin. This was worse when the commercial product was used, while the effect of the distillation residue was much more severe and was accompanied by rapid shallow breathing, apathy, exhaustion and finally death. Subpleural haemorrhages and fatty degeneration of the liver were found post mortem after the instillation of the commercial dichlorohydrin.

TOXIC EFFECTS IN MAN

Two fatal cases have been attributed to the use of dichlorohydrin in industry, but there is some confusion about the one reported by Molitoris (1930, 1931); it is stated by Zernik (1933) that the final decision was that death was due to the monochlorohydrin component of the solution "Enodrin" used, while Kaminski and Seelkopf (1933), reviewing the report of this case, state that ethylene chlorohydrin was the actual toxic agent. The symptoms of the patient, a calico worker, were vomiting, dyspnoea, giddiness, headache and delirium, with a small rapid pulse. Death occurred after a few hours. The post-mortem examination showed blood-stained mucus in the lungs and bronchi, and an enlarged inflamed spleen. The diagnosis was infection with no evidence of poisoning. Molitoris, however, in reviewing the case, pointed out that there were special circumstances on the particular day on which the illness occurred, which had not been present in the previous period, during which it was stated that no symptoms had ever been complained of by workers with "Enodrin", and he insisted that the "new substance" being tested on that day as a solvent of acetyl cellulose was dichlorohydrin and was the cause of death.

In the case reported by Kaminski and Seelkopf (1933) the symptoms followed the bursting of a glass vessel containing dichlorohydrin, and death took place three days later. Nausea, vomiting, disordered action of the heart, quick pulse, and severe dyspnoea were related to the central toxic effect of the dichlorohydrin, while local irritation of the eyes and face were apparently due to contact with fluid when the vessel exploded. On the basis of their animal experiments, however, Kaminski and Seelkopf attributed the more serious toxic effects of the dichlorohydrin used to its content of "neutral organic substances", especially those distilling between 105 and 170° C. rather than to the pure dichlorohydrin itself, though they conclude that the effects of pure dichlorohydrin are similar in nature but of less intensity.

BIBLIOGRAPHY

ANSWER TO QUERY, Journal of the American Medical Association (1931). Use of methanol as an anti-freeze mixture. *J. Amer. med. Ass.*, **96**, 460.
ANSWER TO QUESTION, Industrial Chemist (1930). Use of isopropyl alcohol in perfumery industry. *Industr. Chem. chem. Mfr.*, **6**, 168.
ARKANSAS (1931). Law relating to anti-freeze mixtures. *Publ. Hlth. Rep., Wash.*, **46**, 2254.
ASCHAN, O. (1927). Zur Verwertung des Petroläthers und der Erdol-Residuen. *Chemikerztg.*, **51**, 4.
ATKINSON, H. V. (1925). The toxicity of impurities in wood alcohol. I. Allyl alcohol. *J. Pharmacol.*, **25**, 144.
BAADER, E. W. (1933). Gewerbemedizinische Erfahrungen. *Verh. dtsch. Ges. inn. Med.*, **45**, 318.
BACHEM, C. (1927). Beitrag zur Toxikologie der Halogenalkyle. *Arch. exp. Path. Pharmak.*, **122**, 69.
BAER, G. (1898). Beitrag zur Kenntnis der acuten Vergiftung mit verschiedenen Alkoholen. Dissertation, Berlin.
BARLOW, O. W. (1936). Studies on the pharmacology of ethyl alcohol. I. Comparative study of the pharmacologic effects of grain and synthetic ethyl alcohols. II. A correlation of the local irritant, anaesthetic and toxic effects of three potable whiskeys with their alcoholic content. *J. Pharmacol.*, **56**, 117.
BASKERVILLE, C. (1913). Report on chemistry, technology and pharmacology of, and legislature pertaining to, methyl alcohol. *2nd Rep., Factory Inspection Commission, New York*, **2**, 921.
BENEDICENTI, A. (1935). L'alcool metilico; applicazioni industriali e tossicita. *Ricerca Sci.*, **6**, I. Special suppl. 1–88. See also *Chem. Tr. J.*, **96**, 335.
BERNHARDT, G. (1922). Über Isopropyl alcohol als Mittel zur Handedesinfektion. *Dtsch. med. Wschr.*, **48**, 68.
BERTARELLI, E. (1934). I pericoli pratici dell' alcool metilico. *Ann. Igiene (sper.)*, **44**, 729.
BIRCH-HIRSCHFELD, A. (1901). Experimentelle Untersuchungen über die Pathogenese der Methylalkoholamblyopie. *Arch. Ophthal.*, **52**, 538.
BIRD, C. L. (1932). Auxiliary solvents for dry-cleaning soaps. *J. Soc. Dy. Col., Bradford*, **48**, 256.
BLANC, G. (1922). Auto-oxidation of benzyl alcohol. *Amer. Perfum.*, **17**, 213.
BOGIN, C. (1935). Effect of butanol on the viscosity of lacquer and alkyd resin finishes. *Paint Oil Chem. Rev.*, **97**, 45.

BORUTTAU, H. (1921). Die Verwendung von Isopropylalkohol zur hygienischen und kosmetischen Zwecken. *Dtsch. med. Wschr.*, **47**, 747.
BRACQ, A. (1944). L'intoxication par l'alcool méthylique. Thèse, Paris (cited in *Arch. Mal. prof.*, **6**, 345).
BREZINA, E. (1929). Internationale Übersicht über Gewerbekrankheiten, nach den Berichten der Gewerbeaufsichtsbehörden der Kulturländer über die Jahre 1920 bis 1926. Springer, Berlin.
BRITTON, H. T. S. (1927). Aero dopes and varnishes. *Industr. Chem. chem. Mfr.*, **3**, 116.
BROWN, quoted by Ziegler (1921). *J. Amer. med. Ass.*, **77**, 1160.
BRÜCKNER, H. (1924). Über den gegenwärtigen Stand der Methylalkoholvergiftung mit besonderer Berücksichtigung ihrer Bedeutung für Gewerbe und Industrie. *Zbl. GewHyg.*, **11**, 17.
BUCHNER, H., FUCHS, F. and MEGELE, L. (1901). Wirkungen von Methyl-, Äthyl-, und Propyl-Alkohol auf die arteriellen Blutström bei ausserer Anwendung. *Arch. Hyg., Berl.*, **40**, 347.
BURGER, G. B. C. and STOCKMANN, B. H. (1932). Über Urobilinurie als Folge der Einatmung von organischen Lösungsmitteln in geringer Konzentration. *Zbl. GewHyg.*, **19**, 29.
CARPENTER, T. M. (1929). Ethyl alcohol in fowls after exposure to alcohol vapor. *J. Pharmacol.*, **37**, 217.
CAVALLAZZI, D. (1942). Akute und tödliche Vergiftung durch Äthylenchlorhydrin. *Samml. Vergiftungsf.*, **12**, 79, A 910.
CHAPMAN, A. W. (1920). The chemistry of aeronautics. *Chem. Age (N.Y.)*, **2**, 438.
CLAYTON, E. and CLARK, C. O. (1931). Modern organic solvents: Parts I, II. *J. Soc. Dy. Col., Bradford*, **47**, 183, 247.
COGAN, D. G. and GRANT, W. M. (1945). An unusual type of keratitis associated with exposure to n-butyl alcohol (butanol). *Arch. Ophth., N.S.* **33**, 106.
COGAN, D. G. and GRANT, W. M. (1945). Keratitis due to n-butyl alcohol. *Arch. Ophth., N.S.* **34**, 248.
COLEBATCH, J. H. (1946). Acute methyl alcohol poisoning. *Proc. Australasian Coll. Phys.*, **1**, 57.
COOK, C. A. and SMITH, A. H. (1929). Determination of isopropyl alcohol in the presence of acetone in the urine. *J. biol. Chem.*, **85**, 251.
COOK, W. A. (1945). Maximum allowable concentrations of industrial atmospheric contaminants. *Industr. Med.*, **14**, 936.
CUNNINGHAM, J. G. (1934). Chemical health hazards in industry. *Chem. and Ind.* **12**, 707.
DENNY, F. E. (1928). Chemical treatments for shortening the rest periods of plants. *J. Soc. chem. Ind., Lond.*, **47**, 239.
DIACK, S. L. and LEWIS, H. B. (1928). Studies in the synthesis of hippuric acid in the animal organism: VII. A comparison of the rate of elimination of hippuric acid after the ingestion of sodium benzoate, benzyl alcohol and benzyl esters of succinic acid. *J. biol. Chem.*, **77**, 89.
DIERKER, H. and BROWN, P. G. (1944). Study of fatal case of ethylene chlorohydrin poisoning. *J. industr. Hyg.*, **26**, 277.
DUBLIN, L. I. and LEIBOFF, P. (1922). Occupation hazards and diagnostic signs. *U.S. Department of Labor, Bureau of Labor Statistics Bull.*, **306**.
DURRANS, T. H. (1935). Solvents and plasticisers. *J. Oil Col. Chem. Ass.*, **18**, 340.
DURRANS, T. H. (1950). Solvents. 6th ed. Chapman and Hall, London.
EBERT, F. (1925). Organische Lösungsmittel, ihre Giftwirkung und die Gefahren für die damit beschäftigen Personen, sowie deren Verhütung. *Osterr. Chemikerztg.*, **28**, 176.
EFRON, J. (1885). Beiträge zur allgemeinen Nervenphysiologie. *Pflüg. Arch. ges. Physiol.*, **36**, 467.
EGG, C. (1927). Zur Kenntnis der Methylalkoholvergiftung. *Schweiz. med. Wschr.*, **8**, 5.
ELHARDT, W. F. (1931). Effect of methyl, propyl and butyl alcohol on the growth of white Leghorn chickens. *Amer. J. Physiol.*, **100**, 74.
EYQUEM (1905). Du danger des vapeurs alcooliques dans la fabrication de la poudre sans fumée. *Ann. Hyg. publ., Paris*, 4me sér., **3**, 71.
FILIPPI, E. (1914). Azione fisiologica e compartamento di alcuni derivati del benzene in confronto con quelli di cicloesano. *Arch. Farmacol. sper.*, **18**, 178.
FLEISCHER, R. (1880). Zur Frage der Hautresorption. *Virchows Arch.*, **79**, 558.
FLURY, F. (1928). Moderne gewerbliche Vergiftungen in pharmakologisch toxikologischer Hinsicht. *Arch. exp. Path. Pharmak.*, **138**, 65.
FLURY, F. and WIRTH, W. (1934). Zur Toxikologie der Lösungsmittel (verschiedene Ester, Aceton, Methylalkohol). *Arch. Gewerbepath. Gewerbehyg.*, **5**, 1.
FLURY, F. and ZERNIK, F. (1931). Schädliche Gase, Dämpfe, Nebel, Rauch- und Staubarten. Springer, Berlin.
FÜHNER, H. (1904). Über die Einwirkung verschiedener Alkohole auf die Entwicklung der Seeigel. *Arch. exp. Path. Pharmak.*, **51**, 1.
FÜHNER, H. (1921). Die Wirkungsstärke der Narkotica. *Biochem. Z.*, **120**, 143.
FULLER, H. C. and HUNTER, O. B. (1927). Isopropyl alcohol: an investigation of its physiologic properties. *J. Lab. clin. Med.*, **12**, 326.
GABRIEL, G. L. (1928). Butanol fermentation process. *Industr. Engng. Chem.*, **20**, 1063.

GARDNER, H. A. (1925). Physiological effects of vapors from a few solvents used in paints, varnishes and lacquers. *Paint Manuf. Ass. U.S., Proc. Scientific Sec. Educ. Bur., Circ. No.* 250.
GARLICK, H. S. (1927). Iso-propyl alcohol. *Industr. Chem. chem. Mfr.,* **3,** 392.
GERBIS, H. (1931). Über gewerbliche Gifte. *Z. angew. Chem.,* **44,** 640.
GERBIS, H. (1931). Methylalkoholvergiftung, chronische. *Samml. Vergiftungsf.,* **2,** 1/1, A 160.
GOLDBLATT, M. W. (1944). Toxic effects of ethylene chlorohydrin. Part II: Experimental. *Brit. J. industr. Med.,* **1,** 213.
GOLDBLATT, M. W. and CHIESMAN, W. E. (1944). Toxic effects of ethylene chlorohydrin. Part I: Clinical. *Brit. J. industr. Med.,* **1,** 207.
GOLTDAMMER (1878). Bronchitis durch Holzgeistdämpfe. *Vjschr. gerichtl. Med.,* N.F., **29,** 162.
GRADINESCO, A. (1934). L'action de l'alcool sur le centre respiratoire. *J. Physiol. Path. gén.,* **32,** 363.
GRADINESCO, A. and DEGAN, C. (1934). L'action des alcools méthylique et éthylique sur l'excitabilité du nerf. Determination faite par la méthode de la chronaxie. *J. Physiol. Path. gén.,* **32,** 826.
GRANT, D. H. (1923). Antiseptic and bactericidal properties of isopropyl alcohol. *Amer. J. Med. Sci.,* **166,** 261.
GREENBURG, L., MAYERS, M. R., GOLDWATER, L. J. and BURKE, W. J. (1938). Health hazards in the manufacture of "fused collars". II. Exposure to acetone-methanol. *J. industr. Hyg.,* **20,** 148.
GRIGNOLO, F. (1913). Biochemische Veränderungen im Kammerwasser bei akuten Intoxikationen durch Methylalkohol und durch Toxipeptide. *Klin. Mbl. Augenheilk.,* **51,** 157.
GROSS, E. (unpublished). Cited by Lehmann, K. B. and Flury, F. (1943), Toxicology and hygiene of industrial solvents. Williams and Wilkins, Baltimore.
GRUBER, C. M. (1923). Pharmacology of benzyl alcohol and its esters. *J. Lab. clin. Med.,* **9,** 15, 92.
GRUBER, C. M. (1924). Pharmacology of benzyl alcohol and its esters. *J. Lab. clin. Med.,* **10,** 284.
GRUBER, C. M. (1924). Effect of benzyl alcohol and its esters, benzyl benzoate and benzyl acetate, upon kidney functions. *J. Pharmacol.,* **23,** Proc., 149.
GRUNOW (1912). *Halbmschr.Soz.Hyg. (Med. Reform., Berl.).* Quoted by Koelsch, F. (1921). *Zbl.Gewhyg.,* **9,** 198.
GÜTHERT, H. (1943). Tödliche kombinierte Lösungsmittelvergiftung (Äthylenchlorhydrin-Methylenchlorid-Toluol). *Arch. Gewerbepath. Gewerbehyg.,* **12,** 362.
HAGGARD, H. W. and GREENBERG, L. A. (1934). Quantitative determination of ethyl alcohol in air, blood and urine by means of iodine pentoxide. *J. Pharmacol.,* **52,** 137.
HAGGARD, H. W., MILLER, D. P. and GREENBERG, L. A. (1945). The amyl alcohols and their metabolic fates and comparative toxicities. *J. industr. Hyg.,* **27,** 1.
HALE, A. B. (1901). Case of blindness from drinking bay rum, compared with the reported case due to methyl alcohol and to essence of Jamaica ginger. *J. Amer. med. Ass.,* **37,** 1450.
HÄMÄLAINEN, R. and TERÄSKELI, H. (1928). Ein Beitrag zur Kenntnis der sogenannten Methylalcohol-vergiftung. *Acta ophthal.,* Kbh., **6,** 260.
HAMILTON, A. (1925). Industrial poisons in the United States. Macmillan, New York.
HAMILTON, A. (1934). Industrial toxicology. Harper, New York.
HARNACK, E. (1912). Über die Giftigkeit des Methylalkohols. *Dtsch. med. Wschr.,* **38,** 358.
HARTRIDGE, H. (1920). Economical dehydrating and clearing agents. *J. Physiol.,* **54,** p. viii.
HAWES, A. T. (1905). Amblyopia from the fumes of wood alcohol. *Boston med. surg. J.,* **153,** 525.
HAYHURST, E. R. (1915). Survey of industrial health hazards and occupational diseases in Ohio. Ohio State Board of Health, Columbus, Ohio.
HEFFTER, A. (1920-3). Handbuch der experimentelle Pharmakologie. Springer, Berlin.
HEFFTER, A. and JUCKENACK (1919). Über die Verwendung von Propylalkohol in gesundheitlicher Beziehung. *Vjschr. gerichtl. Med.,* **58,** 1.
HENDERSON, Y. and HAGGARD, H. W. (1927). Noxious gases and the principles of respiration influencing their action. *Amer. Chem. Soc. Monograph.* Ser. 35. Chemical Catalog Co., New York.
HILBERT (1925). Cited by Lewin, L., 1929, in Gifte und Vergiften, 4th Aufl. Springer, Berlin.
HOLDEN, W. A. (1899). The pathology of the amblyopia following profuse hemorrhage, and of that following the ingestion of methyl alcohol, with remarks on the pathogenesis of optic-nerve atrophy in general. *Arch. Ophthal., N.Y.,* **28,** 125.
HOLDEN, H. C. and DOOLITTLE, A. K. (1935). Solvents. *Industr. Engng. Chem.,* **27,** 525.
HUFFERD, R. W. (1932). Toxicity of aliphatic alcohols. *J. Amer. pharm. Ass.,* **21,** 549.
HUMPERDINCK, K. (1941). Zur Frage der chronischen Giftwirkung von Methanoldämpfen. *Arch. Gewerbepath. Gewerbehyg.,* **10,** 569.
HUNT, REID (1902). Toxicity of methyl alcohol. *Johns Hopk. Hosp. Bull.,* **13,** 213.
HUNT, REID (1930). Synthetic methanol; a supplemental statement. *Industr. Engng. Chem.,* **22,** 915.
I. G. FARBENINDUSTRIE AKTIENGESELLSCHAFT (1930). Lösungsmittel und Weichmachungsmittel, p. 61. Frankfurt-am-Main, 1930.

INTERNATIONAL LABOUR OFFICE (1926).
  Brochure No. 48. Alcohol, intoxication by. Also in Occupation and Health, 1, 90. International Labour Office, Geneva, 1930.
  Brochure No. 50. Ethyl alcohol. Also in Occupation and Health, 1, 691. International Labour Office, Geneva, 1930.
  Brochure No. 55. Methyl alcohol. Also in Occupation and Health, 2, 233. International Labour Office, Geneva, 1934.
JAKOBSON, J. (1920). L'alcool bencyligue dans la tuberculose expérimentale (in vitro). Z. Tuberk., 32, 103.
JAMES, V. C. (1931). Acute alcoholic poisoning due to the application of surgical spirit to the legs. Brit. med. J., i, 539.
JELLIFFE, S. E. (1905). Multiple neuritis in wood alcohol poisoning. Med. News, N. Y., 86, 387.
JOIRIS, N.P. (1935). La scotome central positif et transitoire (signe de Weekers) dans la névrite optique rétrobulbaire, au cours de l'intoxication aiguë par l'alcool méthylique. Arch. Ophtal., Paris, 52, 578.
JONES, G. W. (1938). Inflammation limits and new practical application in hazardous industrial operations. Chem. Rev., 22, 1.
KAMINSKI, J. and SFELKOPF, K. (1933). Dichlorhydrin, technisches verursacht tödlichen beruflichen Unfall. Samml. Vergiftungsf., 4, 147, A 350.
KAPLAN, A. and LEVREAULT, G. V. (1945). Methyl alcohol poisoning. Report of forty-two cases. Nav. med. Bull., Wash., 44, 1107.
KASASS, J. J. (1913). Zur Pathologie der Methylalkoholamaurose. Zbl. Biochem. Biophys., 15, 205. (Also in Chem. Abstr., 7, 3794).
KEESER, E. (1931). Ätiologie und therapeutische Beeinflussbarkeit der spezifischen toxischen Wirkungen des Methylalkohols. Arch. exp. Path. Pharmak., 160, 687.
KEITH, H. M. (1932). Effect of diacetone alcohol on the liver of the rat. Arch. Path. Lab. Med., 13, 707.
KEMAL, H. (1927). Beitrag zur Kenntnis der Schicksale des Isopropylalkohols im menschlichen Organismus. Biochem. Z., 187, 461.
KOBER, G. M. and HAYHURST, E. R. (1924). Industrial health. Blakiston, New York.
KOCHMANN, M. (1923). Alkohol; in Heffter, A., Handbuch der experimentellen Pharmakologie. Springer, Berlin.
KOELSCH, F. (1921). Die Gesundheitsschädigungen beim Arbeiten mit denaturiertem Spiritus. Das Polierekzem. Zbl. GewHyg., 9, 203.
KOELSCH, F. (1927). Die Giftigkeit des Äthylenchlorhydrins. Zbl. GewHyg., 14, 312.
KORENMAN, I. M. (1932). Kolorimetrische Bestimmung von Amylalkohol- und Amylacetatdämpfen in der Luft. Arch. Hyg., Berl., 109, 108.
KRÓL, J. (1913). Über das Wesen der Methylalkoholvergiftung. Arch. exp. Path. Pharmak., 72, 444.
KRÜGER, E. (1932). Augenerkrankungen bei Verwendung von Nitrolacken in der Strohhutindustrie. Arch. Gewerbepath. Gewerbehyg., 3, 798.
KUNO, Y. (1913). Über die Wirkung der einwertigen Alkohole auf das überlebende Säugetierherz. Arch. exp. Path. Pharmak., 74, 399.
KUNO, Y. (1914). Über die Wirkung der einwertigen Alkohole auf den überlebenden Kaninchendarm. Arch. exp. Path. Pharmak., 77, 206.
LALLEMAND, S. (1929). Recherches sur la toxicité cellulaire de poisons gazeux et volatils. J. Pharm. Chim., Paris, 9, 380.
LEHMAN, A. J. and CHASE, H. F. (1944). Acute and chronic toxicity of isopropyl alcohol. J. Lab. clin. Med., 29, 561.
LEHMAN, A. J., SCHWERMA, H. and RICKARDS, E. (1944). Isopropyl alcohol: rate of disappearance from the blood stream of dogs after intravenous and oral administration. J. Pharmacol., 82, 196.
LEHMANN, K. B. and FLURY, F. (1943). Toxicology and hygiene of industrial solvents. (Tr. by E. King and H. F. Smyth jr.) Williams and Wilkins, Baltimore.
LENDLE, L. (1928). Untersuchungen über die Narkosegeschwindigkeit homologer und isomerer einwertiger Alkohole. Arch. exp. Path. Pharmak., 129, 85.
LENDLE, L. (1928). Beitrag zur allgemeinen Pharmakologie der Narkose: über die narkotische Breite. Arch. exp. Path. Pharmak., 132, 214.
LEO, H. (1925). Über das Wesen der Methylalkoholvergiftung. Dtsch. med. Wschr., 51, 1062.
LESCHKE, E. (1932). Fortschritte in der Erkennung und Behandlung der wichtigsten Vergiftungen. Münch. med. Wschr., 79, 751, 1755, 1786.
LEWIN, L. and GUILLERY, H. (1913). Die Wirkungen von Arzneimittel und Giften auf das Auge. 2nd Aufl. Hirschwald, Berlin.
LOEWY, A. (1914). Inwieweit ist der gewerbliche Benutzung von vergälltem Branntwein geeignet gesundheitsschädliche Wirkungen hervorzurufen? Vjschr. gerichtl. Med., 48, Suppl. 3, 93.
LOEWY, A. and VAN DER HEIDE, R. (1914). Über die Aufnahme des Methylalkohols durch die Atmung. Biochem. Z., 65, 230.
LOEWY, A. and VAN DER HEIDE, R. (1918). Über die Aufnahme des Äthylalkohols durch die Atmung. Biochem. Z., 86, 125.

LOEWY, A. and MÜNZER, E. (1923). Beiträge zur Lehre von der experimentellen Säurevergiftung. III. Führt Methylalkoholvergiftung zur Acidose? *Biochem. Z.*, **134**, 442.
LOEWY, R. (1912). Über Methylalkohol und Methylalkoholvergiftungen. Dissertation, Berlin.
MCCORD, C. P. (1932). Toxicity of allyl alcohol. *J. Amer. med. Ass.*, **98**, 2269.
MCCORD, C. P. and COX, N. (1931). Toxicity of methyl alcohol (methanol) following skin absorption and inhalation. *Industr. Engng. Chem.*, **23**, 931.
MACFARLAN, J. F. (1855). On methylated spirit and some of its preparations. *Pharm. J. and Trans.*, **15**, 310.
MACHT, D. I. (1918). Pharmacological and therapeutic study of benzyl alcohol as a local anaesthetic. *J. Pharmacol.*, **11**, 263.
MACHT, D. I. (1920). Toxicological study of some alcohols, with especial reference to isomers. *J. Pharmacol.*, **16**, 1.
MACHT, D. I. and NELSON, D. E. (1918). On the antiseptic action of benzyl alcohol. *Proc. Soc. exp. Biol., N.Y.*, **16**, 25.
MARINESCO, G., LISSIEVICI-DRAGANESCO, DRAGANESCO and GRIGORESCO (1929). Fixation de l'alcool éthylique et du méthanol dans les tissus. *C. R. Soc. Biol., Paris*, **101**, 308.
MARIO, S. (1929). L'azione degli alcooli . . . sull' eccitabulita neuromuscolare. *Boll. Soc. ital. Biol. sper.*, **4**, 1043.
MARSHALL, C. R. and HEATH, H. L. (1897). Pharmacology of the chlor-hydrins: a contribution to the study of the relation between chemical constitution and physiological action. *J. Physiol.*, **22**, 38.
MARTINO, G. (1928). Sui rapporti tra fosfogeno e contratture muscolare. *Arch. Fisiol.*, **26**, 362.
MASHBITZ, L. M., SKLIANSKAYA, R. M. and URIEVA, F. I. (1936). Relative toxicity of acetone, methyl alcohol and their mixtures. II. Their action on white mice. *J. industr. Hyg.*, **18**, 117.
MATHEWSON, G. H. and ALEXANDER, B. (1932). Blindness from methyl alcohol successfully treated by lumbar puncture. *Canad. med. Ass. J.*, **26**, 679.
MIDDLETON, E. L. (1930). Fatal case of poisoning by ethylene chlorhydrin. *J. industr. Hyg.*, **12**, 265.
MIESSNER, F. (1891). Über die Wirkung des Allylalkohols. *Berl. klin. Wschr.*, **28**, 819.
MOCARD, C. P. (1933). La toxicité de l'alcool méthylique à la suite de l'absorption par la peau et de l'inhalation. *J. Pharm. Chim., Paris*, **17**, 221.
MOLITORIS, H. (1930). Natürlicher Tod oder Betriebsunfall durch Enodrinvergiftung? *Dtsch. Z. ges. gerichtl. Med.*, **14**, 149.
MOLITORIS, H. (1931). Dichlorhydrin-Vergiftung, gewerbliche? *Samml. Vergiftungsf.*, **2**, 217, A 177.
MONIER WILLIAMS, D. (1929). An account of an outbreak of methyl alcohol poisoning in the Ruhr. Ministry of Health Memorandum.
MOORE, W. (1917). Toxicity of various benzene derivatives to insects. *J. agric. Res.*, **9**, 371.
MORRIS, H. J. and LIGHTBODY, H. D. (1938). Toxicity of isopropanol. *J. industr. Hyg.*, **20**, 428.
MUEHLBERGER, C. W. (1935). Relative toxicological effects of synthetic ethanol and grain fermentation ethanol in blended whiskies. *Amer. J. publ. Hlth.*, **25**, 1132.
MUNCH, J. C. and SCHWARTZE, E. W. (1925). Narcotic and toxic potency of aliphatic alcohols upon rabbits. *J. Lab. clin. Med.*, **10**, 985.
NEIDING, M., GOLDENBERG, N. and BLANK, L. (1932). Die Neurologie der akuten Methylalkoholvergiftung. *Arch. Psychiat. Nervenkr.*, **96**, 24.
NELLER, J. R. (1931). Removal of spray residues from apples; a wax-solvent method. *Industr. Engng. Chem.*, **23**, 323.
NELSON, K. W., EGE, J. F. jr., ROSS, M., WOODMAN, L. E. and SILVERMAN, L. (1943). Sensory response to certain industrial solvent vapors. *J. industr. Hyg.*, **25**, 282.
NEW YORK STATE DEPARTMENT OF LABOR, DIVISION OF INDUSTRIAL HYGIENE (1917). Dangers in manufacture and industrial uses of wood alcohol. *Spec. Bull.*, **86**.
OETTEL, R. (1936). Einwirkung organische Flüssigkeiten auf der Haut. *Arch. exp. Path. Pharmak.*, **183**, 641.
VON OETTINGEN, W. F. (1943). Aliphatic alcohols: their toxicity and potential dangers in relation to their chemical constitution and their fate in metabolism. *Publ. Hlth. Bull., Wash.*, **281**, 122.
OSWALD, A. (1924). Chemische Konstitution und pharmakologische Wirkung ihre Beziehungen zu einander bei den Kohlenstoffverbindungen. Borntraeger, Berlin.
OTTLEY, G. B. (1933). Quick drying ink. *Chem. Abstr.*, **27**, 2830.
OVERTON, E. (1901). Studien über die Narkose, zugleich ein Beitrag zur Allegemeinen Pharmakologie. Fischer, Jena.
PAGLIONI, S. (1927). Effect of alcohols, glycerol and nicotine on survival of leucocytes. *Boll. Soc. ital. Biol. sper.*, **2**, 976.
PATILLO, R. S. and WOOD, C. (1899). Two cases of methyl alcohol amaurosis from the inhalation of vapour. *Ophthal. Rec.*, **8**, 599.
PIAZZA, J. G. (1915). Zur Kenntnis der Wirkung der Allylverbindungen. *Z. exp. Path. Ther.*, **17**, 318.
PIPIK, O. and MEZHEBOVSKAYA, E. (1933). Preparing amyl alcohols from cracked gasoline. Abstr. in *Chem. Abstr.* (1934), **28**, 7504.

POHL, J. (1893). Über die Oxydation des Methyl- und Äthylalkohols im Tierkörper. *Arch. exp. Path. Pharmak.*, **31**, 281.

POHL, J. (1918). Versuche zur Entgiftung des Methylalkohols. *Arch. exp. Path. Pharmak.*, **83**, 204.

POHL, J. (1922). Zur Kenntnis des Methyl- und Isopropyl-alkoholschicksals. *Biochem. Z.*, **127**, 66.

POHL, J. (1925). Über die Giftigkeit einiger aromatischer Hydrierungsprodukte (Tetralin, Hexalin und Methylhexalin). *Zbl. GewHyg.*, **12**, 91.

POINCARÉ, L. (1878). Sur les dangers de l'emploi de l'alcool méthylique dans l'industrie. *C. R. Acad. Sci., Paris*, **87**, 682.

POOLE, J. W. (1929). Solubilities of oils and waxes in organic solvents. *Industr. Engng. Chem.*, **21**, 1098.

PRATT, J. D. (1930). Dangerous properties of ethylene chlorhydrin. *Nature, Lond.*, **126**, 995.

DI PRISCO, L. (1932). Contenuto in glutatione ridotto del sangue nelle intossicazioni croniche sperimentali da solventi. *Minerva med.*, **2**, 423. (Abstr. in *Chem. Abs.*, 1932, **26**, 718.)

REID, E. W. (1934). Modern solvent industry. *Industr. Engng. Chem.*, **26**, 21. See also Modern solvent products. *Chem. Tr. J.*, **94**, 172.

REIF, G. (1923). Über die Giftigkeit des Methylalkohols. *Dtsch. med. Wschr.*, **49**, 183.

RICHARDSON, B. W. (1869). Physiological research on alcohols. *Med. Times and Gazette, Lond.*, **2**, 703.

ROBINSON, J. M. (1918). Blindness from industrial use of a 4 per cent admixture of wood alcohol. *J. Amer. med. Ass.*, **70**, 148.

RØE, O. (1943). Clinical investigations of methyl alcohol poisoning with special reference to the pathogenesis and treatment of amblyopia. *Acta med. scand.*, **113**, 558.

ROGERS, G. W. (1945). Sampling and determination of methanol in air. *J. industr. Hyg.*, **27**, 224.

RÖHRIG, A. (1876). Die Physiologie der Haut, experimentelle und kritisch bearbeitet. Hirschwald, Berlin.

ROMENSKY, A. (1872). Über die physiologischen Wirkungen des Trichlorhydrins. *Pflüg. Arch. ges. Physiol.*, **5**, 565.

ROST, E. and BRAUN, A. (1926). Zur Pharmakologie der niederen Glieder der einwertigen aliphatischen Alkohole. *Arb. ReichsgesundhAmt.*, **57**, 580.

ROST, E. and WOLF, G. (1925). Zur Frage der Beeinflüssung der Nachkommenschaft durch den Alkohol im Tierversuch. *Arch. Hyg., Berl.*, **95**, 140.

RUEMELE, T. (1948). Isopropyl alcohol in perfumes and toilet preparations. *Mfg. Chem.*, **19**, 151.

RUHLE, A. (1912). Tierexperimentelle Befund in Zentral-nervensystem nach Methylalkoholvergiftung. *Münch. med. Wschr.*, **59**, 964.

SALANT, W. (1909). Comparative toxicity of ethyl and amyl alcohol and their effect on blood pressure. *Proc. Soc. exp. Biol., N.Y.*, **6**, 134.

SALANT, W. and KLEITMAN, N. (1922). Pharmacological studies on acetone. *J. Pharmacol.*, **19**, 293.

SANDER, F. (1933). Über den Einfluss von Kohlenstoffverbindungen, insbesondere die aromatischen Reihe, auf Mäuse bei cutaner Applikation. Dissertation, Köln. (Cited by von Oettingen, 1943).

SANDERSON, J. McE. (1933). Solvents used in surface coatings. *Paint Oil Chem. Rev.*, **95**, No. 2, 10.

SATO, K. (1928). Über die pharmakologischen Wirkungen der hydroaromatischen Verbindungen: Cyclohexen, Cyclohexan und Cyclohexanol. *Jap. J. med. Sci., Pharmacol.*, **3**, 1.

SAVIANO, M. (1929). Action of alcohols on excitability of muscle-nerve preparation. *Boll. Soc. ital. Biol. sper.*, **4**, 1043.

SAYERS, R. R. and YANT, W. P. (1930). Observations and notes on the effect of methanol antifreeze on health. *U.S. Bur. Mines, Wash., Inform. Circ. No.* 6415.

SAYERS, R. R., YANT, W. P., SCHRENK, H. H., CHORNYAK, J., PEARCE, S. J., PATTY, F. A. and LINN, J. G. (1944). Methanol poisoning II. *J. industr. Hyg.*, **26**, 255.

SCHWARTZ, L. (1898). Über die Oxydation des Acetons und homologer Ketone der Fettsaurereihe. *Arch. exp. Path. Pharmak.*, **40**, 168.

SCHWARTZ, L. and TULIPAN, L. (1939). Textbook of occupational diseases of the skin. Lea and Febiger, Philadelphia.

SCHWARZMANN, A. (1934). Methylalkohol-Vergiftung, chronische, durch Einatmen von methylalkohol-haltigen Formaldehyddämpfen. *Samml. Vergiftungsf.*, **5**, 129, A 442.

DE SCHWEINITZ, G. E. (1931). Hazards of anti-freeze methanol. *Trans. Coll. Phys., Phila.*, **53**, 221.

SCHWENKENBECHER (1904). Das Absorptionsvermögen der Haut. *Arch. Anat. Physiol., Physiol. Abteil.*, p. 121.

SCOTT, E., HELZ, M. K. and MCCORD, C. P. (1933). Histopathology of methyl alcohol poisoning. *Amer. J. clin. Path.*, **3**, 225.

SHERIDAN, W. F. (1929). Use of (A) normal propyl alcohol and (B) low melting point paraffin in paraffin in infiltration to prevent distortion and hardening of tissues. *J. tech. Meth.*, **12**, 125. Abstr. in *Stain. Tech.*, **5**, 34.

SILVERMAN, L., SCHULTE, H. F. and FIRST, W. W. (1946). Further studies on sensory response to certain industrial solvent vapors. *J. industr. Hyg.*, **28**, 262.

Simon, I. (1933). Cause of toxicity of methanol. *Boll. Soc. ital. Biol. sper.*, **8**, 1376.
Sisley, J. P. (1934). Solvents utilisés dans l'industrie textile. *Rev. gén. Mat. col.*, **38**, 481.
Sklianskaya, R. M., Urieva, F. E. and Mashbitz, L. M. (1936). Relative toxicity of acetone, methyl alcohol and their mixtures. *J. industr. Hyg.*, **18**, 106.
Smith, H. M. (1927). Dewaxing paraffin base crudes. *Oil Gas J.*, **26**, 146.
Smyth, H. F. and Smyth, H. F., jr. (1928). Inhalation experiments with certain lacquer solvents. *J. industr. Hyg.*, **10**, 261.
Smyth, H. F., jr. and Carpenter, C. P. (1945). Note upon the toxicity of ethylene chlorhydrin by skin absorption. *J. industr. Hyg.*, **27**, 93.
Snapper, I., Grünbaum, A. and Sturkop, S. (1924). Over de ontleding en de oxydatie van benzylalcohol en benzylesters in het menschelijk organisme. *Ned. Tijdschr. Geneesk.*, **68**, 3125. (See also: *Biochem. Z.*, 1925, **166**, 73.)
Sollmann, T. (1919). Benzyl alcohol: its anaesthetic efficiency for mucous membranes. *J. Pharmacol.*, **13**, 355.
Sollmann, T. (1920–21). Studies of chronic intoxications on albino rats. II. Alcohols (ethyl, methyl and "wood") and acetone. *J. Pharmacol.*, **16**, 291.
Starrek, E. (1938). Über die Wirkung einiger Alkohole, Glykole und Ester. Dissertation, Würzburg. (Cited by Lehmann and Flury, 1943, in Toxicology and hygiene of industrial solvents. Williams and Wilkins, Baltimore.)
Supniewski, J. V., and Macht, D. I. (1926). Action of opium and related alkaloids on nerve and muscle preparations. *Arch. int. Pharmacodyn.*, **32**, 352.
Tabershaw, I. R., Fahy, J. P. and Skinner, J. B. (1944). Industrial exposure to butanol. *J. industr. Hyg.*, **26**, 328.
Thompson, W. G. Report, Medical Department, Standard Oil Co. (Cited by Morris and Lightbody, 1938).
Treon, J. F., jr. (1949). The Alcohols, in Patty, F. A., Industrial hygiene and toxicology, **2**, p. 831. Interscience Publishers, New York.
Treon, J. F., Crutchfield, W. E., jr. and Kitzmiller, K. V. (1943). The physiological response of rabbits to cyclohexane, methylcyclohexane and certain derivatives of these compounds. I. Oral administration and cutaneous application. *J. industr. Hyg.*, **25**, 199.
Treon, J. F., Crutchfield, W. E., jr and Kitzmiller, K. V. (1943). The physiological response of animals to cyclohexane, methylcyclohexane and certain derivatives of these compounds. II. Inhalation. *J. industr. Hyg.*, **25**, 323.
Trumper, M. (1931). Hazards of anti-freeze methanol. *Trans. Coll. Phys. Phila.*, **53**, 220.
Tyson, H. H. (1912). Amblyopia from inhalation of methyl alcohol. *Arch. Ophthal., N.Y.*, **41**, 459.
Tyson, H. H. and Schoenberg, M. J. (1914). Experimental researches in methyl alcohol inhalation. *J. Amer. med. Ass.*, **63**, 915.
Tyson, H. H. and Schoenberg, M. J. (1915). Changes in the blood and aqueous humor in methyl alcohol inhalation. *Arch. Ophthal., N.Y.*, **44**, 275.
Uri, J. and Jeney, E. (1946). Methyl alcohol poisoning treated by pectins. *Ord. Lapja.*, **2**, 961. (Abstr. in *Brit. J. industr. med.*, 1948, **5**, 35.)
Vayntsvayg, O. M., Kleybes, B. D. and Pasternak, A. E. (1933). Ocular injuries in chronic poisoning from use of stabilin (nitrocellulose glue) in shoe factories. *Sovetsk. Vestnik. Oftal.*, **3**, 157.
Viditz, F. (1933). Zur Pharmakologie des optisch aktiven sekundären Butylalkohols. *Arch. exp. Path. Pharmak.*, **172**, 668.
Vollmer, H. (1931). Fortgesetzte Versuche über die Giftempfindlichkeit von Mäusen und Ratten nach Bestrahlung oder Vorbehandlung mit oxydationssteigernden Substanzen. *Arch. exp. Path. Pharmak.*, **160**, 635.
Vorob'eva, A. A. (1932). Analytical methods for analyzing mixtures of solvents present in "stabilin". Abstr. in *Chem. Abstr.* (1934), **28**, 725.
Walton, D. C., Kehr, E. F. and Loevenhart, A. S. (1928). Comparison of the pharmacological actions of diacetone alcohol and acetone. *J. Pharmacol.*, **33**, 175.
Weber, H. H. and Koch, W. (1933). Zur Methodik der Analyse technischer Lösungsmittel. III. Farbnachweise für n-Propyl, n-Butyl, iso-Butyl- und iso-Amylalkohol. *Chemikerztg.*, **57**, 73.
Weese, H. (1928). Vergleichende Untersuchungen über die Wirksamkeit und Giftigkeit der Dämpfe niederer aliphatischer Alkohole. *Arch. exp. Path. Pharmak.*, **135**, 118.
Weger, A. (1927). Thalamischer Symptomen-komplex bei Formalinintoxikation. *Z. ges. Neurol. Psychiat.*, **111**, 370.
Willcox, W. H. (1929). Toxic effects of methylated spirits and impure forms of alcohol. *Brit. J. Inebr.*, **27**, 64.
Wilson, M. M. and Worster, F. J. (1929). Place of synthetic amyl products among lacquer solvents. *Industr. Engng. Chem.*, **21**, 592.
Winternitz, R. (1891). Zur Lehre von der Hautresorption. *Arch. exp. Path. Pharmak.*, **28**, 405.
Wood, C. A. and Buller, F. (1904). Poisoning by wood alcohol: cases of death and blindness from Columbian spirits and other methylated preparations. *J. Amer. med. Ass.*, **43**, 972.

Yant, W. P., Schrenk, H. H. and Sayers, R. R. (1931). Methanol anti freeze and methanol poisoning. *Industr. Engng. Chem.*, **23**, 551.

Yarsley, V. E. (1933). Health hazards in the lacquer and finishing industries. *Synth. appl. Finish.*, **3**, 80.

Yarsley, V. E. (1934). Solvents and plasticisers. A review of recent progress with special reference to cellulose acetate. *Synth. appl. Finish.*, **5**, 37, 57.

Zangger, H. (1931). Vergällung (Denaturierung) als gewerbepathologisch-hygienisches problem. *Arch. Gewerbepath. Gewerbehyg.*, **2**, 240.

Zangger, H. (1933). Arbeitsunfälle und Arbeitergefährdung bei der Arbeit im Innern von geschlossenen Behältern (Reservoiren, Tanks, Transportwagen usw.). *Arch. Gewerbepath. Gewerbehyg.*, **4**, 117.

Zatman, L. J. (1946). Oxidation of methyl alcohol inhibited by ethyl alcohol. *Biochem. J.*, **40**, lxvii.

Zernik, F. (1933). Neure Erkenntnisse auf dem Gebiete der schädlichen Gase und Dämpfe. *Ergebn. Hyg. Bakt.*, **14**, 139, 212.

Ziegler, S. L. (1921). The ocular menace of wood alcohol poisoning. *J. Amer. med. Ass.*, **77**, 1160.

CHAPTER IV

ETHERS

1. Ethyl Ether

(*Diethyl Ether, Sulphuric Ether*)

$(C_2H_5)_2O$

PROPERTIES

A colourless, mobile, very volatile liquid with a characteristic odour and burning, sweetish taste. Prepared synthetically on a large scale by the action of concentrated sulphuric acid on ethyl alcohol.

Pure Ether. Sp. Gr. 0·7199. B.P. 34·6°C. (Durrans, 1950).

Technical Ether (*British Standard Specification*, No. 579, 1934). Sp. Gr. 0·725 maximum (Durrans, 1950). B.R., 95 per cent below 36°C. Residue, 0·005 per cent maximum. Acidity, 0·002 per cent maximum as sulphuric acid.

Solubility in water at 20° C. 6·9 per cent; dissolves 1·3 per cent of water at 20° C. (Durrans, 1950). Miscible with alcohol, chloroform, benzene, benzine and oils.

EXPLOSIVE RISK

Mixtures of ether and air in the proportion of 75 to 200 mg. ether per l. air explode violently. The presence of impurities in the ether greatly increases the hazard (Hetzel, 1931; Rieche, 1931). It readily acquires high charges of static electricity.

IMPURITIES AND DECOMPOSITION PRODUCTS

As the result of the manufacturing process, ageing or exposure to light (Green and Schietzow, 1933). The chief impurities likely to occur in ether are ether peroxide, acetaldehyde, sulphur compounds (ethyl sulphide and ethyl mercaptan) and ketones. It has been suggested that these substances might be the cause of some of the toxic effects produced by ether inhalation, and even of ether convulsions (Wilson, 1935). An investigation on the effect of these impurities on dogs (Bourne, 1926), when inhaled as an admixture with pure ether, showed that acetaldehyde in concentrations of more than 1 per cent and ether peroxide in concentrations of more than 0·5 per cent caused respiratory embarrassment and fall in blood pressure, while ethyl sulphide (1 per cent) caused also severe gastro-enteritis. Ethyl mercaptan and the ketones had practically no effect. Lower concentrations, however (0·2 per cent of aldehydes and 0·07 per cent of peroxide), were found by Knoefel and Murrell (1935) not to affect the acute toxicity of anaesthetic ether, though their presence slowed down the rate of narcosis in mice, probably by slowing down the rate of entrance of ether into the blood stream.

As the result of the presence of an open flame or the catalytic action of metal. Ether may decompose under the above conditions with the formation of toxic substances (chiefly formaldehyde) which Karber (1930) has shown to be twice as toxic to animals as ether itself. He states, however, that these products are only likely to be found in small unventilated rooms. He failed to detect any in a room of 100 c. m. with an ether concentration of 0·01 per cent.

USES

(1) In the manufacture of smokeless powder.
(2) In the artificial silk industry (Krebs, 1929; Hayhurst, 1930).
(3) In the perfume industry.
(4) As an extraction agent for casein, tannin, etc.
(5) In the celluloid industry.
(6) As a constituent in small proportion of motor fuel.
(7) In the chemical and pharmaceutical industries as a solvent for vegetable alkaloids, resorcinol, etc., and in the synthesis of organic compounds.
(8) As an anaesthetic.

TOXICITY

The acute toxic effects of ether are well known from its use as a surgical anaesthetic, i.e. it is a strong narcotic with some irritant effect on mucous membranes. The metabolic effect of such large (narcotic) doses is much less injurious than that of chloroform, its action on the liver and kidneys being slight and transitory. The long-continued inhalation of fumes of ethyl ether by industrial workers has given rise to a fairly characteristic syndrome of symptoms, but as a rule these have disappeared without serious after-effects on removal from exposure, except in a few recorded cases where nephritis has supervened (e.g. Hamilton, 1925). Whether chronic ether intoxication may lead directly to nephritis appears to be uncertain; it is believed by many authorities that nephritis so caused is a very unusual occurrence, and occurs in susceptible individuals only.

In animals there are few records of any definite injury to the kidneys or of any serious secondary effects from prolonged ether anaesthesia. The experiments of Sand (1910), often quoted as providing conditions comparable with those of industrial exposure, showed fatty degeneration of the liver, but they were carried out with alcohol-ether mixtures.

Maximum Permissible Concentration

Cook (1945) suggests 500 p.p.m. as the maximum allowable concentration, though no actual serious disturbance has been found with concentrations of 2,000 to 3,000 p.p.m., but according to Nelson, Ege, Ross, Woodman and Silverman (1943), nasal irritation occurs at 200 p.p.m., and 300 p.p.m. is considered objectionable.

TOXIC EFFECTS IN ANIMALS.

Absorption and Excretion

The diffusion of ether is sufficiently rapid to ensure absorption of 95 per cent of even large doses in 2 seconds. It is not destroyed or utilized by the body, but is all excreted unchanged. Excretion takes place chiefly through the lungs; approximately 90 per cent can be recovered in the expired air after administration. A small amount is excreted by the skin and in the urine; the concentration in the urine is practically that of the arterial blood passing through the kidneys at the moment of secretion (Haggard, 1924). The amount in the venous blood is always less than that in the arterial even after 3 hours of anaesthesia (Ronzoni, 1923), and the ether content of the venous blood (of the right heart) is an index of the total amount of ether in the body as a whole (Haggard, 1924).

Acute Poisoning

Lethal and Narcotic Concentrations

Lethal dose. The figures for different species of animals as given by different investigators vary a good deal—from about 2 to 10 per cent (65 to 300 mg. per l.). Robbins (1935) in an investigation on dogs found that the concentration was 6·7 to 8 per cent (209 to 250 mg. per l.); he states that the discrepancies found in some of the earlier work can be explained by differences in the induction mixtures used and in the duration of anaesthesia. The concentration of ether in the blood for a lethal effect is 170 to 190 mg. per 100 ml.

Narcotic dose. Robbins' figures are as follows: for anaesthesia with rigidity 3·3 per cent (105 mg. per l.): for anaesthesia with relaxation 4 to 4·5 per cent (125 to 140 mg. per l.). The concentration of ether in the blood for a narcotic effect is 90 to 150 mg. per 100 ml. (Robbins, 1935).

Symptoms

Rapid narcosis sets in after signs of preliminary irritation; if ether is given for long enough and in sufficient concentration death takes place from respiratory paralysis. Recovery on removal from exposure to non-lethal concentrations is rapid and there are no apparent cumulative or after-effects (Bonsmann, 1932). According to Kärber and Lendle (1931), long-continued ether narcosis in animals differs from ether anaesthesia in man in one respect, viz. that in animals the lethal concentration lies very near that which produces loss of reflexes, and that narcosis of very long duration almost invariably leads to death. They suggest as an explanation of this difference that secondary effects may take place in animal metabolism, such as the formation of methaemoglobin observed in cats by Ellinger and Rost (1922), or the chemical (lipoid) changes in the brain by Weil (1929).

Effect on Metabolism

Acidosis. The acidosis occurring in ether anaesthesia was stated by van Slyke, Austin and Cullen (1922) to be a true acidosis with a fall in the alkali reserve, while Chabanier, Lobo-Onell and Lelu (1932) observed in animals a tendency to acidosis during the phase of anaesthesia with a compensatory tendency to alkalosis in the waking period. Schulze (1924) and Minnitt (1933) state that the acidosis is accompanied by ketosis and acetonuria.

Hyperglycaemia. The rise in blood sugar in animals anaesthetized by ether observed by Atkinson and Ets (1922) and Mann (1925) was attributed by Mann to a disturbance of liver function, but Minnitt (1933) suggests that the pancreas may be at fault.

Disturbance of liver function. It appears that, while ether intoxication causes some transitory functional depression of bile secretion by the liver, it leads to no definite pathological changes (Mann, 1925; Rosenthal and Bourne, 1928; Winogradow, 1927). Rosenthal and Bourne found no abnormality in urobilinogen and no increase in bilirubin. Bile secretion was completely normal the next day.

Disturbance of kidney function. While ether narcosis produces some oliguria or even anuria (Stehle and Bourne, 1928) this is only a temporary effect of some diminution of the efficiency of the secretory process (Bonsmann, 1932).

Chronic Poisoning

The only experiments on repeated inhalation of small doses of ether by animals appear to be those of Sand (1910), but it was used in a mixture with alcohol so that it is uncertain which effects were due to the ether itself. Dogs were kept during the night in a closed chamber of 1 c.m. capacity in which 500 g. of alcohol-ether had been evaporated. The chamber was ventilated by a small opening. After some weeks they exhibited signs of intoxication and excitement. At autopsy the lungs showed haemorrhage and emphysema, all the internal organs were congested and there was fatty degeneration of the liver and kidney. Sand himself stated that his conclusions were not definite.

TOXIC EFFECTS IN MAN

Ether intoxication, as exemplified by surgical anaesthesia, is generally regarded as much less serious than that produced by chloroform, since fatal results from large doses and secondary effects are rare. Neither acute nor chronic intoxication by ethyl ether appears to have any very serious or lasting ill-effects; the majority of patients recover completely when removed from exposure. Hayhurst (1930) describes an enquiry made by the Belgian Factory Inspection in three factories making artificial silk by the Chardonnet process. The results showed that the state of health in 723 workers, 388 of whom were women, compared very favourably with that of workers in factories where ether was not used.

Acute Poisoning

The concentration which produces loss of consciousness in man has been variously estimated at values from 2 to 10 per cent. Boothby (1913–14) gave 7 per cent, but later workers including Robbins (1935) and Hirsch and Kappus (1929) consider that this figure is too high and that 3·5 per cent for 30 to 46 minutes is nearer the true value. Higher concentrations are narcotic in a shorter time and longer inhalation of very high concentrations, above 10 per cent according to Lehmann and Flury (1943), may cause death from respiratory paralysis.

Symptoms

Narcosis. Complete narcosis from the industrial use of ether has not often been recorded, but Hamilton in the U.S. Bureau of Labor Statistics Bulletin (1916) mentions two cases, boys who were accidentally exposed to the escape of fumes for 1½ hours. They were unconscious with slow pulse and respiration, but recovered without any after-effects. Frois (1926) quotes a case in which the narcosis lasted 24 hours.

Intoxication and excitement. During an investigation by the U.S. Bureau of Labor statistics in 1916 it was found that the form of intoxication known as "ether jag" was notorious among workers in the smokeless-powder industry, especially in new workers. This condition included all stages of drowsiness, confusion and excitement, dizziness and faintness, and was particularly common in women (Hamilton, 1925; Frois, 1926). One fatal case of this kind is quoted by Hayhurst (1930), where ether was used in perfumery manufacture as an extracting agent. The subject developed acute mania and died in uraemic convulsions.

In 1922 the Home Office examined workers using ether as an extractive in poor ventilation. They complained of "intoxication," especially at first. The after-effects of acute intoxication include nausea, headache, lack of appetite, vomiting, perspiration, mental confusion and irritability. The occurrence of nephritis as a sequel of acute intoxication has been disputed, but it is stated in *Brochure No.* 186, International Labour Office, that in a second study carried out by the National Research Council in 1918, one man had a "large trace" of albuminuria when recovering from a mild, acute attack of "ether jag."

Only one case of disturbance of health attributed to exposure to diethyl ether has been reported to the Factory Department since 1937. The exposure occurred in a dimethylolurea plant, and the workmen affected suffered from weakness, nausea, vomiting and headache, the last three symptoms lasting for several days.

Chronic Poisoning

Symptoms

Many disturbances of health have been reported from long-continued exposure to ether, but there seems to be great variation in susceptibility, some workers having no symptoms other than occasional dizziness and faintness on a hot day, while others, especially women, complain of loss of appetite and distaste for food (the most prominent symptom according to Frois), increased thirst (but vomiting when water was taken), nausea, constipation, lassitude (Hirsch and Kappus quote the case of a chemist using ether as an extracting agent who suffered from "paralysing" fatigue), specks before the eyes and numbness in the fingers and feet. A skin eruption with an itching rash on the side of the face nearest to the cutting machine was observed in some women workers (Hamilton and Minot, 1920).

In the opinion of the physician in charge of the cases investigated by the U.S. Bureau of Labor Statistics in 1916, nephritis was not a frequent sequel, but he stated that it "may occur". One man had symptoms of nephritis, another had albuminuria and puffed eyelids, while amongst the women, 3 out of 55 had slight albuminuria, 2 of these showing no accompanying symptoms, while the third had nausea, vomiting, constipation, loss of appetite and drowsiness.

The most definite case of nephritis recorded is that of Hamilton (1925) in a man in a smokeless-powder factory exposed continuously for 7 years and intermittently for 5 years. He developed severe chronic interstitial nephritis and finally died from pemphigus and pulmonary oedema. In the opinion of the medical examiners the real cause of death was chronic nephritis caused by long inhalation of ether, and the pemphigus was closely related to the nephritis.

Effect on the Blood

In the investigation carried out by Hamilton and Minot (1920), Minot found the chief change to be a polycythaemia, which occurred in about half the women examined, together with a rather high white cell count (red cells over 6 millions, white cells 6,800 to 16,000, averaging 11,000). The polycythaemia appeared to have no clear-cut relation to the symptoms and remained unexplained at the time when Minot's investigation was cut short by the removal of women from work in powder factories. Other cases showed slight anaemia with slight abnormalities of red cells and slightly reduced haemoglobin, a condition which was more marked in the men than polycythaemia.

"Ether Habit"

As has been known for the last hundred years ether-drinking may lead to the "ether habit". Long continued inhalation has the same effect, according to Flury and Zernik (1931). Intoxication is achieved with smaller quantities of ether than alcohol, and the secondary effects are less severe, but the chronic effects appear earlier. These may include inflammation of the respiratory passages, irritability, restlessness, sleeplessness, general debility, headache, noises in the ears, loss of appetite, morning vomiting, tremor, twitching and other nervous symptoms, cardiac irregularity and dilatation of blood vessels with consequent redness of the skin.

2. β β'-Dichloroethyl Ether
(Chlorex)
$C_4H_8OCl_2$, i.e. $(CH_2Cl \cdot CH_2)_2O$

PROPERTIES

A colourless liquid, insoluble in water but soluble in most organic solvents. It has a pungent odour, like that of dichloroethane, and is also irritating to the mucous membranes of the eyes and nose.

B.P. 178°C. Sp.Gr. 1·22 at 20°C. Fl.P. 55°C.

A stable compound, non-corrosive to iron, possessing the combined solvent properties of ethyl ether and ethylene dichloride with a low vapour pressure and fairly high boiling point (Fife and Reid, 1930).

Solvent for many tars, waxes, oils, resins and gums, and for ethyl cellulose, but not for cellulose esters except in conjunction with 10 to 30 per cent alcohol, when it becomes an active solvent.

USES

(1) In the textile industry (*a*) for the removal of paint and tar brand marks from raw wool and of oil and grease spots from cloth; Fife and Reid (1930) state that dichloroethyl ether is indicated where high temperatures are necessary and the loss of the active solvent is an economic consideration; it is often incorporated in scouring and fulling soaps; (*b*) as a melting-out agent and penetrant in combination with diethylene glycol, sulphonated oils, etc.

(2) In the treatment of lubricating oils for wax extraction.

(3) In the synthesis of new chemical compounds (Fife and Reid, 1930).

TOXICITY

Judging from its effects on animals, dichloroethyl ether produces acute toxic effects similar to those of other respiratory irritants such as acid gases, and also appears to produce a delayed response with severe lung lesions. In man, no actual cases of industrial poisoning have been recorded, and the warning properties of its odour and of eye, nose and throat irritation are good, but there appears to be some danger that exposure to low concentrations over long periods may cause considerable respiratory irritation.

Maximum Permissible Concentration

Cook (1945) suggests 15 p.p.m. as the maximum allowable concentration for daily exposure.

TOXIC EFFECTS IN ANIMALS

The investigations of Schrenk, Patty and Yant (1933) have shown that, while dichloroethyl ether is primarily a respiratory irritant, it affects the liver and kidneys also in varying degrees and is to some extent a narcotic.

Lethal Dose

These workers found that it was not possible at room temperature to attain a concentration that would kill in a short time. They found, however, that in guinea-pigs 1,000 p.p.m. for 30 to 60 minutes was dangerous to life; 100 to 200 p.p.m. for 60 minutes and 35 p.p.m. for several hours were the maximum doses which did not produce serious disturbance.

Symptoms

The animals exposed to the higher concentrations showed intense irritation of the conjunctiva and nasal mucous membrane, evidenced by lachrymation and scratching of the nose. They also showed slight unsteadiness or vertigo, gradually decreasing motility, slight retching, slow respiration, gradually becoming shallow and rapid, and loss of consciousness leading to death. Concentrations of 100 p.p.m. produced similar symptoms, but no lachrymation, and unconsciousness occurred only after 13 hours. With 35 p.p.m. there were no symptoms except slight nasal irritation.

Post-mortem Appearances

In those animals which died from the immediate exposure, the chief lesions were those of severe irritation of the respiratory organs—congestion of the lungs, nasal passages, trachea and bronchi, with emphysema, oedema and haemorrhages of the lungs and occasionally complete consolidation. There were also moderate congestion of the brain, slight congestion of the kidney and marked congestion of the liver. A delayed response was shown by animals exposed to 0·1 per cent for 90 minutes. If they were killed immediately the organs showed only light congestion; but if they died 24 hours later the lesions in the lungs were much more severe.

TOXIC EFFECTS IN MAN

The immediate acute effects of inhalation of dichloroethyl ether vapour on human beings were investigated by Schrenk and co-workers (1933) and a summary of their findings is given in Table 40. No toxic effects, either acute or chronic, have been recorded from its industrial use.

TABLE 40
Effect of inhalation of dichloroethyl ether on human beings

Concentration (per cent)	Effects
0·055 0·10	Very irritating to eyes and nose on brief exposure; deep inhalations nauseating and intolerable
0·026	Similar, but in slighter degree; inhalations objectionable but not intolerable
0·01	Slightly nauseating and slightly irritating
0·0035	Easily noticeable but practically free from irritation

3. *iso*Propyl Ether

$(CH_3)_2CH \cdot O \cdot CH(CH_3)_2$

PROPERTIES

A colourless liquid similar in many of its properties to ethyl ether, but has a higher boiling point, evaporates more slowly and is less soluble in water.

B.P. 68·7° C. Sp.Gr. 0·730.

Solvent for ethyl cellulose, but not for other cellulose esters or ethers; solvent for cellulose nitrate when mixed with alcohol; excellent solvent for animal, vegetable and mineral oils and for certain resins and waxes. It is, however, very dangerous to use, because it forms highly unstable peroxides which explode with great violence when heated (Robertson, 1933; Morgan and Pickard, 1936).

USES

(1) In rubber cements.
(2) As a varnish remover (Sanderson, 1933).
(3) In extraction processes, e.g. the de-waxing of oils and de-oiling of waxes (Fife and Reid, 1930) and the extraction of acetic acid from aqueous solutions (Reid, 1934).

TOXIC EFFECTS

No animal experiments have been carried out and no ill-effects in human beings recorded. Flury and Zernik (1931) remark that propyl ether, like ethyl and methyl ethers, has a central paralytic effect leading to complete narcosis, but that the only member of this series of ethers of practical importance is ethyl ether.

Maximum Permissible Concentration

Cook (1945) suggests 500 p.p.m. (2,000 mg. per c.m.) as the maximum allowable concentration for continuous exposure.

4. Propylene Oxide

$$C_3H_6O, \text{ i.e. } CH_3-CH\underset{O}{\overset{}{\diagdown\diagup}}CH_2$$

PROPERTIES

A volatile colourless liquid. B.P. 35° C. Fl.P. <20° F. (−7°C.) On hydration it forms propylene glycol.

TOXIC EFFECTS

No animal experiments have been made, and no injury to man has been recorded.

5. Dioxan
(*Diethylene dioxide*)

$$C_4H_8O_2, \text{ i.e. } O<\genfrac{}{}{0pt}{}{CH_2\cdot CH_2}{CH_2\cdot CH_2}>O$$

PROPERTIES

Dioxan, the cyclic double ether of ethylene glycol, was discovered by Lourenço in 1863. It is a colourless liquid with a faint odour like that of butyl alcohol. It is miscible in all proportions with water and the ordinary organic solvents, and forms a constant boiling mixture with 20 per cent of water (B.P. 87° C.). It is quite stable at ordinary temperatures (Reid and Hofmann, 1929). B.P. of pure substance 101·1° C. B.R. of the technical product 94–110° C. Sp.Gr. at 20° C. 1·0338. Fl.P. 11° C. Its inflammability has been investigated by Jones, Seaman and Kennedy (1933). They found that the lower inflammable limit of dioxan vapour in dry air at ordinary laboratory temperature and pressure was 1·97 per cent by volume, the upper limit at atmospheric pressure and 100–110° C. was 22·25 per cent by volume.

It is a solvent for alcohol-damped cellulose nitrate, cellulose acetate, celluloid and cellulose ethers, also for many oils, fats and resins.

USES

(1) As a degreaser, especially for wool.
(2) In the lacquer industry as a solvent for lacquer and cellulose coatings.
(3) In the celluloid industry.
(4) As a paint remover.
(5) In the manufacture of polishes, pastes, cements, glues, shoe creams, emulsions, detergent and cleaning preparations.
(6) In the manufacture of cosmetic and pharmaceutical preparations.
(7) As a preservative, fumigant or deodorant.
(8) In laboratory preparation of tissues for histology (Weissberger, Young and Carleton, 1934).
(9) In the textile industry.

TOXICITY

From the results of an investigation of five fatal industrial cases which occurred in somewhat exceptional circumstances in 1933, it is clear that dioxan can produce very severe injury to the kidneys. With inhalation of concentrations comparable to those of usage and reasonable exposure, its toxicity to animals is low in relation to that of other compounds of somewhat similar chemical constitution, but lesions of the liver and kidneys have nevertheless been observed, and it has the disadvantage of producing no alarming, lasting or unpleasant effects which might act as a warning to human beings, even in concentrations of 1 in 1,000 and 1 in 500. By subcutaneous, intraperitoneal, or intragastric administration its toxicity to animals is relatively low.

Maximum Permissible Concentration

The maximum allowable concentration suggested by Cook (1945) was 500 p.p.m. According to Silverman, Schulte and First (1946), even this concentration is too high for comfortable working conditions. They found that the majority of workers exposed to varying concentrations found 300 p.p.m. irritating to the eyes, nose and throat, and suggest that 200 p.p.m. is the highest concentration acceptable.

TOXIC EFFECTS IN ANIMALS

Effect of Inhalation

Lethal and narcotic dose. In the experiments carried out by Yant, Schrenk, Waite and Patty (1930), death of guinea-pigs exposed to various concentrations of dioxan (calculated from the quantity vaporized and the quantity of the air in the experimental chamber) only occurred with exposures to 3·0 per cent for 3 hours or more. Narcosis set in within 87 to 141 minutes and was usually preceded by unsteadiness and staggering; respiration was at first dyspnoeic, shallow and rapid, and became shallow and slow later.

In the experiments of Fairley, Linton and Ford-Moore (1934), the material used was a mixture of dioxan and water in the approximate ratio of 80 to 20, since these are the proportions of dioxan and water in their constant boiling mixture, and are those employed in certain commercial processes. The mixture was free from impurities and had a boiling point of 86·5° C. (except for 2 per cent of a higher boiling fraction which eventually proved to be water). The animals used were rats, mice, guinea-pigs and rabbits, and the concentrations to which they were exposed were 1 in 100, 1 in 200, 1 in 500, all obtained by means of a spray, and 1 in 1,000, obtained by heat. With the 1 in 100 concentration, all the animals died after a varying number of exposures of 1½ hours, and in most cases collapse was preceded by convulsions. With the 1 in 200 concentration, the deaths were not so numerous or so rapid, but 8 out of 12 died after exposures varying from 3 hours for a mouse to 43½ hours for a guinea-pig. With the 1 in 100 concentration, in most cases the cause of death was pulmonary—"red-hepatisation" or some degree of oedema—while with the 1 in 200 concentration, the incidence of pulmonary lesions was much smaller.

Irritation. Yant and co-workers (1930) found nasal irritation at all concentrations from 0·1 per cent to 3 per cent by volume, decreasing in intensity with prolonged exposure. Eye irritation (squinting and lachrymation) occurred at all concentrations except the lowest, also decreasing in intensity. Lachrymation was also the immediate effect of the 1 in 100 concentrations applied by Fairley and co-workers (1934). Retching movements were observed by Yant and co-workers with the higher concentrations (1·0 per cent and 3·0 per cent); these were irregular in the time of occurrence and ceased when the animal became unconscious. Pulmonary irritation was observed by Fairley and co-workers with the 1 in 100, and to some extent with the 1 in 200 concentrations, but not with the 1 in 500 and 1 in 1,000 concentrations. Hyperaemia of the lungs and brain was present in the animals which died in Yant's investigation; it was considered uncertain whether death was due to irritation of the lungs or to narcosis.

Lesions of the kidneys and liver. It is of great interest to note that in Yant *et al.*'s investigation, which dealt with the acute effects of inhalation (single exposures), no lesions of the kidneys and liver, other than "paleness of the liver on the surface and cut section" of those animals which died during exposure, were observed, while in Fairley *et al.*'s investigation, where the exposures of low concentrations were continued over long periods, lesions were produced, sometimes of great severity. The lesions were present after exposure to all concentrations, even the non-lethal 1 in 500 and 1 in 1,000. The kidneys showed, in the cortex, marked patchy degeneration of the tubular epithelium, congestion and haemorrhages, both inter- and intra-tubular; in the medulla, only haemorrhages, no degeneration; the glomeruli sometimes showed haemorrhage or degeneration, and sometimes were uninjured. The liver showed degeneration, varying in nature from cloudy swelling to areas of complete necrosis, starting at the periphery of the lobule.

Blood urea. The blood urea of three rabbits, subjected to concentrations of 1 in 200 by inhalation, was estimated at intervals until death. In two of these animals it was nearly doubled, being 35 and 30 mg. per 100 ml. respectively at the beginning of the experiment, and 64 and 61 mg. after 23 days.

Effect of Subcutaneous Injection

Estimated by subcutaneous injection into mice, the toxicity of dioxan appears to be very low, compared with the other ethylene glycol compounds investigated by the same method. The minimum lethal dose for mice was found by von Oettingen and Jirouch (1931) to be more than 10 ml. per kg. body weight, compared with 2·5 ml. for ethylene glycol, 5 ml. for cellosolve and 0·5 ml. for butyl cellosolve. This low toxicity is attributed by these workers to the rapid destruction of dioxan in the organism and in alkaline solutions. Its depressant action on the central nervous system was very slight. Its depressant action on striated muscle and on sensory and motor nerve endings, which were tested with concentrated solutions of dioxan, was found to be distinct, but its action on the heart and on isolated smooth muscle, where more diluted buffered solutions were used, was practically nil. This fact is again attributed to the more rapid decomposition of dioxan in buffered solutions.

Effect of Intraperitoneal, Intravenous and Intragastric Administration

Lethal dose. The minimum lethal dose was 1·5 ml. per kg. body weight intravenously and 2·0 ml. per kg. by stomach tube.

The symptoms of a lethal dose were at first those of drunkenness. They abated after a few hours, but were followed later by the first sign of impairment of renal function—polyuria. This lasted for 24 to 48 hours and was then followed by complete anuria with a rise of blood urea up to or exceeding 300 mg. per 100 ml. Loss of weight, lethargy and finally coma led to death within 2 to 6 days.

Sublethal dose. Toxic symptoms were produced by 1·0 ml. per kg. body weight, the animal becoming limp and dazed with a slow, ataxic, rolling gait and diminished, sluggish reflexes.

Narcotic dose. The narcotic activity of dioxan, when given intraperitoneally to rats and by stomach tube to rabbits, has been compared by Knoefel (1935) to that of some formals and acetals (ethylene formal, ethylene acetal, propylene formal and acetal, butylene formal and acetal, etc.). He found dioxan to possess the lowest narcotic activity of the whole series.

The investigation by de Navasquez (1935) referred to above, was also carried out by intravenous and intragastric administration to rabbits and cats, and the results led him to conclude that dioxan is of a low order of toxicity.

Tolerance. A very characteristic feature with sublethal doses was the tolerance acquired by the animals when given successive and increasing amounts of dioxan. No cumulative action was observed and increasing dosage was necessary to produce either symptoms or changes in renal function. One animal subjected to repeated increasing dosage required 5·0 ml. per kg. to produce a terminal uraemia.

Post-mortem appearances. The appearances observed in the liver and kidneys (the only organs affected) were at variance with those described by Yant and co-workers and in human cases of poisoning (Barber, 1934). Whereas the latter were characterized in the kidney by essentially vascular lesions and in the liver by "central necrosis", the animals examined by de Navasquez showed acute "hydropic" degeneration in the kidney, but the appearances in the liver were regarded as manifestations of glycogen metabolism. According to de Navasquez, dioxan in laboratory animals has a selective action on the secretory epithelium of the kidney tubules, producing a rapid form of degeneration associated with loss of cytoplasmic structure and nuclei and a "ballooning" of the cells with fluid. The tubules are blocked as a result of the swelling, causing anuria and uraemia to develop rapidly. No haemorrhages in the renal cortex, such as were described by Fairley and co-workers, were observed in any of the animals. The liver also showed appearances typical of what is described as "hydropic" degeneration, but the vacuolated cells were shown by special staining to be packed with granules of glycogen. De Navasquez considers that the glycogen accounts for the spurious degeneration which has been described in sections stained with haematoxylin and eosin and that it is not improbable that the "central necrosis" seen in human cases is of a similar nature.

Absorption through the Skin

The occurrence of toxic lesions in the kidneys and liver by absorption through the skin was clearly demonstrated by Fairley and co-workers. Applications of 10 drops in the case of a rabbit and 5 drops in the case of a guinea-pig, eleven times in one week to a clipped area on the nape produced no skin irritation and no apparent disturbance of health, but the kidneys and liver showed, microscopically, severe lesions. In one rabbit, which appeared healthy and which had a normal blood urea, the kidney cortex showed an extreme degree of necrosis.

TOXIC EFFECTS IN MAN
Inhalation

In the investigation by Yant *et al.* concentrations of 0·16 per cent produced immediate slight burning of the eyes with lachrymation, while 0·55 per cent produced also a burning sensation in the throat. Fairley *et al.* used concentrations of 1 in 1,000 and 1 in 500 (produced by vaporizing a dioxan-water mixture in a 10 c.m. chamber), which produced in some of the observers a sensation of constriction in the throat and a desire to breathe rapidly; others felt nothing but a transitory warmth in the chest. The smell diminished as the exposure was prolonged. It was considered, therefore, that concentrations of this order (1 in 1,000 to 1 in 500, capable of producing lesions in animals) could readily be tolerated by workers for prolonged periods without any warning subjective symptoms. However, de Navasquez does not agree with this opinion on the basis of his own experiments and those of Yant and co-workers. He believes it improbable that sufficient dioxan vapour concentrations to cause toxic or lethal effects would escape the notice of workers employed in such processes, and in view of the tolerance easily acquired by his animals, he believes that the five deaths in 1933, among the workers in a factory in which cellulose acetate silk yarn was treated with dioxan, were referable to a few intense exposures and not to chronic poisoning. The process had been in operation for nearly 16 months, no symptoms of poisoning having been observed during this time, but an alteration in the experimental plant took place 2 months before the onset of the illness of these five men, which involved exposure to higher concentrations of the dioxan vapour. All died within a fortnight, but the first two died at home, and although in the second case a post-mortem examination was made, which revealed swollen kidneys with haemorrhages and an enlarged pale liver, no microscopic examination was undertaken. In the other three cases, in which suspicion of the nature of the toxic agent had been aroused, post-mortem examination revealed a condition of necrosis of the liver and of haemorrhage around the glomeruli with some necrosis of the kidney.

Symptoms

An examination of all workers in the firm who were in any way connected with the department in which the men who died had been working (80 in all), as well as enquiry into the symptoms of the five men before death, showed that some of the other men had suffered from anorexia, nausea and vomiting—symptoms similar to those of the "stomach trouble" from which the men who died had complained. No jaundice had been present in either the fatal cases or in the exposed workers examined, but in one of these latter the liver was palpable, and there was a positive indirect van den Bergh reaction, a trace of albumin and a few red blood corpuscles in the urine. There was also a trace of albuminuria in 65 per cent of those "much exposed" and in 50 per cent of the others.

On the basis of this information and the history of the fatal cases, Henry (1933) has divided intoxication by dioxan into three stages:

Stage 1. The earliest effect appears to be irritation of the naso-pharynx (with coughing) and of the eyes (with at times coryza and "misty" vision), but even if the inhalation is continued, this irritation is liable to subside and thus give rise to a false sense of security. Drowsiness, vertigo, headache and moderate gastric symptoms, such as loss of appetite, nausea and even vomiting, follow, but if at this stage a rest of 16 to 24 hours is taken, these ill-effects tend to pass off quickly.

Stage 2. With further continued inhalation the gastric symptoms become more severe and are associated with pain and tenderness in the abdomen and lumbar region, enlargement of the liver and sometimes an indefinite "chilly" feeling. Even at this stage a long rest of one or more weeks enables these ill-effects to pass off gradually, and at present we have no indication as to how far the kidneys are affected, if at all, at this stage.

Stage 3. Still further continued inhalation results in acute haemorrhagic nephritis, leading to suppression of urine, uraemia, coma and death.

Morbid Anatomy

The post-mortem findings on the three autopsies made in hospital were identical. The following account of them is taken from that given by Barber (1934).

Macroscopical. The kidneys were considerably enlarged, with oedematous capsules, under which were large areas of haemorrhage on the kidney surface. The cut surface showed a ring of haemorrhagic tissue with the appearance of a spreading infarct almost all round the outer part of the cortex. The liver was considerably enlarged and pale. The cut surface was uniform, fairly firm and not stained with bile. The liver lobules retained their size and shape because only part of each lobule was involved. The condition was quite distinct from that of acute yellow atrophy. The brain and lungs were congested and oedematous.

Histological. In the kidney a remarkable condition of almost symmetrical necrosis of the outer part of the cortex was found. The destruction was definitely vascular in distribution and obviously dependent on hyaline thrombosis of the vessels in the renal cortex. At the edge of the necrotic zone a good deal of bleeding had taken place. The medulla was on the whole intact, except for some red blood corpuscles in the lumen of the tubules. The liver showed complete necrosis of the inner half, or often two thirds, of each lobule. The line of demarcation between the dead and living cells was very sharp. There was no cellular reaction, no haemorrhage and no sign of regeneration of living cells (e.g. mitotic figures or two nuclei in single cells). Some Kupffer cells in the necrosed areas appeared to have survived, the nuclear staining being good. The absence of fatty degeneration was quite different from that found with any other known liver poison, such as alcohol or chloroform. The spleen in two cases examined showed no abnormality. The lungs in one case examined showed acute oedema.

Effect on the Blood

The blood counts in three of the men who died showed no abnormality of red cells or haemoglobin, but a considerable leucocytosis—24,000, 21,400 and 38,000 respectively, with neutrophils, 80, 94 and 90 per cent. In the eleven cases "much exposed" to the vapour, only one showed less than 10,000 and five more than 13,000; the total number of neutrophils per c.mm. was more than 6,000 in nine of these eleven cases.

The above findings have been extended by Estler (1935), who states that dioxan causes severe kidney and liver poisoning characterized by central liver necrosis without degeneration and by haemorrhagic nephritis, which is finally fatal. He emphasizes the importance of the early symptoms of gastric disturbance as an indication of liver involvement.

6. Methylal

(*Dimethoxymethane, Methylene Glycol Dimethyl Ether, Formal*)

$$CH_2(OCH_3)_2$$

PROPERTIES

A colourless, hygroscopic liquid with an ethereal odour. B.P. 44°C. Sp.Gr. 0·872. Solubility in water, about 33 per cent. Good solvent for cellulose acetate and nitrate.

USES

As a constituent of perfumes and drugs, and as a hypnotic and an anaesthetic.

TOXICITY

Methylal is a fairly powerful narcotic, weaker than chloroform but stronger than diethyl ether (Meyer and Gottlieb-Billroth, 1921).

TOXIC EFFECTS IN ANIMALS

Effect of Subcutaneous Injection

Local inflammation occurs at the site of injection with swelling and a serous exudation (Lewin, 1929). General weakness with polyuria and incontinence of urine followed the injection of 0·3 to 1·0 g.

Effect of Inhalation

The narcotic concentration for mice is 2·8 vols. per cent, or 87 mg. per l. Narcosis is preceded and recovery accompanied by a strong excitation, but with only slight convulsions (Meyer and Gottlieb-Billroth, 1921).

TOXIC EFFECTS IN MAN

No ill-effects have been recorded.

7. Acetal

(*Diethyl Acetal, Ethylidene Diethyl Ether*)

(Impure forms: *Dissolvan C.A.* and *D.N.*, *Lösungsmittel C.*)

$$CH_3 \cdot CH(O \cdot C_2H_5)_2, \text{ i.e. } CH_3 \cdot CH \begin{smallmatrix} O \cdot CH_2 \cdot CH_3 \\ O \cdot CH_2 \cdot CH_3 \end{smallmatrix}$$

PROPERTIES

A liquid with slightly bitter, burning taste, and a cold, peppermint-like aftertaste; it has a sweet ether-like odour. Produced by condensing acetaldehyde with alcohol.

B.P. 102° C. Sp.Gr. 0·83.

Solubility in water, about 5 per cent; miscible with alcohol in all proportions (von Mering, 1882). Forms a constant-boiling mixture with 67 per cent of ethyl alcohol boiling at 79° C. (Durrans, 1944).

USES

In the paint and varnish industry it has been used experimentally as a "latent" solvent for nitrocellulose, that is to say, diethyl acetal is not in itself a solvent for nitrocellulose but on admixture with other substances such as alcohols, it exhibits a high dilution ratio (Durrans, 1935).

TOXICITY

No experiments on inhalation of acetal vapour have been carried out, but from the results of gastric or oral administration to men and animals, and from subcutaneous and intraperitoneal injections in animals, it appears to be a central narcotic, very similar to, though rather more toxic than, paraldehyde when given in equal quantities. It is a skin irritant.

TOXIC EFFECTS IN ANIMALS

Lethal and Narcotic Concentrations

Lethal dose

By intraperitoneal injection. For rats, 7·5 millimols per kg. compared with 10 millimols for paraldehyde.

By gastric administration. For rabbits, 20 millimols per kg., i.e. about the same as paraldehyde (Knoefel, 1934).

Narcotic dose

By subcutaneous injection. For frogs, 0·05 ml. in 5 per cent solution; for rabbits, 2·4 g. (von Mering, 1882).

By intragastric administration. For dogs, 10 g. (von Mering, 1882).

Hypnotic dose

By intraperitoneal injection. For rats, 2·5 millimols per kg. produced 8 minutes sleep; 5·0 millimols, 15 to 23 minutes (Knoefel, 1934).

Acute Poisoning

Symptoms

According to von Mering, diethyl acetal shares with the non-chlorinated narcotics the property of being a central depressant, influencing the respiration more strongly than the cardiac activity. The animals passed from sleep into narcosis with little change in blood pressure and very slight slowing of the heart, even during deep narcosis. Death in fatal cases was due to respiratory failure.

TOXIC EFFECTS IN MAN

Von Mering gave 10 to 12 g. by mouth, which did not induce anaesthesia, but drowsiness and an analgesic effect. No after-effects were observed. It was found by Pereth (1883) that the action of this substance as a hypnotic was uncertain, which Knoefel (1934) states is because of its rapid hydrolysis in the stomach.

No cases of industrial injury from the use of diethyl acetal have been recorded, but it is recognized as a skin irritant (Schwartz, Tulipan and Peck, 1947).

8. Paraldehyde

$(CH_3 \cdot CHO)_3$

PROPERTIES

Produced by the polymerisation of acetaldehyde by acid catalysts. A colourless liquid, soluble in 8 parts of water at 13°C., less soluble in warmer water, so that about half is given off on warming a solution saturated at ordinary temperature (Bau, 1929).

B.P. 124°C. Sp.Gr. 0·998. Water dissolves 12 per cent at 20°C. Fl.P. 27°C.

A "latent" solvent for nitrocellulose, i.e. not in itself a solvent for nitrocellulose, but mixed with other substances such as alcohols it exhibits a high dilution ratio (Durrans, 1935).

USES

(1) In the paint and varnish industry as a diluent; contained in substances called "diluols" (Yarsley, 1934).

(2) In the production of high-grade lubricating oils (Hamilton, 1934).

(3) As a plasticizer with cellulose acetate (Carroll, 1933).

(4) As a combustible (a substitute for denatured ethyl alcohol) and motor fuel (mixed with 40 to 60 per cent of benzene) (Baslini, 1924).

TOXICITY

The properties of paraldehyde as a hypnotic and basal anaesthetic in medicine are well known; it is considered one of the safest of the hypno-anaesthetics with a cortical action. Stewart (1932), for example, states that it has been used as a basal anaesthetic by rectal injection in 500 patients without one fatality, the margin between the hypnotic and the toxic dose being wide.

With regard to its effect in industrial use little is known. Yarsley (1934) states that when it is used as a diluent in paints and varnishes its vapours dissipate rapidly and the possibility of their being sufficiently concentrated to have even a soporific effect is very remote.

In animals, experiments have been carried out only by oral, rectal, subcutaneous or intravenous administration; these have shown that paraldehyde produces depression, preceded by a phase of motor excitement; death from large doses is due to respiratory paralysis.

TOXIC EFFECTS IN ANIMALS

Lethal and Narcotic Concentrations

Lethal dose

By intravenous injection. For dogs, 1·8 ml. per kg. (Nitzescu, 1932).

By oral administration. For dogs, 14 g. per kg. caused death in ¾ hour (Schneider, 1929); after 9 g. per kg. the animal still lived.

Per rectum. For mice, 0·208 ml. per 100 g. (Gage, 1933). Gage found a small proportion of the animals abnormally susceptible.

Narcotic dose

By intravenous injection. For dogs, 0·8 ml. per kg. (Nitzescu, 1932).

By oral administration. For dogs, 1·5 to 2 ml. per kg. Deep anaesthesia was induced within 18 to 30 minutes (Kistler and Luckhardt, 1929).

Per rectum. For mice, 0·102 ml. per 100 g. (Gage, 1933).

Acute Poisoning

Symptoms

The animals undergo a preliminary stage of motor excitation and inco-ordination which disappears as the narcotic dose is approached. The effects appear shortly after administration and last for a comparatively short time. Recovery is complete after non-lethal doses according to Kistler and Luckhardt (1929), though it should be noted that one of their animals died 16 days after a dose of 1 ml. per kg. which had not produced complete narcosis. With lethal doses, pulse and respiration become progressively slower and death occurs from respiratory paralysis, the heart continuing to beat 5 minutes after respiration has ceased (Schneider, 1929).

Changes in the Internal Organs

Congestion of all organs was the only noteworthy change observed by Schneider (1929). There was no fatty degeneration of liver or kidneys or evidence of corrosive action, as shown by hardening of the mucous membrane of the stomach observed in a human suicidal case by Paltauf (1893). Schneider believes that this action takes place only when paraldehyde is in contact with dead tissue.

Chronic Poisoning

According to Bau (1929), earlier workers observed that animals given repeated doses of paraldehyde showed wasting and diarrhoea, albuminuria, lowering of body temperature, respiratory catarrh, and finally died from oedema of the lungs. Post-mortem examination showed fatty degeneration of the heart and liver.

TOXIC EFFECTS IN MAN

Acute Poisoning

Symptoms

A number of cases of acute poisoning, some fatal, have been reported, and though none of them is in connection with the industrial use of the drug, they give some indication of the mechanism of its toxic action.

According to Bau, 20 to 30 g. by mouth will produce a deep coma-like sleep, with dilatation of the pupils and blood vessels, pallor, respiratory disturbance and heart weakness, rigidity of the extremities and loss of reflexes. Recovery may be accompanied by headache, vomiting and nausea. Bau describes four cases of this kind. Epileptic attacks as a residual symptom of acute paraldehyde poisoning were first described by Kraft-Ebbing in 1897, and one of Bau's cases showed this symptom 6 days after recovery from the comatose condition.

In the suicidal case reported by McDougall (1932), death took place 50 hours after 4 oz. had been taken by mouth, while in the often quoted case of Paltauf (1893) the fatal dose was 40 g. in cognac. Bau (1929) states that the fatal dose is from 50 to 100 g. The symptoms in fatal cases are shallow respiration, weak pulse, perspiration, cyanosis, loss of consciousness and death from collapse.

Post-mortem Appearances

The chief abnormalities have been found in the stomach and oesophagus—swelling and redness of the mucous membrane (Bau). In Paltauf's case these appearances amounted to actual corrosion, like that caused by corrosive sublimate, but as mentioned above, Schneider (1929), on the basis of *in vitro* experiments with portions of stomach wall, believes that this is a post-mortem phenomenon rather than a specific corrosive action of paraldehyde on living tissues. The liver and kidneys showed fatty degeneration in McDougall's case;

parenchymatous degeneration of the kidney was present in the fatal case of Bau. The brain showed areas of acute degeneration of the neural cells in McDougall's case.

Chronic Poisoning

According to Flury and Zernik (1931) chronic intoxication from the habitual use of paraldehyde is more frequent than acute. The symptoms resemble those of chronic alcoholism—loss of weight, anaemia, a weak irregular pulse, paraesthesia, muscle weakness, delirium, hallucinations of sight and hearing, loss of intelligence and psychic disturbances. The urine may contain albumin and casts.

BIBLIOGRAPHY

ATKINSON, H. V. and ETS, H. N. (1922). Chemical changes in the blood under the influence of drugs. *J. biol. Chem.*, **25**, 5.
BARBER, H. (1934). Haemorrhagic nephritis and necrosis of the liver from dioxan poisoning. *Guy's Hosp. Rep.*, **84**, 267.
BASLINI, E. (1924). Impiego della paraldeide come combustibile e come carburante. *Atti. Cong. naz. Chim. industr.*, p. 240, Milan.
BAU, S. (1929). Über Paraldehydvergiftung. *Dtsch. Z. ges. gerichtl. Med.*, **13**, 337.
BONSMANN, M. R. (1932). Über Gewöhnungs- und Kumulationserscheinungen unter der Wirkung von Narkoticis beim Hunde. *Arch. exp. Path. Pharmak.*, **165**, 659.
BOOTHBY, W. M. (1913–14). Determination of the anaesthetic tension of ether vapor in man with some theoretical deductions therefrom, as to the mode of action of the common volatile anaesthetics. *J. Pharmacol.*, **5**, 379.
BOURNE, W. (1926). On the effects of acetaldehyde, ether peroxide, ethyl mercaptan, ethyl sulphide, and several ketones—di-methyl, ethyl methyl and di-ethyl—when added to anaesthetic ether. *J. Pharmacol.*, **28**, 409.
CARROLL, S. J. (1933). Use of paraldehyde as a plasticizer with cellulose acetate. *Chem. Abstr.*, **27**, 2809.
CHABANIER, H., LOBO-ONELL, C. and LELU, E. (1932). Modifications de l'équilibre acide-base au cours de l'anesthésie générale par l'éther. *C. R. Soc. Biol., Paris*, **110**, 1282.
COOK, W. A. (1945). Maximum allowable concentrations of industrial atmospheric contaminants. *Industr. Med.*, **14**, 936.
DURRANS, T. H. (1935). Solvents and plasticisers. *J. Oil Col. Chem. Ass.*, **18**, 340.
DURRANS, T. H. (1935). Review of the technical aspects of industrial solvents. *Chem. and Ind.*, **13**, 585.
DURRANS, T. H. (1944). Solvents. 5th ed. Chapman and Hall, London.
DURRANS, T. H. (1950). Solvents. 6th ed. Chapman and Hall, London.
ELLINGER, P. and ROST, F. (1922). Über die Methaemoglobinbildung durch Narkotika. *Arch. exp. Path. Pharmak.*, **95**, 281.
ESTLER, W. (1935). Gewerbliche und experimentelle Vergiftungen mit Dioxan im Schrifttum (Reichsgesundheitsamt., Berlin). *Arztl. Sachverstztg.*, **41**, 119.
FAIRLEY, A., LINTON, E. C. and FORD-MOORE, A. H. (1934). Toxicity to animals of 1:4-dioxan. *J. Hyg., Camb.*, **34**, 486.
FIFE, H. R. and REID, E. W. (1930). New industrial solvents: ethylene dichloride, dichlorethyl ether, and isopropyl ether. *Industr. Engng. Chem.*, **22**, 513.
FLURY, F. and ZERNIK, F. (1931). Schädliche Gase, Dämpfe, Nebel, Rauch- und Staubarten. Springer, Berlin.
FROIS (1926). La santé et le travail des femmes pendant la guerre. Paris. Published by La Dotation Carnégie pour la paix internationale.
GAGE, J. C. (1933). Variations in the susceptibility of mice to certain anaesthetics. *Quart. J. Pharm.*, **6**, 418.
GREEN, L. W. and SCHIETZOW, R. E. (1933). Method for determination of minute amounts of peroxides in ether. *J. Amer. Pharm. Ass.*, **22**, 412.
HAGGARD, H. W. (1924). Absorption, distribution and elimination of ethyl ether. I. The amount of ether absorbed in relation to the concentration inhaled and its fate in the body. *J. biol. Chem.*, **59**, 737.
HAMILTON, A. (1916). U.S. Department of Labor. *Bureau of Labor Statistics Bulletin*, **219**, 54.
HAMILTON, A. (1925). Industrial poisons in the United States. Macmillan, New York.
HAMILTON, A. (1934). Industrial toxicology. Harper, New York.
HAMILTON, A. and MINOT, G. R. (1920). Ether poisoning in the manufacture of smokeless powder. *J. industr. Hyg.*, **2**, 41.
HAYHURST, E. K. (1930). Ether. International Labour Office, *Brochure No.* 186. Also in Occupation and Health, **1**, 684. International Labour Office, Geneva, 1930.

Henry, S. A. (1933). Annual report of Chief Inspector of Factories. H.M. Stationery Office, London, 1934.
Hetzel, K. W. (1931). Warnung beim Arbeiten mit äthylperoxydhältigen Äther. *Z. angew. Chem.*, **44**, 388.
Hirsch, J. and Kappus, A. L. (1929). Über die Mengen des Narkoseäthers in der Luft von Operationssalen. *Z. Hyg. Infekt. Kr.*, **110**, 391.
International Labour Office (1930). *Brochure No.* 186. Ether. Also in Occupation and Health, **1**, 684, International Labour Office, Geneva, 1930.
Jones, G. W., Seaman, H. and Kennedy, R. E. (1933). Explosive properties of dioxan-air mixtures. *Industr. Engng. Chem.*, **25**, 1283.
Kärber, G. (1930). Beiträge zur Toxikologie der bei Gegenwart brennender Flammen auftretenden Zersetzungsprodukte des Äthers. *Klin. Wschr.*, **9**, 1130.
Kärber, G. and Lendle, L. (1931). Quantitative Untersuchungen über die Wirkung des Äthers auf die Atmung. *Arch. exp. Path. Pharmak.*, **160**, 440.
Kistler, G. H. and Luckhardt, A. B. (1929). Pharmacology of some ethylenehalogen compounds. *Curr. Res. Anesth.*, **8**, 65.
Knoefel, P. K. (1934). Narcotic potency of the aliphatic acyclic acetals. *J. Pharmacol.*, **50**, 88.
Knoefel, P. K. (1935). Narcotic potency of some cyclic acetals. *J. Pharmacol.*, **53**, 440.
Knoefel, P. K. and Murrell, F. C. (1935). The rate of production of anesthesia in mice by ether containing aldehyde and peroxide. *J. Pharmacol.*, **55**, 235.
von Kraft-Ebbing, R. (1897). Arbeiten aus dem Gesamtgebiet der Psychiatrie und Neuropathologie. Hft. 2, p. 186. Barth, Leipzig.
Krebs, O. (1929). Recovery of solvents in manufacture of rayon. *Kunstseide*, **11**, 389.
Lehmann, K. B. and Flury, F. (1943). Toxicology and hygiene of industrial solvents. (Tr. by E. King and H. F. Smyth, jr.) Williams and Wilkins, Baltimore.
Lewin, L. (1929). Gifte und Vergiftungen. 4th Aufl. Springer, Berlin.
Lourenço, A.-V. (1863). Recherches sur les composés polyatomiques; éther du glycol, ou anhydride diéthylénique. *Ann. Chim.*, **67**, 3me sér., 288.
McDougall, J. (1932). Fatal case of paraldehyde poisoning with post-mortem findings. *J. ment. Sci.*, **78**, 374.
Mann, F. C. (1925). Investigations of the relation of anesthesia to hepatic function. *Curr. Res. Anesth.*, **2**, 107.
von Mering (1882). Über die hypnotisirende und anästhesirende Wirkung der Acetale. *Berl. klin. Wschr.*, **19**, 648.
Meyer, K. H. and Gottlieb-Billroth, H. (1921). Theorie der Narkose durch Inhalationsanästhetika. *Hoppe-Seyl. Z.*, **112**, 55.
Minnitt, R. J. (1933). Treatment for toxic symptoms from ether anaesthesia. *Brit. J. Anaesth.*, **10**, 106.
Morgan, G. T. and Pickard, R. H. (1936). Explosions arising from di-*iso*propyl ether. *Chem. and Ind.*, **14**, 421.
de Navasquez, S. (1935). Experimental tubular necrosis of the kidneys accompanied by liver changes due to dioxan poisoning. *J. Hyg., Camb.*, **35**, 540.
Nelson, K. W., Ege, J. F., Ross, M., Woodman, L. E. and Silverman, L. (1943). Sensory response to certain industrial solvent vapors. *J. industr. Hyg.*, **25**, 282.
Nitzescu, I. I. (1932). Anesthésie générale par injection intraveineuse de paraldéhyde et d'alcool éthylique. *C. R. Soc. Biol., Paris*, **111**, 337.
von Oettingen, W. G. and Jirouch, E. A. (1931). Pharmacology of ethylene glycol and some of its derivatives in relation to their chemical properties. *J. Pharmacol.*, **42**, 355.
Paltauf, R. (1893). Verhandlungen ärztlicher Gesellschaften und Vereine. *Wien. klin. Wschr.*, **6**, 888.
Pereth (1883). (Cited by Knoefel, J., 1934, in: *J. Pharmacol.*, **50**, 88.)
Reid, E. W. (1934). Modern solvent industry. *Industr. Engng. Chem.*, **26**, 21.
Reid, E. W. and Hofmann, H. E. (1929). 1,4- dioxan. *Industr. Engng. Chem.*, **21**, 695.
Rieche, A. (1931). Ätherexplosionen und "Ätherperoxyd": die Autoxydation der Äther. *Z. angew. Chem.*, **44**, 896.
Robbins, B. H. (1935). Ether anesthesia: concentrations in the inspired air and in the blood required for anesthesia, loss of reflexes and death. *J. Pharmacol.*, **53**, 251.
Robertson, R. (1933). Danger of explosions. *Chem. and Ind.*, **11**, 274.
Ronzoni, E. (1923). Ether anesthesia. II. Anesthetic concentration of ether for dogs. *J. biol. Chem.*, **57**, 761.
Rosenthal, S. M. and Bourne, W. (1928). Effect of anesthetics on hepatic function. *J. Amer. med. Ass.*, **90**, 377.
Sand, R. (1910). L'intoxication expérimentale par l'alcool-éther. Report of second Congrès international des maladies professionnelles, Brussels.
Sanderson, J. McE. (1933). Solvents used in surface coatings. *Paint Oil Chem. Rev.*, **95**, No. 2, 10.
Schneider, P. (1929). Einiges über Paraldehydvergiftung. *Wien. klin. Wschr.*, **42**, 357.
Schrenk, H. H., Patty, F. A. and Yant, W. P. (1933). Acute response of guinea-pigs to vapors of some new commercial organic compounds. VII. Dichlorethyl ether. *Publ. Hlth. Rep., Wash.*, No. **48**, 1389.

Schulze, F. (1924). Zur Frage der postnarkotischer Azidose. *Zbl. Chir.*, **51**, 2688.
Schwartz, L., Tulipan, L. and Peck, S. M. (1947). Occupational diseases of the skin. Kimpton, London.
Silverman, L., Schulte, H. F. and First, W. W. (1946). Further studies on sensory response to certain industrial solvent vapors. *J. industr. Hyg.*, **28**, 262.
van Slyke, D. D., Austin, J. H. and Cullen, G. E. (1922). Effect of ether anesthesia on the acid-base balance of the blood. *J. biol. Chem.*, **53**, 277.
Stehle, R. L. and Bourne, W. (1928). Effects of morphine and ether on the function of the kidneys. *Arch. intern. Med.*, **42**, 247.
Stewart, J. D. (1932). Rectal paraldehyde before operation. *Brit. med. J.*, **ii**, 1139.
U.S. Bureau of Labour Statistics (1916). *Bull.*, **219**, 54. See Hamilton, A. (1916).
Weil, A. (1929). Veränderung in dem histologischen Bilde und dem Lipoidaufbau des Zentralnervensystems in der Äthernarkose. *Pflüg. Arch. ges. Physiol.*, **223**, 351.
Weissberger, A., Young, J. Z. and Carleton, H. M. (1934). Use of dioxan for transference of tissues direct from water to paraffin wax. *Lancet*, **ii**, 1279.
Wilson, W. S. (1935). Convulsions under ether anesthesia. *Curr. Res. Anesth.*, **14**, 281.
Winogradow, A. P. (1927). Die Wirkung von Arzneisubstanzen auf die Absonderung der Galle. *Arch. exp. Path. Pharmak.*, **126**, 17.
Yant, W. P., Schrenk, H. H., Waite, C. P. and Patty, F. A. (1930). Acute response of guinea-pigs to vapors of some new commercial organic compounds: Dioxan. *Publ. Hlth. Rep., Wash.*, **45**, 2023.
Yarsley, V. E. (1934). Solvents and plasticisers. A review of recent progress with special reference to cellulose acetate. *Synth. appl. Finish.*, **5**, 37, 57.

CHAPTER V
ESTERS

1. Methyl Formate
$H \cdot CO \cdot OCH_3$

PROPERTIES

A colourless liquid with a penetrating ethereal odour. B.P. 31·5°C. Sp.Gr. 0·9631 at 25°C. Fl.P. —12°C. Soluble in water, alcohol or ether.

USES

(1) In organic synthesis.
(2) As a solvent for cellulose acetate.
(3) As a fungicide.

TOXICITY

From experiments on animals, methyl formate is an acute irritant of the mucous membranes, and in high dosage has a narcotic effect.

Maximum Permissible Concentration

On the basis of 1,500 p.p.m. as the highest permissible concentration for short exposures, Cook (1945) suggests 400 p.p.m. as the level for prolonged exposure.

TOXIC EFFECTS IN ANIMALS

Lethal Dose

10,000 p.p.m. (25 mg. per l.) after 2 to 3 hours for both guinea-pigs and cats (Schrenk, Yant, Chornyak and Patty, 1936). The maximum amount which could be inhaled for several hours without serious disturbance was 1,500 to 2,000 p.p.m. (3·8 to 5·1 mg. per l.).

Symptoms

Severe irritation of the eyes and nose occurs, and after long exposure, pulmonary oedema; slow deep respiration is followed by inco-ordination, narcosis and death.

TOXIC EFFECTS IN MAN

No cases of injury from its industrial use have been recorded. According to Schrenk *et al.* (1936) inhalation of 1,500 p.p.m. for one minute caused neither irritation of the nose and eyes, nor any other symptoms. They state that "while the odor of methyl formate is distinct and noticeable in concentrations which are relatively safe from the standpoint of producing acute poisoning, owing to its pleasant nature and the occurrence of olfactory fatigue it is doubtful whether the odor of methyl formate will serve as an effective warning of harmful conditions of exposure." Lehmann and Flury (1943) remark that "the greatest care must be taken when using the substance".

2. Ethyl Formate
$H \cdot CO \cdot OC_2H_5$, i.e. $H \cdot CO \cdot O \cdot CH_2 \cdot CH_3$

PROPERTIES

A highly volatile liquid, with an odour similar to that of acetone, but less powerful. Miscible with benzene. Soluble in 9 parts of water at 18°C. B.R. of the commercial product 53–57° C. Sp.Gr. 0·925–0·930. Fl.P. —19° C. Not stable; readily develops acidity. (Technical grade of 83–85 per cent: B.R. 53–76° C. Sp.Gr. 0·898.)

A rapid solvent for cellulose nitrate and acetate. " Formosol" is said to be ethyl formate (Durrans, 1931). Ethyl formate is very often added to other solvents which are not actually sold as a mixture of ethyl formate and other substances (Zangger, 1930).

USES

(1) In the lacquer industry; ethyl formate is not suitable for metal lacquers, but is used to some extent in other lacquers and finishes (Yarsley, 1933).

(2) In artificial silk manufacture as a solvent for cellulose acetate (Cazeneuve, Morel and de Leeuw, 1932).

(3) In the shoe industry as a constituent of the mixture used to dissolve celluloid for covering heels with imitation leather; the mixture is used boiling (Duquenois and Revel, 1934).

(4) In the manufacture of artificial rum.

TOXICITY

While ethyl formate like all the volatile formates has an irritant action on mucous membranes and a paralytic action on the central nervous system, it is apparently less toxic than either butyl or amyl formate, following the rule enunciated by Flury and Zernik (1931) and by Yarsley (1933) that the toxicity of the formates increases with their molecular weight. It should be mentioned here, however, that this rule did not appear to hold clinically in the experience of Duquenois and Revel (1934), who found that the solvent mixtures used in shoe manufacture containing methyl formate, which has a smaller molecular weight than ethyl formate, were responsible for severe intoxication, while those sold as non-toxic and apparently producing no symptoms, consisted chiefly of ethyl formate. According to Flury and Zernik, however, none of the formates can be regarded as entirely harmless; they point out that they can react as aldehydes as well as acids and that the possibility of their intracellular decomposition into aldehydes cannot be excluded.

TOXIC EFFECTS IN ANIMALS

The narcotic effect of ethyl formate is accompanied by excitation; it is short lasting on account of rapid excretion through the lungs, but if continued leads to coma and death. Its irritant effect is only moderate, lachrymation not being produced with concentrations which are severely irritating in the case of methyl formate.

Acute Poisoning

Lethal and Narcotic Concentrations

Lethal dose

By subcutaneous injection. Weber (1902) found 1 ml. per kg. body weight lethal for the rabbit.

By inhalation. In the earlier experiments of Lorenz (1899) and Weber (1902) and in the later researches of Duquenois and Revel (1934), no concentrations are given. Weber merely states that his rabbits inhaled a mixture of air and ethyl formate vapour, which led to deep narcosis after 9 minutes and to death after about 3 hours. In Duquenois and Revel's experiments, frogs inhaled air, heavily impregnated with vapour from the boiling ester, in a closed chamber;

death took place in 5 minutes. With guinea-pigs the inhalation was not pushed to the lethal point. Flury and Zernik (1931) give 32 mg. per l. (10,000 p.p.m.) as the dose which caused death in cats after 90 minutes.

Narcotic dose

According to Flury and Zernik the narcotic dose is the same as the lethal dose (32 mg. per l., 10,000 p.p.m.), but narcosis sets in after 75 minutes.

Symptoms

Weber noted in rabbits deepened respiration and dyspnoea preceding narcosis, which took place after 9 minutes; 2¾ hours later sudden violent twitchings and general convulsions occurred, lasting one minute, followed by coma interrupted by twitchings, and death ¼ hour later. Duquenois and Revel also observed pronounced excitation and rapid respiration in frogs, and twitching of the hind legs preceding narcosis in guinea-pigs. In those animals which were allowed to recover, intoxication and staggering persisted for some time after removal from exposure.

Signs of local irritation are not mentioned by Weber; Duquenois and Revel state that ethyl formate does not produce lachrymation as does methyl formate.

Table 41, summarizing previously unpublished results, is taken from Flury and Zernik, and shows the effects of varying concentrations on different species of animals.

TABLE 41

Effect of different concentrations of ethyl formate on animals

Concentration (mg./l.) (p.p.m.)		Duration	Species		
			Mouse	Cat	Dog
16	5,000	20 min.	Eye irritation, difficult respiration	Eye irritation, salivation	—
32	10,000	20 min.	As above	Slight irritation	—
32	10,000	80 min.	—	Slight irritation, deep narcosis after 75 min., death after 90 min.	—
32	10,000	3 hr.	—	—	Very slight irritation, some vomiting, recovery
32	10,000	4 hr.	—	—	As above; death; P.M. lung oedema

Effect on Nerve and Muscle

According to Duquenois and Revel, on the basis of their experiments on chronaxie in the frog, ethyl formate acts on the nerve before the muscle.

Post-mortem Appearances

Stoppage of the heart in diastole was observed both by Weber and by Duquenois and Revel. The latter note that no exceptional muscle rigidity was found like that occurring after methyl formate. Flury and Zernik note that lung oedema was observed in dogs, and also in cats exposed for 39 minutes to a "moving" mixture of ethyl formate and air of a concentration of 44 mg. per l. (14,000 p.p.m.).

Chronic Poisoning

No experiments on chronic effects in animals have been carried out.

TOXIC EFFECTS IN MAN

According to Flury and Zernik a concentration of 1 mg. per l. (330 p.p.m.) produces slight irritation of the eyes and rapidly increasing irritation of the nose, which subsides only after about 4 hours.

No severe effects of intoxication from the industrial use of ethyl formate have been recorded. Duquenois and Revel (1934) carried out an investigation to determine which solvents used in shoe manufacture were responsible for eye irritation, oppression in the chest and nervous symptoms. Their results did not lead them to ascribe the disturbances to the group containing chiefly ethyl formate, which was sold as non-toxic. Yarsley (1933), while regarding all the volatile formates as having an irritant action on mucous membranes and a paralytic action on the central nervous system, states that the lower aliphatic formates are most common in commercial use and are relatively non-toxic.

3. n-Butyl Formate

$H \cdot CO \cdot OC_4H_9$, i.e. $H \cdot CO \cdot O \cdot CH_2 \cdot CH_2 \cdot CH_2 \cdot CH_3$

PROPERTIES

A colourless liquid.

British Standard Specification. B.R. 96–110° C. Minimum content of esters, 85 per cent. Acidity, 0·02 per cent maximum. Sp.Gr. 0·885.

Pure ester. B.P. 106–107° C. Sp.Gr. 0·9108. Fl. P. 15–20° C. Solvent for some types of cellulose acetate, and for some gums and resins but not for hard copals. Miscible with oils and hydrocarbons, not with water to any extent.

USES

In the lacquer industry for obtaining a film of high strength by using cellulose nitrate of high viscosity. It may replace methyl acetate or acetone in the proportion of about three to eight (Deschiens, 1931).

TOXIC EFFECTS IN ANIMALS

Acute Poisoning

Lethal and Narcotic Dose

As in the case of ethyl formate the lethal dose is the same as the narcotic dose, if narcosis is prolonged for a time. Flury and Zernik (1931) give 43·5 mg. per l. (10,000 p.p.m.) as the concentration producing deep narcosis after 60 minutes and death after 70 minutes in cats.

Symptoms

Concentrations of 17 mg. per l. (4,000 p.p.m.)—moving air mixture—produce lachrymation, salivation and excitation, then stupor, which is followed by slight ataxia (Flury and Zernik, 1931). With larger doses, deep narcosis goes on to death. Table 42 shows the effects on dogs and cats.

TABLE 42
Toxic effects of n-*butyl formate on cats and dogs*

Concentration		Duration	Species	
(mg./l.)	(p.p.m.)	(min.)	Cats	Dogs
43·5	10,000	20	Irritation of eyes, salivation, stupor	—
43·5	10,000	45	Initial excitation followed by quietness; side position after 15 min.; recovery	—
43·5	10,000	60	As above; deep narcosis after 60 min.; death after 70 min.; P.M. lung haemorrhages	Slight irritation, vomiting, drowsiness; side position after 40 min.; recovery

Chronic Poisoning
No experiments on animals have been carried out.

TOXIC EFFECTS IN MAN
Butyl formate is a substance of strongly narcotic and irritant properties. Concentrations of 1 per cent (43·5 mg. per l.) produce such severe irritation of the eyes as to make sight impossible and after a few breaths these concentrations are intolerable (Flury and Zernik).
No cases of injury from industrial use have been reported.

4. Amyl Formate
$$H \cdot CO \cdot OC_5H_{11}$$

PROPERTIES
A colourless liquid with an odour less pronounced than that of amyl acetate. The technical product is a mixture of formates of isomeric and homologous alcohols, but consists chiefly of *iso*amyl formate.
B.R. 110–130° C. Sp.Gr. 0·880–0·885. Fl.P. about 26° C.
Good solvent for cellulose nitrate, some gums and resins, not for cellulose acetate or hard copals. Miscible with castor and linseeed oils, and hydrocarbons, but not with water. Dissolves 0·3 per cent of water at 20° C.

USES
(1) In the artificial leather industry as a substitute for amyl or butyl acetate in the manufacture of leather cloth, its odour being suggestive of leather.
(2) In the manufacture of safety glass.

TOXIC EFFECTS
The only experiments carried out on animals are those of Lehmann (1913), who gives no detailed descriptions of the concentrations used. Lehmann states that the narcotic potency of amyl formate is about three times as great as that of amyl acetate. Since it is also about twice as volatile, its diphasic toxicity

would thus make it an unsuitable substance for industrial use as a substitute for amyl acetate. No injury has, however, been recorded from its employment. Zangger (1930) places the formates in the group of solvents regarded as more toxic than ethyl acetate or ethyl lactate but less than methyl acetate. The irritant effect of amyl formate is presumably similar to that of amyl acetate (Flury and Zernik, 1931).

5. Benzyl Formate
$H \cdot CO \cdot O \cdot CH_2 \cdot C_6H_5$

PROPERTIES

Benzyl formate is similar in character to benzyl acetate but is of somewhat greater volatility (Durrans, 1931). Its odour is less pronounced and is suggestive of leather.

B.R. 200–202° C. Sp.Gr. 1·08–1·09.

Good solvent for cellulose acetate and nitrate, gums and resins. Miscible with castor and linseed oils, and with aromatic and petroleum hydrocarbons.

USES

In the lacquer trade, but only rarely.

TOXIC EFFECTS

No animal experiments have been carried out and no toxic effects from its use in industry recorded. Durrans states that it is non-toxic.

6. Methyl Acetate
$CH_3 \cdot CO \cdot OCH_3$

PROPERTIES

A neutral colourless liquid with a faint ester-like odour. Miscible with about 3 volumes of water.

B.P. 57° C. Sp.Gr. 0·933. Fl.P. —11° C.

British Standard Specification, D 13. Sp.Gr. 0·92–0·94. B.R. 55–60° C. Esters 80 per cent minimum. Acidity 0·1 per cent maximum.

Time of evaporation, 2·2 (compared with ethyl ether, 1); evaporates almost as quickly as acetone, which it closely resembles as a solvent, their boiling points being the same and their solvent powers similar.

An excellent solvent for nitrocellulose (for which it is the chief low-boiling point solvent (Zangger, 1930)), cellulose acetate, celluloid and cellulose ether; also a good solvent for resins, oils and fats. Does not affect rubber. (According to Zangger "E 13" is a mixture of methyl acetate and methyl alcohol; "E 14" of methyl acetate, methyl alcohol and acetone.)

USES

(1) In the lacquer industry. Being an excellent solvent for cellulose acetate it is used in the manufacture of aeroplane lacquers, and in cellulose-acetate coating lacquers, for which purpose it is interchangeable with acetone (Fordyce, 1935),

but the fact that it hydrolyses too easily prevents it being used extensively. It is also used as a thinner for nitrocellulose lacquers (Davidson and Reid, 1928), and in lacquers for spraying motor cars, etc., where rapid drying is desired (Banik, 1929).

(2) In the celluloid industry; for celluloid cements, about 20 per cent solutions of celluloid or film scrap in methyl acetate are used.

(3) In the artificial leather industry; in the manufacture of leather cloth and as an impregnation solvent.

(4) In the shoe industry for softening impregnated shoe puffs and as a varnish for shoe heels.

(5) In the hat industry for painting felt hats, together with amyl acetate, with solutions of camphor and collodion.

(6) In the fur industry skins are sprayed with methyl acetate and methyl alcohol.

TOXICITY

The acute effects of methyl acetate on animals are those of an irritant of mucous membranes (to a slightly greater degree than ethyl acetate, but less than amyl and butyl acetates) and of a narcotic. Its narcotic action is not powerful when compared with that of the other esters, but its general toxic effect is rather different from that of recognized anaesthetics. According to Flury and Wirth (1934), the narcosis produced by methyl acetate is more of the nature of a coma produced by severe acidosis from its decomposition into acid and alcohol. Its chronic effects on animals are not very pronounced, though it has been found to produce loss of weight and some blood disturbances.

In man no severe effects, either acute or chronic, have been observed, but methyl acetate was held responsible by Duquenois and Revel (1934) for comparatively slight symptoms of ocular and nervous disturbance in a group of workers in the shoe industry. It has been suggested, though on the evidence of one case only, that methyl acetate may resemble methyl alcohol in producing atrophy of the optic nerve (Duquenois and Revel, 1934; Lund, 1944).

It is possible that the toxic effects of methyl acetate are due to hydrolysis.

TOXIC EFFECTS IN ANIMALS

Acute Poisoning

In the early experiments of Weber (1902) methyl acetate, when inhaled, was apparently not lethal, either for frogs or rabbits, the latter showing only a short-lasting narcosis from which they rapidly recovered. More recent experiments by Duquenois and Revel (1934) and Flury and Wirth (1934) have shown it to be lethal for frogs, mice and cats when given in sufficiently high concentration. In Duquenois and Revel's experiments, however, temperature was found to be an important factor, the animals dying in 5 minutes, after an initial phase of excitement with nystagmus, when exposed to the vapour from the boiling liquid.

Lethal and Narcotic Concentrations

Lethal dose

For mice 63 mg. per l. (21,000 p.p.m.); for cats 95 mg. per l. (31,000 p.p.m.) (Flury and Wirth, 1934). These workers point out that the lethal dose by subcutaneous injection is considerably greater than that by inhalation. Thus the amount fatal by inhalation to cats in $\frac{1}{2}$ hour, calculated from the concentration and the respiratory volume, is approximately 2 g. per kg., while the subcutaneous lethal dose in the same period is 3 g. per kg.

Narcotic dose

For mice 34 mg. per l. (11,000 p.p.m.); for cats 56 mg. per l. (19,000 p.p.m.). Flury and Wirth found mice more sensitive than cats; rabbits were very insensitive.

Symptoms

In the early stages the animals showed irritation of the eyes, salivation and dyspnoea. Twitching and vomiting and disturbances of breathing occurred during the period of semi-consciousness and the stage of loss of reflexes was extraordinarily difficult to obtain, indicating a general toxaemia rather than a pure narcotic action. Respiration ceased very quickly after the loss of reflexes and many animals died at this stage with symptoms of general poisoning. In survivors the temperature was generally raised, often as much as 2° C. The corneal reflex was restored within $\frac{1}{4}$ to $\frac{1}{2}$ hour after removal from exposure. Recovery was often not complete for several days, but delayed lethal effects were not observed. Duquenois and Revel also observed twitching and trembling in guinea-pigs, with rapid respiration and cardiac action. They believe, on the basis of observations on chronaxie of nerve and muscle, that methyl acetate acts on nerve before muscle.

The effects on mice of increasing concentrations are shown in Table 43, adapted from one given by Flury and Wirth (1934).

TABLE 43

Effect of different concentrations of methyl acetate on mice

Concentration		Interval before apparent effect			After-effects
(mg./l.)	(p.p.m.)	Disturbance of equilibrium	Side position	Deep narcosis	
21	7,000	—	—	—	Recovery
24	8,000	7¼ hr.	7½ hr.	7¾ hr.	Recovery
34	11,000	3–3¾ hr.	3¼–4 hr.	3½–5 hr.	Death after 10 hr.
38	13,000	1¾ hr.	2¼–4 hr.	3¼–4¼ hr.	Recovery
63	21,000	35–40 min.	50–59 min.	52–61 min.	Death after 1–3 min.

Post-mortem Changes

In mice the post-mortem appearances were not characteristic. In cats the lungs showed slight emphysema and there was slight reddening of all respiratory mucous membranes (Flury and Wirth, 1934). In guinea-pigs the heart and lungs were congested (Duquenois and Revel, 1934).

Effect on the Isolated Frog's Heart

Führner's (1921) experiments appear to show that methyl acetate has a less pronounced narcotic action on the isolated frog's heart than ethyl or *n*-propyl acetate. The figures given for the concentrations which inhibit the heart's action are:

methyl acetate	.. .. ..	0·2 mol. per litre
ethyl acetate	.. .. ..	0·13 ,, ,,
n-propyl acetate	.. .. ..	0·03 ,, ,,

Chronic Poisoning

With concentrations of 20 mg. per l. (6,600 p.p.m.) for 8 days, 6 hours a day, cats showed great loss of weight, severe exhaustion followed by very slow recovery, and blood changes, which consisted of an increase of red blood corpuscles, haemoglobin and leucocytes. These changes are attributed by Flury and Wirth to stimulation of the haemopoietic organs.

TOXIC EFFECTS IN MAN

Although Flury and Zernik (1931) state that the irritative effect of methyl acetate on the eyes is weaker than that of ethyl acetate, some observations by Duquenois and Revel (1934) and one case described by Lund (1944) indicate that its industrial use may produce ocular disturbance. Flury and Zernik found that a concentration of about 32 mg. per l. (10,000 p.p.m.) produces irritation of the eyes, nose and throat, that the smell is unpleasant and the taste remains for some time if it is inhaled by the mouth. Zernik (1933) records that a worker who spilt about 30 ml. of methyl acetate on a work-table had a severe headache $\frac{3}{4}$ hour later and drowsiness lasting 6 hours.

Duquenois and Revel, having observed certain symptoms in female workers in shoe manufacture using solvents containing various esters, examined samples of these solvents and classified them according to the severity of the symptoms for which they were responsible. (It should be noted that none of the solvents contained any one ester alone and that all contained a proportion of acetone, varying from 2 to 20 per cent.) Those which contained chiefly methyl acetate were found to cause slight symptoms. These included pricking of the eyes and lachrymation, oppression in the chest, dyspnoea, palpitation and nervous disorders, which are described sometimes as a feeling of intoxication, sometimes as depression and apathy. Methyl acetate was not considered nearly so toxic as, for instance, methyl formate, but Duquenois and Revel state that all the esters examined by them are toxic eventually, especially when used at high temperature. The case of ocular disturbance described by Lund (1944) was that of an old man who had been exposed to the vapours of methyl acetate. He developed general depression, headaches and dizziness, followed by blindness due to atrophy of the optic nerve. Lund suggests that methyl alcohol and formaldehyde may be formed in the organism during the metabolism of methyl acetate.

7. Ethyl Acetate

(*Acetic Ether, Acetic Ester*)

$CH_3 \cdot CO \cdot OC_2H_5$, i.e. $CH_3 \cdot CO \cdot O \cdot CH_2 \cdot CH_3$

PROPERTIES

A colourless neutral liquid with a pleasant fruity odour.
Ethyl acetate is obtainable in several grades:
- (*a*) *Pure*. B.P. 77·1° C. Sp.Gr. 0·901. Fl.P.—5° C.
- (*b*) *British Standard Specification*, No. 553 (1934). Sp.Gr. 0·906–0·909. Esters, 96 per cent minimum. Acidity, 0·01 per cent maximum. B.R. 95 per cent between 74 and 79° C. Residue, 0·01 per cent maximum.
- (*c*) *A.S.T.M. Specification*, D 302–33. Sp.Gr. 0·883–0·888 at 20° C. Esters, 85–88 per cent. Acidity, 0·02 per cent maximum. B.R. up to

70° C. none; to 72° C. 10 per cent minimum; above 80° C. none. Residue, 0·005 per cent maximum.

(d) "*Solvent E*" is said by Durrans to be a mixture of ethyl acetate with either methyl or amyl acetate. Zangger (1930) states that "*E* 13" is a mixture of methyl acetate and methanol.

Ethyl acetate is an excellent solvent for fats, and when mixed with 15 to 20 per cent of alcohol, for cellulose acetate, nitrocellulose, collodion and cellulose.

USES

(1) In the lacquer industry. Ethyl acetate is probably the most commonly used low-boiling cellulose solvent, especially in the manufacture of nitro-cellulose lacquers, as a substitute for acetone, and also, according to Engelhardt (1933), in "Zapon" lacquers. It is also used in cellulose acetate lacquer coatings, in combination with spirit and high-boiling solvents, being specially useful in preventing "blushing" (Fordyce, 1935). It is also a constituent (43 per cent) of a bronze cellulose lacquer (Weingand, 1931).

(2) In the artificial leather industry as a diluent for varnish, in degreasing and in finishing.

(3) In the celluloid industry for joining, being superior for this purpose to acetone or acetone substitutes, since the joints made with ethyl acetate do not "blush".

(4) As an extracting agent.

(5) In the dry-cleaning industry.

(6) In the shoe industry as a constituent of cements and as a softening agent for shoe puffs.

(7) In the hat industry as a constituent of the lacquer (Krüger, 1932).

(8) As a constituent of paint and rust removers.

(9) As a hair wash (its use for this purpose was forbidden in Germany in 1931).

(10) In the pharmaceutical industry (Lloyd, Ostwald and Haller, 1930).

TOXICITY

The local irritative effect of ethyl acetate is about the same as that of methyl acetate; but, according to Flury and Wirth (1934), the irritation of the nose and throat in animals is slightly less than that of butyl acetate and considerably less than that of amyl acetate; its narcotic strength is greater than that of methyl acetate, and deep narcosis, in animals at any rate, is often followed by death. This fact may possibly be correlated with the one fatal case of acute poisoning by ethyl acetate recorded in man (Althoff, 1931).

There is little evidence that much danger is to be apprehended from the chronic effects of ethyl acetate especially as a metabolic poison. Zangger (1930) places it with butyl acetate in the group of substances which he regards as the least toxic of the organic solvents; he considers it essentially less toxic than acetone, amyl acetate or methyl acetate. The substitution of ethyl acetate for acetone in nitrocellulose lacquers is disliked by workmen who have become accustomed to the odour of acetone, although more of the latter may be present in the air because of its higher vapour pressure.

In general ethyl acetate appears to be regarded as a less toxic substance than methyl acetate for industrial use, in spite of its greater acute narcotic effect.

TOXIC EFFECTS IN ANIMALS

Acute Poisoning

Lethal and Narcotic Concentrations

Lethal dose

For mice 31 mg. per l. (8,600 p.p.m.) (Flury and Wirth (1934)). Mice were much more sensitive than cats, some of which died at 56 mg. per l. (16,000 p.p.m.) while others survived 130 mg. per l. (36,000 p.p.m.).

Narcotic dose

For cats 43 mg. per l. (12,000 p.p.m.), and for mice 18 mg. per l. (5,000 p.p.m.). Table 44, adapted from Flury and Wirth, shows the narcotic concentration for mice in increasing doses.

TABLE 44
Narcotic concentrations of ethyl acetate for mice

Concentration (mg./l.) (p.p.m.)	Time before effect		After-effects
	Disturbance of equilibrium (hr.)	Deep narcosis (hr.)	
7·3 2,000	—	—	Recovery
18 5,000	$3-3\frac{3}{4}$	$3\frac{1}{4}-4$	Recovery
26 7,200	$2\frac{1}{2}-3$	$2\frac{3}{4}-3\frac{1}{2}$	Recovery
31 8,600	$1\frac{1}{4}-1\frac{1}{2}$	$1\frac{1}{4}-1\frac{3}{4}$	Death after 3 hr.

Symptoms

The irritative action on the mucous membranes of the eyes and nose was less than that of methyl acetate in the same concentrations. The corneal reflex in cats was lost earlier than skin and pain reflexes and returned later. The fact that most of the mice and about 25 per cent of the cats died during the stage of deep narcosis indicates the high toxicity of narcotic concentrations of ethyl acetate, but those animals which recovered did so more quickly than from methyl acetate narcosis. The narcotic action of ethyl acetate is apparently exerted both peripherally and centrally (Wright, 1935), and it has also a local anaesthetic effect when applied to mucous membranes; Ikebe (1934) found it anaesthetic to the ear of the guinea-pig in 10 to 12 per cent solution. In the frog, Duquenois and Revel (1934) found the phase of excitation less with ethyl than with methyl acetate, and the effect on the eyes was also less, but intoxication and staggering were readily produced and the animals died in coma after 5 minutes.

Post-mortem Changes

The only lesions found in Flury and Wirth's animals were oedema of the lungs, with areas of haemorrhage, and hyperaemia of the respiratory tract. Flury and Wirth (1934) state that these appearances were not sufficient to account for death, which they regard as due purely to the narcotic effect of ethyl acetate.

Effect on the Isolated Frog's Heart

The narcotic strength of ethyl acetate, as measured by its inhibition of the isolated frog's heart, was found by Fühner (1921) to be 0·13 mol. per litre, as compared with 0·2 mol. per litre for methyl, and 0·03 mol. for propyl acetate.

Chronic Poisoning

Concentrations of about 7 mg. per l. (2,000 p.p.m.) were found by Smyth and Smyth (1928) to produce no ill-effects in guinea-pigs after 65 exposures, and they considered ethyl acetate safe for use at this level. Only comparatively slight ill-effects were observed by Flury and Wirth (1934) with higher concentrations than these, e.g. 15 to 16 mg. per l. (4,200 to 4,400 p.p.m.) 6 hours a day for 7 days. The animals refused to eat and lost weight and showed some blood changes, but no changes in the urine which might indicate kidney or liver injury. The blood changes consisted of an increase in red cells, but not of haemoglobin, so that the colour index was below 1. There was no absolute increase in leucocytes, as with butyl acetate, but a relative increase in neutrophils and a diminution in lymphocytes.

TOXIC EFFECTS IN MAN

Acute Poisoning

One fatal case has been reported by Althoff (1931), a workman who was painting the interior of a benzol tank with paint containing 80 per cent of ethyl acetate, the other constituents not being specified. The man was found dead in the tank; post mortem all the organs and tissues smelt strongly of ethyl acetate, as did the interior of the tank and the paint used. The respiratory passages showed much vascular injection, and there were sub-epicardial and sub-pleural punctiform haemorrhages. Spleen and kidneys showed congestion, and the stomach, recent gastritis and extravasation of blood. The blood was fluid. A guinea-pig subjected to parallel conditions died after 80 minutes; post mortem the whole body smelt of ethyl acetate, the blood was dark red but the internal organs showed no gross lesions.

Chronic Poisoning

Such chronic toxic effects as have been observed appear to be limited to irritation of mucous membranes, and it has been suggested (Beintker, 1928; Engelhardt, 1933) that ethyl acetate causes a hypersusceptibility to acute inflammation of mucous membranes and to eczematous conditions of the skin.

Actual irritative effects on the conjunctiva, like those of amyl and butyl acetates, are not recorded, though in Kruger's (1932) cases of conjunctivitis in hat workers ethyl acetate was an occasional constituent of the lacquers used. A severe gingivitis, occurring in a man in whom it was suggested there was a special sensitivity caused by long exposure to ethyl acetate, was reported by Beintker in 1928. The reason for attributing the injury to ethyl acetate, though the man had been exposed to a cellulose lacquer containing also butyl acetate and butyl alcohol, was that after relief of the symptoms by appropriate treatment, they returned with increased severity when a tooth was filled with a "cavity lining and varnish" with a characteristic smell of ethyl acetate. According to Nelson, Ege, Ross, Woodman and Silverman (1943), inhalation of ethyl acetate vapour causes considerable irritation of the nose and throat at 400 p.p.m.

The power to elicit a special susceptibility to eczematous conditions is ascribed to ethyl acetate by Engelhardt (1933). By skin tests he showed that a certain number of workers with "Zapon" and cellulose lacquers, containing mixtures of ethyl acetate, butyl alcohol, toluol and spirit, reacted positively to ethyl acetate.

A permanent smell of ethyl or amyl acetate experienced by workmen using solvents containing these substances, together with loss of perception of other odours is mentioned by Barrios and Devoto (1931). It has a strong odour at 200 p.p.m.

8. n-Propyl Acetate

$$CH_3 \cdot CO \cdot OC_3H_7, \text{ i.e. } CH_3 \cdot CO \cdot O \cdot CH_2 \cdot CH_2 \cdot CH_3$$

PROPERTIES

A liquid, B.P. 101·6° C. Sp.Gr. 0·897. Fl.P. 14·5° C. Solubility in water at 20° C., 1·89 per cent. Evaporation time, 6·1 (compared with ethyl ether, 1). Miscible with castor and linseed oils and hydrocarbons. Solvent for cellulose nitrate, celluloid and resins, but not for cellulose acetate.

n-Propyl acetate is rarely used in industry in a state of purity—the so-called technical product consists of a mixture of n-propyl acetate with other homologues arising from the esterification of fusel oil.

USES

In the lacquer and varnish industry (Alinari, 1930) but not to any great extent.

TOXIC EFFECTS IN ANIMALS
Acute Poisoning

From animal experiments it appears that n-propyl acetate has a moderately irritative effect on mucous membranes, only a little stronger than that of ethyl acetate, a narcotic effect rather stronger than that of methyl or ethyl acetate, but a lethal effect less than that of these two substances.

Lethal and Narcotic Doses

None of Flury and Wirth's mice and only two out of eighteen cats died during deep narcosis, so that propyl acetate cannot be said to be highly toxic to animals in "acute" concentrations.

The narcotic concentration for cats was 38 mg. per l. (9,100 p.p.m.); for mice 25 mg. per l. (6,000 p.p.m.). Rabbits were much less sensitive than cats.

Symptoms

The irritation of mucous membranes, as shown by lachrymation and salivation, was rather greater than that produced by ethyl acetate. Recovery after the stage of loss of reflexes was more rapid than with methyl or ethyl acetate.

Table 45 adapted from Flury and Zernik (1931) shows the effects of increasing concentrations on cats.

TABLE 45
Effect of different concentrations of n-*propyl acetate on cats*

Concentration (mg. l.) (p.p.m.)		Duration of exposure	Effect
22	5300	6 hr. on 5 consecutive days	Only slight eye irritation and salivation
38	9,100	$5\frac{1}{2}$ hr.	Stupor after $\frac{1}{2}$–$\frac{3}{4}$ hr., side position after $4\frac{1}{2}$ hr., deep narcosis $4\frac{1}{4}$–$5\frac{1}{4}$ hr., one died after $5\frac{1}{2}$ hr.
102	24,500	$\frac{1}{2}$ hr.	Side position 5–16 min., deep narcosis 13–18 min., one died after 4 days

Effect on the Isolated Frog's Heart

Judged from its inhibitory effect on the isolated frog's heart, the narcotic potency of *n*-propyl acetate would appear to be less than that of *iso*propyl acetate but more than that of ethyl acetate (Fühner, 1921). The figures for the narcotic concentrations are:

n-propyl acetate	.. ..	0·03 mol. per litre
*iso*propyl acetate	.. ..	0·06 ,, ,,
ethyl acetate	.. ..	0·13 ,, ,,

Chronic Poisoning

In long continued non-narcotic concentration propyl acetate produces respiratory irritation and some liver injury according to Flury and Wirth (1934).

The lower degree of toxicity of propyl acetate than that of methyl and ethyl acetates when inhaled for long periods was shown by the fact that the animals tolerated a concentration of more than half the narcotic level (22 mg. per l., 5,300 p.p.m.) with no ill-effects other than exhaustion, from which they recovered in a few days. In one animal which died, slight tracheitis, bronchitis and emphysema of the lung and also an increased fat content of the liver were observed.

TOXIC EFFECTS IN MAN

No ill-effects have been reported.

9. *iso*Propyl Acetate

$$CH_3 \cdot CO \cdot OC_3H_7, \text{ i.e. } CH_3 \cdot CO \cdot O \cdot CH {<} {CH_3 \atop CH_3}$$

PROPERTIES

A liquid similar in character to *n*-propyl acetate, with properties intermediate between ethyl and butyl acetates. Its evaporation rate is about one third that of ethyl acetate.

Pure grade. B.P. 92° C. Sp.Gr. 0·93. Fl.P. about 0° C. Solubility in water at 20° C. 3·1 per cent by weight.

Technical grade. B.R. 82–89° C. Esters 85–100 per cent. Sp.Gr. 0·865–0·90. Miscible with benzene in all proportions.

Good solvent for cellulose nitrate and some gums and resins; partly for shellac; but not for cellulose or hard copals.

USES

In the lacquer trade (Hackett, 1932).

TOXIC EFFECTS

The only experiments on the toxic action of *iso*propyl acetate appear to be those of Fühner (1921) on the isolated frog's heart. Its narcotic potency judged by this method would seem to be less than that of *n*-propyl and *iso*butyl acetates but more than that of ethyl or methyl acetates. The concentrations which stop the contraction of the heart are as follows:

*iso*propyl acetate	..	0·06 mol. per litre
n-propyl acetate	..	0·03 ,, ,,
ethyl acetate	..	0·13 ,, ,,

No ill-effects from its industrial use have been recorded, but according to Silverman, Schulte and First (1946), it causes eye irritation at 200 p.p.m. They recommend 100 p.p.m. as the upper limit for sensory disturbance.

10. *n*-Butyl Acetate

$CH_3 \cdot CO \cdot OC_4H_9$, i.e. $CH_3 \cdot CO \cdot O \cdot CH_2 \cdot CH_2 \cdot CH_2 \cdot CH_3$

PROPERTIES

n-Butyl acetate is a neutral, colourless liquid with a pleasant ester-like odour; not miscible with water. B.P. 126° C. Evaporation rate, 12 (compared with ethyl ether, 1).

British Standard Specification, No. 551 (1934). Sp.Gr. 0·883–0·886. Esters, 97 per cent minimum. B.R. 124–128° C. 95 per cent minimum. Residue, 0·01 per cent maximum.

A.S.T.M. Specification. D 303-33. Sp.Gr. 0·872–0·878 at 20° C. Esters, 88–92 per cent. B.R. up to 110° C. none; 110–120° C. 15 per cent maximum; above 127° C. 30 per cent maximum; above 145° C. none.

It is an excellent solvent for nitrocellulose, celluloid, cellulose ethers, resins and oils, and one of the most important solvents for nitrocellulose lacquers, especially the 85 per cent grade, which contains 15 parts of butyl alcohol (I. G. Farbenindustrie Aktiengesellschaft, 1930). It is used often as a substitute for amyl acetate, when, if the lower volatility of amyl acetate is desired, larger quantities of butyl alcohol (up to 30–40 per cent) may be added. It may be diluted to a considerable extent with spirit, benzol, toluol, benzine, etc., without the risk of causing precipitation or impeding the formation of the film.

USES

(1) In the lacquer industry; chiefly used for nitro-cellulose lacquers, at any rate in the U.S.A., since amyl acetate is more expensive and ethyl acetate less efficient (Hamilton, Bricker and Smyth, 1929). It is also used in the production of combination lacquers containing drying oils as well as nitro-cellulose and possibly resins. It is the chief constituent of many high polish lacquers and varnishes.

(2) In the straw hat industry; Krüger (1932) states that the use of butyl acetate as a constituent of lacquers for this industry is steadily increasing.

(3) In the leather industry especially in the manufacture of patent leather (Enna, 1930).

(4) In the motor industry. It is stated by Holden and Doolittle (1935) that butyl acetate is the ideal solvent for making a protective coating vehicle for low viscosity pyroxylin.

(5) In the manufacture of cleansing and stain-removing agents, spirit varnishes, etc., as a perfuming agent especially in shoe-cleaning pastes (Zangger, 1930).

TOXICITY

Butyl acetate apparently stands close to amyl acetate in both irritant and narcotic action. Whether it is more dangerous, from an industrial point of view, than amyl acetate appears to be uncertain; Banik (1929) points out that since it is 25 per cent more volatile than amyl acetate it might be expected to be more dangerous, but on the other hand, the technical product commonly used is purer. Flury and Wirth (1934) claim that on the basis of animal experiments it is both less narcotic and less irritant than the amyl ester. These authors state that the difference in toxic action between methyl, ethyl and propyl acetates on the one hand and amyl and butyl acetates on the other is much greater than the differences between individual members of these two groups. Amyl and butyl acetates produce a sensation of pressure in the chest, cough and exhaustion, as well as the slight irritation of nose, throat and eyes produced by the other three. According to Nelson et al. (1943), the capacity of butyl acetate for throat irritation is about the same as that of amyl acetate. Zangger (1930) on the other hand places butyl acetate with ethyl acetate in the group of solvents which he regards as least toxic of all the organic solvents.

Evidence of any organic injury, either in man or animals is very slight and inconclusive. The evidence produced by Burger and Stockmann (1932) of irritation of the liver is weakened by the fact that the butyl acetate used in the cases quoted was mixed with several other constituents, while in the case of injury reported by Führner and Pietrusky (1934), the patient had been exposed to a mixture of butyl acetate and toluol. One of the lacquers which was known to produce eye lesions was analysed by Krüger and was found to contain collodion cotton, shellac, butyl acetate and butyl alcohol.

Maximum Permissible Concentration

The maximum allowable concentration is given as 400 p.p.m. in America, but 50 p.p.m. in Germany.

TOXIC EFFECTS IN ANIMALS
Acute Poisoning
Lethal and Narcotic Doses

The lethal effect of butyl acetate, judged by subcutaneous injection, is less than that of ethyl, methyl or amyl acetates (Flury and Wirth, 1934). The narcotic effect of butyl acetate when inhaled appears to be slightly greater than that of amyl acetate (Gardner, 1925; Flury and Wirth, 1934), while its irritative effects on mucous membranes, especially the cornea, are also slightly more pronounced.

The narcotic concentration for cats according to Emrich (1932) is 33 mg. per l. (7,000 p.p.m.) and for mice 30 mg. per l. (6,300 p.p.m.). Table 46 shows Flury and Wirth's results with increasing concentrations in mice.

TABLE 46

Effect of different concentrations of n-butyl acetate on mice

Concentration		Time before effect			After-effects
(mg./l.)	(p.p.m.)	Disturbance of equilibrium (hr.)	Side position (hr.)	Deep narcosis (hr.)	
20	4,200	—	—	—	Recovery
28	5,900	3½–4	—	—	Recovery
30	6,300	2½–3	4½–5	4½–5	Recovery
35	7,400	1¼	2–3½	3¼–3½	Recovery

Sayers, Schrenk and Patty (1936) found no symptoms or only very slight disturbances in guinea-pigs at concentrations of 3,300 p.p.m.

Symptoms

There was more irritation of the mucous membranes of the eyes and nose in Flury and Wirth's animals than with amyl acetate. Central irritation was slight, only one animal showing slight clonic spasms.

Secondary Effects

In mice the narcosis was very well tolerated, recovery being rapid and the reflexes returning within about half an hour. Gardner (1925) also observed that, while the vapour of butyl acetate inhaled in large quantities by rabbits produced profound anaesthesia, they recovered with no impairment of health. In his experiments narcosis appeared to be produced more rapidly with butyl than with amyl acetate. This was the case also when the ester was given by stomach tube in a 25 per cent mixture with alcohol and water.

Chronic Poisoning

The investigations of the Pennsylvania Department of Labor and Industry (1926) gave somewhat inconclusive results. A series of animals exposed to 500 p.p.m. for 36 3-hour exposure periods seemed to show some harmful effects, as evidenced by early depression and by some renal changes. A later investigation by Smyth and Smyth (1928), however, using 900 p.p.m. over 65 exposure periods failed to confirm any ill-effects other than fatigue. Flury and Wirth observed slightly more toxic effects than these, with concentrations of 14·3 to 20 mg. per l. (3,100 to 4,200 p.p.m.) over periods of 6 days for 6 hours a day. In cats these exposures produced exhaustion, loss of weight and some blood changes, which consisted of a progressive increase in erythrocytes, leucocytes and haemoglobin, but no pathological cell formation.

Post-mortem Changes

One animal which died after exposure to the higher concentrations showed slight conjunctivitis, laryngitis, tracheitis and slight emphysema of the lungs. No mention is made of any liver changes such as are caused by amyl acetate.

TOXIC EFFECTS IN MAN

Symptoms

None of the cases of poisoning by butyl acetate recorded in the literature is from the use of this substance alone. The symptoms complained of vary from conjunctival irritation (Krüger, 1932), and bronchial catarrh (Führer and Pietrusky, 1934), to headache, stupor and gastric disturbance (Weber and Gueffroy, 1932). A very unusual description of the effects of butyl acetate (unconfirmed and not annotated) is given by Faucett (1935). He states that it "lengthens the focus of the eyes, gives the members of the body a greater apparent length and affects the hearing", but adds that it has only a very temporary effect on the body even when used in large amounts and that in his 7 years' experience of its use in the hat industry he has never heard of anyone being affected by products containing the usual small amounts, about 15 per cent.

Conjunctival irritation. In Krüger's investigation (1932), butyl acetate was a constituent of one of the hat lacquers with which severe conjunctival and corneal irritation were associated. The lacquer contained also butyl alcohol and spirit. It is stated by Flury and Zernik (1931) that concentrations of 34 to 50 mg. per l. (7,200 to 10,000 p.p.m.) give rise to marked eye and nose irritation in man, and Krüger points out that the volatility of butyl acetate is 25 per cent greater than that of amyl acetate, to which symptoms of eye irritation were very definitely attributed in her series of cases.

Respiratory irritation. In the case reported by Fühner and Pietrusky (1934) as being one of intoxication by butyl acetate and toluol, it is by no means clear that any of the symptoms complained of were due directly to butyl acetate. The nitrocellulose lacquer used was a mixture of 48 per cent of aromatic hydrocarbons (chiefly toluol) and 52 per cent of esters and alcohols, including ethyl and butyl acetates and ethyl and butyl alcohols. The subjects suffered from anaemia, cough, bronchial catarrh and nervous symptoms, and it was suggested that the latter might be a psychogenic reaction caused by over-sensitivity to the smell of the lacquer.

Headache, stupor and gastro-intestinal disturbance. These symptoms, complained of by workmen using lacquers containing butyl acetate, were described by Weber and Gueffroy (1932), but it is to be noted that the lacquers contained also benzol, toluol and xylol.

Urobilinuria. A greater frequency of urobilinuria (regarded as evidence of liver damage) in a group of workers using a wallpaper paint containing butyl acetate as compared with a control group, is mentioned by Burger and Stockmann (1932), but here again the paint contained a mixture of amyl and butyl alcohols, acetone and amyl acetate as well as butyl acetate.

11. *Secondary* Butyl Acetate

$$CH_3 \cdot CO \cdot OC_4H_9, \text{ i.e. } CH_3 \cdot CO \cdot O \cdot CH {<} {{CH_3}\atop{CH_2 \cdot CH_3}}$$

PROPERTIES

A colourless liquid with a fruity odour. The technical product contains isomers and homologues.

B.R. of the commercial product 107–114° C. Sp.Gr. 0·861. Contains a minimum of 85 per cent esters. Rate of evaporation considerably faster than that of *n*-butyl acetate (Metzinger, 1935). Clearly miscible with benzene, castor and linseed oils. Solubility in water 3 per cent; dissolves 1 per cent of water at 25° C.

Solvent for cellulose nitrate and many gums and resins, but not for cellulose acetate.

USES

sec.-Butyl acetate is used in the lacquer industry, but only to a limited extent, since its evaporation rate, being faster than that of *n*-butyl acetate, causes difficulty in making substitutions (Metzinger, 1935).

TOXIC EFFECTS

No animal experiments have been carried out and no ill-effects from its industrial use reported.

12. *iso*Butyl Acetate

$$CH_3 \cdot CO \cdot OC_4H_9, \text{ i.e. } CH_3 \cdot CO \cdot O \cdot CH_2 \cdot CH {<} {{CH_3}\atop{CH_3}}$$

PROPERTIES

A colourless liquid of mild pleasant odour. Occurs with other homologues to a considerable extent in the so-called "technical" amyl acetate and is seldom used in a state of purity.

Commercial quality. B.R. 110–118° C. Sp.Gr. 0·868–0·875. Fl.P. 18° C. Contains 95–100 per cent of esters. Miscible with castor and linseed oils and hydrocarbons and partly with water. Good solvent for cellulose nitrate and for some gums and resins; not for cellulose acetate, copals or copal ester.

USES

(1) In the lacquer trade (Hackett, 1932).
(2) In the preparation of flavouring essences (Vandevelde, 1906).

TOXIC EFFECTS

The only experiments are those of Vandevelde (1906) on its haemolytic action, and those of Führner (1921) on its action on the isolated frog's heart.

Its haemolytic "coefficient of toxicity" as estimated by its capacity for haemolysing sterile defibrinated blood, compared with that of ethyl alcohol taken as 100, is 4·34 according to Vandevelde. The corresponding value for ethyl acetate was 11·31. Judged by its action on the isolated frog's heart *iso*butyl acetate would also appear to be more toxic than ethyl acetate, the narcotic concentrations being given by Führner as follows: *iso*butyl acetate, 0·01 mol. per l.; ethyl acetate, 0·13 mol. per l.

No experiments on inhalation of the vapour have been carried out, and no toxic effects from its industrial use have been reported.

13. Amyl Acetate

$$CH_3 \cdot CO \cdot OC_5H_{11}$$

PROPERTIES

A colourless liquid of a very characteristic "peardrop" odour. It is prepared by the acetylation of amyl alcohol obtained either from fusel oil or by synthetic methods, being called "fusel-oil" or "synthetic" amyl acetate accordingly.

British Standard Specification, No. 552 (1934). Sp.Gr. 0·872–0·880. Esters, 95 per cent minimum. Residue, 0·01 per cent maximum. B.R. 120–145° C. 95 per cent minimum; above 135° C. 33 per cent minimum.

A.S.T.M. Specifications. Fusel oil amyl acetate. Sp.Gr. 0·860–0·865 at 20° C. Esters, 85–88 per cent. B.R. below 110° C. none; 120° C. 15 per cent maximum; 130° C. 50 per cent maximum; 140° C. 60 per cent minimum; above 150° C. none.

Synthetic amyl acetate. Sp.Gr. 0·860–0·870 at 20° C. Esters 85–88 per cent. B.R. below 126° C. none; 130° C. 5 per cent maximum; 135° C. 25 per cent maximum; 140° C. 75 per cent minimum; above 155° C. none.

Secondary amyl acetate. Also commercially available. B.R. 128°–134° C. Sp.Gr. 0·863. Dissolves 0·2 per cent of water at 25° C. This was the grade tested on animals by Patty, Yant and Schrenk (1936) (p. 301).

Amyl acetate is a good solvent for cellulose nitrate, celluloid, many resins, oils and fats. It does not dissolve cellulose acetate. The technical grade effects solution more rapidly and gives solutions of lower viscosity than the B.S.I. grade and the latter differs comparably from the pure grade.

USES

(1) In the lacquer industry, where it tends to replace butyl acetate.
(2) In artificial pearl manufacture, the beads are dipped into fish scale emulsified with amyl acetate and then dried on a rotating rack. The temperature

of the room must be above 65° F. (18° C.) for satisfactory lustre of the nacre.

(3) In the textile industry for the printing of fabrics with nitrocellulose varnish and dry-dyeing. Clayton and Clark (1931) state that amyl acetate enters into the composition of certain wetting-out agents, but these are not to be recommended for use in hot liquors or in standing baths.

(4) In the manufacture of photographic and cinematographic films.

(5) In the leather industry for preparing patent and buff leathers, and in special processes for cleaning and graining leather.

(6) In the boot and shoe industry for pastes used in manufacture and finishing.

(7) In artificial glass manufacture.

(8) In electrical industries in the manufacture of lamps, dry batteries and accumulators.

(9) In the preparation of fruit essences, etc.

(10) In the straw hat industry amyl acetate is one of the chief constituents both of the lacquer and of the stiffening solutions (Krüger, 1932). It was contained in six out of eleven lacquers analysed in Krügers' investigation.

(11) In furniture polishes. According to Zangger (1930) amyl acetate is much used for improving the odour of shoe polishes.

ESTIMATION OF VAPOUR IN AIR

A method of estimating the amount of amyl acetate vapour present in a sample of room air is described by Korenman (1932). The method, which is based on Komarowsky's reaction for amyl alcohol, the production of a rose-pink, red or violet-red coloration on the addition of furfural and concentrated sulphuric acid, is apparently identical for both amyl acetate and amyl alcohol, but the colour is less intense with amyl acetate, and in Korenman's modification is increased by heating for 3 minutes in a boiling waterbath. In this method the air to be tested is passed over drying agents, and drawn into an evacuated glass flask of known volume. By means of a pipette, 20 ml. of ethanol is introduced and diluted with an equal volume of water. The flask is tightly closed and shaken for 2 to 3 hours. The solution of amyl acetate vapour so obtained is filtered, if necessary, and 0·1 ml. of 1 per cent furfural solution is added to 1 ml. of the filtrate, and 1·5 ml. of concentrated sulphuric acid is poured carefully over the walls of the flask. The contents are well mixed while being cooled under a stream of water. Both the test flask and a flask containing a standard solution are placed for 3 minutes in a boiling waterbath to increase the intensity of the colour, and after complete cooling the colours are compared in a colorimeter.

TOXICITY

Amyl acetate when inhaled in high concentrations possesses the irritative and narcotic action common to almost all the esters. There is also some evidence that repeated long-continued inhalation of non-narcotic concentrations produces injury not only to the respiratory organs, but to the liver, kidneys and digestive organs (Flury and Wirth, 1934; Holstein, 1935; Burger and Stockmann, 1932). From their animal experiments, Flury and Wirth (1934) summarize their conclusions by stating that, although amyl acetate is more strongly narcotic when inhaled than methyl or ethyl acetate, it is better tolerated; pure amyl acetate, and still more the impure product, is more toxic than butyl acetate. Some experiments by Blina (1933), however, appear to indicate that amyl acetate is more toxic to animals than either methyl or ethyl acetate.

ESTERS

Maximum Permissible Concentration

On the basis of irritation of mucous membranes, Cook (1945) recommends 200 p.p.m. (1,000 mg. per c.m.) as the maximum allowable concentration for continued exposure.

TOXIC EFFECTS IN ANIMALS
Acute Poisoning

While the experiments of Flury and Wirth (1934) and of Patty, Yant and Schrenk (1936) have confirmed the earlier statements of Lehmann (1913), Koelsch (1912) and Grimm, Heffter and Joachimoglu (1914) that inhalations of amyl acetate, even in high concentrations, are practically never immediately lethal to animals, Flury and Wirth find it more strongly narcotic than did Lehmann, while Patty, Yant and Schrenk, found it dangerous to life in concentrations of 0·5 to 1·0 per cent by volume after several hours.

Lethal and Narcotic Concentrations

Lethal dose

By inhalation. In Lehmann's experiments inhalation of concentrations up to 37 mg. per l. (approx. 7,000 p.p.m.) did not cause death in rabbits, and Flury and Wirth (1934) were unable to kill mice with still higher concentrations. They state that on account of the low volatility of amyl acetate it is not possible to measure with certainty concentrations above 37 mg. per l., but for comparison with spraying processes they used concentrations probably amounting to 56 mg. per l. (10,500 p.p.m.). Even under these conditions the animals recovered the next day with no symptoms other than fatigue. Patty, Yant and Schrenk (1936), however, found concentrations of 1·0 per cent fatal to guinea-pigs after 300 minutes.

By subcutaneous injection. Flury and Wirth found amyl acetate more toxic to animals than butyl acetate, 3 g. of amyl acetate per kg. body weight proving lethal after 6 to 12 days, while 3 to 5 g. of butyl acetate was tolerated.

Narcotic dose

According to Flury and Wirth the concentrations necessary to produce narcosis were somewhat lower than those given by Lehmann and by Koelsch. Lehmann, for instance, found that only after 9 hours did inhalation of 28 mg. per l. (4,600 p.p.m.) produce deep narcosis in cats, while Koelsch produced complete loss of reflexes with 24 mg. per l. within ½ hour. Table 47 shows Flury and Wirth's results with increasing concentrations in mice.

TABLE 47
Effect of different concentrations of amyl acetate on mice

Concentration		Time before effect			After-effects
(mg./l.)	(p.p.m.)	Disturbance of equilibrium	Side position	Deep narcosis	
10	1,900	—	—	—	Recovery
20	3,800	3¼–4¾ hr.	3¾–5¼ hr.	4¼–6¼ hr.	Recovery
30	5,600	18–30 min.	40-90 min.	1 –2½ hr.	Recovery
37	7,000	4–20 min.	16-40 min.	1¼–1½ hr.	Recovery

These results agree closely with those of Patty, Yant and Schrenk (1936), who produced unconsciousness in guinea-pigs with concentrations of 5,000 p.p.m. after 300 to 540 minutes and with 10,000 p.p.m. in 20 minutes.

Symptoms

The immediate effects of narcotic concentrations of amyl acetate are not characterized by extreme irritation of mucous membranes. Lehmann stated that the irritative effects were slight; Flury and Wirth found them about the same, or slightly less, than with butyl acetate, producing lachrymation, and in some animals salivation at first. Only one animal showed severe conjunctivitis. Convulsions were observed in some of Flury and Wirth's animals; this evidence of central irritation was stronger than with the butyl ester.

The secondary effects of narcosis are apparently slight; fatigue and exhaustion are usually the only after-effects, and recovery from even deep narcosis is rapid. Koelsch's (1912) guinea-pigs which were exposed to concentrations of about 0·5 to 1 per cent for 10 hours a day showed nervous symptoms, ataxia and motor paralysis, which subsided on removal from exposure. After long periods, however, they lost appetite and weight and two died. Smyth and Smyth (1928) found that guinea-pigs tolerated inhalations of 5·3 mg. per l. (1,000 p.p.m.) for 2 to 3 hours without any ill-effects. In a few of Flury and Wirth's animals diarrhoea and albuminuria were observed.

The principal changes in the animals examined by Patty, Yant and Schrenk (1936) were congestion of the brain, lungs, liver and kidneys. None of Flury and Wirth's acutely intoxicated animals appear to have been examined post mortem. Koelsch's experiments were not strictly quantitative, but he appears to have used high concentrations with prolonged exposure which must be regarded as acutely toxic, and his two animals which died showed oedema and pneumonic infiltration of the lungs and fatty degeneration of the liver.

Action on the Vasomotor System

Amyl acetate at ordinary temperatures exerts no paralysing action on the vasomotor system of animals, but when it is slightly warmed, so that the inspired air mixture is at a temperature of 28 to 30° C., vasomotor paralysis occurs in less than 2 minutes (Dautrebande, Feced, Philippot, Charlier and Bodson, 1935).

The paralytic action is apparently exerted on the muscle rather than the nerve, since fragments of isolated organs cease to contract in Ringer's solution containing amyl acetate and cellulose varnish and since amyl acetate greatly reduces the hypertensive action of vasomotor excitants like lobeline, adrenaline, ephedrine and pituitrin (Dautrebande *et al.*, 1935). The toxic action at the level of smooth muscle is like that exerted by benzol, but becomes irreversible after 1 minute, compared with 3 minutes for benzol.

Chronic Poisoning

The results of repeated inhalations of non-narcotic concentrations in Flury and Wirth's experiments show a more definite chronic toxic action than did those of Grimm, Heffter and Joachimoglu (1914) and of the Pennsylvania Department of Labor and Industry in 1926. Grimm, Heffter and Joachimoglu observed only slight eye irritation and transient loss of weight; the animals recovered rapidly and those killed on the ninth and eleventh days showed no anatomical changes. The Pennsylvanian authors found no evidence of harm with concentrations of 500 and 1,000 p.p.m. for 36 3-hourly and 2- to 3-hourly periods of inhalation. Flury and Wirth, using concentrations of 10 mg. per l. (1,900 p.p.m.) 8 hours daily for 6 days, observed albuminuria, which they interpret as evidence of a toxic effect upon the kidneys. A feature of their observations was the rapidly acquired tolerance to the irritative action on mucous membranes.

Post-mortem Changes

Flury and Wirth found inflammation of the respiratory mucous membranes (trachea, bronchi and lungs) and an increased fat content of the liver.

TOXIC EFFECTS IN MAN

Dautrebande *et al.* (1935) have found that amyl acetate at ordinary temperatures exercises no paralysing action on the vasomotor system, but if it has been warmed so that the temperature of the inspired air mixture reaches 28 to 30° C., it induces vasomotor paralysis in less than 2 minutes, as judged by the vasomotor response to the occlusion and release of the two common carotids. He states that this fact is of special industrial importance in the motor industry, where a workman may be spray-painting in or near a room, the temperature of which has been raised for the purpose of rapid drying.

In an investigation by the Home Office in 1932 of a workshop used for making artificial pearls, the actual room temperature was 67° F. (19° C.), and there was a moderate odour of amyl acetate, but, apart from the common initial sensations from such an atmosphere, there were no evidences of ill-health among the workers.

Brightman (1930) mentions the danger of fire and explosion from the generation of static electricity during high-temperature processes of drying and impregnating fabrics using amyl acetate.

The dangers of amyl, methyl and ethyl acetates when used in sprays for colouring leather goods are emphasized by Holtzmann (1935). Hamilton (1925) quotes a Report of the German Factory Inspection describing the unpleasant effects of substituting a mixture of amyl acetate, amyl and propyl alcohols for ethyl alcohol in furniture polishes.

Certain symptoms, referable chiefly to the irritative effect of amyl acetate on mucous membranes, are fairly commonly reported, but the evidence for a specific metabolic toxic effect of amyl acetate is slight. Hamilton (1934) remarks "most workmen dislike its sweet sickish odour and insist that it causes 'dopiness', which is a convenient term for the sensations produced by a mild anaesthetic". Even in those reports which describe a condition characterized by nervous disturbance, eye irritation, oppression in the chest, cough, gastro-intestinal and vasomotor disturbance, quick pulse and tremor, occasionally with pharyngitis and bronchitis, it is uncertain whether this syndrome has been related to the effect of amyl acetate alone, the solvents used being usually a mixture of amyl acetate with other esters, acetone, amyl and butyl alcohols, etc.

A case with a fatal outcome was reported by Crecelius (1930) as "injury due to amyl acetate", when the lacquer used consisted of a solution of cellulose in acetone, amyl acetate and ether; similarly the workers showing eye lesions reported by Krüger (1932) were exposed to the vapours of lacquers of varying constitution. It is therefore difficult to prove with any certainty in the majority of the reported cases that the symptoms observed were specifically due to amyl acetate.

Hamilton (1925), during her investigation of aeroplane "doping" during the 1914–18 war, concluded that the workers might be expected to complain of smarting and running eyes, dryness of the throat, tightness in the chest and inclination to cough, and occasionally drowsiness, fatigue and vague nervousness. She did not consider nausea, headache, vomiting, gastro-intestinal disturbance or anaemia characteristic of amyl acetate intoxication, nor did she believe that any organic disturbance followed continued exposure. Similarly, Legge in 1924 stated (Home Office correspondence) that he had never found any constitutional disturbance traceable to amyl acetate, and Bridge in 1929, during an examination of a "cold lacquer" factory where there was a smell of amyl acetate but no evidence of ill-health among the workers, stated that he had examined a very large number of workers so employed but had never been able to find any injury to health except occasional complaints of soreness of throat and conjunctivae when commencing work. According to Patty, Yant and Schrenk (1936), the warning properties of amyl acetate are likely to take effect in concentrations well below those found harmful to guinea-pigs; 0·5 and 0·1 per cent vapour in air produces strong irritation to eyes and nasal passages and even 0·02 per cent is unpleasant.

Some authors, however, especially in Germany, have reported symptoms of gastro-intestinal disturbance (Baader, 1933; Holstein, 1935), while some observations of Burger and Stockmann (1932) appear to indicate that amyl acetate may produce irritation of the liver.

Acute Poisoning

The immediate effects of inhalation of amyl acetate vapour have been well described by Lehmann (1913) and Koelsch (1912) from personal experience. Koelsch gave no details of the amount inhaled, but reported that inhalation was followed by a sensation of heat, giddiness, slight drowsiness, a feeling of intoxication, fatigue, rapid respiration and increased pulse rate, and a cough which lasted for a month. Lehmann, after inhalation for half an hour of a concentration of 5 mg. per l. (950 p.p.m.) observed no change in the pulse rate, but headache, fatigue, oppression in the chest, and irritation of the conjunctivae and of the mucous membranes of the nose and throat with excessive secretion of mucus.

Similar symptoms were observed, though not frequently, by Koelsch in workers in the hat industry, but they were usually only experienced at the beginning of the work and passed off after a time without permanent injury. This was not so, however, in the one case in which a fatal outcome was attributed, or at any rate related, to amyl acetate intoxication. This case occurred also in the hat industry (Crecelius, 1930), in a man who, having glycosuria and pneumonia which were not definitely related to the effects of amyl acetate, eventually died of acute oedema of the glottis. Crecelius states that the man showed typical symptoms of amyl acetate poisoning, stupor, giddiness, pain in the chest, dyspnoea and cough, and during the course of his disease developed laryngitis leading to oedema of the glottis. He believed that the laryngitis was originally due to the irritation caused by the amyl acetate.

No cases of acute narcosis by amyl acetate have been recorded, with the possible exception of a man who was reported to the Home Office in 1932, as having had a fainting attack from painting with cellulose paint containing both amyl acetate and petroleum.

Chronic Poisoning

Symptoms

The chief symptoms observed in workers exposed over long periods to atmospheres containing amyl acetate are those of irritation of mucous membranes, but slight disturbances of general health have also been reported.

Nervous symptoms

Krüger (1932) noted headache as one of the symptoms occurring among hat workers; Heim de Balsac, Agasse-Lafont and Feil (1922) observed it as a transitory symptom in celluloid workers; and one girl employed in lacquering fountain pens was reported to the Home Office in 1931 as complaining of headache. Giddiness, drowsiness and palpitation were rarely complained of in Koelsch's (1912) investigation.

Gastro-intestinal disturbances

These seem to be comparatively rare, though Holstein (1935) states that they do occur, and Baader (1933) describes them as the chief manifestation in a group of men using a lacquer, of which amyl acetate was the chief constituent, for wagons and motor cars. The symptoms consisted of loss of appetite, a sensation of fullness, eructations, dyspepsia and loss of weight.

Urobilinuria

Urobilinuria was observed by Burger and Stockmann (1932) in a girl aged sixteen who had been working with amyl acetate for 3 months. The excess of urobilin, tested by Schlesinger's fluorescent reaction, was present in the morning urine, and was much reduced 4 to 8 weeks after removal from exposure. Burger and Stockmann found urobilinuria also in groups of workers using mixtures of alcohols and acetates, and interpret it as evidence of liver injury.

Irritation of mucous membranes

Respiratory. Cough, throat irritation, oppression in the chest, and occasionally dyspnoea, are the commonest initial effects of amyl acetate, but are generally transitory (Koelsch, 1912; Hamilton, 1934, etc.). Barrios and Devoto (1931) mention a permanent smell of amyl (and ethyl) acetate experienced by workers using lacquers containing these solvents, together with loss of perception of other odours.

Conjunctivitis. "Smarting and running eyes" were amongst the usual symptoms noted by Hamilton in aeroplane dope workers, and one case, which cleared up when precautions were taken, was reported to the Home Office in 1927 from a celluloid factory, but the only severe and intractable cases of eye irritation are those reported by Krüger in 1932. During an intensive season's work the workers complained of pain in the eyes, dimness of vision, especially in the mornings, and lachrymation. On examination they showed swelling of the lids, diffuse reddening of the conjunctivae and injection of the vessels. In a few cases the inflammation extended to the cornea. No improvement resulted from treatment until the end of the season, and a characteristic feature of these cases was that no tolerance seemed to be acquired, the symptoms returning on re-exposure.

Effect on the Blood

Anaemia. As stated above, Hamilton (1925) did not consider anaemia a characteristic symptom of amyl acetate intoxication. Baader (1933), however, records secondary anaemia as one of the findings in his group of "Zapon" lacquer workers.

Eosinophilia. A toxic eosinophilia, observed as a constant sign by Heim de Balsac and co-workers (1922) in women using a mixture of equal parts of amyl acetate and acetone, is regarded by them as an indubitable sign of intoxication, or at any rate a reaction of the organism to the solvents, but they do not distinguish between them in this particular group of cases.

14. *Secondary* Hexyl Acetate

$$CH_3 \cdot CO \cdot OC_6H_{13}, \text{ i.e. } CH_3 \cdot CO \cdot O \cdot \underset{\underset{}{|}}{CH} \cdot CH_2 \cdot CH {<}_{CH_3}^{CH_3}$$

with the middle CH bearing a CH_3 substituent.

PROPERTIES

A colourless mobile liquid, with characteristic odour more pleasant than those of the normal esters. B.R. 146°–156° C. Sp.Gr. 0·863. Water dissolves less than 0·1 per cent at 25° C. Content of esters, 85–88 per cent.

Solvent for cellulose nitrate and most gums, but not for cellulose acetate or shellac.

USES

In the lacquer industry, either as solvent of nitrocellulose or as a thinner (Park and Hopkins, 1930).

TOXIC EFFECTS

No animal experiments have been carried out and no toxic effects from its industrial use reported. Park and Hopkins (1930) include *secondary* hexyl acetate in the group of secondary acetates which they state do not seem to affect the lower part of the throat to the same extent as normal acetate and consequently produce less gagging effect.

15. *cyclo*Hexyl Acetate
(*Adronol acetate, Hexalin acetate*)

$$C_6H_{11}O \cdot CO \cdot CH_3, \text{ i.e. } CH_2{<}_{CH_2 \cdot CH_2}^{CH_2 \cdot CH_2}{>}CH \cdot O \cdot CO \cdot CH_3$$

PROPERTIES

A neutral, colourless, oily liquid, non-miscible with water, with a characteristic ester-like odour resembling amyl acetate.

B.R. of the technical product 170–176° C. Sp.Gr. 0·947–0·950 at 15° C. Fl.P. 57·5° C. Esters, 90 per cent minimum.

Evaporation time, 77 (compared with ethyl ether, 1).

A good all-round solvent for resins and gums, and one of the best solvents for collodion cotton.

USES

(1) In the lacquer industry, sometimes as a solvent in combination lacquers of collodion cotton and drying oils, since it prevents separation of the oil in the drying lacquer, but must be used sparingly as its characteristic odour persists for a long time.

(2) In the textile industry, according to Clayton and Clark (1931), it has largely replaced amyl acetate owing to its higher flash point.

(3) In the leather industry for improving the adhesion of leather varnishes.

(4) In the manufacture of raincoats as a constituent of the waterproofing composition (Clayton and Clark).

TOXICITY

*cyclo*Hexyl acetate generally produces effects similar to those of amyl acetate i.e. it is a local irritant and mild narcotic. As a narcotic for animals it is in fact about three times as potent as amyl acetate, but owing to the fact that it is about six times less volatile, it is less hazardous in actual use (Lehmann, 1913). No toxic effects from its industrial use have been recorded, though personal experiments by Lehmann showed it to be highly irritating.

TOXIC EFFECTS IN ANIMALS
Acute Poisoning

Lethal and Narcotic Concentrations

Lethal dose

By subcutaneous injection. For cats, 0·8 g. per kg. (Lehmann).

By inhalation. For cats, 10 mg. per l. (1,700 p.p.m.) after 1 day; 9 mg. per l. (1,550 p.p.m.) after 11 to 24 days. (28 mg. per l. of amyl acetate under the same conditions was not fatal.)

Narcotic dose

By inhalation. 10 mg. per l. (1,700 p.p.m.) after 10 hours, compared with 28 mg. per l. (5,250 p.p.m.) of amyl acetate.

Symptoms

Only slight irritation of mucous membranes follows exposure to low concentrations. Flury and Zernik (1931) state that white mice and guinea-pigs exposed for 1 to 2 hours to an atmosphere containing up to 20 per cent of *cyclo*hexyl acetate vapour showed only slight signs of irritation, while dogs tolerated 7 mg. per l. (1,200 p.p.m.) without any other symptoms. In cats 8 to 9 mg. per l. (1,400 to 1,550 p.p.m.) produced successively ataxia and slight narcosis but not deep narcosis.

Table 48 summarizes Lehmann's findings for cats.

TABLE 48

Effect of different concentrations of cyclohexyl acetate *on cats*

Concentration		Exposure	Effect
(mg./l.)	(p.p.m.)	(hr.)	
Up to 5	Up to 860	9	Only slight irritation
8	1,400	8½	Side position after 7 hr.; recovery
9	1,550	10½	Ataxia after 3–4½ hr.; side position after 8¾ hr.; no complete narcosis; death after 11–24 days
10	1,700	10	Ataxia after 5 hr.; side position and light narcosis after 6½ hr.; deep narcosis after 10 hr.; death after 1 day

Chronic Poisoning

Drowsiness and stupor, with no special after-effects, were the only symptoms observed by Lehmann in cats after exposure for 8½ hours daily for 5 days to concentrations of 9·5 mg. per l. (1,650 p.p.m.). Much higher concentrations than these, 30 mg. per l. (5,150 p.p.m.) for 8 to 9 hours daily for 30 days in dogs and cats, also produced no significant symptoms.

TOXIC EFFECTS IN MAN

A suffocating odour producing immediate cough were the only effects noted by Lehmann on inhalation.

Blood changes in the form of lymphocytosis, were found by Meyer (1931) to occur in men working with mixtures of *cyclo*hexyl acetate, methyl glycol acetate, butanol, etc., but as Meyer also observed lymphocytosis with the use of butanol alone it is not possible to attribute this change specifically to *cyclo*hexyl acetate.

16. Methyl*cyclo*hexyl Acetate
(*Sextate, Methylhexalin Acetate*)
$$CH_3 \cdot C_6H_{10}O \cdot CO \cdot CH_3$$

PROPERTIES

A mixture of three isomeric esters. A neutral liquid with solvent properties somewhat similar to those of *cyclo*hexyl acetate, but its evaporation rate is lower (Clayton and Clark, 1931), and it is slower in action.

B.R. of the technical product 172–192° C. Sp.Gr. 0·94–0·95. Fl.P. 65° C. (Flury and Zernik, 1931).

Good solvent for cellulose nitrate, some resins, raw rubber, waxes and bitumen.

USES

Chiefly in the lacquer industry in nitrocellulose lacquers for automobiles (Deschiens, 1928).

TOXICITY

No animal experiments have been carried out. An investigation by Browning (unpublished results) of a number of workers on a process, printing on plastic materials, in which "Sextate" was one of the constituents of the solvent mixture, showed the only symptoms to be slight drowsiness and irritation of the throat. Since other *cyclo*hexane derivatives were also in use, it was not possible to attribute these effects entirely to the "Sextate". Blood examinations of these workers showed a fairly high incidence of anaemia with a low colour index as the only abnormality.

17. Butoxyl
(*Methoxybutyl Acetate*)
$$CH_3 \cdot O \cdot CH(CH_3) \cdot CH_2 \cdot CH_2 \cdot O \cdot CO \cdot CH_3$$

PROPERTIES

A colourless liquid with a weak and somewhat pleasant odour. B.R. 167–171°. Sp.Gr. 0·956. Fl.P. 60° C. Volatility seventy-five times less than that of ether (Lehmann and Flury, 1943). Soluble in water to the extent of nearly

10 per cent but will only dissolve traces of water. Solvent for cellulose nitrate, cellulose ethers, ester gum, gutta percha, vinyl and other resins, and oils; not a solvent for cellulose acetate or rubber; has a high tolerance for diluents.

USES

Chiefly in the lacquer industry. It can be added to spray lacquers in amounts of 5 to 15 per cent without injuring their drying capacity. It gives a smooth surface with a high polish.

TOXICITY

Except for a moderate irritative effect on mucous membranes, butoxyl is practically innocuous (Lehmann and Flury, 1943).

TOXIC EFFECTS IN ANIMALS

Local Effect

According to Lehmann and Flury (1943), irritation of mucous membranes occurs with exposure to almost saturated vapour-air mixtures at room temperature.

Skin Application

No irritation is produced by the undiluted substance or by a 10 per cent lanoline salve, even after 24 hours of application (Lehmann and Flury, 1943).

Oral and Subcutaneous Administration

No injury to cats, rabbits or guinea-pigs followed either single doses of 0·1, 0·5 or 1·0 g. per kg. or 6 repeated doses of 0·1 g. per kg.

Inhalation

Only slight irritation of mucous membranes was caused by inhalation for 6 hours of almost saturated vapour-air mixtures.

TOXIC EFFECTS IN MAN

No toxic effects have been reported.

18. Benzyl Acetate

$CH_3 \cdot CO \cdot O \cdot CH_2 \cdot C_6H_5$

PROPERTIES

A liquid of powerful but pleasant odour, suggestive of jasmine flowers. Its volatility is very low (evaporation time, 393, compared with ethyl ether, 1).

B.P. 215° C. Sp.Gr. 1·060–1·062. Fl.P. 93° C. Solubility in water about 1 part in 4,000 parts.

Solvent for cellulose acetate and nitrate, some gums, resins and oils, and, in conjunction with alcohol, for glyceryl phthalate resins and shellac.

USES

(1) In the lacquer industry, chiefly for exerting a temporary softening effect on cellulose films; "Plastolin" is reputed to be benzyl acetate.

(2) In the perfumery industry (Herold, 1931).

TOXICITY

Although from experiments on animals benzyl acetate appears to have greater irritative and narcotic effects than ethyl, methyl, *n*-propyl, *n*-butyl and amyl acetates, according to Flury and Wirth (1934), these are of little practical importance unless the ester is used in a vaporized or nebulized form. On account of its very slight volatility a narcotic effect is rarely observed from air mixtures. Long-continued inhalation of air saturated with benzyl acetate was, however, found to be highly toxic to mice, though fairly well tolerated by cats, and apparently produced kidney injury, as evidenced by albuminuria and microscopical appearances.

TOXIC EFFECTS IN ANIMALS
Acute Poisoning

Lethal and Narcotic Concentrations
  Lethal dose

In white mice, a condition similar to narcosis was produced after a period longer than 10 hours, and proved fatal to many of the animals, some dying during the experiment and some after apparent recovery. The concentrations producing death during the experiment, or at the latest 10 hours after the end of it, were from 1·3 mg. per l. (210 p.p.m.) for constant air mixtures, where the animal was moved to a fresh test chamber containing the same concentration every 8 hours in order to avoid the effects of accumulation of carbon dioxide, to 1·5 mg. per l. (250 p.p.m.) in "moving" air mixtures.

The lethal dose was 3 g. per kg. body weight for rabbits and guinea-pigs subcutaneously, and 4 to 5 g. per kg. orally. The animals showed disturbances of equilibrium, convulsions, rapid respiration and paralysis of the hind limbs.

Narcotic dose

For mice (constant air mixture) 0·5 mg. per l. (80 p.p.m.). after 47 to 90 hours; 1·3 mg. per l. (210 p.p.m.) after 9 to 17 hours.

Owing to its slight volatility the narcotic action of benzyl acetate was not exerted on cats even after 10 hours.

Symptoms

The effects were both irritative and narcotic, though the irritative effects were most prominent when the benzyl acetate was vaporized by means of a compressed air apparatus. When this was done there was much salivation, lachrymation and increase of nasal secretion in cats, rabbits and guinea-pigs, but no other disturbances. Flury and Wirth (1934) point out that with this procedure a good deal of the vapour condensed on the glass walls of the test chamber, so that only part could have actually exerted any toxic effect.

Subcutaneous and oral administration emphasized the toxic action of benzyl acetate on the central nervous system.

It may be noted here that, according to Gruber (1924), injection of benzyl acetate into animals produces a great acceleration in the excretion of urine.

Chronic Poisoning

In cats subjected to inhalation of air-ester mixtures (practically saturated, corresponding to about 1·1 mg. per l. or 180 p.p.m.) for 8 to 9½ hours for 7 days, the immediate effects were those of irritation of mucous membranes, with rapidly acquired tolerance. Towards the end of the experiment the animals showed fatigue, drowsiness, tremor, weak and rapid respiration, sometimes urination and defaecation. After-effects took the form of loss of appetite, fatigue, drowsiness, slight loss of weight and slight stupor; in some cases death took place after 8 days. The urine contained a trace of albumin. Slight tracheitis and bronchitis, and hyperaemia of the kidneys were found post mortem.

Blood changes similar to those observed with the other esters investigated by Flury and Wirth were not found after inhalation, but with sub-lethal subcutaneous injections a slight leucocytosis occurred.

TOXIC EFFECTS IN MAN

No injurious effects from the industrial use of benzyl acetate have been recorded, but in describing the report of Müller (1932), (incorporated in Flury and Wirth's results) on the irritative effects of vaporized benzyl acetate, Zernik (1933) adds a warning note on spray processes. Zangger (1930), without making special reference to benzyl acetate, describes the additional dangers attending spray processes in which the solvent is inhaled in the form of pure vapour or fine drops, rather than as a mixture with air.

19. n-Butyl Propionate

$C_2H_5 \cdot CO \cdot OC_4H_9$, i.e. $CH_3 \cdot CH_2 \cdot CO \cdot O \cdot CH_2 \cdot CH_2 \cdot CH_2 \cdot CH_3$

PROPERTIES

A colourless liquid with an apple-like odour less pronounced than that of amyl acetate. Manufactured in U.S.A. from the residue, chiefly calcium propionate, from the production of acetone from seaweed.

The pure product has B.P. 146° C. Sp. Gr. 0·883.

A.S.T.M. Specification, D 320–33. Sp.Gr. 0·872–0·878 at 20° C. Esters, 90–93 per cent. B.R. up to 115° C. none; 130° C. 20 per cent maximum; 138° C. 60 per cent maximum; above 168° C. none. Residue, 0·005 per cent maximum.

Good solvent for cellulose nitrate and some gums and resins, but not for cellulose acetate or for hard copals. Miscible with oils and hydrocarbons but not with water. Dissolves 0·576 per cent of water at 22·5° C. (Durrans, 1950).

USES

In the lacquer industry as a substitute for amyl acetate when it is to be applied under humid conditions. Specially used in mixtures for white brushing nitrocellulose lacquer.

TOXIC EFFECTS

No animal experiments have been carried out and no toxic effects from its use in industry recorded.

20. Amyl Propionate

$CH_3 \cdot CH_2 \cdot CO \cdot OC_5H_{11}$

PROPERTIES

A colourless liquid with an odour milder and more like apples than amyl acetate, it evaporates more slowly than amyl acetate; time of evaporation, 53 minutes compared with amyl acetate, 50.

The composition of technical amyl propionate varies considerably, but a high grade product has a B.R. 140–170° C. Sp.Gr. 0·807–0·873. Fl.P. 40° C. Content of esters 85–100 per cent.

Solvent for cellulose nitrate, gums and resins and all fats, but not for cellulose acetate, hard copals or shellac.

USES

In the lacquer industry, for brushing lacquers where a slower evaporation than with amyl acetate is required. It gives a more brilliant gloss to varnish, and has a more pronounced anti-blush action; from this point of view it is the best of all the esters (Tersand, 1935). It is also used in metallic lacquers since it gives no deposit of verdigris with metallic pigments.

TOXIC EFFECTS

No animal experiments have been carried out and no toxic effects from its use in industry recorded.

21. *n*-Butyl Butyrate
(said to be "*Butol*" (Durrans, 1931))

$C_3H_7 \cdot CO \cdot OC_4H_9$

PROPERTIES

A colourless liquid of strong apple-like odour. A mixture of the *n*-butyl esters of *n*- and *iso*butyric acids.

B.R. 160°–165° C. Sp.Gr. 0·870–0·880. Dissolves 0·497 per cent water at 25° C.

Solvent for cellulose nitrate and for some gums and resins, but not for cellulose acetate or hard copals.

USES

In the lacquer industry, where slower evaporation than with amyl acetate is required.

TOXIC EFFECTS

No animal experiments have been carried out and no toxic effects from its industrial use recorded.

22. Methyl Benzoate
(*Oil of Niobé*)

$C_6H_5 \cdot CO \cdot OCH_3$

PROPERTIES

A liquid of less pronounced odour than the ethyl ester.

B.R. of the technical product 198–200° C. B.P. of pure ester 198·1° C. Sp.Gr. 1·093–1·094.

Solvent for cellulose acetate and nitrate, ester gum, coumarone and rubber. Miscible with oil but not with water.

USES

In the lacquer industry.

ESTERS

TOXIC EFFECTS

No animal experiments have been carried out and no ill-effects from its use recorded. Durrans states that it is non-toxic.

23. Ethyl Benzoate

$$C_6H_5 \cdot CO \cdot OC_2H_5$$

PROPERTIES

A liquid of pleasant aromatic odour, which boils for the most part without decomposition.

B.R. of the technical product 212–215° C. Sp.Gr. 1·051–1·053.

The technical product is usually of a high degree of purity, with an ester content of 98–100 per cent.

Solvent for cellulose acetate and nitrate, ethyl cellulose, many gums and resins, but not for hard copals except on prolonged boiling. Miscible with castor and linseed oils and hydrocarbons, but not with water.

USES

Ethyl benzoate is chiefly used in the lacquer industry for imparting good brushing properties to cellulose lacquers and high gloss to finishing films.

TOXIC EFFECTS

No animal experiments have been carried out and no toxic effects reported from its use in industry. Durrans states that it is non-toxic.

24. Ethyl Lactate

(*Actylol, Estisol, Eusolvan, Solactol*)

$$C_2H_4(OH) \cdot CO \cdot OC_2H_5, \text{ i.e. } CH_3 \cdot CH(OH) \cdot CO \cdot O \cdot CH_2 \cdot CH_3$$

PROPERTIES

A colourless liquid. When pure it is nearly odourless, but technical products often have a weak odour like that of ethyl hydroxybutyrate. Evaporation time, 80 (compared with ethyl ether, 1).

British Standard Specification, No. 663 (1936). Sp.Gr. 1·032–1·040. Esters, 95 per cent minimum. B.R. 135–160° C. 95 per cent minimum. Residue, 0·01 per cent maximum.

A.S.T.M. Specification, D 321–33. Sp.Gr. 1·020–1·036 at 20° C. Esters, 96–100 per cent. B.R. up to 102° C. none; 139° C. 10 per cent maximum; 155° C. 90 per cent minimum; above 173° C. none.

Its solvent power for cellulose nitrate and acetate is very high, but its action on cellulose nitrate is slower than that of amyl and butyl acetates. It is a solvent also for many gums, resins, basic dyes, etc., and will dissolve hard copals on boiling. Miscible with linseed oil, castor oil, water and aromatic hydrocarbons, but not with petroleum hydrocarbons.

314   TOXICITY OF INDUSTRIAL ORGANIC SOLVENTS

USES

(1) In the lacquer industry; owing to its slow rate of evaporation ethyl lactate is good for brushing lacquers, and for those with a high gloss, but tends to slow drying.

(2) In the manufacture of safety glass.

TOXIC EFFECTS

Although no actual animal experiments appear to have been recorded, Lewin (1929) states that ethyl lactate is narcotic to both cold- and warm-blooded animals including man, and that large doses are lethal to animals causing disturbance of respiration and paralysis of the respiratory centre. In actual industrial use, however, ethyl lactate appears to be considered relatively harmless. Zangger (1930) places it in the group of relatively harmless solvents, and Yarsley (1933) also states that all lactic acid esters used in lacquers and finishes are perfectly harmless.

25. Butyl Lactate

$C_2H_4(OH) \cdot CO \cdot OC_4H_9$, i.e. $CH_3 \cdot CH(OH) \cdot CO \cdot O \cdot CH_2 \cdot CH_2 \cdot CH_2 \cdot CH_3$

PROPERTIES

A colourless liquid when pure; the technical product is often slightly brown. Practically devoid of odour.

Sp.Gr. of the technical product, 0·984–0·988. Esters, 95 per cent minimum. B.R. 185–195° C. Evaporation time, 443 (compared with ethyl ether, 1). Solvent for nitrocellulose and for fats, oils, waxes and resins.

USES

Chiefly in the lacquer industry, especially in nitrocellulose brushing lacquers, with diluents of petroleum hydrocarbon mixtures having a greater evaporation rate than butyl lactate (Bogin, 1933).

TOXIC EFFECTS

No animal experiments have been carried out and no toxic effects recorded from its industrial use.

26. Amyl Lactate

$CH_3 \cdot CH(OH) \cdot CO \cdot OC_5H_{11}$

PROPERTIES

A colourless liquid with a faint brandy-like odour.
B.P. about 210° C. Sp.Gr. 0·968–0·972.

Solvent for cellulose nitrate, some gums and resins, but not for cellulose acetate or hard copals. Miscible with castor and linseed oils and hydrocarbons, but not with water.

USES

In the lacquer industry, chiefly in nitrocellulose brushing lacquers (Bogin, 1933) when it is desirable to prolong the drying period.

TOXIC EFFECTS

No animal experiments have been carried out and no toxic effects from its use in industry have been recorded.

27. Ethyl Hydroxy*iso*butyrate
(*Ethyl oxybutyrate*)

$$C_3H_6(OH)\cdot CO\cdot OC_2H_5, \text{ i.e. } \overset{CH_3}{\underset{CH_3}{>}}C(OH)\cdot CO\cdot O\cdot CH_2\cdot CH_3$$

PROPERTIES.

A colourless liquid with a mild pleasant odour.

B.R. of the commercial product 142–146° C. Esters, 96–100 per cent. Sp.Gr. 0·978–0·986.

Solvent for both cellulose acetate and nitrate, but slow in action. Solutions of the nitrate in it are about twice as viscous as those of the corresponding concentration in ethyl lactate but less viscous than those in amyl nitrate.

USES

Ethyl oxybutyrate is chiefly used in the lacquer industry. It is mentioned by Hamilton *et al.* (1929) as one of the new solvents for lacquer in the United States. Although it is not widely used, according to Durrans (1937), it tends to be preferred to ethyl lactate, as being more homogeneous and of somewhat higher volatility.

TOXIC EFFECTS

No animal experiments have been carried out, but Vandevelde in 1906 examined its haemolytic action and determined the co-efficient of toxicity as compared with that of ethyl alcohol (100) by this method. So judged it appears to be slightly more "toxic" than *iso*butyl acetate and considerably less so than ethyl acetate, the actual figures being ethyl acetate 11·31, ethyl oxybutyrate 4·85, *iso*butyl acetate 4·34.

No actual cases of injury from its industrial use have been reported, but Hamilton *et al.* (1929) state that although it is advertised as non-toxic, masterpainters say that men using it complain of dryness in the throat, a feeling of tightness in the chest, and "dopiness".

28. Diethyl Carbonate
(*Diatol*)

$(C_2H_5)_2CO_3$, i.e. $(CH_3\cdot CH_2O)_2CO$

PROPERTIES

A colourless liquid with a weak odour resembling that of ethyl oxybutyrate.

B.R. of the technical product 120°–130° C. (B.P. 126°). Sp.Gr. 0·976–0·988. Fl.P. 48° C.

Miscible with castor oil, aromatic hydrocarbons and most organic solvents and with some petroleum hydrocarbons, but not with water. Soluble in water to the extent of about 10 per cent.

Not a good solvent for cellulose acetate or nitrate but addition of an alcohol or ester increases its solvent power for the latter. Its solutions of cellulose nitrate are more viscous than solutions of corresponding concentrations in butyl acetate and have a poor tolerance for hydrocarbons; this is increased by the addition of a small proportion of ethyl lactate. Solvent for some gums, especially with the addition of 5 to 10 per cent of anhydrous alcohol. "Diatol" is a commercial material containing 90 per cent of diethyl carbonate.

USES

In the lacquer industry; brushing lacquers containing diethyl carbonate are more or less devoid of any objectionable odour (Clayton and Clark, 1931). "Diatol" is specially used in lacquers for brass and copper (Sanderson, 1933).

TOXIC EFFECTS

No animal experiments have been carried out, and no toxic effects from its industrial use recorded, but Durrans states that it has a faintly lachrymatory after-effect similar to that of ethyl chloroacetate.

29. Dialkyl Carbonates

R_2CO_3 (R = Alkyl)

PROPERTIES

	Sp. Gr.	B.P.
Methyl	1·076	90·5° C.
n-Propyl	0·946	168·5
isoPropyl	0·921	147·2
n-Butyl	0·928	207·5
isoButyl	0·919	189·8
isoAmyl	0·911	233

USES

In the lacquer industry.

TOXIC EFFECTS

No animal experiments have been carried out and no toxic effects from the industrial use of dialkyl carbonates have been reported.

30. Diethyl Oxalate

$CO \cdot OC_2H_5$
$CO \cdot OC_2H_5$

PROPERTIES

Pure ester. B.P. 185° C. Sp.Gr. 1·084–1·087. Fl.P. 75° C. Dissolves 0·5 per cent water at 25°C. Good solvent for cellulose acetate and nitrate, ester gum and benzyl abietate; since it is unstable in the presence of moisture it is not suitable for a lacquer solvent (Durrans, 1935). In France, however, it has been used for spray-painting in a mixture of 5 parts oxalate and 20 parts syrup of nitrocellulose and butyl acetate (Desoille, Truffert, Assouly and Bidegarray, 1947).

TOXICITY

Only one set of workers, Desoille *et al.* (1947) attribute a toxic action to diethyl oxalate and describe it as a "benzene-like" effect of mild degree.

TOXIC EFFECTS IN ANIMALS

Desoille *et al.* (1947) state that guinea-pigs exposed to the vapour of ethyl oxalate showed a slight anaemia.

TOXIC EFFECTS IN MAN

In a chemical works where the concentration of ethyl oxalate was found to be 0·76 mg. per l. without ventilation, and 0·46 mg. per l. with it, a number of workmen showed both symptoms and blood changes similar to those of benzene poisoning. The symptoms were fatigue, headache, nausea, dyspepsia, pallor and epistaxis. The blood picture showed anaemia, with red cells down to 3,200,000 per c.mm., a slight fall in total white cells, neutropenia and a mild eosinophilia (7 to 8 per cent). These changes developed slowly, sometimes after 1 month of exposure, but more usually after 2 or 3 months; when exposure ceased they disappeared (Desoille *et al.*, 1947).

BIBLIOGRAPHY

ALINARI, E. (1930). Miscele entettiche fra alcooli ed eteri acetici. *Ann. Chim. appl., Roma*, **20**, 159.
ALTHOFF (1931). Tod durch Einatmung von Essigäther. *Z. MedBeamt.*, **15**, 246. (Abstr. in *Zbl. GewHyg.*, 1932, **9**, 67).
BAADER, E. W. (1933). Gewerbemedizinische Erfahrungen. *Verh. dtsch. Ges. inn. Med.*, **45**, 318.
BANIK (1929). Die Gesundheitsgefahren beim Arbeiten mit Zaponlack und ihre Verhütung. *Zbl. GewHyg.*, **6**, 197.
BARRIOS, D. L. and DEVOTO, J. S. (1931). Commentarios sobre la acion toxica del "Barniz", "Tini", y "Pomada". *Sem. méd., B. Aires*, **1**, 252.
BEINTKER (1928). Überempfindlichkeit gegen Aethylazetat. *Dtsch. med. Wschr.*, **54**, 528.
BLINA, L. V. (1933). Experimental investigation of toxic action of acetic acid esters. Amyl acetate. *Med. d. Lavoro*, **24**, 166. (Abstr. in *J. industr. Hyg.*, **15**, 125).
BOGIN, C. (1933). Use of lactates in brushing lacquers. U.S. Patent 1,927,539. Sept. 19. *Chem. Abstr.*, **27**, 5995.
BRIGHTMAN, R. (1930). Industrial safety: chemical risks in the rubber trade. *Industr. Chem. chem. Mfr.*, **6**, 252.
BURGER, G. E. C. and STOCKMANN, B. H. (1932). Über Urobilinurie als Folge der Einatmung von organischen Lösungsmitteln in geringer Konzentration. *Zbl. GewHyg.*, **19**, 29.
CAZENEUVE, P., MOREL, A. and DE LEEUW, H. (1932). L'hygiène et l'industrie de soie artificielle. *Chim. et Industr.*, **28**, 473.
CLAYTON, E. and CLARK, C. O. (1931). Modern organic solvents: Parts I, II. *J. Soc. Dy. Col., Bradford*, **47**, 183, 247.
COOK, W. A. (1945). Maximum allowable concentrations of industrial atmospheric contaminants. *Industr. Med.*, **14**, 936.
CRECELIUS, A. (1930). Schädigung durch Amylacetat. *Klin. Wschr.*, **9**, 452.
DAUTREBANDE, L. (1935). L'action sur le système vasomoteur de certains solvents volatils utilisés dans l'industrie (benzol, éther de pétrole, acétate d'amyle, vernis cellulosique. *Médecine*, **16**, 202.
DAUTREBANDE, L., FECED, J. A., PHILIPPOT, E., CHARLIER, R. and BODSON, M-Th. (1935). Paralysie du système vasomoteur par les vapeurs d'acétate d'amyle et de vernis dit cellulosique. *C. R. Soc. Biol., Paris*, **119**, 314.
DAVIDSON, J. D. and REID, E. W. (1928). Thinners for nitrocellulose lacquers. *Industr. Engng. Chem.*, **20**, 199.

DESCHIENS, M. (1928). Les vernis et peintures cellulosiques pour automobiles. 7me Congrès, Paris. *Chim. et Industr.* Special No., p. 673.
DESCHIENS, M. (1930). Mélanges solvants et vernis cellulosiques. 10me Congrès, Paris. *Chim. et Industr.* Special No. p. 692.
DESOILLE, H., TRUFFERT, L., ASSOULY and BIDEGARRAY (1947). À propos de quelques cas d'intoxication par l'oxalate d'éthyle. *Arch. Mal. prof.*, **8**, 265.
DUQUENOIS, P. and REVEL, P. (1934). Intoxications professionnels par vapeurs de quelques esters employés comme solvants. *J. Pharm. Chim., Paris*, 8me sér., **19**, 590.
DURRANS, T. H. (1931). Solvents. 2nd ed. Chapman and Hall, London.
DURRANS, T. H. (1935). Solvents and plasticisers. *J. Oil Col. Chem. Ass.*, **18**, 340.
EMRICH (1932). Dissertation, Würzburg. (Cited by Flury and Wirth, 1934, in *Arch. Gewerbepath. Gewerbehyg.*, **5**, 1.)
ENGELHARDT, W. E. (1933). Überempfindlichkeitserkrankungen der Haut durch Alkoholesterverbindungen der aliphatischen Alkoholreihe im Lackierberuf. *Arch. Derm. Syph.*, **169**, 236.
ENNA, F. G. A. (1930). Processes of patent leather manufacture. *Leather World*, **22**, 412.
FAUCETT, P. H. (1935). Spirit proof finishes. *Paint Oil Chem. Rev.*, **97**, No. 5, 5.
FAUCETT, P. H. (1935). From shellac to oil lacquer for wood sealers. *Paint Oil Chem. Rev.*, **97**, No. 14, 12.
FLURY, F. and WIRTH, W. (1934). Zur Toxikologie der Lösungsmittel (Verschiedene Ester, Aceton, Methylalkohol). *Arch. Gewerbepath. Gewerbehyg.*, **5**, 1.
FLURY, F. and ZERNIK, F. (1931). Schädliche Gase, Dämpfe, Nebel, Rauch- und Staubarten. Springer, Berlin.
FORDYCE, C. R. (1935). Cellulose acetate lacquers. *Paint Oil Chem. Rev.*, **97**, No. 9, 36.
FÜHNER, H. (1921). Die Wirkungsstärke der Narkotica. *Biochem. Z.*, **120**, 143.
FÜHNER, H. and PIETRUSKY, F. (1934). Butylazetat-Toluol-Vergiftung, chronische, berufliche Spätfolgen. *Samml. Vergiftungsf.*, **5**, 1, B40.
GARDNER, H. A. (1925). Physiological effects of vapors from a few solvents used in paints, varnishes and lacquers. *Paint Manuf. Ass. U.S., Proc. Scientific Sec. Educ. Bur. Circ.* No. 250.
GRIMM, V., HEFFTER, A. and JOACHIMOGLU, G. (1914). Gewerbliche Vergiftungen in Flugzeugfabriken. *Vjschr. gerichtl. Med.*, **48**, Suppl. 2, 161.
GRUBER, C. M. (1924). Effect of benzyl alcohol and its esters, benzyl benzoate and benzyl acetate, upon kidney functions. *J. Pharmacol.*, **23**, Proc., 149.
GRUBER, C. M. (1924). Pharmacology of benzyl alcohol and its esters. *J. Lab. clin. Med.*, **10**, 284.
HACKETT, J. D. (1932). Lacquers and their hazards. *Industr. Hyg. Bull.*, **11**, 223.
HAMILTON, A. (1925). Industrial poisons in the United States. Macmillan, New York.
HAMILTON, A. (1934). Industrial toxicology. Harper, New York.
HAMILTON, A., BRICKER, E. B. and SMYTH, H. F. (1929). Volatile solvents used in industry. *Amer. J. publ. Hlth.*, **19**, 523.
HEIM DE BALSAC, F., AGASSE-LAFONT, E. and FEIL, A. (1922). Enquête sur les manifestations morbides présentées par les ouvriers peintres en voitures. *Progr. méd., Paris*, **49**, 157.
HEIM DE BALSAC, F., AGASSE-LAFONT, E. and FEIL, A. (1922). Morbid disturbances due to handling of celluloid. *J. Amer. med. Ass.*, **79**, 146.
HEIM DE BALSAC, F., AGASSE-LAFONT, E. and FEIL, A. (1922). Manifestations morbides chez les ouvriers maniant le celluloid et ses solvants. *Paris méd.*, **i**, 477.
HEROLD, I. (1931). Benzylacetat und Homologe. *Dtsch. ParfümZtg.*, **17**, 362.
HOLDEN, H. C. and DOOLITTLE, A. K. (1935). Solvents. *Industr. Engng. Chem.*, **27**, 525.
HOLSTEIN, E. (1935). Die Gesundheitsgefahren des Malerberufes. III. Binde-Lösungsmittel und Lacke. *Med. Welt*, **9**, 302.
HOLTZMANN, F. (1935). Die Lederindustrie und ihre gesundheitlichen Gefahren. *Med. Welt*, **9**, 671.
I. G. FARBENINDUSTRIE AKTIENGESELLSCHAFT (1930). Lösungsmittel und Weichmachungsmittel. Frankfurt-am-Main.
IKEBE, S. (1934). Experimental study on surface anesthesia of outer ear passage in guinea pigs. *Folia pharm. japon.*, **19**, 62. (See also *Chem. Abstr.*, **29**, 2599).
KOELSCH, F. (1912). Gesundheitsschädigungen durch Amylacetat. *Concordia* (Berlin), **19**, 246. (Cited by Krüger, E., 1932, in *Arch. Gewerbepath. Gewerbehyg.*, **3**, 798.)
KORENMAN, I. M. (1932). Kolorimetrische Bestimmung von Amylalkohol- und Amylacetatdämpfen in der Luft. *Arch. Hyg., Berl.*, **109**, 108.
KRÜGER, E. (1932). Augenerkrankungen bei Verwendung von Nitrolacken in der Strohhutindustrie. *Arch. Gewerbepath. Gewerbehyg.*, **3**, 798.
LEHMANN, K. B. (1913). Experimentelle Studien über den Einfluss technisch und hygienisch wichtiger Gase und Dämpfe auf den Organismus. Amylazetat und Cyclohexanolazetat. *Arch. Hyg., Berl.*, **78**, 260.
LEHMANN, K. B. and FLURY, F. (1943). Toxicology and hygiene of industrial solvents. (Tr. by E. King and H. F. Smyth, jr.) Williams and Wilkins, Baltimore.
LEWIN, L. (1929). Gifte und Vergiftungen. 4th Aufl. Springer, Berlin.
LLOYD, J. V., OSTWALD, W. and HALLER, W. (1930). A study in pharmacy. *J. Amer. pharm. Ass.*, **19**, 1076.

Lorenz (1899). Über die physiologische Wirkung einiger Ester der Fettreihe. Dissertation, Würzburg.
Lund, A. (1944). Et Til faeld af Tobsisk amblyopia efter inhalation ef methylacetat. *Ugeskr. Laeg.*, **106**, 308.
Metzinger, E. F. (1935). Lacquer solvents. Recent introduction of secondary solvents and ketones, with their comparative values and substitutions for normal solvents. *Paint Oil Chem. Rev.*, **97**, No. 9, 42.
Meyer, S. (1931). Über Blutveränderungen bei gewerblichen Schädigungen. *Arch. Gewerbepath. Gewerbehyg.*, **2**, 516, 553.
Müller, W. (1932). Benzylacetat. Beitrag zur Pharmakologie und Toxikologie der aromatischen Ester. Dissertation, Würzburg. (Cited by Zernik, 1933, in *Ergebn. Hyg. Bakt.*, **14**, 139).
Nelson, K. W., Ege, J. F., Ross, M., Woodman, L. E. and Silverman, L. (1943). Sensory response to certain industrial solvent vapors. *J. industr. Hyg.*, **25**, 282.
Park, J. G. and Hopkins, M. B. (1930). Secondary esters and their use in lacquers. *Industr. Engng. Chem.*, **22**, 826.
Patty, F. A., Yant, W. P. and Schrenk, H. H. (1936). Acute response of guinea-pigs to vapors of some new commercial organic compounds. XI. Secondary amyl acetate. *Publ. Hlth. Rep., Wash.*, **51**, 811.
Pennsylvania Department of Labor and Industry (1926). Spray painting in Pennsylvania. *Dept. Lab. Industr., Spec. Bull. No.* **16**.
Sanderson, J. McE. (1933). Solvents used in surface coatings. Group V: Esters, part I. *Paint Oil Chem. Rev.*, **95**, No. 2, 10.
Sanderson, J. McE. (1933). Solvents used in surface coatings. Group V: Esters, part II. *Paint Oil Chem. Rev.*, **95**, No. 17, 8.
Sayers, R. T., Schrenk, H. H. and Patty, F. A. (1936). Acute response of guinea-pigs to vapors of some new commercial organic compounds. XII. Normal butyl acetate. *Publ. Hlth. Rep., Wash.*, **51**, 1229.
Schrenk, H. H., Yant, W. P., Chornyak, J. and Patty, F. A. (1936). Acute response of guinea-pigs to vapours of methyl formate. *Publ. Hlth. Rep., Wash.*, **51**, 1329.
Silverman, L., Schulte, H. F. and First, W. W. (1946). Further studies on sensory response to certain industrial solvent vapors. *J. industr. Hyg.*, **28**, 262.
Smyth, H. F. and Smyth, H. F., jr. (1928). Inhalation experiments with certain lacquer solvents. *J. industr. Hyg.*, **10**, 261.
Smyth, H. F. and Smyth, H. F., jr. (1928). Spray painting hazards as determined by the Pennsylvania and the National Safety Council Surveys. *J. industr. Hyg.*, **10**, 163.
Tersand, R. (1935). Les propionates et l'oxide de mésityle, solvants de la nitrocellulose. *Rev. Prod. chim.*, **38**, 97.
Vandevelde, A. J. J. (1906). Über die Anwendung von biologischen Methoden zur Analyse von Nahrungsstoffen. *Biochem. Z.*, **1**, 5.
Weber, S. (1902). Über die Giftigkeit des Schwefelsäure-dimethylesters; Dimethylsulfates und einiger verwandter Ester der Fettreihe. *Arch. exp. Path. Pharmak.*, **47**, 113.
Weber, H. H. and Gueffroy, W. (1932). Über einige Beiz-, Lackier- und Poliermittel, ihre Zusammensetzung und physiologische Wirkung. *Schr. GesGeb. GewHyg.*, **40**, 38.
Weingand, R. (1931). Neuer Zelluloseesterbronzelack. *Farbe u. Lack.*, p. 28.
Wright, Samson (1935). Mode of action of certain drugs which stimulate respiration. *J. Pharmacol.*, **54**, 1.
Yarsley, V. E. (1933). Health hazards in the lacquer and finishing industries. *Synth. appl. Finish.*, **3**, 80.
Zangger, H. (1930). Über die modernen organischen Lösungsmittel. *Arch. Gewerbepath. Gewerbehyg.*, **1**, 77.
Zangger, H. (1930). Weitere Mitteilungen über Vergiftungen durch flüchtige Gifte und deren Beziehung zu gewerblichen Vergiftungen. *Schweiz. med. Wschr.*, **60**, 1193.
Zernik, F. (1933). Neuere Erkenntnisse auf dem Gebiete der schädlichen Gase und Dämpfe. *Ergebn. Hyg. Bakt.*, **14**, 139, 220.

CHAPTER VI

KETONES

1. Acetone

(*Propan-2-one, Dimethyl Ketone*)

CH$_3$·CO·CH$_3$

PROPERTIES

A colourless liquid with a peculiar pungent odour, very volatile and inflammable. Manufactured from acetic acid by many different routes with variations of catalysts and conditions; also made by fermentation of starch by bacteria, yielding *n*-butyl alcohol and acetone.

Soluble in water, miscible with alcohol and ether.

British Standard Specification, No. 509 (1933). Sp.Gr. 0·796–0·798. B.R. 55·5–56·5° C. 95 per cent. Acidity, 0·002 per cent maximum. Residue, 0·005 per cent maximum. Withstands permanganate test for 30 min.

A.S.T.M. Specification, D 329–33. Sp.Gr. 0·791–0·799 at 20° C. 98 per cent dimethyl ketone. B.R. up to 55° C. none; above 57·5° C. none. Residue, 0·005 per cent maximum. Withstands permanganate test for 30 min.

Evaporation time, 2·1 (compared with ethyl ether, 1).

Solvent for resins, fats, oils, collodion cotton, celluloid, cellulose acetate, etc.

USES

(1) In the manufacture of smokeless powder and explosives (Hamilton, 1925).

(2) In the lacquer and varnish trade especially for "Zapon" lacquer and metallic lacquers. "Blush" difficulties with acetone are serious, and consequently its use is avoided where possible.

(3) In the celluloid industry especially in the manufacture of small portable accumulator boxes.

(4) In the rubber industry.

(5) In the chemical industry, e.g. in the preparation of chloroform, ketones, iodoform and sulphonal (Kagan, 1924).

(6) In the dyeing industry.

(7) In the manufacture of boots and shoes as a constituent of cements.

(8) In the manufacture of aeroplane dope.

(9) In the manufacture of artificial silk and artificial leather, etc. (Clayton and Clark, 1931).

(10) As a solvent for acetylene (Flury and Wirth, 1934).

(11) In the production of lubricating oils (Smoley and Kraft, 1935).

ESTIMATION IN AIR

A method of estimating small amounts of acetone vapour in air is described by Spatz (1930); it is similar to that used by Kagan (1924) in his calculation of the amount of inhaled acetone absorbed. The method depends on the fact that when acetone is passed into an alkaline iodine solution (10 ml. of $N/10$ iodine solution and 10 ml. 20 per cent caustic soda) iodoform is produced, 1 molecule of acetone binding 6 atoms of iodine. The solution is acidified with sulphuric acid and titrated for free iodine with sodium thiosulphate (see also Elliott and Dalton's method, cited by Flury and Zernik).

TOXICITY

By injection, oral administration and inhalation, acetone is a narcotic of more rapid action than methyl alcohol, but less toxic as regards its lethal and secondary effects; its effect is similar to but greater than that of ethyl alcohol. It also has an irritating effect on mucous membranes. In animals it produces a progressive general narcosis characterized by depression of the body temperature, respiratory and heart rate, and by abolition of corneal, auditory and equilibratory reflexes. Several authors have considered, however, that it has some toxic action upon the kidneys when inhaled (Flury and Wirth, 1934), as well as when ingested (Albertoni and Pisenti, 1887; Poliak, 1925), while Lewin (1907) described capillary hyperaemia of the mucous membrane of the stomach and small intestine and fatty infiltration of the liver.

Later experiments on inhalation seem to show that the acute toxicity of acetone by this route is not so great as was asserted by the U.S. Bureau of Mines in 1921. This authority stated that, with regard to the immediate acute effects, acetone was more toxic than chloroform and only a little less toxic than benzol, and that inhalation of 50 mg. per l. (21,000 p.p.m.) was fatal to mice. Later experiments by Kagan (1924), Schultze (1932), Flury and Wirth (1934) and Sklianskaya, Urieva and Mashbitz (1936) have shown a considerably lower degree of toxicity for acetone vapour; concentrations of 130 mg. per l. are tolerated by mice (Flury and Wirth), while according to Sklianskaya *et al.*, 100 to 200 mg. per l. are attended by a mortality of only 15 per cent as compared with 45 per cent for the same concentrations of methyl alcohol. Flury and Wirth suggest that the acetone used in the U.S. Bureau of Mines experiments may have been impure. It is believed by Haggard, Greenberg and Turner (1944) that the toxicity of acetone, as expressed in terms of severe symptoms or of death, is not a true criterion of its basic toxicity. They have correlated the symptoms of intoxication with the blood concentration, and on this basis find the toxicity of acetone much lower than the figures given by many earlier workers. In rats exposed to lethal concentrations of vapour, the actual concentration in the blood required to bring about respiratory failure was on the average 9,190 mg. per l. of blood—very nearly identical with that for ethyl alcohol (9,300 per mg. l.).

Dangerous Concentrations

The explosive risks of acetone are described by Jones, Harris and Miller (1933). The lower limit for acetone-air mixtures is 2·55 per cent acetone and the upper 12·80 per cent at laboratory temperatures.

The maximum permissible concentration suggested by Cook (1945) is 500 p.p.m. (1,000 mg. per c.m.).

TOXIC EFFECTS IN ANIMALS

Lethal and Narcotic Concentrations

Lethal dose

By intravenous injection. For rats, 6 to 8 ml. per kg. body weight; for rabbits, 4 ml. per kg. (Walton, Kehr and Loevenhart, 1928).

By stomach tube. For rabbits, 10 ml. per kg.; the animals died after 3 hours (Walton *et al.*).

By mouth. For dogs, 8 g. per kg. (Albertoni, 1884).

By inhalation. For mice, 100 to 200 mg. per l. (42,200 to 84,400 p.p.m.). The mortality was 15 per cent in the experiments of Sklianskaya *et al.*, remaining at about the same level when the duration of exposure was increased; with lower concentrations (40 to 100 mg. per l.) the toxicity increased sharply when exposure was prolonged. In the experiments of Schultze (1932) death took place a few minutes after the end of an hour's exposure to 110 mg. per l. (46,000 p.p.m.).

Haggard, Greenberg and Turner (1944) found 300 mg. per l. fatal to rats in $1\frac{3}{4}$ to $2\frac{1}{4}$ hours, and 100 mg. per l. in $4\frac{1}{2}$ to $5\frac{1}{2}$ hours.

Narcotic dose

By intravenous injection. For rabbits, 2 ml. per kg. body weight.

By intramuscular injection and by stomach tube. Rabbits were depressed but not unconscious with 5 and 7 ml. per kg. respectively (Walton *et al.*).

By mouth. For dogs, 4 g. per kg. body weight (Albertoni, 1884).

By inhalation. For mice, 48 mg. per l. after $1\frac{1}{2}$ hours (Flury and Wirth, 1934); 20 mg. per l. after $7\frac{3}{4}$ hours (Schultze, 1932); 40 mg. per l. after nearly 3 hours (Sklianskaya *et al.*). For cats, 125 mg. per l. (Kagan, 1924); 178 mg. per l. with preliminary convulsions (Schultze, 1932). For rats, loss of corneal reflex, 100 mg. per l. in 105 to 115 minutes (Haggard, Greenberg and Turner, 1944). For guinea-pigs, loss of reflexes, 2 per cent in 8 to 9 hours (Specht, Miller and Valaer, 1939).

Acute Poisoning

Symptoms

A feature of the narcosis produced by all forms of administration is the appearance of irritative effects when the narcosis is pushed to the stage of unconsciousness. Sklianskaya *et al.* observed inco-ordination in mice after 25 to 28 minutes with rhythmical clonic contractions of the hind legs and abdominal muscles, features also observed by Kagan (1924) in cats; Kagan noticed twitchings and convulsions during narcosis, while Flury and Wirth (1934) record symptoms of choking and vomiting at this stage. Di Prisco (1936) states that irritation of the nasal mucosa causes a temporary cessation of breathing as a protective reflex.

Effect on Blood Pressure and Respiration

The results of inhalation and of intravenous injection of acetone have been somewhat inconstant in the hands of different observers, but the consensus of opinion seems to be that in non-lethal concentrations it produces a specific irritative effect on the respiratory centre (Gollwitzer-Meier, 1927) followed by respiratory paralysis (von Tappeiner, 1884; Schlomovitz and Seybold, 1924). After intravenous and intramuscular injection, it causes a fall in blood pressure, which is regarded by Walton and co-workers (1928) and by Salant and Kleitmann (1922) as due primarily to a decrease in cardiac output.

In this connection may be mentioned the findings of Sklianskaya and co-workers with respect to the action of acetone on the isolated frog's heart. It produces a more powerful depression of cardiac activity in the absorptive phase than an equal concentration of methyl alcohol, but recovery in the elimination phase is much better.

Table 49 (adapted from Flury and Zernik, 1931, and from Specht, Miller, Valaer and Sayers, 1940), summarizes the most important results of inhalation experiments.

TABLE 49
Summary of results of inhalation of acetone on some animals

Species	Concentration (mg./l.)	Concentration (p.p.m.)	Duration (hr.)	Effects	Author
Mice	20	8,300	$7\frac{3}{4}$	Side position after 4–$7\frac{3}{4}$ hr.; deep narcosis in only a few animals after $7\frac{3}{4}$ hr.	Schultze (1932)
	48	20,000	$1\frac{1}{2}$	Side position after 60–70 min., deep narcosis after $1\frac{1}{2}$ hr.	
	110	46,000	1	Side position after 20–30 min.; deep narcosis after 40–60 min.; death a few minutes after end of experiment	
Cats	8–10	3,370–4,220	5	Initial salivation, irritation of nose and eyes; slight stupor and drowsiness after 5 hr.	Kagan (1924)
	20–50	8,440–21,100	3–4	Usually drowsy in first $\frac{1}{2}$ hr.; later sleepy; increased sensitivity to pain	
	80–100	33,700–42,200	4	No drowsiness; marked irritation of central nervous system; giddiness, ataxia, narcosis, twitchings and convulsions during narcosis	
	125	52,750	$1\frac{1}{3}$	As above	
Cats	40	17,000	$4\frac{1}{2}$	Side position after $3\frac{3}{4}$–4 hr.; recovery	Schultze (1932)
	114	48,000	3	Side position after $1\frac{1}{2}$ hr.; no deep narcosis; recovery	
	178	75,000	$1\frac{1}{2}$	Side position after $\frac{1}{2}$–1 hr.; deep narcosis after 1–$1\frac{1}{4}$ hr., with preliminary convulsions	
Guinea-pigs	50	21,100	25 min.	Lachrymation only	Specht et al. (1940)
			4–$8\frac{1}{2}$	Loss of auditory reflex; side position	
			9–$23\frac{1}{2}$	Coma; death	

Chronic Poisoning

Repeated exposure to low concentrations of acetone vapour apparently causes very little injury to animals. Both Kagan and Schultze gave repeated inhalations of 3 to 5 mg. per l. (1,265 to 2,110 p.p.m.) to cats and observed no ill-effects, with the exception of very slight irritation of the eyes and nose. Kagan noted an increased tolerance to low concentrations in animals which had been previously narcotized by higher concentrations.

Absorption through the Skin

Lazarew, Brussilowskaja and Lawrow (1931) estimated the amount absorbed by immersion of the foot of the animal in acetone and measurement of the amount exhaled and of that present in the blood; it was absorbed more slowly than ethyl ether or chloroform. This is explained by the fact that acetone is more water-soluble than ether or chloroform and therefore penetrates more slowly into the epidermal cells. Lazarew and co-workers regard the danger of absorption by this route as very small and very unlikely to occur in industry.

Changes in the Internal Organs

The earlier workers differed as to their statements of the injury caused by acetone to the internal organs, especially the kidneys, when taken by mouth. While Albertoni and Pisenti (1887) recorded very definite changes in the kidney, consisting of granular degeneration in the less severe intoxications, necrosis of the tubular epithelium accompanied by albuminuria in the more severe, Baginsky (1888) and Schwartz (1898) found no evidence of injury after repeated administration to dogs. Later experiments by Poliak (1925) have however confirmed the findings of Albertoni and Pisenti. Poliak found lesions in the kidneys, especially in the convoluted tubules of the cortex, increasing in severity with the amount of acetone administered. He suggests, therefore, that acetone is specifically excreted by the convoluted tubules, which are only injured when considerable absorption takes place. Baginsky's dogs were reported to have vomited frequently and since the method of feeding acetone would give rise to little absorption and therefore little excretion, this might explain his negative results.

Kagan found the kidneys enlarged with some fatty infiltration in one cat subjected to inhalation of 180 mg. per l., while Flury and Wirth observed albuminuria following inhalation in some of their animals.

TOXIC EFFECTS IN MAN

Acetone taken by mouth in doses of 15 to 20 g. daily for several days was shown by Albertoni to produce no ill-effects other than slight drowsiness, but when inhaled it has more serious consequences. Kagan (1924), in personal experiments, found it impossible to inhale concentrations of 22 mg. per l. (8,500 p.p.m.) for longer than 5 minutes owing to acute irritation of the throat. According to Flury and Zernik (1931), longer inhalation of small amounts produces irritation of the upper respiratory passages and bronchi, headache, heaviness in the head, a feeling of oppression and bad dreams, and they state that many workmen show individual susceptibility to its effects.

Accounts of acute industrial poisoning from inhalation of acetone are rare; the most outstanding one being that described by Sack, 1940 (see p. 325).

Two non-industrial cases of acute poisoning from absorption through the skin have been reported by Cossman in 1903 and by Strong in 1944. Two cases of eye injury were attributed by Halbertsma (1926) to the use of acetone containing acetaldehyde, and a series of cases showing slight symptoms of poisoning are mentioned by Heim de Balsac, Agasse-Lafont and Feil (1922) where the solvent used contained equal parts of amyl acetate and acetone. Two cases of fainting and collapse, where butanone as well as acetone was used, were recorded by Smith and Mayers in 1944.

The following statement in the International Labour Office's *Brochure No.* 33 (1930) is apparently based upon the opinions given in earlier text-books, such as Roth's *Kompendium der Gewerbekrankheiten* (1909), Eulenberg's *Handbuch der Gewerbehygiene* (1876), without illustrations from actual cases observed: "Numerous cases of illness have been noticed among persons occupied in trades using lacquers, varnishes, cements, colours, etc., in which the solvent contains acetone; in the manufacture of indiarubber, of boots, of aeroplane wings, of metal articles (process of painting) etc. In these same trades cases of eczema and conjunctivitis caused by acetone vapour are also noticed fairly frequently." This opinion is not confirmed by Hamilton (1934) who states that she encountered no ill-effects from acetone in her experience during the war when enormous quantities were used. It may be mentioned here that an additional danger arises when acetone is mixed with chlorinated hydrocarbons, as under the influence of light there may be a transference of chlorine leading to the formation of the highly irritant chloroacetone (Gerbis, 1931).

Absorption and Elimination

That acetone is absorbed into the blood stream when inhaled, partly eliminated in the expired air and urine and partly metabolized in the body has been shown by the researches of Kagan (1924), Greenburg, Mayers, Goldwater and Burke (1938) and by Haggard, Greenberg and Turner (1944).

According to Kagan, 71 to 77 per cent of inhaled acetone is absorbed. He inhaled acetone from a flask containing a known percentage of acetone in water, and exhaled it into a series of flasks containing water and sodium iodide. The amount inhaled was estimated by the loss of acetone from the inhalation flask and the amount exhaled by the acetone content of the sodium iodide solution. The amount inhaled in 5 minutes from a flask containing a 10 per cent solution was 910 mg.; 264 mg. were expired; thus the amount absorbed was 646 mg., or 71 per cent.

That acetone vapour is absorbed by workers inhaling it during industrial processes has also been demonstrated by Greenburg *et al.* (1938) who found a positive reaction for acetone in the urine of every worker examined.

The phenomenon of "absorptive concentration" with progressive accumulation as described by Haggard, Greenberg and Turner (1944) sheds a somewhat different light on the hitherto accepted ideas on the absorption of acetone. They found that at the end of each working day there is a small residual accumulation of acetone in the tissues of a man inhaling acetone for 8 hours. This amount, however, over a period of days, never reaches the saturation level. Thus, a man inhaling an atmosphere containing 2,100 p.p.m. of acetone for 8 hours accumulated 330 mg. per l. of blood, an amount well below the saturation level and producing no significant symptoms.

Acute Poisoning

Among the few cases of acute intoxication from the industrial use of acetone is one reported by Sack in 1940, two by Smith and Mayers in 1944, and one by Strong in 1944.

Sack's case was that of a man engaged in cleaning a tank in which a solution of artificial silk in acetone had been filtered. He became unconscious, but recovered following an injection of coramine, showed excitement and vomited several times. Complete elimination of acetone was slow, the blood level on the day after the acute poisoning being still 18 mg. per cent (normal, 1 to 2 mg. per cent) and acetone still detectable in the urine. The presence of slight albuminuria, red and white blood corpuscles in the residue after centrifugation, a high level of urobilin and an early rise of bilirubin in the blood, symptoms of gastric irritation and an increased blood sugar level are suggestive of slight injury to the kidneys and liver and probably to the islets of Langerhans, according to Sack, as well as of a disturbance of the motor nerve mechanism of the stomach. Previous symptoms of slight intoxication had included headache, lassitude, drowsiness, loss of appetite, nausea and vomiting, with intolerance to alcohol.

The two cases reported by Smith and Mayers (1944) occurred in the process of water-proofing seams of raincoats, the solvent containing butanone (atmospheric concentration 398 to 561 p.p.m.) as well as acetone (330 to 495 p.p.m.). The symptoms shown by the two patients were gastric distress, fainting and collapse. The cases reported by Cossman (1903) and by Strong (1944), if accepted as true cases of acetone poisoning, would indicate that this substance can be absorbed through the skin in sufficient quantities to exert a narcotic effect. Cossman's case was a 12-year-old boy suffering from a tubercular hip to which, after long immobilization, a plastic bandage of "celluloid mull" made with acetone was applied. Within 24 hours he exhibited symptoms similar to those of diabetic coma, mydriasis, dilatation of the pupils, small pulse, deep slow respiration and sleep of 24 hours' duration. Acetone was present in the urine, the vomit and the expired air. An attempt to reproduce the condition in a dog by a similar application was only partly successful, the dog became semi-narcotized, but no acetone was present in the urine. The case reported by Strong was somewhat similar, though of even greater severity. The symptoms followed the application of a plaster cast, a compound of polymerized vinyl acetate, pyroxylin, boric acid and surgical gauze, to which was further applied a lacquer containing acetone. During the following 4 days the patient developed nausea, with vomiting of blood; sugar, albumin and acetone appeared in the urine, and there were intervals of semi-consciousness. Improvement began after the lacquer was removed.

Chronic Poisoning

There seems to be a general impression that exposure to acetone produces effects similar to those of amyl acetate—sensation of heat, vertigo, slight fainting attacks, irritation of throat, attacks of coughing, etc. (Lehmann, quoted by Carozzi in the International Labour Office's *Brochure No.* 33), but, as previously stated, there are few actual cases of such effects recorded.

Headache in workers making celluloid boxes and using a glue containing a mixture of equal parts of amyl acetate and acetone is described by Heim de Balsac, Agasse-Lafont and Feil (1922), but they do not distinguish between the effects of acetone and those of amyl acetate. According to the International Labour Office's Brochure it is in this industry that the greatest danger from acetone arises.

An examination by Greenburg *et al.* (1938) of workers in a shirt factory using acetone and methyl alcohol as a stiffening fluid showed no significant incidence of disturbance of health. Haggard, Greenberg and Turner (1944), pointing out that the general physiological reaction to acetone, based on concentration in the blood, is virtually the same as that for ethyl alcohol, consider that the mildest effects of intoxication might lead to some loss of judgement and concentration, and thus increase the likelihood of industrial accidents. With concentrations in air below 0·5 mg. per l., the blood concentrations would rise only to some 30 mg. per l. for one exposure of 8 hours. Such low blood concentrations would cause no intoxication and the amount of acetone absorbed would be completely, or nearly completely, eliminated during the 16 hours following exposure, and would not accumulate with daily exposure. Above the level of 0·5 mg. per l., however, the blood would not be free of acetone at any time during the working week, and though the low blood concentrations given with exposures up to 5 mg. per l. would cause no obvious symptoms, it is considered by these observers that the value of 5 mg. per l. approaches the upper limit of safety for men performing moderate exertion. This limit is in close agreement with the standards already proposed by the California State Department of Industrial Relations (1939).

Local irritation of the eyes. Halbertsma (1926) describes two cases of severe irritation of the cornea and conjunctiva, occurring in a driving-belt factory where the men came into close and frequent contact with acetone. Both showed lachrymation, photophobia, and infiltration of the corneal epithelium, while one had corneal opacities. Since such lesions had never been observed before, the acetone was examined and found to contain acetaldehyde, to which Halbertsma considers the effects may have been due.

Effect on the Blood

No specific blood changes following exposure to acetone have been reported in the literature except the eosinophilia described by Heim de Balsac, Agasse-Lafont and Feil (1922), who state that they consider eosinophilia "a proof of the reaction to saturation of the organism by fumes of acetone and amyl acetate".

No eosinophilia was observed by Browning (unpublished) in a small series of workers employed on a cellulose acetate process, nor by Rosgen and Mamier (1944) in a much larger series (45 men and 39 women) in a dry-cleaning process. The latter investigation was undertaken following the death of a female worker from haemolytic icterus and complaints of anaemia from several other workers. No clinical symptoms suggesting intoxication were found in any of the operatives examined. Haematological examinations included enumeration of red and white corpuscles, estimation of haemoglobin, differential count, sedimentation rate and coagulation time. No abnormality was found except for a few cases of rather low haemoglobin percentage among the women, believed to be of nutritional origin. Rosgen and Mamier conclude that the possibility of any injurious effect of acetone on the blood system can be excluded.

2. Methyl Acetone

CONSTITUTION

An impure fraction of variable composition obtained during the manufacture of acetone and methyl alcohol from wood. Its usual constituents are methyl acetate 20 to 30 per cent, acetone 35 to 60 per cent, methyl alcohol 20 to 40 per cent.

"Methone" is a form of methyl acetone in which the alcohols are replaced by an ester, thus enhancing the solvent properties considerably.

PROPERTIES

British Standard Specification 2D1. Almost colourless liquid, clearly miscible with water in all proportions.

B.R. 50–70°C. Sp.Gr. 0·830. Fl.P. below 0°C.

Excellent solvent for cellulose acetate and nitrate, many gums and resins, castor and linseed oils.

USES

(1) In the lacquer industry in nitrocellulose lacquers, in combination with benzol as a diluent (Symons, 1931).

(2) As a varnish remover in place of acetone (Sanderson, 1933).

TOXIC EFFECTS

No animal experiments have been carried out and no toxic effects in human beings reported.

3. Acetone Oils
(Ketone Oils, Ketols)

CONSTITUTION

Acetone oils are by-products in the manufacture of acetone from wood distillates. They consist of complex mixtures of a large number of aliphatic and polymethylene ketones. Flury and Zernik (1931) describe two varieties:
(1) Light or white acetone oil. B.R. 70–120° C.
(2) Heavy or yellow acetone oil. B.R. 60–250° C.

USES

(1) In the lacquer industry; not such rapid solvents as acetone, but can be used in humid atmospheres.
(2) In the manufacture of secondary aliphatic alcohols and esters such as secondary butyl acetate.

TOXICITY

While no direct experiments on the effects of acetone oils on animals have been performed, the effects of diethyl ketone, methyl ethyl ketone and acetone either alone or in a mixture with ether have been tested by Bourne (1926). Ether mixtures containing up to 5 per cent of these substances had practically no effect upon the blood pressure and respiration of a dog already anaesthetized with ether.

No toxic effects in human beings have been recorded.

4. Methyl Ethyl Ketone
(Butanone, M.E.K.)

$$CH_3 \cdot CH_2 \cdot CO \cdot CH_3$$

PROPERTIES

A clear colourless liquid with an odour resembling that of acetone. Produced for many years in limited quantity as a by-product in the distillation of wood, but by this method it was uncertain both in quantity and quality. By more recent synthetic methods it is produced from secondary alcohols formed during the large-scale production of ethyl and higher alcohols by olefine absorption processes, and also from butylene glycol. By the use of special catalysts (alloys of copper, zinc, or aluminium oxide) yields of methyl ethyl ketone up to 96 per cent are obtained from secondary butyl alcohol (Durrans, 1935).

According to Zangger (1930) the "special solvent of Hiag", which is a mixture of acetone with formates and acetates, contains a certain amount of methyl ethyl ketone.

Technical. B.R. 68–85°C. Sp.Gr. 0·810–0·815.

Pure. B.P. 79·6°C. Sp.Gr. 0·8051 to 0·8095. V.P. at 20°C. 77·5 mm. Hg (International Critical Tables, 1928).

Good solvent for cellulose acetate and nitrate and for some gums and resins.

The explosive hazard of methyl ethyl ketone is according to Patty, Schrenk and Yant (1935) not to be ignored, though it has good warning properties in mixtures considerably below the flammable range, the lower limit of which is about 2 per cent.

USES

(1) In artificial leather manufacture often mixed with acetone, methanol, etc. (Zangger, 1930).

(2) In the lacquer and varnish industry (Despartmet, 1928) as a solvent for cellulose acetate; it can be used as a substitute for ethyl acetate (Metzinger, 1935), and is often mixed with acetone, alcohol, triphenyl phosphate, etc., especially for aeroplane lacquering (Zangger).

(3) As a paint remover (Lougovoy, 1933).

(4) In the manufacture of pharmaceuticals and cosmetics, where its use is increasing (Holden and Doolittle, 1935).

(5) In the production of lubricating oils as a refining and dewaxing solvent (Smoley and Kraft, 1935).

ESTIMATION OF VAPOUR IN AIR

The method used by Patty, Schrenk and Yant was similar to that used by Kagan and Spatz for acetone (p. 320), i.e. a known volume of vapour-air was passed into a solution of $1N$ sodium hydroxide and an excess of $N/10$ iodine added; the solution was then neutralized with $2N$ sulphuric and a slight excess added; the excess iodine was determined by titration with $N/20$ sodium thiosulphate solution (see also Cassar, 1927).

TOXICITY

While methyl ethyl ketone has a narcotic effect on animals, with a greater effect than acetone in depressing the bodily functions, its odour and its irritating effect on mucous membranes are so strong that concentrations harmful to animals are apparently intolerable to man (Patty, Schrenk and Yant, 1935). Ten per cent vapour in air was the highest concentration obtainable by Patty, Schrenk and Yant in a closed chamber by extended re-circulation of air over wicks wet with the liquid. From their experiments on the effects of inhalation by men of 0·3, 1 to 3·3, and approximately 10 per cent vapour in air, it appears that concentrations apparently harmless to guinea-pigs after several hours' exposure give warning of their presence by their odour and by causing irritation. The higher concentrations were intolerable even with momentary exposure owing to irritation of the eyes and nasal passages; 1 per cent vapour had a strong odour and almost intolerably irritating effects after several inhalations, while 0·33 per cent vapour had a moderately strong odour and was moderately irritating to the eyes and nose.

Maximum Permissible Concentration

On the basis of the sensory response to the vapour of methyl ethyl ketone, Nelson, Ege, Ross, Woodman and Silverman (1943) suggest 200 p.p.m. (500 mg. per c. m.) as a suitable concentration for prolonged exposure, though animals tolerated 3,000 p.p.m.

TOXIC EFFECTS IN ANIMALS

Lethal and Narcotic Concentrations
  Lethal dose
  By inhalation. For guinea-pigs, 147·5 to 295 mg. per l. (5 to 10 per cent by volume in air) in 30 to 60 minutes.
  Narcotic dose
  29·5 mg. per l. in 4 to $4\frac{3}{4}$ hours or 97 mg. per l. in 48 to 90 minutes.

Acute Poisoning

Symptoms

Exposure to concentrations higher than 0·33 per cent by volume were followed by irritation of nose and eyes, lachrymation, inco-ordination, narcosis, gasping type of respiration, and death with concentrations higher than 1 per cent, according to the duration of exposure. The time for occurrence of these symptoms decreased rapidly with increases in concentration, and death was produced by 45 and 200 minutes' exposure to 10·0 and 3·3 per cent vapour in air respectively. Table 50 summarizes the findings of Patty, Schrenk and Yant, which have been confirmed by the experiments of Specht and co-workers (1940).

TABLE 50
Effect of different concentrations of methyl ethyl ketone on some animals

Concentration		Duration of exposure (min.)	Effects
(mg./l.)	(p.p.m.)		
9·7	3,300	810	No abnormal signs
29·5	10,000	810	Irritation of eyes and nose after 2–4 min.; narcosis after 240–280 min.; recovery
97·3	33,000	200	Inco-ordination after 18–30 min.; narcosis after 48–90 min.; death
295	100,000	45–55	Narcosis after 10–11 min.; death
	(per cent)		
	5–10	30–60	Dangerous to life
	1·0	60	Maximum amount without serious disturbance
	0·3	Several hours	Maximum amount without serious disturbance

Changes in the Internal Organs and Cornea

Congestion and emphysema of the lungs and congestion of the liver, kidneys and brain were found in animals subjected to high concentrations. Opacity of the cornea was also present in animals exposed to 10 per cent vapour for 30 minutes or more. The cornea gradually improved in animals living 4 to 8 days after exposure and was nearly normal at the end of 8 days. When death occurred it was not clear whether it was due to irritation of the lungs or to narcosis, but all the animals either died during exposure or recovered, indicating that irritation of the respiratory centre was seconuary to the narcotic action.

Absorption by the Skin

In the experiments of Lazarew and co-workers (1931) methyl ethyl ketone was so slowly absorbed by the skin of animals that they consider its toxicity by this route very small and unlikely to be of any importance in industry.

TOXIC EFFECTS IN MAN

No injury from the use of methyl ethyl ketone in industry has been recorded except the production of occupational dermatitis mentioned by Schwartz, Tulipan and Peck (1947).

5. Methyl *iso*Butyl Ketone
(Hexone, 4-Methyl-2-Pentanone)

$$CH_3 \cdot CO \cdot C \cdot H_9, \text{ i.e. } CH_3 \cdot CO \cdot CH_2 \cdot CH {<}^{CH_3}_{CH_3}$$

PROPERTIES

A clear liquid with characteristic ketone odour, but less penetrating than acetone oil.

B.R. of the technical product 112–118°C. Sp.Gr. 0·802. Fl.P. 23°C. (Doolittle, 1935). V.P. at 30°C. 26·2 mm. Hg. Evaporation rate slower than that of acetone, approximating to that of butyl acetate. Only slightly soluble in water (1·9 per cent, von Heuckenroth, 1934). Solvent for ester gum and some modified phenol-type resins which are only partially soluble in acetone (Sanderson, 1933); solvent for nitrocellulose, but disperses cellulose acetate only in the presence of a small quantity of a definite solvent such as acetone (Metzinger, 1935; Yarsley, 1934).

USES

Chiefly in the lacquer industry, where it is specially suitable for reducing the viscosity of cellulose lacquers and as an anti-blushing agent (Yarsley, 1934; Despartmet, 1928; von Heuckenroth, 1934; Reid, 1934, etc.). It can be substituted for *secondary* butyl acetate.

TOXICITY

In high concentrations methyl *iso*butyl ketone is narcotic to animals, causing depression of bodily functions and body temperature. It is an irritant to nasal and conjunctival mucous membranes in both men and animals.

Maximum Permissible Concentration

Maximum allowable concentration suggested by Cook (1945) is 200 p.p.m. (500 mg. per c. m.)

TOXIC EFFECTS IN ANIMALS

At concentrations above 0·1 per cent marked lachrymation and salivation occur. The lethal dose is 1·0 per cent in about 4 hours (Specht, 1938). Complete recovery follows if the animals are removed in any but the terminal stage.

TOXIC EFFECTS IN MAN

No cases of industrial poisoning have been recorded, but concentrations below 0·1 per cent cause irritation of conjunctival and nasal mucosa, though this concentration is tolerated by guinea-pigs (Specht *et al.*, 1940; Jacobs, 1944).

6. Mesityl Oxide
(2-Methyl-2-pentenone-4)

$$C_6H_{10}O, \text{ i.e. } {}^{CH_3}_{CH_3}{>}C{:}CH \cdot CO \cdot CH_3$$

An unsaturated ketone, occurring in acetone oils; obtained by dehydrating diacetone alcohol or direct from acetone by means of alkalis.

PROPERTIES

A liquid of strong odour, suggestive of "mice and peppermint" according to Durrans (1933). It is more stable than diacetone alcohol, but its colour tends to darken on storage (Sanderson, 1933). The commercial variety may contain traces of aldehydes, especially crotonaldehyde (Zangger, 1930).

B.P. 129·5° C. Sp.Gr. 0·861. Solubility in water 2·9 per cent by weight. V.P. 9·2 mm. Hg at 20° C.; 14·3 mm. Hg at 30° C. Fl.P. 32° C. Excellent solvent for nitrocellulose, acetylcellulose, celluloid, many resins, oils and fats. Smyth, Seaton and Fischer (1942) state that it is specially suitable for resin coatings. According to Tersand (1935) its solvent power is equal to that of ethyl acetate, but its evaporation time is like that of a heavy solvent; it can tolerate large dilutions with benzol, xylol, toluol or alcohol.

USES

(1) In the lacquer industry (Sanderson, 1933).
(2) In the leather industry (Tersand, 1935).
(3) As a paint remover, suggested by Durrans (1933).

TOXICITY

Mesityl oxide is a strong narcotic with some irritative effect, but it apparently has no cumulative effect on the internal organs. No ill-effects from its industrial use have been recorded.

TOXIC EFFECTS IN ANIMALS

The results of administration of mesityl oxide by inhalation to animals show that both single and repeated exposures produce toxic effects by its narcotic action and that it does not primarily exert any cumulative effect on the internal organs. It is also an irritant, and in its acute narcotic and irritant effects it is slightly more toxic than *cyclo*hexanone or methyl *iso*butyl ketone.

Acute Poisoning

Lethal Dose

In high dosage (13,000 p.p.m.) inhalation of mesityl oxide is fatal in a few minutes, while 5,000 p.p.m. is fatal in 30 to 60 minutes.

Narcotic Dose

According to Specht *et al.* (1940), inhalation of 2,000 p.p.m. for 6 hours has a strong narcotic effect.

Lesions of the Internal Organs

Congestion of the lungs, kidneys, spleen, adrenals and brain, with some dilatation of the kidney tubules was observed by Specht and co-workers. They considered that the kidney lesions showed a selective action, but, from the evidence provided by repeated exposures, this opinion is not shared by Smyth *et al.*, who believe that the toxic effect of mesityl oxide is largely due to its action on the circulation and respiration, and is not primarily a kidney injury.

Chronic Poisoning

With repeated exposures to smaller concentrations Smyth and co-workers observed no toxic effect following inhalation of 50 p.p.m., but noted a high mortality with 500 p.p.m. Mild albuminuria occurred with 250 and 550 p.p.m. Only congestion and cloudy swelling of the kidney were observed. Smyth and co-workers suggest that mesityl oxide is probably not rapidly eliminated and that with frequent exposures the blood concentration reaches an anaesthetic level. The lesions are therefore largely due to anaesthetic action on the circulation and respiration and do not represent a selective injury to the kidneys.

KETONES

TOXIC EFFECTS IN MAN

While no ill-effects from the industrial use of mesityl oxide have been recorded, Silverman, Schulte and First (1946) have examined the sensory response of human subjects exposed to various concentrations of this solvent. They found that a majority experienced some degree of eye irritation at 25 p.p.m., and nasal irritation at 50 p.p.m.; at the latter level half the subjects found the odour objectionable. An unpleasant taste remained in many cases for 3 to 6 hours after the exposure. Silverman *et al.* therefore suggest that 25 p.p.m. is the highest concentration which would be satisfactory for an 8-hour day.

7. cycloHexanone

(*Sextone, Hexanon, Anon*)

$$C_6H_{10}O, \text{ i.e. } CH_2{<}{{CH_2 \cdot CH_2}\atop{CH_2 \cdot CH_2}}{>}CO$$

PROPERTIES

A colourless, neutral liquid, non-miscible with water, of peculiar ketone-like odour.

B.R. of the technical product 154–156°C. Sp.Gr. 0·945. Evaporation time, 40·6 (compared with ethyl ether, 1).

Solvent for oils, fats, rubber and many resins; also for celluloid, cellulose ethers and cellulose acetate; specially good solvent for collodion cotton. The dilution ratios are greater than those of amyl acetate solutions, so that very large quantities of cheap diluents may be incorporated.

USES

(1) In the lacquer industry where it is particularly adaptable as a slowly evaporating improver added sparingly to nitrocellulose and cellulose acetate lacquers, also for lacquers which are insulators for heat and electricity (Loehr, 1931) and for automobile lacquers (Deschiens, 1928).

(2) In the textile industry as a spotting agent, except for cellulose acetate silk, but it is said to be a re-lustring agent for this (Clayton and Clark, 1931); also used in the spray-painting of textiles.

(3) In the leather industry as a thinner for fast-coating finishes on light and fancy leathers, also for wet and dry degreasing processes in this trade; it improves the adhesion of varnishes especially on greasy leather. Good for removing chlorophyll stains (Lamb and Gilman, 1932).

(4) As a degreaser especially in removing grease from nickel sheets.

TOXICITY

*cyclo*Hexanone is a narcotic, causing irritation of mucous membranes, narcosis and death, in high dosage. It is less toxic than *cyclo*hexanol and the "six carbon" ketones, methyl *iso*butyl ketone or mesityl oxide, but more toxic than *cyclo*hexane or methyl*cyclo*hexanol.

Maximum Permissible Concentration

100 p.p.m. has been suggested by Cook (1945) as the maximum allowable concentration.

TOXIC EFFECTS IN ANIMALS

Lethal and Narcotic Concentrations

Lethal dose

Repeated exposure to the narcotic dose of 12·12 mg. per l. caused death to a number of animals in the experiments of Treon, Crutchfield and Kitzmiller (1943).

Narcotic dose

Gross (cited by Lehmann and Flury, 1943) found 15 mg. per l. (3,800 p.p.m.) produced deep narcosis after 1 hour, a result in fairly close agreement with that of Treon *et al.* (1943), 12·12 mg. per l., and Specht *et al.* (1940), 16 mg. per l. No toxic effects were observed by Treon *et al.* with concentrations of 0·75 mg. per l., and they regard this as the maximum safe concentration for prolonged exposure for rabbits. According to Frey (1939) no abnormality other than a trace of albuminuria was found in a guinea-pig exposed to a total of 0·31 ml. of liquid *cyclo*hexanone vaporized in a desiccator.

Symptoms

Apart from slight salivation, conjunctival irritation and lethargy, no other symptoms preceded the onset of narcosis (Treon *et al.*, Gross); Specht *et al.* (1940) noted corneal opacity in guinea-pigs and slow recovery from narcosis.

Lesions of the Internal Organs

No specific lesions were found following exposure to lethal doses, the toxic effects being those of general vascular injury and inflammation. With repeated exposure to smaller concentrations (0·75 mg. per l. or 190 p.p.m.) there were barely demonstrable changes in the liver or kidneys (Treon *et al.*).

Effect on the Blood

No significant changes in the blood picture were observed.

Effect on Metabolism

Some conjugation of urinary sulphates and some increase in the excretion of glucuronic acid were observed by Treon *et al.*

TOXIC EFFECTS IN MAN

No serious industrial poisoning has been reported. A number of workers employed in degreasing nickel sheets were examined by the Home Office in 1932, but no ill-effects were observed, nor were any complaints of such received. During a recent investigation by Browning (unpublished) of twenty operatives employed in a silk-screen printing process where the solvent contained about 25 per cent of *cyclo*hexanone, with varying admixtures of *cyclo*hexyl acetate and methyl*cyclo*hexanone, most of the workers complained of drowsiness. Blood examinations showed no abnormality of the white cell or differential count, but there was some anaemia with a low colour index among the women.

8. Methyl*cyclo*hexanone
(Sextone B, Methyl Anon)

$CH_3 \cdot C_6H_9O$

PROPERTIES

A permanently neutral, almost colourless oily liquid (darkening on storage, especially if exposed to light), non-miscible with water, with a peculiar peppermint-like odour.

KETONES

90 *per cent technical product*. B.R. 160–175°C. Sp.Gr. 0·925–0·930. Fl.P. 53°C. Water dissolves 3 per cent. Evaporation time, 47 (compared with ethyl ether, 1).

Solvent properties similar to those of *cyclo*hexanone but not such a good solvent for cellulose acetate. Good solvent for chlorophyll (Lamb and Gilman, 1932).

USES

(1) In the lacquer industry as a thinner for cellulose lacquers, replacing acetone and methyl acetate (Deschiens, 1928); minimizes blushing, as it keeps the nitrocellulose in solution when the other solvents and diluents which favour blushing have already evaporated (I. G. Farbenindustrie Aktiengesellschaft, 1930). Has particular advantage in crystal lacquers since it allows formation of crystals of unusual size and regularity (Sanderson, 1933).

(2) In the leather industry particularly suitable for improving the adhesion of cellulose finishes for light and fancy leathers.

(3) As a rust remover for loosening tenacious rust in screws and bolts.

(4) As a slow-setting varnish remover (Sanderson, 1933).

Maximum Permissible Concentration

The maximum allowable concentration suggested by Cook (1945) is 1,000 p.p.m.

TOXIC EFFECTS IN ANIMALS

The only animal experiments carried out by Gross (cited by Lehmann and Flury, 1943) and by Treon *et al.* show that while methyl*cyclo*hexanone has a local irritating effect on the eyes and nose, and in high concentrations is slightly narcotic, its vapour is not lethal at concentrations which can be reached at ordinary temperatures. Subcutaneous injections of 0·5 ml. per kg. body weight caused only acceleration of respiration and no perceptible injury to the circulation (Gross). Inhalation of 8·19 and 5·12 mg. per l. (1,822 and 1,139 p.p.m.) in the experiments of Treon *et al.* and of 10 mg. per l. (2374 p.p.m.) in those of Gross caused only lethargy.

Lesions of the Internal Organs

Only slight damage to the liver and kidneys was caused by repeated exposure to 2·31 mg. per l. (514 p.p.m.). Treon *et al.* give the safe maximum concentration for repeated exposure of rabbits as between 0·82 and 2·31 mg. per l. No significant blood changes were observed.

TOXIC EFFECTS IN MAN

No toxic effects have been recorded, but, according to Gross, the vapour has a relatively strong irritating effect on mucous membranes. Sanderson (1933) refers to its "non-toxicity" as a varnish remover.

11. Isophorone

(1:1:3-*Trimethyl*cyclo*hex*-3-*en*-5-*one*)

$C_9H_{14}O$, i.e. $HC \genfrac{<}{>}{0pt}{}{C(CH_3) \cdot CH_2}{CO\text{------}CH_2} C(CH_3)_2$

PROPERTIES

A liquid with an unpleasant odour, but less obnoxious than that of mesityl oxide. B.P. 215·2°C. Sp.Gr. 0·9229. Fl.P. 96°C. (205°F.)

Good solvent, especially for vinyl resins.

USES

Chiefly in the lacquer industry.

TOXICITY

Isophorone is a narcotic and an irritant of mucous membranes, but its characteristic toxic action, as judged from animal experiments, is that of cumulative injury to the kidneys. Its toxicity to human beings is unknown.

Maximum Permissible Concentration

Isophorone has twice the cumulative toxicity of mesityl oxide but is 30 times less volatile at room temperatures, so that for protection against chronic toxic effects, 15 times less local ventilation would be needed than for mesityl oxide.

Nevertheless, a level of 20 to 40 p.p.m. has been suggested as a desirable limit for an eight-hour working day of continuous exposure.

TOXIC EFFECTS IN ANIMALS
Acute Poisoning

Effect of Inhalation

According to Smyth and Seaton (1940), the vapour of isophorone is about 4 times as toxic as methyl *iso*butyl ketone in single exposures, but owing to its lower vapour pressure it has actually half the hazard.

Chronic Poisoning

Lethal Concentrations

By inhalation. Rats and guinea-pigs subjected by Smyth, Seaton and Fischer (1942) to repeated inhalation (30 eight-hour exposures) of concentrations between 25 and 500 p.p.m. showed no injurious effect below a level of 100 p.p.m. At that level 12 per cent of the animals died; at 200 p.p.m. 17 per cent died; and at 500 p.p.m., 45 per cent. Only those animals inhaling 500 p.p.m. showed albuminuria, though the kidneys later showed some damage microscopically, which was more severe than that inflicted by the same concentration of mesityl oxide.

Injury of Kidneys and Lungs

The kidneys showed congestion, and the lungs showed desquamation of bronchial epithelium, microscopic haemorrhage in alveoli and bronchioles, and sometimes pneumonia. Smyth *et al.* (1942) suggest that death resulting from repeated inhalation of isophorone is due primarily to kidney and lung injury from a cumulative effect.

TOXIC EFFECTS IN MAN

No records of a toxic effect from the industrial use of isophorone have been recorded. Vapour concentrations of 200 to 500 p.p.m. are irritating to the human eyes and nose.

BIBLIOGRAPHY

ALBERTONI, P. (1884). Die Wirkung und die Verwandlungen einiger Stoffe im Organismus in Beziehung zur Pathogenese der Acetonämie und des Diabetes. *Arch. exp. Path. Pharmak.*, **18**, 219.
ALBERTONI, P. and PISENTI, G. (1887). Uber die Wirkung des Aceton und der Acetessigsäure auf die Nieren. *Arch. exp. Path. Pharmak.*, **23**, 393.
BAGINSKY, A. (1888). Über Acetonurie bei Kindern. *Arch. Kinderheilk.*, **9**, 1.
BOURNE, W. (1926). On the effects of acetaldehyde, ether peroxide, ethyl mercaptan, ethyl sulphide, and several ketones, dimethyl, ethyl methyl and diethyl, when added to anaesthetic ether. *J. Pharmacol.*, **28**, 409.
CALIFORNIA STATE DEPARTMENT OF INDUSTRIAL RELATIONS (1939). *Min. Ind. Accident Comm.*
CAROZZI, L. (1930). Acetone. International Labour Office, *Brochure No.* 33, also in Occupation and Health, **1**, International Labour Office, Geneva, 1930.
CASSAR, H. A. (1927). Determination of isopropyl alcohol in presence of acetone, and of methylethylketone in presence of secondary butyl alcohol. *Industr. Engng. Chem.*, **19**, 1061.
CLAYTON, E. and CLARK, C. O. (1931). Modern organic solvents: Parts I, II. *J. Soc. Dy. Col.*, Bradford, **47**, 183, 247.
COOK, W. A. (1945). Maximum allowable concentrations of industrial atmospheric contaminants. *Industr. Med.*, **14**, 936.
COSSMANN (1903). Acetonvergiftung nach Anlegung eines Zelluloid-Mullverbandes. *Münch. med. Wschr.*, **50**, 1556.
DESCHIENS, M. (1928). Les vernis et peintures cellulosiques pour automobiles. 7me Congrès, Paris. *Chim. et Industr.* Special No., p. 673.
DESPARTMET, E. (1928). Les cétones et leurs applications dans l'industrie des matières plastiques et des vernis nitrocellulosiques. 7me Congrès, Paris. *Chim. et Industr.* Special No., p. 697.
DOOLITTLE, A. K. (1935). Lacquer solvents in commercial use. *Industr. Engng. Chem.*, **27**, 1169.
DURRANS, T. H. (1933). Solvents. 3rd ed. Chapman & Hall, London.
DURRANS, T. H. (1935). Review of the technical aspects of industrial solvents. *Chem. and Ind.*, **13**, 585.
DURRANS, T. H. (1935). Solvents and plasticisers. *J. Oil Col. Chem. Ass.*, **18**, 340.
EULENBERG, H. (1876). Handbuch der Gewerbehygiene. Hirschwald, Berlin.
FLURY, F. and WIRTH, W. (1934). Zur Toxikologie der Lösungsmittel (Verschiedene Ester, Aceton, Methylalkohol). *Arch. Gewerbepath. Gewerbehyg.*, **5**, 1.
FLURY, F. and ZERNIK, F. (1931). Schädliche Gase, Dämpfe, Nebel, Rauch- und Staubarten. Springer, Berlin.
FREY, J. (1939). The effect of cyclohexanon on the hematopoietic system. *Haematologica*, **20**, 725.
GERBIS, H. (1931). Über gewerbliche Gifte. *Z. angew. Chem.*, **44**, 640.
GOLLWITZER-MEIER, K. (1927). Zur Frage der spezifischen Wirkung der Ketonkörper auf die Atmung. *Arch. exp. Path. Pharmak.*, **125**, 278.
GREENBURG, L., MAYERS, M.R., GOLDWATER, L. J. and BURKE, W. J. (1938). Health hazards in the manufacture of "fused collars". II. Exposure to acetone-methanol. *J. industr. Hyg.*, **20**, 148.
GROSS, E. Cited by Lehmann, K. B. and Flury, F. (1943). Toxicology and hygiene of industrial solvents. Williams and Wilkins, Baltimore.
HAGGARD, H. W., GREENBERG, L. A. and TURNER, J. McC. (1944). The physiological principles governing the action of acetone together with determination of toxicity. *J. industr. Hyg.*, **26**, 133.
HALBERTSMA, K. T. A. (1926). Hornhautschädigung bei Anwendung von Aceton. *Dtsch. Z. ges. gerichtl. Med.*, **10**, 109. (See also *Ned. Tijdschr. Geneesk.*, **70**, 1593.)
HAMILTON, A. (1925). Industrial poisons in the United States. Macmillan, New York.
HAMILTON, A. (1934). Industrial toxicology. Harper, New York.
HEIM DE BALSAC, F., AGASSE-LAFONT, E. and FEIL, A. (1922). Enquête sur les manifestations morbides présentées par les ouvriers peintres en voitures. *Progr. méd.*, Paris, **49**, 157.
HEIM DE BALSAC, F., AGASSE-LAFONT, E. and FEIL, A. (1922). Morbid disturbances due to handling of celluloid. *J. Amer. med. Ass.*, **79**, 146.
HEIM DE BALSAC, F., AGASSE-LAFONT, E. and FEIL, A. (1922). Manifestations morbides chez les ouvriers maniant le celluloid et ses solvants. *Paris méd.*, **i**, 477.
VON HEUCKENROTH, A. W. (1934). Note on some newer lacquer raw materials. *Natl. Paint Varn. Lacq. Ass., Proc. Sci. Sec.*, Circ. No. 472, 364.
HOLDEN, H. C. and DOOLITTLE, A. K. (1935). Solvents. *Industr. Engng. Chem.*, **27**, 525.
I. G. FARBENINDUSTRIE AKTIENGESELLSCHAFT (1930). Lösungsmittel und Weichmachungsmittel. p. 36. Frankfurt-am-Main.
INTERNATIONAL CRITICAL TABLES (1938). **3**, 218. National Research Council. McGraw-Hill, New York.
INTERNATIONAL LABOUR OFFICE. *Brochure No.* 33. Acetone. Also in Occupation and Health, **1**, 37. International Labour Office, Geneva, 1930.

JACOBS, M. B. (1944). Analytical chemistry of industrial poisons, hazards and solvents. Interscience Publishers, New York.

JONES, G. W., HARRIS, E. S. and MILLER, W. E. (1933). Explosive properties of acetone-air mixtures. *U.S. Bur. Mines, Wash., Technical Paper*, No. **544**.

KAGAN, E. (1924). Experimentelle Studien über den Einfluss technisch und hygienisch wichtiger Gase und Dämfpe auf den Organismus. XXXVI. Aceton. *Arch. Hyg., Berl.*, **94**, 41.

LAMB, M. C. and GILMAN, J. A. (1932). Note on the solution of chlorophyll by cellulose solvents and diluents. *Leather World*, **24**, 708.

LAZAREW, N. W., BRUSSILOWSKAJA, A. J. and LAWROW, J. N. (1931). Quantitative Untersuchungen über die Resorption einiger organischer Gifte durch die Haut ins Blut. *Arch. Gewerbepath. Gewerbehyg.*, **2**, 641.

LEWIN, L. (1907). Ueber das Verhalten von Mesityloxyd und Phoron im Tierkörper im Verglieche zu Aceton. *Arch. exp. Path. Pharmak.*, **56**, 346.

LOEHR, O. (1931). Acetylcellulose composition suitable for films, lacquers etc. U.S.P. 1,783,176. (See also *Chem. Abstr.*, **25**, 414.)

LOUGOVOY, B. N. (1933). Paint and varnish removers. U.S.P. 1,884,772. (See also *Chem. Abstr.*, **27**, 1118.)

MASHBITZ, L. M., SKLIANSKAYA, R. M. and URIEVA, F. I. (1936). Relative toxicity of acetone, methyl alcohol and their mixtures. II. Their action on white mice. *J. industr. Hyg.*, **18**, 117.

METZINGER, E. F. (1935). Lacquer solvents. Recent introduction of secondary solvents and ketones, with their comparative values and substitutions for normal solvents. *Paint Oil Chem. Rev.*, **97**, No. 9, 4.

NELSON, K. W., EGE, J. F., ROSS, M., WOODMAN, L. E. and SILVERMAN, L. (1943). Sensory response to certain industrial solvent vapors. *J. industr. Hyg.*, **25**, 282.

PATTY, F. A., SCHRENK, H. H. and YANT, W. P. (1935). Acute response of guinea-pigs to vapors of some new commercial organic compounds. VIII. Butanone. *Publ. Hlth. Rep., Wash.*, **50**, 1217.

POLIAK, B. (1925). Anatomische Veränderungen bei der experimentellen Azetonvergiftung. *Arch. exp. Path. Pharmak.*, **105**, 220.

DI PRISCO, L. (1936). A respiratory protective reflex during inhalation of acetone. *Folia med., Napoli*, **24**, 669.

REID, E. W. (1934). Modern solvent industry. *Industr. Engng. Chem.*, **26**, 21.

ROSGEN and MAMIER (1944). Sind azeton Gase blutschädigend? *Der offentliche Gesundheitskranke Dienst*, **10**, A83.

ROTH, E. (1909). Kompendium der Gewerbekrankheiten und Einführung in die Gewerbehygiene. 2nd Aufl. Schoetz, Berlin.

SACK, G. (1940). Ein Fall von Azeton Vergiftung. *Arch. Gewerbepath. Gewerbehyg.*, **10**, 80.

SALANT, W. and KLEITMAN, N. (1922). Pharmacological studies on acetone. *J. Pharmacol.*, **19**, 293.

SANDERSON, J. McE. (1933). Solvents used in surface coatings. *Paint Oil Chem. Rev.*, **95**, No. 2, 10.

SANDERSON, J. McE. (1933). Solvents used in surface coatings: Group VI: Ketones. *Paint Oil Chem. Rev.*, **95**, No. 18, 6.

SCHLOMOVITZ, B. H. and SEYBOLD, E. G. (1924). The toxicity of the "acetone bodies". 1. Acetone administered intravenously. *Amer. J. Physiol.*, **70**, 130.

SCHULTZE, D. (1932). Dissertation, Würzburg. (Cited by Flury and Wirth, 1934, *Arch. Gewerbepath. Gewerbehyg.*, **5**, 1.)

SCHWARTZ, L. (1898). Über die Oxydation des Acetons und homologer Ketone der Fettsaurereihe. *Arch. exp. Path. Pharmak.*, **40**, 168.

SCHWARTZ, L., TULIPAN, L. and PECK, S. M. (1947). Occupational diseases of the skin. Kimpton, London.

SILVERMAN, L., SCHULTE, H. F. and FIRST, W. W. (1946). Further studies on sensory response to certain industrial solvent vapors. *J. industr. Hyg.*, **28**, 262.

SKLIANSKAYA, R. M., URIEVA, F. E. and MASHBITZ, L. M. (1936). Relative toxicity of acetone, methyl alcohol and their mixtures. *J. industr. Hyg.*, **18**, 106.

SMITH, A. R. and MAYERS, M. R. (1944). Poisoning and fire hazards of butanone and acetone. *Industr. Bull.*, **23**, 174.

SMOLEY, E. R. and KRAFT, W. W. (1935). Production of lubricating oils. *Industr. Engng. Chem.*, **27**, 1418.

SMYTH, H. F., jr. and SEATON, J. (1940). Acute response of guinea pigs and rats to inhalation of the vapors of isophorone. *J. industr. Hyg.*, **22**, 477.

SMYTH, H. F., jr., SEATON, J. and FISCHER, L. (1942). Response of guinea-pigs and rats to repeated inhalation of vapors of mesityl oxide and isophorone. *J. industr. Hyg.*, **24**, 46.

SPATZ, R. (1930). Die quantitative Bestimmung kleiner Mengen von Alkohol- und Acetodämpfen in Luft. *Arch. Hyg., Berl.*, **91**, 315.

SPECHT, H. (1938). Acute response of guinea pigs to inhalation of methyl iso-butyl ketone. *Publ. Hlth. Rep., Wash.*, **53**, 292.

SPECHT, H., MILLER, J. W. and VALAER, P. J. (1939). Acute response of guinea-pigs to inhalation of dimethyl ketone (acetone) vapor. *Publ. Hlth. Rep., Wash.*, **54**, 944.

SPECHT, H., MILLER, J. W., VALAER, P. J. and SAYERS, R. R. (1940). Acute response of guinea-pigs to the inhalation of ketone vapor. *Nat. Inst. Hlth. Bull.*, No. 176, 36.
STRONG, G. F. (1944). Acute acetone poisoning: case report. *Canad. med. Ass. J.*, 51, 359.
SYMONS, P. S. (1931). Nitrocellulose film structure. *Paint Varn. Lacq. Manuf.*, 1, 21.
VON TAPPEINER, H. (1884). Über die giftigen Eigenschaften des Acetons. *Dtsch. Arch. klin. Med.*, 34, 450.
TERSAND, R. (1935). Les propionates et l'oxide de mésityle, solvants de la nitrocellulose.' *Rev. Prod. chim.*, 38, 97.
TREON, J. F., CRUTCHFIELD, W. E., jr. and KITZMILLER, K. V. (1943). The physiological response of rabbits to cyclohexane, methyl cyclohexane and certain derivatives of these compounds. I. Oral administration and cutaneous application. *J. industr. Hyg.*, 25, 199.
TREON, J. F., CRUTCHFIELD, W. E., jr. and KITZMILLER, K. V. (1943). The physiological reponse of animals to cyclohexane, methyl cyclohexane and certain derivatives of these compounds. II. Inhalation. *J. industr. Hyg.*, 25, 323.
U.S. BUREAU OF MINES (1921). Technical paper, No. 272. U.S. Dept. of Interior. (Cited by Flury and Wirth, 1934, *Arch. Gewerbepath. Gewerbehyg.*, 5, 1.)
WALTON, D. C., KEHR, E. F. and LOEVENHART, A. S. (1928). Comparison of the pharmacological actions of diacetone alcohol and acetone. *J. Pharmacol.*, 33, 175.
YARSLEY, V. E. (1934). Solvents and plasticisers. Parts I, II. A review of recent progress with special reference to cellulose acetate. *Synth. appl. Finish.*, 5, 37, 57.
ZANGGER, H. (1930). Über die modernen organischen Lösungsmittel. *Arch. Gewerbepath. Gewerbehyg.*, 1, 77.
ZANGGER, H. (1930). Weitere Mitteilungen über Vergiftungen durch flüchtige Gifte und deren Beziehung zu gewerblichen Vergiftungen. *Schweiz. med. Wschr.*, 60, 1193.

CHAPTER VII

GLYCOLS AND THEIR DERIVATIVES

1. Ethylene Glycol

$$HO \cdot CH_2 \cdot CH_2 \cdot OH$$

PROPERTIES

A colourless, odourless liquid, with a bitter-sweet taste. It is somewhat viscous, as hygroscopic as glycerol, and miscible in all proportions with water, ethyl alcohol, methyl alcohol, glycerol and acetone. When it is mixed with water, heat is evolved. It is non-inflammable.

It is observed by Hanzlik, Seidenfeld and Johnson (1931) that many specimens of supposedly pure ethylene glycol contain impurities, as indicated by darkening of the product on heating; according to Lunge and Berl (1933), commercial ethylene glycol contains no appreciable quantity of impurities apart from water; but sometimes its higher homologues, propylene or butylene glycols, may be present, as indicated by a low specific gravity of the dry rectified glycol.

B.P. 197–198° C. Sp.Gr. 1·116–1·118.

USES

Ethylene glycol is a solvent for some dyes and gelatinizes cellulose nitrate, but is not applicable to the preparation of lacquers and varnishes. It readily dissolves salts, many heavy metallic salts, barbital, alkaloids, iodine, etc.

In 1916, Bachem suggested it as a substitute for glycerine in the preparation of drugs, while Curme and Young (1923) described its uses as a solvent, preservative and organic base. Haag and Bond (1927) have used it as a solvent for chloretone for an intraperitoneal anaesthetic for laboratory animals.

Its chief applications in industry at present appear to be:

(1) As a constituent of ethylene glycol electrolyte for electrolytic condensers.

(2) As a cosmetic added to "stanniform powder" in skin lotions used for burns, sunburn, etc. The combination is a physical not a chemical one, the ethylene glycol being unchanged in the lotion.

(3) As a constituent of anti-freeze mixtures for car radiators.

(4) As a constituent of flavouring essences.

TOXICITY

Numerous experiments with oral, subcutaneous, intramuscular, intravenous and intraperitoneal administration have established the fact that its effect on the animal organism is that of a depressant of the central nervous system with severe toxic action upon the kidneys. Although this effect has further been observed in several cases of accidental ingestion by human beings, there is little evidence that the use of ethylene glycol in industry is attended by danger to health. Reid Hunt (1932) remarks that "it is so slightly volatile that there is no danger of poisonous doses being inhaled and there is no evidence that it is absorbed from the skin. . . . Used legitimately, this compound is a far less dangerous substance than aniline, benzene, carbon disulphide, turpentine, carbon tetrachloride, and many other chemicals extensively employed in the industries." Dyson (1934), however, also referring to the fact that most of the investigations indicating the toxicity of ethylene glycol have been based on its administration by injection, observes: "It is clear that continued exposure to vapour, or continued taking of small quantities in beverages of flavoured foodstuffs must be treated as a different problem involving a greater possibility of chronic poisoning." An investigation carried out some years ago by the Factory Inspection Department of the Home Office gave only slight support to this possibility (p. 345). It is also of interest to note that in the list of occupational diseases in New York for 1934, published by the Division of Industrial Hygiene (1935), one case is mentioned under the heading of "Ethylene Glycol" as having been allowed compensation. No details of this case are given.

Some recent experiments on the effects of inhalation of the vapour from ethylene glycol (Flury and Wirth, 1934; Wiley, Hueper and von Oettingen, 1936) indicate that, with concentrations which would be encountered under normal conditions, little, if any, pathological effect is to be expected. With regard to its inclusion in any food or drug preparation, however, its toxicity is not negligible, according to Laug, Calvery, Morris and Woodard (1939). These workers advise that it should be omitted entirely from food and drug preparations.

Relative Toxicity

Laug and co-workers (1939), determining the relative toxicity to animals of several members of the glycol group by stomach-tube introduction, found that all except propylene glycol produced injuries to the liver and kidneys. Ethylene glycol was found to be more toxic than diethylene glycol, but less toxic than the monoethyl ether of both these glycols. That ethylene glycol is much more toxic than propylene glycol as judged by oral administration and injection into animals, has been shown by Reid Hunt (1932) and confirmed by Hanzlik and co-workers (1931). Reid Hunt found that young rats grew at the normal rate and reached maturity when the only liquid they received was a 5 per cent solution of propylene glycol; rats receiving a similar solution of ethylene glycol died within 2 or 3 days. The same order of toxicity for the three glycols is shown by the results (L.D. 50) of stomach-tube administration (Laug, Calvery, Morris and Woodard, 1939); the results of their examination of some glycol derivatives by this method show some slight difference from the minimum lethal dose (M.L.D.) determined by von Oettingen and Jirouch (1931) in mice by subcutaneous administration, as can be seen in Table 51.

TABLE 51

Relative toxicity of glycols and their derivatives

Solvent	Lethal dose (ml./kg.)	
	Laug et al. (L.D.50)	Von Oettingen and Jirouch (M.L.D.)
Propylene glycol	23·9	—
Diethylene glycol	23·7	5
Ethylene glycol	13·1	2·5
Diethylene glycol monoethyl ether	6·58	2·5–5·0
Dioxan	5·66	>10
Ethylene glycol monoethyl ether	4·31	5
Ethylene glycol monoethyl ether monoacetate	—	5
Ethylene glycol monobutyl ether		0·5

TOXIC EFFECTS IN ANIMALS

Lethal Dose

By oral administration. Fairly large doses of ethylene glycol are apparently well tolerated. Page (1927) gave 9 ml. per kg. of body weight (as a 30 per cent solution) to dogs without harm, while Bachem (1917) also found little effect from the administration of 20 g. diluted with water to rabbits (approximately the same amount per kg.). Larger doses than this, however, produced haemoglobinuria, collapse and death. Hanzlik, Seidenfeld and Johnson (1931) found 4 to 7 g. per kg. fatal for the majority of rats. Reid Hunt (1932) found that the addition of 5 per cent of ethylene glycol to the drinking water of rats caused death within 3 days, while with the addition of 1 per cent many died within 7 days.

By stomach tube introduction. Laug et al. (1939) found the L.D. 50 for mice to be 13·1 ml. per kg. as compared with 23·7 ml. for diethylene glycol.

By subcutaneous injection. For mice, von Oettingen and Jirouch (1931) found 2·5 ml. per kg. body weight a lethal dose. Apparently mice are more susceptible than other animals to this compound, for Page (1927) observed in dogs only slight irritation after subcutaneous injection of 3 ml. per kg., while Bachem (1917) found that only doses larger than 10 to 15 ml. per kg. produced haematuria and collapse in rabbits.

By intravenous injection. For rats, 2·2 g. per kg.; for rabbits, 3·3 g. per kg. (Hanzlik et al., 1931).

By intramuscular injection. For rats, 4·4 g. per kg.; for rabbits, 6·6 g. per kg. (Hanzlik et al.).

These values agree fairly closely with those of Reid Hunt (1932), though the latter's figures indicate a slightly higher degree of toxicity.

By intraperitoneal injection. The results obtained by Haag and Bond (1927) from intraperitoneal injection led them to regard ethylene glycol as having a low toxicity, since they found that 5 ml. per kg. of body weight in dogs produced no effect. Page (1927) also found no effect from 6 ml. per kg., but at 8 ml. per kg. depression, weakness and muscle twitching occurred and with 10 ml. death took place within 12 hours.

By inhalation. Air containing 0·5 mg. per l. is, according to Flury and Wirth (1934), to be regarded as saturated, and this concentration was not found to be lethal to rats subjected to it for 28 hours during 5 days. It produced only slight narcosis with recovery.

Acute Poisoning

Symptoms

Central nervous effects. The symptoms preceding death, after intravenous or intramuscular injection of a lethal dose, were those of respiratory acceleration, loss of equilibrium, and general motor depression (Hanzlik and co-workers, 1931). The effect of intravenous and intramuscular injections differed only in that the former produced immediate muscular excitation (twitching, tremors and convulsions). Central depression was also observed by von Oettingen and Jirouch (1931) in frogs, after injection of 1 ml. of a 25 per cent aqueous solution into the lymph sac, together with twitching of the muscles and spastic extension which was not stopped by decapitation and pithing of the spinal cord and which was therefore partly peripheral in origin. A fall in blood pressure occurred after the intravenous injection of a 50 per cent solution, presumably due to central or cardiac depression. This was also observed by Page (1927). It was followed by an increase in the amplitude of the heart beat with either no change or a slight slowing of the heart rate.

Effects on the kidneys. The earlier investigations by Pohl (1896), Mayer (1902, 1917), Dakin (1907) and Bachem (1917) showed that an increase in the oxalate content of the urine is possible after the administration of ethylene glycol, since oxalic acid and glycollic acid are intermediate products in its final oxidation to carbon dioxide and water. Experimentally, these workers found that changes in urinary oxalate occurred only after large doses—1 to 2·6 g. per kg. body weight—the conversion of glycol to oxalic acid being only from 0·4 to 3 per cent. Although severe injury to the kidneys in animals has been demonstrated after administration of ethylene glycol, the relation of this injury to the formation of oxalic acid does not appear to have been definitely proved.

A condition of acute nephrosis in rats was produced by von Oettingen and Jirouch (1931) by the subcutaneous injection of 2·5 to 5·0 ml. per kg. of a 50 per cent solution, while 10 ml. per kg. resulted in filling of the intracapsular spaces and tubules with blood, and in degenerative processes which these workers regard as an indication of the production of pathological processes in the kidney. The production of oxaluria was not investigated. Mayer (1917) claims to have observed renal injury and the formation of calculi after the hypodermic injection into animals of 13 to 15 g., but the results of the investigations of Hanzlik and co-workers (1931) were somewhat irregular. They examined the effects of long-continued administration of small amounts, both subcutaneously and orally, on the urine and kidneys of rats. Subcutaneously, 0·05 ml. (0·055 g.) was given daily for 40 days. No abnormalities were found. Orally, 10, 5, 3, 2, 1 and 0·5 per cent solutions in water were given, the lowest concentrations daily for from 120 to 130 days. With 0·5 per cent no oxaluria or renal changes were found. With 1 per cent there was a variable presence of calcium oxalate crystals and red corpuscles in the urine. In one rat renal calculi consisting of calcium phosphate, carbonate and oxalate were found, in the others abnormalities occurred occasionally or not at all.

Reid Hunt (1932) at first attributed the toxic action on the kidneys to the production of oxalic acid, but later thought that it might rather be due to other toxic intermediates which are destroyed within the body. A later investigation of this aspect of the toxicity of ethylene glycol by Wiley, Hueper, Bergen and Blood (1938) appears to indicate that while ethylene glycol does cause an increase in urinary oxalic acid, the amount formed is scarcely sufficient to account for the actual injury to the kidneys. The animals investigated all showed degenerative changes in the kidneys, but no oxalic acid deposits in the renal parenchyma. There was a high incidence of degenerative changes in the pyramidal tissue, which these workers suggest may indicate the existence of some systemic organic injury by the actual substance injected.

Local irritant effects. The irritant action of undiluted ethylene glycol on mucous surfaces and on motor nerve fibres is weak, compared with that of some of its derivatives, and, according to Hanzlik and co-workers, is due to hyper-tonicity; 10 and 2·1 per cent solutions of glycol in water produced no demonstrable effects. Von Oettingen and Jirouch found it to have practically no inhibitory effect on the reflex of time of withdrawal of the foot of the decapitated frog as measured by the Turck method, while its inflammatory effect on the cornea was very slight, producing only hyperaemia.

Chronic Poisoning

Effects on Internal Organs of Inhalation of Vapour

Concentrations of 0·398 mg. per l. (average) were obtained by Wiley, Hueper and von Oettingen (1936) by warming ethylene glycol in a cylinder to 62° C. and passing air over it at a rate of 9 l. per minute. The inhalations (by mice and rats) were continued, 8 hours a day, 5 days a week for 16 weeks. Although some of the animals died during the investigation their deaths were attributed to other causes. Their weight showed slight variations, but in mice the resulting average was an increase. The only pathological changes found were evidences of intestinal infection (ulcerations of the caecum and cystic lymph nodes in the peri-caecal tissue). The internal organs were all healthy. These workers concluded, therefore, that ethylene glycol vapour, in concentrations of the order of 300 mg. per 1,000 l. repeated over a period of 3½ months, has no injurious effect. It appears unlikely that higher concentrations will be encountered under normal conditions.

TOXIC EFFECTS IN MAN

The only serious cases of poisoning by ethylene glycol reported have occurred from the accidental drinking of a comparatively large quantity.

The exact dose which will produce acute symptoms is not known, but Dyson (1934), on the basis of the amount supposed to have been drunk in the two almost fatal cases reported by Hansen (1930), suggests about 100 ml. as the fatal dose. In the eighteen fatal cases reported by Pons and Custer (1946) the dosage was known in only one case, 200 ml. Page (1927) drank 15 ml. diluted without any ill-effect; Bachem (1917) took larger quantities than this and though feeling no subjective ill-effects had an increased excretion of oxalic acid in the urine.

According to Hanzlik, Seidenfeld and Johnson (1931), the effect of swallowing a moderate amount of ethylene glycol is a sensation of warmth on the tongue and in the oesophagus.

The effect of intramuscular injection of ethylene glycol was tested to some extent by these workers, by treating patients with "Iodobismitol", the bulk of which was glycol. Out of fourteen patients so treated, eleven showed no oxaluria.

Acute Poisoning

Symptoms

In fatal cases, two of which have been described in the Answer to a Query in the *Journal of the American Medical Association* (1930) and eighteen by Pons and Custer (1946), death appears to be due to respiratory failure and convulsions. All these deaths resulted from drinking anti-freeze radiator fluid, and the symptoms before death consisted of vomiting, cyanosis, extreme prostration and coma within the first few hours. The post-mortem findings in Pons and Custer's cases indicated severe damage to the renal epithelium in one case only, though in other cases the urine before death contained oxalate crystals, and there were crystals of calcium oxalate in the kidneys in others. Lesions in the central nervous system varied from congestion and oedema to exudative meningo-encephalitis.

In non-fatal cases, two of which have been recorded by Hansen (1930) and also by Brekke (1930), the symptoms appear to confirm the finding in experimental animals of severe injury to the kidneys. The patients, two young men, drank ethylene glycol from a tin found in a garage. Two hours later they complained of a feeling of intoxication, and soon passed into a condition of stupor, somnolence and coma. Complete double abducens paralysis was present and the pupils were dilated and failed to react. Death was apparently only averted by surgical decapsulation of one kidney, which was found to be enlarged and dark red and in a condition of haemorrhagic nephritis.

Chronic Poisoning

Symptoms

Careful examination of workers in factories in England where ethylene glycol has been used has revealed very few definite symptoms of intoxication, but several cases have shown slight urinary abnormalities. After fairly long periods of exposure the only symptoms complained of were slight eye irritation and "phlegm" in one case (after 2 years), and loss of appetite and slight "dopiness" in another (after 9 months). The urines of four of these workers were examined: in two of them there was found considerable albuminuria and increased urobilin; and in one of these two, more red blood corpuscles than would be expected in a normal specimen; in the third, a trace of albumin but no red blood corpuscles or casts; the fourth was normal. (The patient with considerable albuminuria had recently had influenza.)

2. Ethylene Glycol Monomethyl Ether
(*Methyl Cellosolve, 2-Methoxyethanol*)

$CH_3 \cdot O \cdot CH_2 \cdot CH_2 \cdot OH$

PROPERTIES

A colourless, permanently neutral, very stable liquid, with a very faint pleasant odour. Miscible with water, with which it forms a constant boiling mixture; miscible with aromatic and light paraffin hydrocarbons, but not with heavy paraffins or linseed oil. Hygroscopic and of fairly high volatility, B.R. 124–125° C. Sp.Gr. 0·967–0·975. Fl.P. 36° C.

Good solvent for cellulose acetate and nitrate, but not such a good solvent for most resins as is the monoethyl ether (cellosolve).

USES

(1) In the lacquer industry, especially for white lacquers, its hygroscopic quality tending to produce "blushing" in coloured lacquers, unless used with higher glycol ethers or other solvents.
(2) As a stiffener for shirt collars.
(3) As a solvent for rotogravure inks.

TOXICITY

While methyl cellosolve appears from the results of animal experiments to be a potential toxic agent for the kidneys, in human beings its effects have been exerted predominantly on the nervous system, especially with chronic exposure. It has also been found, in both animals and man, to produce anaemia of a macrocytic type.

Concentrations in Atmosphere

The actual concentrations in the shirt-collar factory in which cases of chronic intoxication occurred (p. 347) were 25 p.p.m. with the windows open and 76 p.p.m. when closed. In a cement spraying and printing process the concentrations estimated by Elkins, Stortazzi and Hammond (1942) were approximately 25 and 20 p.p.m. respectively. No ill-effects were observed in these factories where the process was intermittent.

Greenburg, Mayers, Goldwater, Burke, Moskowitz and Moskowitz (1938) suggest 25 p.p.m. as the maximum safe limit.

The method of estimation used by Elkins *et al.* is a differential oxidation by potassium dichromate. This is effective in the presence of ethyl and methyl alcohols, but is interfered with by *iso*propyl alcohol, ethyl cellosolve, and probably by some other water-soluble solvents and by acetone, but the latter can be removed by aeration.

TOXIC EFFECTS IN ANIMALS

Large doses of methyl cellosolve appear to exert an injurious effect on the kidneys similar to that of ethylene glycol and its other monoalkyl derivatives. Repeated smaller dosage has a less marked effect on the kidneys. Changes in the blood can apparently be correlated with a similar effect observed in human beings.

Acute Poisoning

Like other members of the monoalkyl ether series, methyl cellosolve is lethal to animals in fairly small dosage by injection, oral administration or inhalation.

By oral administration. According to Smyth, Seaton and Fischer (1941) methyl cellosolve is more toxic than cellosolve and butyl cellosolve for guinea-pigs, but less toxic for rats than the butyl derivative.

By injection. Wiley *et al.* (1938) state that methyl cellosolve, given by injection, is "rather toxic" to dogs and rabbits, one dog becoming anuric after three injections of 6 ml. daily, and one rabbit dying after two injections of 2 ml.

By inhalation. The minimum lethal dose for mice (L.D. 50) is given by Werner, Mitchell, Miller and von Oettingen (1943) as 4·6 mg. per l. (1,480 p.p.m.) for methyl cellosolve, and 3·4 mg. per l. (700 p.p.m.) for the most toxic of the series, butyl cellosolve. Their figure for 100 per cent mortality in 32 hours (10·5 mg. per l.) is practically the same as that given by Starrek (1938), cited by Lehmann and Flury (1943), 10 mg. per l.

No typical narcotic action was observed following either injection or inhalation. Dyspnoea and weakness were the most constant results; death usually occurred between 7 and 32 hours from the beginning of exposure.

Lesions of the Internal Organs

In the animals subjected to injection by Wiley *et al.*, marked degenerative lesions of the kidneys, hyperaemic changes in the liver and degeneration of the spermatic epithelium were found. These findings were not so marked or constant in mice subjected to inhalation. The spleen showed the most consistent evidence of toxic effect, though a few animals showed interstitial nephritis, and a few showed a typical picture of broncho-pneumonia.

Chronic Poisoning

The results of repeated injections or repeated exposure to inhalation have given somewhat varying results in the hands of different observers and with different species of animals. With repeated injection, evidence of kidney injury is more prominent than with inhalation. Lehmann and Flury found red blood corpuscles, albumin and cylindrical casts in the urine of rabbits, and glomerulitis at autopsy after two to seven injections of 0·1 to 2 ml. per kg. After repeated inhalations of 2·5 to 5 mg. per l. (800 to 1,600 p.p.m.) they mention slight temporary albuminuria in one animal only. In the inhalation experiments of Werner, Nawrocki, Mitchell, Miller and von Oettingen (1943) and Werner, Mitchell, Miller and von Oettingen (1943), using rats and dogs respectively, dogs showed much more marked evidence of toxicity to the blood than did rats following exposure to methyl cellosolve. Injury to the kidneys, as evidenced by pathological changes in the urine, or in the kidneys themselves, was not of predominant importance with the concentrations used (300 to 400 p.p.m. for rats and 800 p.p.m. for dogs).

Effect on the Blood

Changes in the blood picture were more striking in the dogs than in the rats and were indicative of anaemia and a shift to the left, the percentage of young granulocytes in the circulating blood increasing between the first and eighth weeks of exposure. The actual site of action of this effect upon the blood was not completely definable. It appears that the bone marrow was not seriously damaged nor was there any definite indication of a haemolytic action since there was no significant increase of haemosiderin in the spleen; also methyl cellosolve, which is less haemolytic *in vitro* than cellosolve or butyl cellosolve, produced the greatest alteration in the red cells.

The variations in response of the animals are, in the opinion of Werner *et al.*, due mainly to species differences, and they indicate that toxic effects in man might manifest themselves in different ways. In view of the fact, however, that similar blood changes have been observed in human beings (Parsons and Parsons, 1938; Greenburg *et al.*, 1938) it is suggested that blood examinations might offer a means of detecting toxicity.

TOXIC EFFECTS IN MAN
Acute Poisoning

Only one fatal case of poisoning by methyl cellosolve has been recorded and this followed its accidental ingestion in an industrial process. The case is described by Young and Woolner (1946). About half a pint of fluid, which on analysis showed the constants characteristic of methyl cellosolve, had been ingested; the man was comatose and died 5 hours later without regaining consciousness. The chief post-mortem findings were haemorrhagic gastritis and toxic changes in the kidneys and liver.

Chronic Poisoning

With the concentrations likely to be present during the industrial use of methyl cellosolve, it appears that injury to the kidneys and liver is less probable than a toxic action on the nervous system and the blood.

Effect on the Nervous System

A form of "toxic encephalopathy" described by Donley (1936) occurred in a woman employed in the manufacture of shirt-collars. She suffered from headache, drowsiness and general loss of interest. She showed irregular and unequal pupils and some ataxia. The two cases reported by Parsons and Parsons (1938) were similar, but more severe in character. The symptoms of these two young men were lassitude, sleepiness, giddiness, severe frontal headache, burning and weakness of the eyes, and a complete change of personality from intelligence and quickness to stupidity and lethargy. They showed a general hypersensitivity to light and accommodation of the pupils, moderate ataxia and a positive Romberg reaction. The more severe case also had gastro-intestinal disturbance and both men showed loss of weight and some nocturia. Both recovered with apparently no permanent damage to the nervous system.

Abnormal neurological findings, in the form of abnormal reflexes and tremor of the hands, were present in four out of nineteen workers in the same factory examined by Greenburg *et al.*, and four other workers, in addition to these signs, showed the mental retardation described by Parsons.

Effect on the Blood

Anaemia of a macrocytic type was present in Parsons' two cases, and in eight of the nineteen examined by Greenburg *et al.* Parsons also observed granulopenia, but Greenburg found no granulopenia or leucopenia, but a general immaturity of the leucocytes (shift to the left) in every case. The high colour index and relative lymphocytosis persisted in Parsons' cases even after removal from exposure and suitable anti-anaemic treatment.

3. Ethylene Glycol Monoethyl Ether
(Cellosolve, Solvulose, 2-Ethoxyethanol)

$C_2H_5 \cdot O \cdot CH_2 \cdot CH_2 \cdot OH$

PROPERTIES

A colourless, permanently neutral, very stable liquid, nearly odourless in low concentrations, rather disagreeable in high concentrations. Taste at first sweet, followed by a burning sensation. Completely miscible with water. B.R. 126–138° C. (pure 134·8° C.) Sp.Gr. 0·934–0·938. Fl.P. 40° C. Its vapour is approximately three times heavier than air. Vapour pressure at 20° C. 4·5 mm. Hg.

A.S.T.M. Specification, D 331–35. Sp.Gr. 0·927–0·933 at 20° C. B.R. up to 128° C. none; 128–132° C. 5 per cent maximum; below 136° C. 95 per cent minimum; above 137° C. none. Residue, 0·005 per cent maximum. Acidity, 0·02 per cent acetic acid, maximum.

Cellosolve is a more powerful solvent for nitrocellulose and stands dilution with non-solvents better than any solvent of similar boiling range that has been examined (Davidson, 1926; Smith, 1928). Its ease of application and lack of residual odour make it specially suitable for interior decoration.

USES

Its chief industrial use is in the lacquer industry, especially for lacquers for refrigerators, kitchen cabinets, etc.

TOXICITY

Examination of workers using lacquer containing cellosolve has revealed no definite symptoms, and only rarely, slight physical signs which might indicate changes in the liver or kidneys. Experiments carried out on animals have given results which suggest a possible toxic action on the central nervous system and the kidneys; even with subcutaneous injection cellosolve appears to be less toxic than ethylene glycol itself, while, with inhalation, serious acute poisoning could only be produced at room temperatures by long exposure to saturated air on account of the low vapour pressure. At this concentration there are warning signs of danger—a disagreeable odour and some irritation of the eyes. It is pointed out by Hamilton (1934), however, that if cellosolve is heated, as in drying or baking coated goods, a much higher concentration in the room air may be produced.

TOXIC EFFECTS IN ANIMALS
Acute Poisoning
Effect of Inhalation

The effect of exposure to varying concentrations of the vapour of cellosolve has been investigated in guinea-pigs by Waite, Patty and Yant (1930), and, with a view to establishing the comparative toxicity of this substance with other monoalkyl ethers of ethylene glycol, in mice by Werner, Mitchell *et al.* (1943). Waite *et al.* (1930) found that it required 24 hours' continuous exposure to 0·6 per cent by volume of the vapour (the highest concentration obtainable at room temperature) to kill the animals at the time of the exposure; or a similar period of exposure to half-saturated air to kill them 24 hours after removal from exposure. Werner *et al.* took as their criterion of toxicity the minimum concentration which would kill 50 per cent of the animals, basing this figure on all deaths within 3 weeks of exposure. They found this dosage to be 6·7 mg. per l. (1,820 p.p.m.), as compared with 4·6 mg. per l. (1,480 p.p.m.) for methyl cellosolve, and with 3·4 mg. per l. (700 p.p.m.) for butyl cellosolve. The order for toxicity is therefore: cellosolve < methyl cellosolve < butyl cellosolve. Werner *et al.* point out that the extent of industrial hazard is related to the amount of toxic substance possible in the atmosphere as well as to its actual toxicity. From this fact they have constructed a "hazard index" for each substance by dividing the average saturation concentration by the minimum lethal concentration (Table 52). Translating the above values into the actual industrial hazard of the three solvents, therefore, the order of toxicity becomes altered to butyl cellosolve < cellosolve < methyl cellosolve.

TABLE 52
Relative toxicities of cellosolve, methyl and butyl cellosolve

	Methyl cellosolve	Cellosolve	Butyl cellosolve
Saturation concentration (mg./l.)	31·6	19·0	4·7
Minimum lethal concentration (mg./l.)	4·6	6·7	3·4
Hazard index	6·9	2·8	1·4

This apparently indicates that the butyl derivative is least hazardous with regard to acute toxicity, but Werner *et al.* point out that a relatively low index does not necessarily indicate an insignificant hazard; it may indicate one which can be relatively easily eliminated.

Effect of Subcutaneous Injection

As estimated by von Oettingen and Jirouch (1931) by subcutaneous injections into mice, cellosolve appears to be less toxic than ethylene glycol itself, considerably less toxic than butyl cellosolve, but more toxic than dioxan. The minimum lethal dose was found to be 5·0 ml. per kg. body weight, compared with 2·5 ml. for ethylene glycol, 0·5 ml. for the butyl ether, and above 10 ml. for dioxan. As a central nervous and cardiac depressant, the action of cellosolve was less than that of any of the other ethylene glycol derivatives tested. Its depressant effect on motor nerve fibre, on excised skeletal and smooth muscle, while less than the corresponding effect of butyl cellosolve, was greater than that of ethylene glycol. On heart muscle it had practically no effect, in contrast to the diethylene glycol ether (carbitol) which was comparatively depressant, and to butyl cellosolve which was extremely depressant.

The low toxicity of this compound especially for the central nervous system, as compared with other ethylene glycol derivatives, is correlated by von Oettingen and Jirouch with its slight effect in precipitating protein, which is based on its comparatively high surface tension compared with that of ethylene glycol and butyl cellosolve, and with its low lipoid partition coefficient.

Symptoms

No narcotic action is exerted by lethal or near lethal concentrations of cellosolve. In the experiments of Werner *et al.* dyspnoea was the most marked symptom; they and Waite *et al.* also noted extreme weakness. A few animals showed lens or corneal opacities after exposure to very high concentrations.

Lesions of the Internal Organs

The pathological changes in the series of animals examined by Waite *et al.* consisted of lung congestion and oedema, hyperaemia of the kidneys and occasional haemorrhage into the gastric mucosa, and, according to Waite *et al.*, were less with cellosolve than with the other derivatives. There was no evidence of liver damage, and no evidence of interstitial nephritis, but some congestion of lung tissue, and the spleen showed the most marked toxic change—follicular phagocytosis.

Effects on the Kidneys

Like the other ethylene glycol compounds studied by von Oettingen and Jirouch, large doses (10 ml. per kg. of a 50 per cent solution) of cellosolve injected subcutaneously into rats produced acute nephrosis, with degenerative processes and filling of the intracapsular spaces and tubules with blood.

Chronic Poisoning

Exposure of both rats and dogs to repeated inhalations of cellosolve has been shown by Werner, Mitchell *et al.* (1943) to have less toxic effect than the other derivatives, particularly with regard to the effect upon the blood picture. Concentrations of 300 to 400 p.p.m., however, did produce a slight increase in the percentage of young granulocytes in the circulating blood and a decrease in the density of the cytoplasm of liver cells, which did not persist after exposure was discontinued.

TOXIC EFFECTS IN MAN

During the investigation carried out by Waite *et al.* (1930), two men breathed air containing 0·6 per cent by volume of cellosolve for a few seconds. They found it irritating to the eyes and of a disagreeable odour. According to Davidson (1926), however, operators using cellosolve with a spray gun for large scale work can work all day without interruption or discomfort. Actual examinations of workers in the lacquer and paint manufacture by the Factory Department have revealed very little evidence of any injury to health from the use of cellosolve. In a cellulose paint factory no effects of any kind were noted. No symptoms were observed in three lacquer workers, one of whom had been so employed for 6 years and two for 13 years, but in one case the sclerotics had a "muddy (scarcely icteroid) tinge," and in another the urine showed a very faint trace of albumen, and the blood a slightly increased content of bilirubin, "which may indicate some slight degree of liver damage".

4. Ethylene Glycol Monoethyl Ether Monoacetate

(*Cellosolve Acetate, Ethyl Glycol Acetate*)

$$C_2H_5 \cdot O \cdot CH_2 \cdot CH_2 \cdot O \cdot CO \cdot CH_3$$

PROPERTIES

A colourless liquid with a weak, pleasant, ester-like odour, stronger than that of the simple monoalkyl ether. Taste first sour and then burning. B.R. of the technical product 140–160° C. (pure 153° C.). Sp.Gr. 0·975–0·982. Fl.P. 47° C. Vapour pressure at 20° C. 1·25 mm.Hg. Slightly soluble in water (22 per cent at 20° C.).

A.S.T.M. *Specification*, D343–35. Sp.Gr. 0·968–0·976 at 20° C. B.R. up to 145° C. none; 150–160° C. 90 per cent minimum; above 165° C. none. Residue, 0·03 per cent maximum. Esters, 95–96 per cent minimum. Acidity, 0·024 per cent maximum.

Solvent for cellulose nitrate, ester gum, coumarone, mastic, kauri, shellac. Durrans (1931) states that it is not a solvent for cellulose acetate, but Davidson (1926) states that it dissolves both cellulose nitrate and acetate; the difference probably lies in the varying degree of purity. It does not dissolve zanzibar or hard copal.

USES

Chiefly in the lacquer industry.

TOXIC EFFECTS IN ANIMALS

No experiments with inhalation of the vapour of this compound have been carried out. When administered by subcutaneous and intravenous injection, cellosolve acetate appears to be less acutely toxic than ethylene glycol, more toxic than dioxan, and of about the same toxicity as cellosolve. The minimum lethal dose for mice was found by von Oettingen and Jirouch (1931) to be 5 ml. per kg. body weight as compared with 2·5 ml. for ethylene glycol and 5 ml. for cellosolve. As a central nerve depressant, however, cellosolve acetate had a stronger action than either ethylene glycol or the monoethyl ether, 1 ml. of a 25 per cent solution producing very rapid central nervous paralysis and death within 5 minutes, but this action was considerably less than that of butyl cellosolve (q.v.) It produced great depression of motor nerves and heart muscle, while its action on excised skeletal and smooth muscle was stronger than that of any of the other glycol derivatives examined, except butyl cellosolve, to which it was about equal in this respect. Like the other compounds of the series, subcutaneous injection of large doses produced acute nephrotic changes in the kidneys. In the investigations of Wiley *et al.* (1938), repeated daily subcutaneous injection of 8·5 ml. to dogs produced no symptoms of toxicity, but did produce lesions of the kidneys. Animals killed 5 to 8 days after the last injection showed some calcium casts, mild nephrotic lesions and some degenerative lesions of the testes and of the brain. Some increase in urinary oxalic acid was found, as in the case of ethylene glycol, but Wiley *et al.* do not consider that the amount was sufficient to explain the toxicity of cellosolve acetate.

TOXIC EFFECTS IN MAN

While no toxic effects of cellosolve acetate on human beings have been recorded, it appears, on the basis of animal experiments, to have an action on the kidneys and the central nervous system similar to that of other glycol derivatives.

5. Ethylene Glycol Diethyl Ether
(*Diethyl Cellosolve*)

$C_2H_5 \cdot O \cdot CH_2 \cdot CH_2 \cdot O \cdot C_2H_5$

PROPERTIES

A colourless liquid of sweetish odour. B.P. 121°C. Sp.Gr. 0·853. Refractive index, 1·3914. It is considerably less soluble in water than the monoethyl ether (Davidson, 1926). At 22° C. it is 29·2 times less volatile than ether.

Solvent for cellulose nitrate, but not cellulose acetate; solvent for ester gum, elemi, zanzibar, coumarone, dammar, mastic, kauri, shellac; not for hard copal or sandarac.

USES

Chiefly in the lacquer industry, but not of such wide application as the monoethyl ether.

TOXICITY

No injury to human beings has been recorded from the use of diethyl cellosolve. From the results of animal experiments it appears to exert a mild narcotic effect and to produce some injury to the kidneys and some irritation of the respiratory tract.

TOXIC EFFECTS IN ANIMALS
Acute Poisoning

Lethal and Narcotic Doses

By oral administration. Cats, the most susceptible species, showed narcosis followed by death, after four doses of 1·0 ml. per kg. body weight, though similar doses were well-tolerated by dogs and rabbits.

By subcutaneous injection. For guinea-pigs, 0·5 to 1·0 ml. per kg. was lethal.

By inhalation. 50 mg. per l. (10,550 p.p.m.) was not lethal and caused only a tendency to narcosis.

Lesions of the Tissues

Animals which died after repeated subcutaneous injection of 0·5 to 1 ml. per kg. showed parenchymatous and interstitial nephritis.

Chronic Poisoning

Lethal and Narcotic Doses

By subcutaneous injection. 9·5 ml. daily produced no symptoms of intoxication in dogs (Wiley, Hueper, Bergen and Blood, 1938).

By inhalation. Repeated inhalations of 2·5 mg. per l. (520 p.p.m.) were lethal to rabbits and cats, but well tolerated by mice and guinea-pigs (Lehmann and Flury, 1943).

Lesions of the Internal Organs

After repeated subcutaneous injections, Wiley *et al.* found degenerative lesions of the kidneys and a mild degeneration of the liver cells and brain tissues. After inhalation, Lehmann and Flury found injury to the kidneys, and in one cat, purulent inflammation of the trachea.

TOXIC EFFECTS IN MAN

No ill-effects have been recorded.

6. Ethylene Glycol Mono-*n*-butyl Ether
(*Butyl Cellosolve*)

$C_4H_9 \cdot O \cdot CH_2 \cdot CH_2 \cdot OH$

PROPERTIES

A colourless liquid of mild odour. Taste at first sour, but later produces a burning sensation, followed by marked numbness of the tongue, indicating paralysis of sensory nerve endings (von Oettingen and Jirouch, 1931).

A.S.T.M. Specification, D330–35. Sp.Gr. 0·899–0·905 at 20° C. B.R. up to 163° C., none; to 167° C. 5 per cent maximum; to 171° C. 95 per cent minimum; Above 174° C., none. Residue, 0·005 per cent maximum. Acidity, 0·02 per cent acetic acid, maximum.

Good, but slow solvent for cellulose nitrate; solvent for shellac, sandarac, mastic, kauri, coumarone, ester gum, but not for copal or cellulose acetate. Miscible in all proportions with water at room temperatures but two phases may separate above 49° C. depending upon the proportions.

USES

(1) In the lacquer industry for brushing lacquers and for reducing their viscosity.

(2) In the cotton thread industry as a constituent of the "wetting out" solution.

TOXICITY

From the results of animal experiments it appears that butyl cellosolve is the most toxic of the monoalkyl ethers of ethylene glycol, though evidence of any injuries to man from its industrial use is very slight. Chronic exposure to the vapours may, according to Werner *et al.* (1943), have a slight though specific effect on the blood picture which may possibly be of value in its industrial use as an assessment of early toxic action.

TOXIC EFFECTS IN ANIMALS
Acute Poisoning

Lethal Dose

By oral administration. According to Lehmann and Flury (1943) the administration of 1 and 2 ml. per kg. body weight was lethal to rabbits within 22 or 30 hours; Smyth, Seaton and Fischer (1941) found butyl cellosolve the most toxic of the three cellosolves to rats, though less toxic than methyl cellosolve to guinea-pigs.

By subcutaneous injection. Von Oettingen and Jirouch (1931) consider the minimum lethal dose for mice to be 0·5 ml. per kg. body weight, as compared with 2·5 ml. for ethylene glycol, 5 ml. for cellosolve, and above 10 ml. for dioxan. For rabbits, Lehmann and Flury (1943) estimated the lethal dose at 0·4 ml. per kg., and for cats, 2·0 ml.

By inhalation. Butyl cellosolve has a higher toxicity than either cellosolve or methyl cellosolve, though, according to Werner *et al.* (1943), its "hazard index" (the relation between saturation concentration and minimum lethal concentration) is the lowest. This, as already explained (p. 349), does not necessarily indicate an insignificant hazard, but rather one which is relatively easily eliminated. The minimum lethal concentration in these experiments was found to be 3·4 mg. per l. (700 p.p.m.). The discrepancy between this result and that of Lehmann and Flury (1943), who state that "the inhalation of concentrations up to the saturation value (about 10 mg. per l. air, or 0·21 per cent) for 1 hour showed no effect on mice, guinea-pigs, rabbits or cats", may be explained by the fact that the results of Werner *et al.* (1943) were based on all deaths occurring within 3 weeks of exposure. It may also be noted that the latter give the saturation concentration of butyl cellosolve as 4·7 mg. per l.

Symptoms

In the experiments of von Oettingen and Jirouch (1931) the action of butyl cellosolve as a nervous depressant was very marked; rapid paralysis, followed by death within 1 minute, being produced by doses of 1 ml. of a 25 per cent aqueous solution. No twitching of the muscles, or spastic extension, such as were produced by ethylene glycol, preceded death. With regard to depression of nerve endings, smooth muscle and cardiac muscle, this compound was also by far the most toxic of those investigated. Its action on the kidneys was likewise very marked and, in the experiments of Lehmann and Flury (1943), extremely rapid. After oral administration they found the urine contained both blood and albumin with many granulated and epithelial casts and red and white corpuscles; the kidneys showed acute nephritis. Haemorrhage into the intracapsular spaces and tubules, as well as degenerative nephrotic changes, were also observed by von Oettingen and Jirouch (1931).

Marked dyspnoea and weakness were the most constant symptoms after inhalation, and, with lethal or near-lethal concentrations, haemoglobinuria. There was no narcotic action (Werner *et al.*, 1943).

Lesions of the Internal Organs

The kidneys showed only occasional interstitial nephritis. The liver showed definite damage in only one animal, which also had the most severe inflammatory changes in the lungs. The spleen showed follicular phagocytosis and congestion of the cavernous veins. The lungs showed a typical picture of broncho-pneumonia in a few cases.

Chronic Poisoning

Lehmann and Flury (1943) record that repeated subcutaneous injections of 0·5 to 0·1 ml. per kg. body weight produced no toxic effects in rabbits, but that some guinea-pigs died after several injections and showed albumen and red corpuscles in the urine.

In Lehmann and Flury's (1943) investigations, repeated inhalation of rather high concentrations (2·5 mg. per l. or 500 p.p.m.) produced the same signs of kidney injury as were observed from the more acute toxic action of butyl cellosolve. Smaller dosage (1·54 mg. per l. or 300 p.p.m.) produced no severe kidney lesions, but there was a moderate retention of urea in the blood. Changes in the blood picture observed by Werner et al. (1943) were slight, anaemia being much less severe than with methyl cellosolve, but there was a shift to the left in the white cell differential count. These changes occurred also in dogs exposed to 800 p.p.m. of butyl cellosolve.

TOXIC EFFECTS IN MAN

The only case giving evidence of possible injury from the industrial use of butyl cellosolve came under the Factory Inspectorate in 1934. It was that of a man who had two isolated attacks, with 5 months' interval, of haematuria. During the latter attack casts were present in the urine, but there was no albumen. A month later the urine was clear. The man also suffered from nasal catarrh, and the question arose of a possible absorption by way of the nasal passages. Since butyl carbitol may also have been used it is difficult to assess whether it or butyl cellosolve was the direct toxic agent. Irritation of the mucous membranes (in one case, eyes, and in another, the nose) and headache were complained of. In 1934 two girls, who had worked for a year on enamel splattering in which the enamel contained butyl cellosolve, reported similar symptoms.

7. Ethylene Glycol Monoacetate
(*Solvent G.C.*)

$$HO \cdot CH_2 \cdot CH_2 \cdot O \cdot CO \cdot CH_3$$

PROPERTIES

An odourless, colourless liquid, miscible with water or aromatic hydrocarbons, not miscible with benzine, paraffins or linseed oil, neutral and of low volatility. On prolonged storing it may develop slight traces of acetic acid, especially in the presence of water. B.R. 178–195° C. (about 181° C.). Sp.Gr. 1·109. Fl.P. 102° C.

Good solvent for cellulose nitrate and acetate, dyestuffs, and some resins and oils; not for shellac, kauri, sandarac, dammar, zanzibar, hard copals, ester gum or coumarone.

USES

(1) In the lacquer industry it has a very limited application, owing to its low volatility.

(2) In the textile industry in solutions of cellulose acetate intended for printing on fabrics.

(3) In the manufacture of essences and cosmetics.

TOXIC EFFECTS IN ANIMALS

An investigation by Wiley et al. (1938), primarily to determine whether the toxicity of ethylene glycol and some of its derivatives is due to increased excretion of oxalic acid, showed that ethylene glycol monoacetate is very similar in effect and toxicity to ethylene glycol itself. Injections of 8·5 ml. daily to dogs while producing no actual symptoms of toxicity did produce degenerative lesions of the kidneys, the testes and the brain. Like ethylene glycol also, the monoacetate produced an increase in the urinary excretion of oxalic acid, but not in sufficient amount to explain its toxicity.

TOXIC EFFECTS IN MAN
No ill-effects have been recorded.

8. Ethylene Glycol Diacetate
$$CH_3 \cdot CO \cdot O \cdot CH_2 \cdot CH_2 \cdot O \cdot CO \cdot CH_3$$

PROPERTIES
A colourless liquid, with slight odour resembling ethyl acetate (Durrans, 1931). B.P. 186–190° C. Sp.Gr. 1·11–1·15. Fl.P. 102° C. (Wolff, 1927). Esters, 98 per cent minimum. Not miscible with water, petroleum or linseed oil.

Solvent for cellulose acetate and nitrate, and some gums and resins; not for shellac, kauri, sandarac, dammar, zanzibar, hard copal, ester gum or coumarone.

USES
(1) In the lacquer industry; a very limited use.
(2) In developing blue-prints to some extent.

TOXIC EFFECTS
No animal experiments appear to have been carried out with this substance, and no toxic effects in human beings have been recorded.

9. Diethylene Glycol
(*Dihydroxydiethyl Ether*)

$$HO \cdot CH_2 \cdot CH_2 \cdot O \cdot CH_2 \cdot CH_2 \cdot OH$$

PROPERTIES
A colourless, almost odourless liquid, more viscous and more hygroscopic than ethylene glycol. It has a sweetish taste, with a distinctly bitter after-taste. B.P. 244·5° C. Sp.Gr. 1·121. Fl.P. 124° C. It is non-inflammable in air at ordinary temperatures, but if slowly heated in a shallow dish becomes inflammable at 130° C. (Renkenbach and Aaronson, 1931). It is miscible with water, alcohols, glycols, acetone, furfural, chloroform, esters, etc.; not miscible with ether, benzol, toluol, carbon tetrachloride, linseed oil or petroleums.

Solvent for cellulose nitrate and dyes, but not for ester gum, shellac, coumarone, copals or cellulose acetate.

USES
(1) In the manufacture of lacquers and thinners it has some application.
(2) In the manufacture of face creams.
(3) As a substitute for glycerine as a hygroscopic agent for tobacco (Flinn, 1935).
(4) As a vehicle for medicinal preparations in America, at one time with disastrous results.

TOXICITY

Though no toxic effects from the industrial use of diethylene glycol or from its use as a hygroscopic agent in tobacco have been recorded, the fatal results of its accidental ingestion suggest that it is potentially a severe toxic agent towards the kidneys; this is confirmed by the results of animal experiments.

TOXIC EFFECTS IN ANIMALS

Acute Poisoning

In the series of ethylene glycol derivatives investigated by von Oettingen and Jirouch (1931), subcutaneous injection into animals showed diethylene glycol to be less lethal than ethylene glycol (minimum lethal dose, 5·0 ml. per kg. body weight as compared with 2·5 ml. for ethylene glycol). The symptoms were those of central nervous depression with twitching of muscles and spastic extension, and the nephrotic and degenerative changes in the kidneys were like those produced by ethylene glycol. There was no depression of reflex nerve excitability, nor of the excitability of motor nerve fibres, but there was slightly more depression of excised skeletal muscle than with ethylene glycol and very slight depression of smooth and heart muscle. Its inflammatory action on the cornea was slight, producing only hyperaemia. This latter effect has been confirmed by Mulinos and Osborne (1935); they used cigarettes, in which diethylene glycol served as the hygroscopic agent, smoked in a mechanical puffer designed to simulate human smoking of cigarettes, and tested the effect of the smoke on the conjunctival sac of rabbits. Diethylene glycol was found in this respect much less irritating than glycerine, the average oedema production for glycerine being 2·7 as compared with 0·8 for diethylene glycol.

Chronic Poisoning

By oral administration. 4 per cent concentrations, according to Fitzhugh and Nelson (1946), produce extensive lesions in the urinary tract, including hydronephrotic changes with focal tubular atrophy and hyaline cast formation in the kidneys, bladder stones and tumours. Some of the tumours showed varying degrees of malignancy. Moderate liver damage in the form of hydropic degeneration was also caused in over half the rats at this level. According to Laug et al. (1939), Nelson and Morris (1941) and Morris, Nelson and Calvery (1942) in these respects it is slightly less toxic than ethylene glycol itself.

By skin application. According to Hanzlik, Lawrence, Fellows, Luduena and Laqueur (1947a) diethylene glycol is the only one of the glycols to cause liver injury by skin absorption. This injury occurred after ten daily applications of 1·0 ml. per kg. body weight.

TOXIC EFFECTS IN MAN

A serious non-industrial effect of diethylene glycol occurred in 1937. It was used as a constituent (about 72 vols. per cent) of a sulphonamide-elixir and caused over sixty deaths. In ten of the fatal cases the amount of the elixir ingested was between 14 and 198·5 g. The symptoms were drowsiness, pallor, slight oedema of the face, scanty urine with albuminuria, or in some cases, anuria, and later, oedema, ascites and coma (Geiling and Cannon, 1938).

No effects from its inhalation have been reported. When inhaled from the smoke of cigarettes to which it has been added, Flinn found that it produced neither symptoms nor irritation of mucous membranes, in fact, in habitual smokers suffering from congestion of the pharynx or larynx, this condition disappeared when diethylene glycol cigarettes were smoked exclusively.

10. Diethylene Glycol Monoethyl Ether
(Carbitol)

$$C_2H_5 \cdot O \cdot CH_2 \cdot CH_2 \cdot O \cdot CH_2 \cdot CH_2 \cdot OH$$

PROPERTIES

A colourless, nearly odourless, slightly hygroscopic liquid with a sweetish taste, but bitter after-taste. Miscible with water in all proportions. B.R. of the technical product 180–200° C. (pure, 201·9° C.). Sp.Gr. 0·9898. Volatility at room temperature practically negligible. Good solvent for cellulose nitrate, shellac, kauri, mastic, sandarac, copal, coumarone, colophony and dye stuffs; partial solvent for ester gum, zanzibar and dammar, but not for cellulose acetate. The trade product "carbitol solvent" or "technical carbitol" contains about 30 per cent ethylene glycol.

USES

(1) In textile and soap manufacture and dye printing.
(2) As a base or vehicle for cosmetics, and to a limited extent, in dermatological formulae (Harry, 1940; de Navarre, 1941).
(3) In the manufacture of safety glass.

TOXICITY

While no toxic effects in human beings from the use of carbitol have been recorded, the results of animal experiments suggest that care should be taken with regard to its use as a cosmetic constituent. Its general toxic effect is to depress the central nervous system, and to damage the kidneys.

TOXIC EFFECTS IN ANIMALS
Acute Poisoning

The acute toxicity of carbitol appears to be only slightly less than that of ethylene glycol or "technical carbitol". Symptomatically it differs from ethylene and diethylene glycols in not producing an initial effect on the central nervous system.

Lethal Dose

By subcutaneous, gastric, or intravenous injection. The acute toxicity of carbitol, as measured by L.D.50, was found by Hanzlik, Luduena, Lawrence and Hanzlik (1947c) to be about 6 ml. per kg. for rats and mice; for other animals it was apparently more toxic. Laug et al. (1939) also estimated the L.D.50 for guinea-pigs as about twice that for rats and mice. The M.L.D. for mice was found by von Oettingen and Jirouch (1931) to be 2·5 to 5 ml. per kg.

By external application. For one single large application to an area of 100 sq. cm., after clipping the hair, Hanzlik et al. (1947a) found the L.D.50 to be 8·5 ml. per kg. body weight. This result agrees fairly closely with that of Calvery (1944) who found the acute fatal dose to be 7·8 ml. per kg. after contact for 24 hours. The technical product was slightly less toxic, according to Hanzlik et al.

The cause of death in acute poisoning from carbitol may be respiratory or circulatory failure or both, circulatory failure being marked with rapid intravenous injection.

Effect on the Central Nervous System

Central nervous depression without the initial twitching, tremors or convulsions produced by ethylene and diethylene glycols is the characteristic effect of carbitol, followed by death.

Effect on Nerve Endings and Muscle

According to von Oettingen and Jirouch (1931), carbitol produced no inhibition of reflex excitability in frogs. Its slight depressant action on motor nerve fibre and excised skeletal muscle was about the same as that of ethylene glycol and cellosolve, but its depressant action on heart muscle was greater than these, though very much less than that of butyl cellosolve.

Effect on the Internal Organs

From the investigations of Hanzlik, Lawrence and Laqueur (1947b) and of Laug et al. (1939), it appears that the early injuries, especially those of the kidneys, are qualitatively similar to, but quantitatively somewhat less than, those from ethylene and diethylene glycols. Renal tubular injuries with some haemoglobinuria appeared only after large doses, and recovery from medium doses (2 to 3 ml. per kg.) was more likely to occur after carbitol than after the glycols.

Chronic Poisoning

Both by long continued ingestion and by repeated external application, animal experiments appear to show that pure carbitol produces less demonstrable renal damage, especially in the form of oxaluria and oxalate calculi, than the other glycols. This finding agrees with the hypothesis that from the chemical composition of these compounds the incomplete oxidation of ethylene and diethylene glycols should tend to produce oxalate compounds. Nevertheless, the fact that even external application can cause death, probably from uraemia with impairment of renal functional efficiency in animals (Hanzlik et al., 1947a), suggests that the toxicity of carbitol when applied to the skin in the form of cosmetic ointments, creams or lotions cannot be ignored. This applies with even greater force to "technical carbitol" in which the presence of ethylene glycol makes the occurrence of damage to the kidneys more probable.

Effect of Ingestion

Three sets of workers, Morris, Nelson and Calvery (1942), Smyth, Carpenter and Shaffer (1945), and Hanzlik et al. (1947b), investigated the results in mice of continued drinking of 1 per cent solutions, and eating for long periods food containing 5 per cent carbitol. Their observations showed negligible impairment of health and only slight damage to the kidneys, except when "technical carbitol" was used. Delayed death, probably due to uraemia resulting from renal damage, did occur with some of the animals.

Effect of External Application

Continued application of carbitol to the skin for 1 hour daily for not more than 30 days was found by Hanzlik et al. to have a L.D.50 of 0·32 ml. per kg., and they consider doses down to 0·08 ml. per kg. may be injurious. Only a mild transient dermatitis was produced, but definite kidney injury—necrosis of tubular epithelium—was observed after repeated daily applications of 0·04 ml. per kg.

Effect of Inhalation

Inhalation of carbitol by animals is well tolerated. Lehmann and Flury (1943) observed no injury to animals inhaling almost saturated concentrations for 12 days.

TOXIC EFFECTS IN MAN

So far no toxic effects have been reported, but the findings of Hanzlik *et al.* (1947a) on the effects of external application of carbitol to animals are of special importance with regard to the growing use, in America at any rate, of diethylene glycol monoethyl ether in cosmetic and dermatological mixtures. Experiments with dermatological formulae used in the Stanford Skin Clinic, such as rose-water ointment containing 10 per cent carbitol, and calamine lotion containing 50 per cent, showed that mixture with other ingredients did not hinder the absorption of carbitol since the majority of animals died. Actually the doses prescribed, such as 180 ml. of a 5 per cent solution applied daily for 14 days, which would be equivalent to 9 ml. of carbitol daily, are reasonably safe and could be given without harm in concentrations of 25 per cent for liquids or 50 per cent for ointment, but a dosage of 0·16 ml. per kg. should not be exceeded and 0·08 ml. per kg. would be safer, especially if applied to broken skin or to people with renal disorders. The pure rather than the technical product should be used in cosmetics.

11. Diethylene Glycol Mono-*n*-butyl Ether
(*Butyl Carbitol*)

$$C_4H_9 \cdot O \cdot CH_2 \cdot CH_2 \cdot O \cdot CH_2 \cdot CH_2 \cdot OH$$

PROPERTIES

A colourless liquid of mild odour. B.R. of the technical product 165–230° C. Sp.Gr. 0·969.

Solvent for cellulose nitrate, kauri, sandarac, mastic, ester gum, coumarone, colophony; not for cellulose acetate.

USES

Chiefly as "wetting out" solution in the thread industry.

TOXIC EFFECTS

No animal experiments on the toxicity of this substance have been carried out.

The case of haematuria which came to the notice of the Home Office in 1934, in connection with the use of either butyl carbitol or of the ethylene glycol monobutyl ether (butyl cellosolve) has already been described under the latter (p. 355).

12. Diethylene Glycol Monoacetate

$$CH_3 \cdot CO \cdot O \cdot CH_2 \cdot CH_2 \cdot O \cdot CH_2 \cdot CH_2 \cdot OH$$

PROPERTIES

A colourless liquid. B.P. 198° C.

Solvent for cellulose acetate and nitrate, colophony and gum camphor; partial solvent for coumarone, mastic, elemi and kauri, but not for ester gum, shellac, copals or sandarac. Miscible with water and aromatic hydrocarbons, but not with linseed oil or petroleum.

USES

Chiefly in the lacquer industry.

TOXIC EFFECTS

No animal experiments on the toxicity of this compound have been carried out, and no injurious effects have been reported from its use in industry.

13. Dipropylene Glycol

$$CH_3 \cdot CHOH \cdot CH_2 \cdot O \cdot CH_2 \cdot CHOH \cdot CH_3$$

PROPERTIES

An odourless liquid, highly germicidal (Robertson, Puck, Lemon and Loosli, 1943).

USES

(1) Chiefly as an anti-freeze agent.
(2) As an insecticide.

TOXICITY

As an acute depressant of the central nervous system dipropylene glycol in large doses is more toxic than ethylene glycol. As shown by animal experiments and, in conformity with its chemical composition, possessing an ether linkage between two glycol molecules, in large doses it has the same capacity as dioxan, methyl carbitol, butyl carbitol and diethylene glycol to injure the kidneys. From the point of view of chronic toxicity, that is to say, in smaller repeated doses, it is relatively non-toxic to the central nervous system, kidneys and liver.

TOXIC EFFECTS IN ANIMALS
Acute Poisoning

Lethal Dose

By oral administration. For rats, 10 per cent in drinking water, in 10 to 30 days (Kesten, Mulinos and Pomerantz, 1939).

By intravenous injection. For dogs, 11·5 ml. per kg. body weight compared with 25 ml. per kg. for propylene glycol.

Anaesthetic Dose

5·9 ml. per kg. body weight compared with 20 ml. per kg. for ethylene glycol.

Effect on Internal Organs

Some animals which died from either oral or intravenous doses showed lesions of the kidneys identical with those produced by diethylene glycol, i.e. extensive hydropic degeneration of the renal epithelium and occasional casts. In some rabbits (in one animal on the third day) early lesions were produced by non-lethal dosage.

Liver changes were inconstant, and on the evidence of results of perfusion of isolated cats' livers, dipropylene glycol is less toxic in this respect than either ethylene or diethylene glycols (Newman, van Winkle, Kennedy and Morton, 1940).

Chronic Poisoning

In small repeated doses, either by mouth or intravenously, dipropylene glycol is considerably less toxic than ethylene glycol.

Effect on Central Nervous System

A decrease in the running activity of rats was produced by the substitution of dipropylene glycol for one quarter of the carbohydrate of their diet. In this respect it was more toxic than propylene, but less toxic than ethylene glycol (van Winkle and Kennedy, 1940). Dogs given less than 11·5 ml. per kg. body weight intravenously showed no evidence of injury, nor did cats given 1 to 5 per cent in their drinking water for 33 to 77 days.

Effect on Kidneys

Some animals which survived repeated oral dosage (six doses of 2 ml. per kg.) showed slight degenerative changes in the convoluted tubules of the kidneys (Hanzlik, Newman, van Winkle, Lehman and Kennedy, 1939).

TOXIC EFFECTS IN MAN

No ill-effects have been reported, and it appears that in any dosage which might conceivably be ingested by human beings it is relatively non-toxic.

BIBLIOGRAPHY

ANSWER TO QUERY, Journal of the American Medical Association (1930). Possible death from drinking ethylene glycol ("prestone"). *J. Amer. med. Ass.*, **94**, 1940.
BACHEM, C. (1916). Ein neuer brauchbarer Glyzerinersatz. *Münch. med. Wschr.*, **63**, 1475.
BACHEM, C. (1917). Pharmakologische Untersuchungen über Glykol und seine Verwendung in der Pharmazie und Medizin. *Med. Klinik*, **13**, 7.
BACHEM, C. (1917). Schlusswort zu vorstehenden "Bemerkungen" des Herrn. P. Mayer (Karlsbad). *Med. Klinik*, **13**, 312.
BREKKE, A. (1930). To tilfelle av forgiftung med etylenglykol. *Norsk. Mag. Laegevidensk*, **91**, 381. (Abstr. in *Chem. Abstr.*, **26**, 772.)
CALVERY, H. O. (1944). Safeguarding foods and drugs in wartime. *Amer. Sci.*, **32**, 103.
CURME, G. O. and YOUNG, C. O. (1923). Ethylene glycol: its uses and properties. *Chem. metall. Engng.*, **28**, 169.
DAKIN, H. D. (1907). Experiments bearing upon the mode of oxidation of simple aliphatic substances in the animal organism (acetic acid, glycollic acid, glyoxylic acid, oxalic acid, glycocoll and glycol). *J. biol. Chem.*, **3**, 57.
DAVIDSON, J. D. (1926). The glycol ethers and their use in the lacquer industry. *Industr. Engng. Chem.*, **18**, 669.
DIVISION OF INDUSTRIAL HYGIENE (1935). Occupational poisons and diseases in New York State, 1934. *Industr. Bull.*, **14**, 4.
DONLEY, D. E. (1936). Toxic encephalopathy and volatile solvents in industry. *J. industr. Hyg.*, **18**, 571.
DURRANS, T. H. (1931). Solvents. 2nd ed. Chapman & Hall, London.
DYSON, G. M. (1934). Toxicity of ethylene glycol, propylene glycol and diethylene dioxide (dioxan). *Industr. Chem. chem. Mfr.*, **10**, 102.
ELKINS, H. B., STORTAZZI, E. D. and HAMMOND, J. W. (1942). Determination of atmosphere contaminants. II. Methyl cellosolve. *J. industr. Hyg.*, **24**, 229.
FITZHUGH, O. G. and NELSON, A. A. (1946). Comparison of the chronic toxicity of triethylene glycol with that of diethylene glycol. *J. industr. Hyg.*, **28**, 40.
FLINN, F. B. (1935). Some clinical observations on the influence of certain hygroscopic agents in cigarettes. *Laryngoscope, St. Louis*, **45**, 149
FLURY, F. and WIRTH, W. (1934). Zur Toxikologie der Lösungsmittel (Verschiedene Ester, Aceton, Methylalkohol). *Arch. Gewerbepath. Gewerbehyg.*, **5**, 1.
GEILING, E. M. K. and CANNON, P. R. (1938). Pathologic effects of elixir of sulfanilamide (diethylene glycol) poisoning. A clinical and experimental correlation; final report. *J. Amer. med. Ass.*, **111**, 919.
GREENBURG, L. M., MAYERS, M. R., GOLDWATER, L. J., BURKE, W. J., MOSKOWITZ, C. E. and MOSKOWITZ, S. (1938). Health hazards in the manufacture of "fused collars". I. Exposure to ethylene glycol monomethyl ether. *J. industr. Hyg.*, **20**, 134.

HAAG, H. B. and BOND, W. R. (1927). Some notes on glycol; glycol chloretone anaesthesia. *J. Lab. clin. Med.*, **12**, 882.
HAMILTON, A. (1934). Industrial toxicology. Harper, New York.
HANSEN, K. (1930). Äthylenglykol-Vergiftung. *Samml. Vergiftungsf.*, **1**, 175, A77.
HANZLIK, P. J., LAWRENCE, W. S., FELLOWS, J. K., LUDUENA, F. P. and LAQUEUR, G. L. (1947a). Epidermal application of diethylene glycol monoethyl ether (carbitol) and some other glycols. *J. industr. Hyg.*, **29**, 325.
HANZLIK, P. J., LAWRENCE, W. S. and LAQUEUR, G. L. (1947b). Comparative chronic toxicity of diethylene glycol monoethyl ether (carbitol) and some related glycols; results of continued drinking and feeding. *J. industr. Hyg.*, **29**, 233.
HANZLIK, P. J., LUDUENA, F. P., LAWRENCE, W. S. and HANZLIK, H. (1947c). Acute toxicity and general systemic actions of diethylene glycol monoethyl ether (carbitol). *J. industr. Hyg.*, **29**, 190.
HANZLIK, P. J., NEWMAN, H. W., van WINKLE, W., jr., LEHMAN, A. J. and KENNEDY, N. K. (1939). Toxicity, fate and excretion of propylene glycol and some other glycols. *J. Pharmacol.*, **67**, 101.
HANZLIK, P. J., SEIDENFELD, M. A. and JOHNSON, C. C. (1931). General properties, irritant and toxic actions of ethylene glycol. *J. Pharmacol.*, **41**, 387.
HARRY, R. G. (1940). Modern cosmetology, the principles and practice of modern cosmetics. Chem. Publishing Co., New York.
HUNT, REID (1932). Toxicity of ethylene and propylene glycols. *Industr. Engng. Chem.*, **24**, 361.
KESTEN, H. D., MULINOS, M. G. and POMERANTZ, L. (1939). Pathologic effects of certain glycols and related compounds. *Arch. Path.*, **27**, 447.
LAUG, E. P., CALVERY, H. O., MORRIS, H. J. and WOODARD, G. (1939). Toxicology of some glycols and derivatives. *J. industr. Hyg.*, **21**, 173.
LEHMANN, K. B. and FLURY, F. (1943). Toxicology and hygiene of industrial solvents. (Tr. by E. King and H. F. Smyth, jr.) Williams and Wilkins, Baltimore.
LUNGE, G. and BERL, E. (1933). Chemisch-technische Untersuchungsmethoden. 8th Aufl., Hefte III. Springer, Berlin.
MARSHALL, C. R. and HEATH, H. L. (1897). Pharmacology of the chlor-hydrins: a contribution to the study of the relation between chemical constitution and physiological action. *J. Physiol.*, **22**, 28.
MAYER, P. (1902). Experimentelle Untersuchungen über Kohlenhydratsäuren. *Z. klin. Med.*, **47**, 68.
MAYER, P. (1917). Bemerkungen zu der Arbeit von C. Bachem. *Med. Klinik*, **13**, 312.
MOLITORIS, H. (1931). Dichlorhydrin-Vergiftung, gewerbliche? *Samml. Vergiftungsf.*, **2**, 217, A177.
MORRIS, H. J., NELSON, A. A. and CALVERY, H. O. (1942). Observations on the chronic toxicities of propylene glycol, ethylene glycol, diethylene glycol, ethylene glycol monoethyl-ether and diethylene glycol mono-ethyl ether. *J. Pharmacol.*, **74**, 266.
MULINOS, M. G. and OSBORNE, R. L. (1935). Irritating properties of cigarette smoke as influenced by hygroscopic agents. *N.Y. St. J. Med.*, **35**, 590.
DE NAVARRE, M. G. (1941). Chemistry and manufacture of cosmetics, with a chapter on the Federal Food, Drug and Cosmetic Act of 1938, by R. J. Mill. van Nostrand, New York.
NELSON, A. A. and MORRIS, H. J. (1941). Reticulum cell lymphosarcoma in rats. *Arch. Path.*, **31**, 578.
NEWMAN, H. W., VAN WINKLE, W., jr., KENNEDY, W. K. and MORTON, N. C. (1940). Comparative effects of propylene glycol, other glycols and alcohol on the liver directly. *J. Pharmacol.*, **68**, 194.
VON OETTINGEN, W. F. and JIROUCH, E. A. (1931). Pharmacology of ethylene glycol and some of its derivatives in relation to their chemical properties. *J. Pharmacol.*, **42**, 355.
PAGE, I. H. (1927). Ethylene glycol—a pharmacological study. *J. Pharmacol.*, **30**, 313.
PARSONS, C. E. and PARSONS, M.E.M. (1938). Toxic encephalopathy and "granulopenic anemia" due to volatile solvents in industry. *J. industr. Hyg.*, **20**, 124.
POHL, J. (1896). Über die Oxydation Abbau der Feltkörper in thierischen Organismus. *Arch. exp. Path. Pharmak.*, **37**, 413.
PONS, C. A. and CUSTER, R. P. (1946). Acute ethylene glycol poisoning: a clinico-pathologic report of eighteen fatal cases. *Amer. J. med. Sci.*, **211**, 544.
RENKENBACH, W. H. and AARONSON, H. A. (1931). Properties of diethylene glycol. *Industr. Engng. Chem.*, **23**, 160.
ROBERTSON, O. H., PUCK, T. T., LEMON, H. F. and LOOSLI, C. G. (1943). Lethal effect of triethylene glycol vapor on air-borne bacteria and influenza virus. *Science*, **97**, 142.
SMITH, S. (1928). The cellulose lacquers. Pitman, London.
SMYTH, H. F., jr., SEATON, J. and FISCHER, L. (1941). Single dose toxicity of some glycols and derivatives. *J. industr. Hyg.*, **23**, 259.
SMYTH, H. F., jr., CARPENTER, C. P. and SHAFFER, C. B. (1945). Subacute toxicity and irritation of polyethylene glycols of approximate molecular weights of 200, 300, and 400. *J. Amer. pharm. Ass.*, **34**, 172.

STARREK, E. (1938). Über die Wirkung einiger Alkohole, Glykole und Ester. Dissertation, Würzburg. (Cited by Lehmann and Flury, 1943, in Toxicology and hygiene of industrial solvents. Williams and Wilkins, Baltimore.)

WAITE, C. P., PATTY, F. A. and YANT, W. P. (1930). Acute response of guinea-pigs to vapors of some new commercial organic compounds. III. Cellosolve (mono-ethyl ether of ethylene glycol). *Publ. Hlth. Rep., Wash.*, **45**, 1459.

WERNER, H. W., MITCHELL, J. L., MILLER, J. W. and von OETTINGEN, W. F. (1943). The acute toxicity of vapors of several monoalkyl ethers of ethylene glycol. *J. industr. Hyg.*, **25**, 157.

WERNER, H. W., MITCHELL, J. L., MILLER, J. W. and von OETTINGEN, W. F. (1943). Effects of repeated exposure of dogs to monoalkyl ethylene glycol ether vapors. *J. industr. Hyg.*, **25**, 409.

WERNER, H. W., NAWROCKI, C. Z., MITCHELL, J. L., MILLER, J. W. and VON OETTINGEN, W. F. (1943). Effects of repeated exposures of rats to vapors of mono-alkyl ethylene glycol ethers. *J. industr. Hyg.*, **25**, 374.

WILEY, F. H., HUEPER, W. C., BERGEN, D. S. and BLOOD, E. R. (1938). Formation of oxalic acid from ethylene glycol and related solvents. *J. industr. Hyg.*, **20**, 269.

WILEY, F. H., HUEPER, W. C. and VON OETTINGEN, W. F. (1936). Toxicity and potential dangers of ethylene glycol. *J. industr. Hyg.*, **18**, 123.

VAN WINKLE, W. and KENNEDY, N. K. (1940). Voluntary activity of rats fed propylene glycol and other glycols. *J. Pharmacol.*, **69**, 140.

WOLFF, H. (1927). Die Lösungsmitteln der Fette, Öle, Wachse und Härze. Stuttgart.

YOUNG, E. G. and WOOLNER, L. B. (1946). A case of fatal poisoning by 2-methoxyethanol. *J. industr. Hyg.*, **28**, 267.

ZERNIK, F. (1933). Neuere Erkenntnisse auf dem Gebiete der schädlichen Gase und Dämpfe. *Ergebn. Hyg. Bakt.*, **14**, 139.

CHAPTER VIII

AMINES AND COAL TAR BASES

1. cycloHexylamine

$C_6H_{11} \cdot NH_2$

PROPERTIES

A water-white liquid with a strong fishy odour. B.P. 134·5°C. Sp.Gr. 0·8647 at 25°C. Completely miscible with water and all common solvents. It is a strong base and forms salts with all acids, and with fatty acids it reacts to form soaps. Attacks copper alloys and lead (Carswell and Morrill, 1937).

USES

(1) As an insecticide (Kearns and Flint, 1937).
(2) As a hardening and condensing agent in the plastic industry.
(3) As an inhibitor of corrosion in steam-heating systems.
(4) As a vulcanizer.
(5) As an emulsifier for soaps.
(6) In the dyeing industry as an intermediate and also as a solvent for dyestuffs.

TOXICITY

The results of animal experiments have shown it to be a skin irritant and also a nerve poison. No toxic effects in man have been recorded.

TOXIC EFFECTS IN ANIMALS

Effect of Oral Administration

After a period of 82 days, repeated doses of 100 mg. per kg. body weight in water were not lethal to rabbits, guinea-pigs or rats, and autopsy showed no injury to internal organs (Flinn, cited by Carswell and Morrill, 1937).

Effect of Subcutaneous Injection in Olive Oil

Doses of 0·5 g. per kg. body weight proved fatal to rabbits. Convulsions, delayed for several hours, preceded death which occurred within 3 to 4 hours. According to Flinn (1937) the toxic action is on the motor centres of the spinal column and on the medulla. Doses of 0·25 g. per kg. body weight, repeated daily for 10 days, had no ill-effects; if left on the skin for any time it produced a dermatitis, although it was not absorbed through the skin in lethal quantities. The danger of accidental ingestion is slight because solutions containing amounts which do not cause injury to animals are so bitter as to be completely unpalatable.

2. Dicyclohexylamine

$(C_6H_{11})_2NH$

PROPERTIES

A water-white liquid with an odour like that of *cyclo*hexylamine. B.P. 255–8° C. Sp.Gr. 0·914. Only slightly soluble in water but soluble in all common solvents and miscible with *cyclo*hexylamine. Does not form an azeotropic mixture on distillation. Slightly more basic than *cyclo*hexylamine.

USES
Similar to those of *cyclo*hexylamine.

TOXICITY
From animal experiments di*cyclo*hexylamine appears to be more toxic than *cyclo*hexylamine. Death occurs sooner, and convulsions and temporary paralysis of the hindquarters appear with a smaller dose. No toxic effects in man have been recorded, but Flinn (1937) states that di*cyclo*hexylamine, if left on the skin, will produce a dermatitis. Unlike *cyclo*hexylamine, di*cyclo*hexylamine can be absorbed through the skin in lethal quantities.

3. Ethanolamines

Of the three ethanolamines, mono-, di- and tri-, triethanolamine is of the greatest commercial importance. The properties of all three are very similar and therefore may be described under one heading.

PROPERTIES
The ethanolamines are formed by the reaction of ethylene oxide with aqueous ammonia.

Monoethanolamine $\quad NH_2 \cdot C_2H_4 \cdot OH$
B.P. at 150 mm. Hg, 171°C. Sp. Gr. 1·04.

Diethanolamine $\quad NH(C_2H_4 \cdot OH)_2$
B.P. 268°C. Sp. Gr. 1·098.

Triethanolamine $\quad N(C_2H_4 \cdot OH)_3$
B.P. at 150 mm. Hg, 277°C. Sp. Gr. 1·124.

They are soluble in gasoline or carbon tetrachloride; form fatty acid soaps with oleic, linoleic and stearic acids.

USES
Monoethanolamine is used chiefly for dry-cleaning and degreasing.
Triethanolamine has a wide range of application:

(1) In the soap, fat, oil and wax industries as an emulsifier. By virtue of its mild alkalinity, surface tension and depressant action, water solutions of the oleate approach closely the properties of a theoretically "neutral soap". The stearate forms a hard soap. Wax emulsions of triethanolamine may be used for coating paper, leather and linoleum, and in the manufacture of shoe polishes and metal cleansers. These emulsions are not injurious to lacquer finishes.

(2) In the paint and lacquer industry (Fischer, 1935).

(3) In the textile industry; a scouring compound containing 60 per cent triethanolamine and 40 per cent ethylene dichloride has a good detergent and solvent action (Wilson, 1930).

(4) In the cosmetic industry as a constituent of face creams, rouge, lipsticks, etc.

TOXICITY
The results of some animal experiments in the I. G. Elberfeld Toxicology Index (1931) indicate that triethanolamine is practically non-toxic except for

some local effect at the site of subcutaneous injections. In the experiments of Henry (1944), intraperitoneal injections of triethanolamine tartrate were also non-toxic to animals, but it was found to be a powerful vasodilator. Diethanolamine acetate has been found to have a diuretic effect on dogs by intravenous injection; monoethanolamine has only a slight action (Chabrol, Cottet and Sallet, 1939).

TOXIC EFFECTS IN ANIMALS

Effect of Skin Applications

Applications of 5 or 10 per cent solutions to the scarified dorsal skin of rabbits and rats produced no ill-effects.

Effect of Oral Administration

No ill-effects were observed from the administration to rats of 0·5 g. per kg. body weight of a 10 per cent solution (I. G. Elberfeld Toxicology Index). The L.D.50 for rats, according to Kindsvatter (1940), is 8 g. per kg. body weight.

Effect of Subcutaneous Injection

0·1 g. per kg. body weight of a 10 per cent solution produced swellings at the site of the injection.

Effect of Intraperitoneal Injection

In the experiments of Henry (1944) guinea-pigs survived intraperitoneal injection of 0·7 g. per kg. body weight of ethanolamine tartrate but showed marked vaso-dilation.

Effect on Internal Organs

Kindsvatter found only slight reversible effects on the livers and kidneys of rats and guinea-pigs. He states that the presence of mono- and di-ethanolamines in the commercial grade of triethanolamine does not enhance its toxicity.

TOXIC EFFECTS IN MAN

The results of application of pure monoethanolamine to the skin of human beings (I. G. Elberfeld Toxicology Index, 1931) showed that only when applied on gauze for 1½ hours was marked redness and infiltration of the skin produced. Human beings remaining 24 hours in the close neighbourhood of the liquid, and therefore possibly near vapour emanating from it, showed nothing more than very slight redness of the skin. According to Downing (1944), a protective cream used in industry and containing 1·5 per cent triethanolamine is non-irritating.

4. Pyridine

$$C_5H_5N, \text{ i.e. } CH \leqslant _{CH \cdot CH} ^{CH:CH} \geqslant N$$

PROPERTIES

A clear colourless liquid with a bitter taste. Readily soluble in water, alcohol, ether and chloroform. It is a weak base. It is present in coal-tar and may be extracted from the crude "benzole" by acid. It is present as an impurity in turpentine made by the distillation of wood (Hamilton, 1934), and is a constituent of tobacco-smoke (Vohl and Eulenberg, 1871).

Pure Pyridine. B.P. 115·5°C. Sp.Gr. 0·9893.
"Pure Pyridine 90 per cent." B.R. 114·4–116·8°C. (Ludwig, 1934).
Commercial Pyridine. B.R. 94–160°C.

The commercial product contains impurities, chiefly higher homologues with a higher boiling point, including picoline (monomethylpyridine), lutidine (dimethylpyridine) and collidine (trimethylpyridine). Picoline is found in tobacco-smoke (Vohl and Eulenberg, 1871).

USES

(1) In the dyeing industry as a "leveller," to obtain more even distribution of the dye.
(2) In denaturing alcohol.
(3) In purifying anthracene.
(4) In the manufacture of chemicals and explosives.

TOXICITY

Pyridine is an irritant of mucous membranes and to some extent a narcotic, and in animals has been found to produce severe liver and kidney damage.

In man, injurious effects on the central nervous system, producing a symptomatology resembling that of Wernicke's polioencephalitis or tabes, have been reported but are believed to be rare and to occur only in sensitive individuals.

Gastro-intestinal disturbances from the use of pure pyridine have also been recorded (Holtzmann, 1936).

TOXIC EFFECTS IN ANIMALS

Lethal and Narcotic Concentrations

Most of the earlier workers found that warm-blooded animals were very insensitive to pyridine, whether given orally, intraperitoneally or by inhalation, though frogs were easily narcotized and died from respiratory paralysis (narcotic dose 0·1 g., lethal dose 0·2 to 0·4 g., Brunton and Tunnicliffe, 1894).

Lethal dose

By intraperitoneal injection the lethal dose for guinea-pigs was found to be 0·087 g. per 100 g. body weight (Brunton and Tunnicliffe). No lethal effect from inhalation has been observed in animals.

Narcotic dose

In warm-blooded animals the exact narcotic dose has not been recorded, but Bochefontaine (1883) stated that while inhalation of the vapour given off by pyridine at ordinary temperatures produced stupor, drowsiness and increased respiration in frogs, guinea-pigs were practically unaffected after 3 hours. Ludwig (1936) found that cats, while becoming slightly less active after being kept in a chamber containing pyridine vapour in a concentration of 100 p.p.m., showed no real stupor, while His (1887) gave 1 g. daily by stomach tube to dogs for a week with no narcotic effect.

Acute Poisoning

Symptoms

In the frog, according to Rohde (1920), pyridine has a curare-like action on **nerve endings**, while in guinea-pigs after intraperitoneal injection Brunton and

Tunnicliffe (1894) found that it paralysed the sensory nerve endings and sensory spinal centres, leading to reduction of peripheral sensation. The occurrence of convulsions during narcosis has been disputed. Harnack and Meyer (1880) stated that pyridine produced convulsions in frogs, but both Bochefontaine (1883) and Brunton and Tunnicliffe (1894) found it purely paralytic and non-convulsive. Brunton and Tunnicliffe observed a decided but transient fall in blood pressure, due to its action on the heart-muscle.

The question whether these effects are due to pyridine itself or to the presence of impurities has also been disputed. Certainly the higher homologues, especially picoline, have been shown to be more toxic than pyridine itself. Vohl and Eulenberg (1871) found the pyridine bases distilled from tobacco very toxic. Pure picoline killed a rabbit in doses of 30 drops injected subcutaneously and the vapour of 5 g. was lethal to pigeons. Cohn (1894) also examined the action of pure picoline and produced toxic symptoms in dogs, including convulsions, albuminuria and casts, by subcutaneous injection of 0·1 to 1 g. In Brunton and Tunnicliffe's experiments the pyridine used contained some picoline.

Chronic Poisoning

The contention of earlier workers (Rambousek, 1911; Lehmann, 1919, etc.) that inhalation of pyridine vapour in less than narcotic dosage is comparatively innocuous is supported by Ludwig (1936) and has not been disproved; however, pyridine has been shown by later investigators (Baxter, 1946, 1947, 1949, and Coulson and Brazda, 1948) to be most toxic by oral administration, causing extensive liver and kidney damage in animals. The extent of this damage was found to be considerably lessened by the administration of "protective" substances such as methionine and cystine.

Effect on the Liver and Kidneys

The slight fatty degeneration of the liver and kidneys observed by Ludwig (1936) in cats after inhalation of 20 ml. of pyridine, vaporized in a chamber of 0·2 c.m. capacity for an hour at a time, were not definitely attributed by him to the pyridine.

In the experiments of Baxter (1946), and of Coulson and Brazda (1948), however, the liver and kidneys of rats to whose diet pyridine had been added showed cirrhosis or necrosis, or both, of the liver, and degenerative changes in the kidneys which were roughly but not absolutely proportional to those of the liver.

When the high protein diet was supplemented by cystine or methionine, the changes in the liver were largely prevented.

According to Coulson and Brazda (1948), pyridine given by mouth is more toxic than many of its derivatives and must be included in the list of substances known to produce severe liver and kidney damage.

METABOLISM

According to the early workers, His (1887) and Cohn (1894), pyridine when given by mouth is not oxidized into products such as glycuronic acid derivatives but is excreted in the form of methyl pyridinium hydroxide.

Baxter (1946) agrees that pyridine is methylated in the body, probably at the expense of lipotropic substances which act as methyl donors.

TOXIC EFFECTS IN MAN

Acute Poisoning

The only instance of pyridine producing acute narcosis in man is that reported to the Home Office in 1934, when a man employed in cleaning a tank wagon, which had contained pyridine used in the manufacture of chemicals and explosives, became temporarily semi-conscious. During therapeutic inhalation of pyridine for asthma, Lublenski (1885) observed tremor and nausea in several cases and in one case vomiting, headache and giddiness.

Chronic Poisoning

Symptoms

The symptoms attributed to the long-continued use of pyridine by Ludwig (1934), Holtzmann (1936) and Teisinger (1947) fall into two classes, those produced by its effect on the central nervous system and on the gastro-intestinal tract respectively.

Nervous symptoms. The severe symptoms shown by one out of eight men employed as chemists and laboratory assistants using pyridine were attributed by Ludwig (1934) to an injury, probably of the nature of a haemorrhagic polio-encephalitis, of the cerebral blood vessels. Five of the other men complained of headache, giddiness, fatigue and gastro-intestinal disturbance, and two younger men were unaffected. The one severely affected patient had these slighter symptoms at first, but later, after a transitory attack of unconsciousness, developed facial paresis, conjugate deviation and horizontal nystagmus, ptosis of the left eye, loss of palate reflex, anisocoria and ataxia—a condition which Ludwig considered very similar to Wernicke's pseudo-encephalitis. It is to be noted that the commercial pyridine used had a boiling range of 94–160°C. and contained picoline, lutidine and collidine. Slighter nervous symptoms, including giddiness, noises in the ears, disturbance of equilibrium and neuritic pains in the arms were present in the case recorded by Holtzmann (1936) where pure pyridine was used. Seven cases of chronic pyridine poisoning, in which the symptoms included headache, vertigo, nervousness, sleeplessness and in one case, lack of concentration and deterioration of memory, are recorded by Teisinger (1947). The pyridine, boiling at 115°C., was presumably pure, though small amounts of ammonia were present in the air samples. Meyer (1950) describes a case, with a history of 13 years' exposure to pyridine, in which nervousness, insomnia, giddiness, pain and weakness of the limbs and generalized tremor were followed by sudden anuria, due to paralysis of the bladder. Babinski and Romberg tests were both positive, and the symptomatology resembled that of tabes, but there was no optic atrophy and the Wassermann reaction was negative. Treatment with vitamin B (thiamine) produced considerable improvement and Meyer suggests that pyridine may act as an "anti-vitamin", injuring the metabolism of thiamine.

Gastro-intestinal symptoms. The symptoms observed in Ludwig's cases were loss of appetite, thirst, sickness and diarrhoea. In the case recorded by Holtzmann (1936), also a laboratory worker, the symptoms were specially referable to the digestive tract. Vomiting was the chief complaint, and the gastric contents showed hypo-acidity and reduction of free hydrochloric acid. There was also great loss of weight, viz. 23 lb. in a few months. Occasional digestive troubles were present in Teisinger's cases.

5. Picoline
(*Methylpyridine*)

C_6H_7N, i.e. $CH_3 \cdot C_5H_4N$

The picolines are a series of homologues of pyridine obtained by the dry distillation of bones and coal. They are:

        α-picoline (2-methylpyridine)
        β-picoline (3-methylpyridine)
        γ-picoline (4-methylpyridine)

They occur as impurities in commercial pyridine and in tobacco smoke (Vohl and Eulenberg, 1871).

PROPERTIES

α-picoline—a colourless liquid. B.P. 128°C. Sp.Gr. 0·952. Soluble in alcohol or water.

β-picoline—a colourless liquid. B.P. 144°C. Sp.Gr. 0·977. Soluble in alcohol.

γ-picoline—a colourless liquid. B.P. 147°C. Sp.Gr. 0·974. Soluble in water.

USES

Similar to those of pyridine. α-Picoline is used medicinally as a nerve sedative.

TOXICITY

It has been mentioned on p. 370 that the toxic effects on the nervous system observed by Ludwig (1934) in two workers exposed to pyridine were probably attributable to the picoline content of the crude pyridine used. The picolines are regarded as more toxic than pyridine, their potential toxicity increasing with their boiling point.

TOXIC EFFECTS IN ANIMALS

In the animal experiments of Vohl and Eulenberg (1871) pure picoline was used; thirty drops subcutaneously was sufficient to kill a rabbit, and a pigeon was killed by inhaling the vapour given off by 5 g. Repeated subcutaneous injections of 0·5 to 1 g. of α-picoline (Cohn, 1894) produced albuminuria and cramp in rabbits after one week. Hyperaemia of the brain was observed by Vohl and Eulenberg (1871) in their animals. This was not confirmed by Ludwig (1934) using crude pyridine in his experiments on cats, and he suggests that cats may be relatively insensitive.

TOXIC EFFECTS IN MAN

No definite cases of injury following exposure to picolines have been recorded.

BIBLIOGRAPHY

BAXTER, J. H. (1946). The mechanisms of the liver and kidney injury produced by toxic substances. I. Some effects of pyridine and their prevention by methionine. *J. clin. Invest.* (*Proc.*), **25**, 908.

BAXTER, J. H. (1947). Studies of mechanisms of liver and kidney injury; methionine protects against damage produced in rat by diets containing pyridine. *J. Pharmacol.*, **91**, 345.

BAXTER, J. H. (1949). Pyridine liver and kidney injury in rats; influence of diet, with particular attention to methionine, cystine, and choline. *Bull. Johns Hopk. Hosp.*, **85**, 138.

BOCHEFONTAINE, M. (1883). Expériences pour servir à l'étude des propriétés physiologiques de la pyridine. *C. R. Soc. Biol., Paris*, **5**, 5.

BRUNTON, T. L. and TUNNICLIFFE, F. W. (1894). On the physiological action of pyridine. *J. Physiol.*, **17**, 272.
CARSWELL, T. S. and MORRILL, H. L. (1937). Cyclohexylamine and dicyclohexylamine. *Industr. Engng. Chem.*, **29**, 1247.
CHABROL, E., COTTET, J. and SALLET, J. (1939). Recherches expérimentales sur l'action diurétique des éthanolamines et de leurs sels. *C. R. Soc. Biol.*, **131**, 637.
COHN, R. (1894). Über das Verhalten einiger Pyridin- und Naphthalinderivate in thierischen Stoffwechsel. *Hoppe-Seyl. Z.*, **18**, 112.
COULSON, R. A. and BRAZDA, F. G. (1948). Influence of choline, cystine and methionine on toxic effects of pyridine and certain related compounds. *Proc. Soc. exp. Biol.*, **69**, 480.
DOWNING, Z. G. (1944). A new protective cream. *Arch. Derm. Syph., N.Y.*, **49**, 436.
FISCHER, E. J. (1935). Triäthanolamin in der Industrie der Harze, Lacke, Anstrich- und Isoliermittel. *Farbe u. Lack*, **519**, 533.
FLINN, F. B. (1937). Quoted by Carswell, T. S. and Morrill, H. L. (1937). *Industr. Engng. Chem.*, **29**, 1247.
HAMILTON, A. (1934). Industrial toxicology. Harper, New York.
HARNACK, E. and MEYER, H. (1880). Untersuchungen über die Wirkungen der Jaborandi-Alkaloide. *Arch. exp. Path. Pharmak.*, **2**, 394.
HENRY, R. (1944). Sur quelques propriétés physiologiques du tartrate de triéthanolamine. *C. R. Soc. Biol., Paris*, **138**, 745.
HIS, W. (1887). Über das Stoffwechselprodukt des Pyridins. *Arch. exp. Path. Pharmak.*, **22**, 254.
HOLTZMANN, F. (1936). Pyridinvergiftung. *Zbl. GewHyg.*, **23**, 8.
I. G. ELBERFELD TOXICOLOGY INDEX (1931). Report of LK Division of I.G. Ludwigshafen.
KEARNS, C. W. and FLINT, W. B. (1937). Contact mechanical properties of various derivatives of cyclohexylamine. *J. econ. Ent.*, **30**, 158.
KINDSVATTER, V. H. (1940). Acute and chronic toxicity of triethanolamine. *J. industr. Hyg.*, **22**, 207.
LEHMANN, K. B. (1919). Kurzes Lehrbuch der Arbeits- und Gewerbehygiene. Hirzel, Leipzig.
LUBLENSKI, W. (1885). (Cited by Ludwig, H., 1934, in *Arch. Gewerbepath. Gewerbehyg.*, **5**, 654.)
LUDWIG, H. (1934). Zur Toxikologie des Pyridins und seiner Homologen. (Pseudoencephalitis Wernicke bei Pyridinarbeiter). *Arch. Gewerbepath. Gewerbehyg.*, **5**, 654.).
LUDWIG, H. (1936). Über die Wirkung von Pyridin-Einatmung bei Katzen. *Arch. exp. Path. Pharmak.*, **182**, 178.
MEYER, A. (1950). Über die Pyridinvergiftung. *Z. Unfallmed. u. Berufskrank.*, **43**, 144.
RAMBOUSEK, J. (1911). Gewerbliche Vergiftungen deren Vorkommen, Erscheinungen, Behandlung Verhütung. Veit, Leipzig. (Tr. by T. M. Legge, Industrial poisoning. Arnold, London, 1913).
ROHDE, E. (1920). Pyridin, Chinolin, Chinin und Chininderivate. Heffter's Handbuch der experimentellen Pharmakologie, **2**, 1. Springer, Berlin.
TEISINGER, J. (1947). Mild chronic intoxication with pyridine. *Čas. Lék. čes.*, **86**, 1185.
VOHL, H. and EULENBERG, H. (1871). Über Tabak in toxicologischen Beziehung. *Berl. klin. Wschr.*, **8**, 385.
WILSON, A. L. (1930). Triethanolamine emulsions. *Industr. Engng. Chem.*, **22**, 143.

CHAPTER IX

NITRO-COMPOUNDS

1. Nitromethane

$$CH_3 \cdot NO_2$$

PROPERTIES

A colourless liquid with a moderately strong, disagreeable odour. B.P. 100°C. at 760 mm. Sp. Gr. 1·139.

USES

(1) As a solvent for nitrocellulose, cellulose acetate, resins, waxes, fats and dyestuffs.
(2) As a mixed solvent with chlorinated hydrocarbons.
(3) In chemical synthesis.

TOXICITY

Like the other nitroparaffins, nitromethane is an irritant of mucous membranes and has some narcotic action. Its irritative effect is less than that of nitroethane but its effect on the central nervous system is somewhat more severe.

Maximum Permissible Concentration

The maximum practical working level suggested by Cook (1945) is 200 p.p.m.

TOXIC EFFECTS IN ANIMALS

Lethal Dose

By oral administration. For rabbits, 0·75 to 1 g. per kg. body weight (Machle, Scott and Treon, 1940).

By inhalation. For rabbits and guinea-pigs, 50,000 p.p.m. (124·8 mg. per l.) for 1 hour or 30,000 p.p.m. (74·9 mg. per l.) for 2 hours.

Symptoms

Conjunctival irritation, salivation, audible râles in the lung, twitching, jerking and sometimes generalized convulsions.

Lesions of the Internal Organs

The brain showed some congestion, oedema and occasionally small haemorrhages. The livers in fatal cases showed some injury, e.g. oedema, cloudy swelling and occasionally necrosis. There was some oedema of the kidneys.

TOXIC EFFECTS IN MAN

No cases of industrial injury have been recorded.

According to Sterner (1949) slight irritation to the eyes and upper respiratory tract follow short exposures to nitromethane, as they do to the other nitroparaffins.

2. Nitropropanes

Nitropropane in two isomeric forms (1- and 2-nitropropane) is a nitroparaffin produced by vapour-phase nitration of lower aliphatic hydrocarbons at high temperatures (Hass, Hodge and Vanderbilt, 1936).

A. 1-Nitropropane

$$\text{n-}C_3H_7\cdot NO_2$$

PROPERTIES

A colourless, non-hygroscopic liquid of a somewhat disagreeable odour. B.P. 131·6° C. Sp.Gr. 1·1003. Fl.P. 49° C. Solubility in water 1·4 per cent. Solvent for cellulose nitrate, gum, coumarone, oils and petroleum, but not for resins, tar, waxes, gelatin or shellac.

USE

In the chemical industry as a raw material for synthesis.

TOXICITY

Like other mono-nitroparaffins, 1-nitropropane is an irritant, and in high dosage, either by oral administration or by inhalation, is lethal to animals. With regard to liver injury, it appears to be the most toxic of the nitroethane, nitromethane and nitropropane series.

Maximum Permissible Concentration

According to Machle, Scott and Treon (1940), the odour of 1-nitropropane is so disagreeable in low concentrations that it may be the best guide to harmful concentrations. Sterner (1949) suggests a "maximum practical working level of 100 p.p.m.", a concentration which would be produced in 100 c. ft. of air by volatilization and equal distribution of 10·2 ml. of liquid at 25° C.

TOXIC EFFECTS IN ANIMALS

Lethal Dose

By oral administration. 0·25 to 0·5 g. per kg.

By inhalation. Concentrations of 1·5 per cent or greater killed all animals exposed, and concentrations of 1 per cent were fatal to some (Machle *et al.*, 1940).

Symptoms

After oral administration, progressive weakness and collapse, unsteadiness of gait and inco-ordination occurred, ending in ataxia; respiration was at first slow but rapidly increased.

After inhalation, there was restlessness, slow, full respiration with audible râles, conjunctival irritation with lachrymation and only slight evidence of either anaesthesia or of hyper-irritation of the central nervous system.

Lesions of the Internal Organs

General visceral and cerebral congestion in all animals that died from exposure to 1-nitropropane was similar to that caused by the other nitroparaffins, but the severe liver damage associated with fatty infiltration in animals exposed to inhalation of 1-nitropropane indicated that the compound is a more potent liver poison than any other member of the series.

TOXIC EFFECTS IN MAN

No injury to man has been recorded.

B. 2-Nitropropane

$$CH_3 \cdot CH(NO_2) \cdot CH_3$$

PROPERTIES

B.P. 120·3° C. Sp.Gr. 0·992. Fl.P. 39 °C. Solubility in water 1·7 per cent.

USES

Similar to that of 1-nitropropane, but also in the coating industry (Bogin and Dampier, 1942).

TOXICITY

Though 2-nitropropane has not been so closely studied as 1-nitropropane, it appears to be slightly less toxic to animals, but causes essentially the same injuries as 1-nitropropane. Skinner (1947) reports some digestive disturbance in man.

Maximum Permissible Concentration

Skinner suggests 25 p.p.m. as the maximum allowable concentration.

TOXIC EFFECTS IN ANIMALS

Lethal Dose

By oral administration. 0·5 to 0·75 g. per kg. for rabbits (Machle *et al.*, 1940).
By inhalation. No animal experiments have been made.

TOXIC EFFECTS IN MAN

Skinner (1947) reports symptoms which include anorexia, nausea, vomiting and diarrhoea, and severe headache occurring in men employed in dipping articles into a tank when the solvent used was a mixture of xylol and 2-nitropropane. The concentration of 2-nitropropane in the atmosphere was between 20 and 45 p.p.m. No such disturbance of health had arisen in two men who had used a spray lacquer containing 20 per cent 2-nitropropane where the concentration in the breathing area was 10 to 25 p.p.m. These men had been exposed for only 4 hours a day, 3 days a week, whereas the affected men were working all day and every day. The symptoms ceased when methyl ethyl ketone was substituted for 2-nitropropane.

3. Nitrobutanes

PROPERTIES

A. 1-Nitrobutane

$$n\text{-}C_4H_9 \cdot NO_2$$

Produced by the nitration of butane.

A liquid, B.P. 151° C. Sp.Gr. 0·9774. Solubility in water 0·5 ml. per 100 ml. (Machle *et al.*, 1940). Miscible with ethyl alcohol or ether (Sterner, 1949).

B. 2-Nitrobutane

$$CH_3 \cdot CH(NO_2) \cdot CH_2 \cdot CH_3$$

A liquid, B.P. 139° C. Sp.Gr. 0·9728. Solubility in water 0·9 ml. per 100 ml.

USES

As solvents for cellulose acetate.

TOXICITY

Like other nitroparaffins, nitrobutanes are primarily moderate irritants to the eyes and upper respiratory tracts, with secondary general toxic action and a slight narcotic effect. With lethal dosage some degree of liver and kidney damage is found.

TOXIC EFFECTS IN ANIMALS

Lethal Dose

By oral administration. For rabbits, 0·50 to 0·75 mg. per kg. (Machle *et al.*, 1940), which is less than that of 1-nitropropane.

Symptoms

There was initial restlessness and discomfort with respiratory and conjunctival irritation and increased salivation.

TOXIC EFFECTS IN MAN

No cases of injury from industrial use have been recorded.

4. Nitrobenzene
(*Oil of Mirbane*)
$C_6H_5 \cdot NO_2$

PROPERTIES

A yellow, oily liquid with a strong odour of bitter almonds. B.P. 210·85° C. Sp.Gr. 1·198. Soluble in alcohol, ether, benzine or oils; very slightly soluble in water.

USES

(1) In the manufacture of aniline, quinoline, etc.
(2) As a constituent of shoe and floor polishes.
(3) In the perfume industry, and as a substitute for essence of almond.
(4) As a preservative in certain spray paints.

TOXICITY

Nitrobenzene is an acute poison to the central nervous and blood systems, and in high concentrations causes the formation of methaemoglobin, leading to death from respiratory paralysis. In less severe poisoning the nervous symptoms are accompanied by a characteristic lilac cyanosis and often by jaundice. Subacute and chronic poisoning is associated with anaemia, fatigue, headache and loss of appetite.

Formation of Methaemoglobin

The primary toxic action of nitrobenzene consists in its ability to convert the haemoglobin of the blood into methaemoglobin. Since methaemoglobin cannot transport oxygen, anoxia of the tissues results, and in this respect nitrobenzene is said to be fifty times more toxic than aniline (Patty, 1949).

It was first suggested by Heubner (1913) that the initial step in the metabolism of nitrobenzene to methaemoglobin was its reduction to *p*-aminophenol, and that another intermediate compound was phenylhydroxylamine. Although *p*-aminophenol is itself a powerful producer of methaemoglobin, this effect is relatively transient, and the fact that re-conversion to haemoglobin takes place rapidly may explain why many non-fatal cases of nitrobenzene poisoning recover within 24 hours. According to Clark, van Loon and Morrissey (1943), it may also explain the delay in the clinical appearance of methaemoglobinaemia following absorption of nitrobenzene and the relatively slight variations in pulse and respiratory rate and blood pressure.

It is suggested that even when there is a marked deficiency of available oxygen in venous blood, the oxygen tension of arterial blood is not sufficiently disturbed to set up compensatory responses in the circulatory and respiratory centres.

Estimation of Methaemoglobinaemia

Methaemoglobin is demonstrable in the blood by means of the spectrometer (Evelyn and Malloy, 1938).

TOXIC EFFECTS IN ANIMALS

Animal experiments have been somewhat misleading, especially with regard to the formation of methaemoglobin, because many results have been based on the reactions of the guinea-pig to aniline and its derivatives. It is now known that herbivorous animals, such as the guinea-pig and rabbit, are resistant to methaemoglobin formation and have entirely different metabolic reactions to these compounds from those of carnivorous animals (Lester, 1943).

The toxic effects on the nervous system have been studied by several investigators, including Dresbach and Chandler (1917) and Smyth (1931). Dresbach and Chandler found that, though rabbits, rats and cats reacted to inhalation of nitrobenzene chiefly by depression and paralysis, dogs showed an initial stimulation (vomiting), loss of muscle co-ordination, rigidity and nystagmus, followed by paralysis and ultimately death from respiratory failure. These symptoms were sometimes delayed from a few hours to 4 days after inhalation. Smyth (1931) states that nitrobenzene is fatal in a dosage of 1 g. per kg. body weight. The blood changes in animals have been very little studied, and the results are indefinite.

TOXIC EFFECTS IN MAN

Acute Poisoning

Nitrobenzene is readily absorbed by the skin as well as by inhalation, and some of the most severe cases of poisoning have been caused by accidental ingestion. Ravault, Bourret and Roche (1946) report a fatal case in which nitrobenzene had been drunk. The immediate symptoms were a burning sensation in the chest, followed by dyspnoea and syncope which passed into deep coma with cyanosis. Chapman and Fox (1945) reported a less severe case following accidental ingestion of furniture cream containing 4 to 5 per cent nitrobenzene (the total amount of nitrobenzene was about 15 mg.). The first symptoms were vomiting and drowsiness, followed by marked cyanosis $6\frac{1}{2}$ hours later.

Subacute Poisoning

Most of the cases occurring after contact with the skin, by inhalation or a combination of both, as in the two instances reported by Ravault, Bourret and Roche (1946), have an insidious onset. In these cases, which occurred in the course of textile printing, the two men used a rag soaked with nitrobenzene and worked in a small room where normally there was natural ventilation but in winter the windows were kept closed. They complained of headache, weakness and drowsiness, and later became unconscious with deep cyanosis. Under suitable treatment both recovered completely. According to Hamblin (1949) the onset of the methaemoglobinaemia is often symptomless, cyanosis being unaccompanied by any feeling of illness.

Nervous disturbance. This is a prominent feature of sub-acute poisoning and is chiefly due to the tissue anoxia, but also to a direct toxic action on the nerve centres; the symptoms include excessive fatigue, headache, vertigo, tinnitus and numbness of the limbs.

Methaemoglobinaemia. The condition is indicated by cyanosis, the face and lips being of a characteristic grey-blue (lilac) colour. Hamblin (1949) states that, as the methaemoglobinaemia increases, the headache becomes more severe, that cyanosis is usually recognizable when the methaemoglobin content of the blood reaches 10 to 15 per cent, but that headache and nausea are not experienced until the level has reached 40 per cent, and gastro-intestinal symptoms not even at 70 per cent or above. This is at variance with the earlier statements of Darling and Roughton (1942) and others, that more than 30 per cent will produce serious tissue anoxia. Hamblin (1949) bases his statement on a series of methaemoglobin estimations by means of a sensitive spectrophotometer.

Hamblin and Mangelsdorff (1938) state that the haemoglobin-methaemoglobin reaction is reversible, and that, if the sufferer is removed from exposure, reversion to haemoglobin is usually complete within 24 to 48 hours.

Effect on the Blood

In spite of the methaemoglobinaemia, in the few cases investigated changes in the blood picture were not marked, especially in sub-acute cases. In the case of a man who swallowed some laundry ink (Wirtschafter and Walpaw, 1944), although there was marked cyanosis, the chief abnormality of the blood picture was an initial leucocytosis (28,100 per c.mm.) with no anaemia, and later a relative lymphocytosis, but even these changes were transitory. Rejsek (1947) records that a man who suffered severe poisoning while exposed to dinitrobenzene had shown no change in his blood picture when previously employed for a year in the production of nitrobenzene, or during a later period when he was re-employed on nitrobenzene. Two industrial cases recorded by Gaultier, Benguigui and Smagghe (1949) showed no significant changes, and in the two cases reported by Ravault, Bourret and Roche (1946) the only blood changes were slight, anaemia (4·3 millions) with a raised colour index (1·1). These workers remark, however, that the anaemia is of a haemolytic type, and that the red blood corpuscles may show a marked irregularity of shape and size, and that there is sometimes an intense polynucleosis. This has not been the experience of Hamblin (1949), who states that he has frequently observed an initial rise in both the number of red cells and haemoglobin, the values returning to normal within 24 hours.

Effect on the Urine

The urine is usually a dark mahogany colour which Hamblin (1949) suggests is due to the presence of *p*-aminophenol which darkens rapidly as it oxidizes. Bile pigments in the urine have been demonstrated by Ravault, Bourret and Roche (1946) and by Wright Smith (1929), whose patient's urine showed a positive result for the indican test.

Treatment

Removal from exposure, rest, warmth, thorough cleansing of the skin and administration of oxygen are the most adequate, generally approved measures of treatment.

There is less agreement on the subject of intravenous administration of 1 per cent solution of methylene blue. This mode of therapy is based on the suggestion of Wendel (1939) that methylene blue helps to bring about the reversion of methaemoglobin to haemoglobin. A case so treated by Williams and Challis (1933) recovered, but in addition, the patient had received injections of glucose; Gaultier et al. (1949) report beneficial results in two relatively slight cases. Hamblin (1949), however, states that he has knowledge of several cases in which the administration of methylene blue has been followed by an alarming increase in cyanosis. Glucose therapy (1,000 ml. of a 5 per cent solution) is recommended by him. Alcohol in any form is strongly contra-indicated.

BIBLIOGRAPHY

BOGIN, C. and DAMPIER, H. L. (1942). Nitroparaffins as solvents in the coating industry. *Industr. Engng. Chem.*, **34**, 1091.
CHAPMAN, W. A. and FOX, C. G. (1945). Nitrobenzene poisoning from aniline cream. *Brit. med. J.*, **i**, 557.
CLARK, B. B., VAN LOON, E. J. and ADAMS, W. L. (1943). Respiratory and circulatory responses to acute methoglobinemia produced by aniline. *Amer. J. Physiol.*, **139**, 64.
CLARK, B. B., VAN LOON, E. J. and MORRISSEY, R. W. (1943). Acute experimental aniline intoxication. *J. industr. Hyg.*, **25**, 1.
COOK, W. A. (1945). Maximum allowable concentrations of industrial atmospheric contaminants. *Industr. Med.*, **14**, 936.
DARLING, R. C. and ROUGHTON, F. J. W. (1942). Effect of methemoglobin on equilibrium between oxygen and hemoglobin. *Amer. J. Physiol.*, **137**, 56.
DRESBACH, M. and CHANDLER, W. L. (1917). Some physiological disturbances induced in animals by nitrobenzol fumigation. *Amer. J. Physiol.*, **42**, 604.
EVELYN, K. A. and MALLOY, H. T. (1938). Micro-determination of oxyhemoglobin, methemoglobin and sulfhemoglobin in a single sample of blood. *J. biol. Chem.*, **126**, 655.
GAULTIER, M., BENGUIGUI, M. and SMAGGHE, G. (1949). À propos de 33 cas d'intoxications aiguës méthemoglobinisantes légères. *Arch. Mal. prof.*, **10**, 368.
HAMBLIN, D. O. (1949). Nitro and amino compounds of the aromatic series: In Patty, F. A., Industrial hygiene and toxicology. Part II, p. 987. Interscience Publishers, New York.
HAMBLIN, D. O. and MANGELSDORFF, A. F. (1938). Methaemoglobinaemia and its measurement. *J. industr. Hyg.*, **20**, 523.
HASS, H. B., HODGE, E. B. and VANDERBILT, B. M. (1936). Nitration of gaseous paraffins. *Industr. Engng. Chem.*, **28**, 339.
HEUBNER, W. (1913). Studien über Methämoglobinbildung. *Arch. exp. Path. Pharmak.*, **72**, 241.
LESTER, D. (1943). Formation of methemoglobin; species differences. *J. Pharmacol.*, **77**, 154, 160.
MACHLE, W., SCOTT, E. W. and TREON, J. (1940). The physiological response of animals to some simple mononitroparaffins and to certain derivatives of these compounds. *J. industr. Hyg.*, **22**, 315.
PATTY, F. A. ed. (1949). Industrial hygiene and toxicology. Interscience Publishers, New York. Parts I & II.
RAVAULT, P. P., BOURRET, J. and ROCHE, L. (1946). Deux intoxications par le nitrobenzène. *Arch. Mal. prof.*, **7**, 305.
REJSEK, K. (1947). M-dinitrobenzene poisoning. *Acta med. scand.*, **127**, 179.
SKINNER, J. B. (1947). Toxicity of 2-nitropropane. *Industr. Med.*, **16**, 441.
SMYTH, H. F. (1931). The toxicity of certain benzene derivatives and related compounds. *J. industr. Hyg.*, **13**, 87.
STERNER, J. H. (1949). Aliphatic nitro, diazo, and amino compounds: in Patty, F.A., Industrial hygiene and toxicology. Part II, p. 969. Interscience Publishers, New York.
WENDEL, W. B. (1939). Control of methemoglobinemia with methylene blue. *J. clin. Invest.*, **18**, 179.
WILLIAMS, J. R. and CHALLIS, F. E. (1933). Methylene blue as an antidote for aniline dye poisoning. *J. Lab. clin. Med.*, **19**, 166.
WIRTSCHAFTER, Z. T. and WALPAW, R. (1944). Case of nitrobenzene poisoning. *Ann. intern. Med.*, **21**, 135.
WRIGHT-SMITH, R. Y. (1929). Poisoning by nitrobenzene or "essence of mirbane"; with recovery. *Med. J. Aust.*, **i**, 867.

CHAPTER X
MISCELLANEOUS
1. Carbon Disulphide
CS_2

PROPERTIES

A colourless liquid in the pure state, becoming yellow on standing under the influence of light, with an ethereal odour when pure, very unpleasant if impure. B.P. 46° C. Sp.Gr. 1·2661. Fl.P. below −20°C.

EXPLOSION RISKS

Carbon disulphide has an extremely low auto-ignition temperature (125–135° C.) so that even contact with a hot steam pipe or electric lamp bulb may be sufficient to cause ignition of the vapour. The minimum explosive mixture in air is 0·063 mg. per l. or 19 volumes in 1,000 (Home Office Memorandum, 1935). It also generates static electricity in pipes, and even by friction on rubber (Barthelemy, 1939).

USES

(1) In the rubber industry in the manufacture of "dipped" rubber goods and as a solvent for sulphur in the cold vulcanization process.

(2) In artificial silk manufacture in the "viscose" process.

(3) In extraction processes, e.g. the extraction and purification of fats, recovery from waste materials, rags, etc. (Mattei and Sédan, 1924), extraction of sulphur in gas purification, etc.

(4) In pharmaceutical processes, e.g. in the manufacture of carbon tetrachloride, camphor, mustard plasters, etc. (Terrien, 1920).

(5) In the manufacture of waterproof cements, e.g. "Plastiline" (Flury and Zernik, 1931).

(6) In the manufacture of transparent paper.

(7) In the perfume industry sometimes as a constituent of hair lotions.

(8) As an insecticide for vines, tobacco plants, etc.

(9) In the manufacture of matches as a solvent for phosphorus (Mattei and Sédan, 1924).

TOXICITY

In high concentrations carbon disulphide may be a narcotic, causing loss of consciousness preceded by delirium or acute mania and followed by death from respiratory failure. Its better known effects, however, are those of a severe, chronic, general nerve poison, producing a great variety of symptoms depending on the part affected, individual susceptibility and the conditions of exposure, and ranging from slight fatigue and giddiness to profound psychic disturbances, failure of vision and complete paralysis. The variety of symptomatology apparently depends upon its affinity for lipoid tissue, particularly the cells of the central and peripheral nervous systems, and also the lipoid substance of the endocrine organs, especially the adrenals.

Concentrations in Air producing Symptoms

According to Lehmann (1894) 3·5 mg. per l. (1,100 p.p.m.) will produce severe symptoms with unconsciousness in half an hour, while higher concentrations (7 or even 10 mg. per l.) are followed by the same symptoms with very slow recovery. With 2·5 mg. per l. headache comes on immediately and lasts many hours, while 1 to 1·2 mg, per l. (320 to 380 p.p.m.) after a few hours causes headache and dullness, or confusion.

In the report of the Second International Congress on Industrial Diseases at Brussels in 1910, Constensoux and Heim noted the difference in severity of symptoms in two groups of workers exposed to different concentrations. The first group, before improvement in the conditions, worked in atmospheres containing 1 to 1·5 mg. per l. and showed severe symptoms, including cerebral disturbance, polyneuritis and amblyopia. The second group, exposed to atmospheres of 0·133 to 0·218 mg. per l., showed much slighter disturbance; the cerebral symptoms which did arise (headache, torpor and irascibility) disappeared in fresh air. In the factories making artificial silk examined by Weise in 1933, concentrations of 0·16 to 0·18 mg. per l. were accompanied by such severe symptoms that it was found necessary to place the containers under glass, while Kranenberg and Kessener (1925) found the amounts to vary considerably in different parts of the factory and under varying conditions of ventilation, e.g. 0·3 mg. per l. in one room, falling to 0·18 mg. after a rest pause, 0·6 mg. near the mixing drum in the viscose process, 0·4 mg. in transport from the drum, etc.

These results were summarized by the Department of Scientific and Industrial Research (1939), and are shown in Table 53.

TABLE 53
Toxic concentrations of carbon disulphide

Concentration of vapour in air		Duration of exposure	Effect
(parts by vol.) (approx.)	(mg./l.) (approx.)		
1 in 300– 1 in 500	10–6	About ½ hr.	Serious illness and danger of mania and coma
1 in 3,000	1	Single exposure for a few hours	Severe headache and mental dullness or confusion
		Daily exposures	Increasingly severe symptoms with neuritis, distorted vision and mental disturbance
1 in 15,000– 1 in 30,000	0·2–0·1	Repeated daily exposures	General condition of ill-health with headache, drowsiness and hysterical outbursts

Maximum Permissible Concentration

The following table (cited by Paluch, 1948) gives the maximum allowable concentration for carbon disulphide in various countries:

France	47 p.p.m.
Germany	47–63 p.p.m.
American Standards Association	20 p.p.m.
U.S.S.R. (Official)	3 p.p.m.

DETERMINATION IN AIR

The method suggested by the Department of Scientific and Industrial Research, (1939, Leaflet No. 6) depends upon the formation of cupric diethyl-dithiocarbamate from the interaction of carbon disulphide with diethylamine and copper acetate. A known volume of air containing carbon disulphide is drawn by means of a hand-pump through 14 ml. of "mixed reagent" (10 ml. of absolute alcohol, 2 ml. of a solution of 2 ml. of diethylamine in 100 ml. benzene, and 2 ml. of a solution of 0·1 g. of copper acetate in 100 ml. of absolute alcohol). The carbon disulphide is determined by matching the yellow colour produced against standards prepared by the same procedure from alcoholic solutions of carbon disulphide of known concentration. Concentrations down to 1 in 120,000 (0·025 mg. per l.) can be estimated in this manner with 20 strokes, or less, of the hand-pump.

TOXIC EFFECTS IN ANIMALS

Lethal and Narcotic Concentrations
  Lethal dose
  By inhalation. For cats, 23 mg. per l. (7,400 p.p.m.) (Lehmann, 1894); 20 mg. per l. (6,450 p.p.m.) (Luig, 1913).
  Narcotic dose
  By inhalation. For mice, 34 mg. per l. (11,000 p.p.m.) (Flury and Zernik, 1931); for cats, 10 mg. per l. (3,220 p.p.m.) (Luig, 1913); for rabbits, 7·6 mg. per l. (2,450 p.p.m.) (Lehmann, 1894).

Acute Poisoning

Symptoms

The animals become at first quiet and semi-somnolent, then show staggering, inco-ordination and convulsions, sometimes with vomiting, and finally paralysis with loss of reflexes; death takes place from respiratory paralysis. According to Flury and Zernik (1931) impure carbon disulphide is more toxic to animals than pure; young animals are more easily and earlier affected than older.

Post-mortem Appearances

Hyperaemia and slight oedema of the lungs and acute inflammatory lesions of the gastro-intestinal tract have been the chief pathological changes observed. Tomassia (1882) noted injection of the whole alimentary tract and Lehmann (1894) punctiform ecchymoses of the mucous membrane. Degeneration of the optic nerve in dogs subjected to repeated inhalations of concentrations which usually caused death were demonstrated by Offret in 1907. Richter (1945) found bilateral degenerative changes in the basal ganglia of monkeys.

Chronic Poisoning

The effects of repeated exposure of animals to carbon disulphide are regarded by most observers as similar to those recorded in human beings, particularly with regard to the changes in the nervous system.

Symptoms

Most of the earlier workers from Delpech (1856) and Lehmann (1894) onwards are agreed that concentrations in the neighbourhood of 1 mg. per l. (320 p.p.m.) are badly tolerated by animals. Many of these earlier investigations are described by Ranelletti (1931), and his own results generally confirm and amplify them. He exposed rabbits to concentrations of 1·28 mg. per l. for 8 hours a day (in two spells of 4 hours each), with a pause of 1½ hours for free ventilation, over a period of 5 months. The first symptoms appeared after 2 weeks' exposure—restlessness, irritability and increased sexual activity. These disappeared rapidly on removal from exposure. Later, however, they increased and were accompanied by muscular spasms of the limbs and throat, rapid heart action and difficulty in breathing. No tolerance to the effects was acquired; the symptoms occurred sooner after exposure in the later than in the earlier stages of the experiment. Growth and development were retarded. Even when acute symptoms (general convulsions and paralysis) appeared, recovery took place after 3 days in fresh air. More detailed investigations by Wiley, Hueper and von Oettingen (1936), Alpers and Lewey (1940) and Lewey and his collaborators (1941) have tended to show that experimental animals subjected to repeated exposure to various concentrations of carbon disulphide develop clinical and pathological signs analogous to those of human beings in similar conditions.

The dogs investigated by Lewey et al. (1941) which were exposed for 8 hours daily for 5 days a week to 400 p.p.m. (1·25 mg. per l.) showed marked behaviour changes, rigidity and tremor (Parkinsonian), choreiform movements, motor weakness, flaccid paralysis and nerve tenderness caused by polyneuropathy and degeneration of axis cylinders.

Pathological Changes

Effect on the nervous system

Chromatolysis of the cells of the cerebral cortex and increase of neuroglia cells were found in Ranelletti's animals. Audo-Gianotti's (1932) observations correlated the nervous lesions more closely with the special Parkinsonian syndrome occurring in some cases of chronic carbon disulphide poisoning in man, and which, by some authorities, is attributed to a special affinity of carbon disulphide for the central (strio-pallidal) nuclei. Audo-Gianotti found atrophy of the cerebral nerve-cells and dendrites especially in these areas and in the locus niger.

In the experiments of Wiley, Hueper and von Oettingen (1936), mice and rats exposed to 37 to 350 p.p.m. for 8 hours a day for 20 weeks showed no entirely uniform changes. Only one animal showed marked oedema and perivascular haemorrhages in the brain with degeneration of the Purkinje cells. Alpers and Lewey (1940) noted changes in dogs which were much more definite. They found swelling of the cytoplasm of ganglion cells, extensive damage to the Purkinje cells and basal ganglia, proliferation of the vascular epithelium, and thickening and hyalinisation of the cortical arterioles, swelling and fragmentation of the axis cylinders and myelin sheaths of the peripheral nerves.

Eye changes

A marked decrease in the corneal and pupillary reflexes and retinal angiospasm were observed by Lewey et al. (1941).

Effect on the blood

Some of the early observers and Davis (1929) reported changes in the size, shape and appearance of the erythrocytes of animals exposed to carbon disulphide which might suggest that their disintegration and haemolysis was a definite feature of carbon disulphide intoxication. With this view Brieger (1941) does not agree. He found no haemolysis in his animals subjected to intravenous, intramuscular or hypodermic injection of carbon disulphide, nor was anaemia frequent enough to suggest that decrease of total red cells or haemoglobin was an outstanding symptom. Alpers and Lewey (1940) found no definite changes in the blood picture of their dogs. Definite anaemia was observed by Binet and Bourlière (1944) in dogs subjected to inhalation of concentrations varying from 320 to 3,200 p.p.m. daily. The fall in red cells amounted to 1 to 2 millions per c. mm. The white cells apparently show no constant change. Brieger found a slight increase in neutrophil polymorphonuclear leucocytes and a few immature cells of this series, while Binet and Bourlière noted a slight increase in monocytes, but not to the degree of "toxic monocytosis" postulated by Creskoff in 1938.

Effect on the liver

Some suggestion of deranged liver function was indicated in the investigation of Lewey *et al.* (1941). They found some variations in the total blood cholesterol and a decrease in liver fat.

Effect on Sulphur Metabolism

Little effect on the sulphur metabolism, as estimated by the amount of sulphur in the blood and urine and the urinary sulphates, was observed by Binet and Bourlière (1944) during chronic intoxication, though a slight increase in the glutathione content of the blood occurred when the intoxication became acute. There was, however, an increase in urinary sulphates lasting for some days after the cessation of exposure.

Connexion with Vitamin B_1 Deficiency

According to Lewey (1939) the neuropathy and encephalopathy of chronic carbon disulphide poisoning in animals may have some association with the interference caused to the enzyme systems of respiration of the nervous system, which in its turn is associated with vitamin B_1 deficiency. These workers found that dogs, exposed to carbon disulphide and receiving a diet without added thiamine, had a pathologically low thiamine excretion, died earliest, and were the earliest to be electrically under-excitable. Thiamine appears to give some protection in delaying the pathological nervous and muscular reaction and the death of the animals.

TOXIC EFFECTS IN MAN

Absorption and Excretion

The principal mode of entry into the body is by inhalation of the vapour, but the liquid may be absorbed through the skin producing a sensation of burning and of anaesthesia. Actual burns with blistering and a local neuritis have been reported from prolonged contact with the skin (Hueper, 1936). It may also be absorbed by the gastro-intestinal tract after ingestion, and typical symptoms of poisoning have occurred in these conditions (Zangger, 1930; Floret, cited by Krause, 1931).

According to Lehmann (1908), about 23·7 per cent of the carbon disulphide inhaled is absorbed. (These later results differ slightly from his earlier estimation (1894) that 30 to 35 per cent is absorbed and 65 to 70 per cent exhaled.) Elimination, which is slow, takes place mainly by the lungs (strong odour in the breath), and a small fraction by the sweat, faeces and urine (Mattei and Sédan, 1924). According to Wiener (1908), 40 per cent of the carbon disulphide in the blood disappears within an hour after removal from exposure, a further 25 per cent during the second hour.

The presence of carbon disulphide in the urine is shown by the formation of a brownish-black precipitate with Fehling's solution.

Estimation in the Blood

Diluted blood is boiled with 0·5 per cent alcoholic potash solution; the potassium xanthogenate formed is titrated with a $N/100$ copper sulphate solution (Rodenacker, 1931).

Acute Poisoning

Symptoms

Inhalation of concentrated fumes of carbon disulphide causes a narcotic condition similar to that of chloroform anaesthesia—loss of consciousness, preceded or followed by delirium and maniacal symptoms, and accompanied by dilatation of the pupils, loss of reflexes, complete paralysis in the severest cases and respiratory failure leading to death. Two such cases with a fatal outcome were reported by Harmsen in 1905, from cleaning a reservoir and a pump through which carbon disulphide was delivered.

Slighter cases are characterized by headache, giddiness, breathlessness, vomiting, precordial and abdominal pain, palpitation, etc. These symptoms may subside rapidly on removal of the patient into fresh air, or some may persist for several weeks or even months after the acute attack. In a case reported to the Home Office in 1931, where the onset of symptoms occurred after the patient had inhaled fumes while cleaning a churn used in the manufacture of transparent paper, sickness, abdominal pains, loss of appetite, trembling of limbs and exaggerated knee jerks persisted for 2 months. Of the eighteen cases reported to the Home Office since carbon disulphide poisoning became compulsorily notifiable in December, 1924, eight were of the acute variety and occurred chiefly during the process of cleaning tanks containing carbon disulphide.

Only eight cases of "gassing" by carbon disulphide have been reported to the Factory Department since 1939, and all were of very short duration; only one showed loss of consciousness for a short period. The chief symptoms were headache, vomiting, general weakness, loss of power in the legs and abdominal pain, and in two cases (one in the rayon industry and one in paper-manufacture) mental confusion.

Of late years the mental symptoms of recorded cases appear to have become less severe than, for example, those described in Laudenheimer's survey of fifty cases in 1899, or those described in the Report of the Departmental Committee on Dangerous Trades in 1899, where a factory was mentioned in which the windows of the vulcanizing room had been barred to keep men in an acute stage of poisoning from leaping out during attacks of mania. Cases of this kind are rare in recent literature, though one case of maniacal type was reported to the Home Office in 1929, and in 1946, out of a hundred cases in Italy reported by Vigliani, there were several of severe psychotic type. The decrease in the number of cases as well as their severity is well shown by some statistics in France for 1946 (Auffret, 1946), where the figures show a fall from 252 in 1942, and 166 in 1943, to 66 in 1944 and 62 in 1945, with a corresponding decrease in the number of severe cases.

The persistence of the after-effects of acute carbon disulphide poisoning has been specially noted by Floret (cited by Krause, 1931) in three cases who had inhaled concentrated carbon disulphide vapour for short periods. One patient had anorexia, vomiting and tremor of the hands for a year, and another epileptiform seizures under excitement.

Post-mortem Changes

Very few fatal cases have come to autopsy. Foreman (1886) merely records congestion of the cortical veins. Redaelli (1925) noted congestion and oedema of the brain, but made no histological examination. The only detailed study of the nervous system that is available is that of Quensel (1904) who found diffuse lesions of the cortical ganglion cells, swelling of the endothelium and scattered perivascular haemorrhages.

Chronic Poisoning

Chronic carbon disulphide poisoning is much more common than acute, but acute exacerbations may arise during the course of intoxication. The clinical manifestations are so varied that special "nervous" or "gastro-intestinal" forms can scarcely be distinguished, each type showing symptoms arising from a widespread organic disturbance.

Carbon disulphide is, however, a severe nerve poison, affecting both the brain and the peripheral nerves. In some cases it produces psychosis of manic or depressive character; in others a syndrome similar to Parkinsonism and multiple sclerosis is found, and in still others, evidence of polyneuritis and of visual disturbance. It may also produce gastro-intestinal disturbance and an anaemia characterized by diminution of red cells and haemoglobin.

The most severe outbreak of chronic poisoning reported has been in Poland (Paluch, 1948). In one factory there were 148 cases among the 600 workers exposed. The majority of these cases (52 per cent) were of the polyneuritic type, the chief site of paresis being the peripheral nerves, but in some patients the nerves of the upper extremities and the facial nerves were affected. Psychotic symptoms were present in 18 per cent of the cases, and neurotic or pre-psychotic in 30 per cent.

Pre-clinical Poisoning

A pre-clinical stage of poisoning in workers exposed to carbon disulphide is postulated by some authorities including Paluch, who emphasizes the poor mental and physical condition which may exist before actual symptoms appear.

Among the six cases (three of which are referred to as "so-called acute") reported by Gordy and Trumper (1938) the tendency to psychotic episodes of a periodic or cyclic character is stressed; these episodes were accompanied by a variety of paraesthetic or hallucinatory phenomena, which were sequelae in four of the cases.

Mechanism of Action

The mechanism of action of carbon disulphide in producing psychic disturbance has for long been believed to be that of a lipoid solvent. In 1938, for example, Gordy and Trumper stated that the mode of action of carbon disulphide is clear; it is lipotrophic and therefore neurotoxic. A modification of this view is offered by Lewey (1941). He points out that there is, in animal experiments at any rate, a delay in the development of injury to the peripheral nerves, in contrast to the immediate effect on the cerebral hemispheres, and that there is no correlation between the appearance in man of the psychic symptoms and signs of peripheral nerve injury. He suggests that the mechanism of chronic carbon disulphide intoxication is different from that of acute poisoning. Lewey also found that thiamine has a protective and curative effect in experimental animals (p. 384), and that human beings suffering from chronic carbon disulphide poisoning have a lowered thiamine excretion; he believes that thiamine deficiency may therefore be associated with chronic carbon disulphide absorption, possibly by way of liver damage and by direct poisoning of the co-enzymes of nerve metabolism and respiration.

Symptoms

The mild chronic form of poisoning is characterized by fatigue, giddiness, heaviness and pain in the limbs, increasing pain in the forehead and temples, listlessness, difficulty in concentration, general "nervousness", often accompanied by numbness and weakness, gastro-intestinal disturbance and tingling of the legs in the "polyneuritic" type. A typical case of this kind was described by Walshe in 1929, and six were reported to the Home Office between 1934 and 1935.

Nervous symptoms

Psychoses. The psychosis of carbon disulphide poisoning is most frequently of the maniacal type with acute confusion, delirium and hallucinations. The onset of acute symptoms may be fairly slow, after weeks or even months of exposure, and may be preceded by a phase of irritability with headache and sleeplessness and a phase of depression. The disorder may subside after a few months or may pass into incurable dementia (Laudenheimer, 1899; Peterson, 1892, etc.). As already mentioned, such cases have become rare but a case of severe poisoning of 13 years' duration was reported by Abe in 1933. This was a young man, employed in mending rubber shoes, in whom psychosis of the Korsakoff type was eventually superimposed on neurasthenic symptoms of long standing. He showed loss of memory, disturbance of speech, weakness of the legs, positive Romberg sign and positive Babinski reflex, muscular cramp, marked tremor, oedema of the legs, high fever and albuminuria. Death occurred 10 months after development of severe symptoms. Post-mortem examination revealed regressive changes in the nervous system, especially in the grey matter of the cerebrum, cerebellum and medulla, with marked degeneration of the pyramidal tracts in the pons, extending into the cord.

Rodenacker (1931) has put forward an interesting theory in connection with the psychic manifestations of carbon disulphide poisoning. He states that it has a special affinity for individuals of the "cyclothymic" type and that carbon disulphide poisoning is not in itself a cyclic psychosis, but merely an intoxication during a phase of mania or depression. Rodenacker believes that the schizophrenic type of individual is comparatively insensitive to carbon disulphide poisoning, and quotes the case of a workman, who had been treated for insanity of the schizophrenic type for 7 years, and who felt well when working exposed to carbon disulphide in concentrations which caused considerable discomfort to his co-workers.

Hysterical manifestations. In women these were often recorded by the earlier workers. Pierre-Marie (1888), for example, held that carbon disulphide was only the exciting agent in individuals already disposed to hysteria, while Bignami (1925) states that it may bring latent hysteria to a crisis, sometimes in epidemic form, in a factory.

Some observations by André (1947) would appear to suggest that some of the minor manifestations which follow long exposure to low concentrations of carbon disulphide may in fact have a purely functional origin. He found a number of workers at an artificial silk factory exposed to "permitted" concentrations of carbon disulphide suffering from certain symptoms, of which muscular weakness without atrophy was the most pronounced, and which included asthenia and a tendency to emotion with suggestibility. The severe functional disability had apparently no organic basis and no physiopathological basis was discovered. Nevertheless, only workers exposed to carbon disulphide in this factory were affected and one case, who was lost sight of for several years, was later found employed as a diamond cutter, with complete recovery of all muscular power and capacity for fine movements.

Strio-pallidal or Parkinsonian syndrome. A number of cases of the Parkinsonian syndrome are characterized by tremor, ataxia, disturbance of speech, amnesia, muscular spasticity (Schramm, 1940) and mental depression (Quarelli, 1930; Negro, 1930; Audo-Gianotti, 1932; Gordy and Trumper, 1940). These symptoms are believed to be caused by the action of carbon disulphide, either directly or indirectly through nutritional and circulatory disturbance, on the nerve cells of the corpus striatum, the globus pallidus and the locus niger. The detailed microscopic examination by Abe (1933) of the nervous system of a patient who died after slow development of these symptoms strongly supports this view. In a typical case, the face is pale and greasy-looking, there is a history of a slow onset of general malaise, weakness of the limbs, and a change in mental disposition towards melancholy and irritability. Sometimes, as in a case investigated by the Home Office in 1932, there are also headache, indigestion, numbness, tingling and pains in the arms and legs. The most noticeable physical sign is the tremor of the face, tongue, eyelids, arms and hands, very suggestive of paralysis agitans. The nerve reflexes are exaggerated, the gait is unsteady and the speech slow (Abe, 1933; Negro, 1930). Romberg's sign and the Babinski reflex are usually positive (Abe, etc.). An interesting feature of the Home Office case was the occurrence of a mild attack of this syndrome with recovery on removal from exposure, followed by a much more severe attack when the patient returned to the same work 2 months later.

In some rarer cases, such as that described by Zeglio (1942), the syndrome appears to be "striate" rather than "pallidal", the symptoms including choreic and athetotic movements rather than the slowness and rigidity of the Parkinsonian variety of poisoning. In Zeglio's case the intense involuntary muscular movements became general, and later there was formiculation, cold in the extremities and lack of muscular power. Recovery took place 6 months after removal from exposure and Zeglio regarded this as an unusual manifestation in a young man showing individual predisposition with slight endocrine dysfunction, including hypogenitalism.

Polyneuritic syndrome. The chief symptoms in this group of cases are loss of power in the muscles of the upper and lower limbs, so that the gait is unsteady, even simulating that of pseudo-tabes (as in the case recorded by Mutschlechner, 1924), and the grip weak. Paresis of isolated muscles may occur e.g. of the vagus, giving rise to paresis of the soft palate, disturbances of swallowing, and a nasal voice (Ranelletti, 1931). Loss of sensation may also occur. In Ranelletti's series of 100 cases, of which 10 per cent were the polyneuritic type, loss of sensation was less frequent than loss of motor power, but in a typical, severe case of this kind reported to the Home Office in 1925, there was almost complete anaesthesia from the chin downwards.

Peripheral nerve lesions are generally of slow development, after exposure lasting not less than a month, according to Schramm (1940), though earlier workers such as MacGregor (1892) have recorded cases of peripheral neuritis after exposure as short as 4 days.

The electrical reaction of the nerve-muscle system of the extremities is regarded by Bashore and Staley (1938) as an early diagnostic criterion of carbon disulphide poisoning. Usually electrical irritability is decreased, and in only a few cases is it increased.

The prognosis of carbon disulphide polyneuritis appears to be less favourable than was generally believed. Zeglio (1946) states that, out of 20 cases reviewed after a period of 4 to 8 years since the beginning of the disease, only 1 had recovered, 4 had improved, 5 had remained unchanged and 10 had become worse. One of 2 cases observed by Bourret and Kohler (1946) showed only slight improvement after 3 months' treatment with strychnine.

Eye disturbances. Eye lesions due to affections of the optic nerve, the pupils and the retina, and kerato-conjunctivitis still constitute a considerable problem. In 1945, 600 cases were observed in France (Auffret, 1946).

Two cases of severe amblyopia in men employed in making mustard plasters were recorded by Terrien in 1920, and 2 more in men employed in the extraction of fats from waste materials by Mattei and Sédan in 1924. Ocular disturbances, (diminution of visual acuity, retraction of the visual field and some disturbance of colour vision), going on to amblyopia in a few cases, were fairly prominent among the cases investigated by Constensoux and Heim in 1910, but did not occur in Ranelletti's case, and were not common among the 18 cases reported to the Home Office and the Factory Department of the Ministry of Labour since 1925; these included one case of slight failure of vision, one in which the eyes were easily fatigued and 12 cases of conjunctivitis and keratitis in 1930. Fatigue of accommodation after exposure to 0·172 mg. per l. was observed by Haas and Heim in 1910.

Cases of progressive failure of vision due to retrobulbar neuritis have, however, been described by Monbrun, Richet and Facquet (1932) and by Nectoux and Gallois (1931). In the 4 cases of Nectoux and Gallois (all young women) the intoxication was slow and progressive, lasting from 1 to 3 years, and all showed a bilateral central scotoma for colours. In the cases reported by Monbrun *et al.*, which occurred as a result of the use of a hair lotion containing carbon disulphide, frontal headache, loss of memory, intellectual apathy, fatigue and loss of power in the lower limbs accompanied the decrease in visual acuity, which was below one tenth normal. Recovery was more rapid than with alcohol or nicotine amblyopia, which it resembled.

In 3 cases described by Baader (1932), including one already recorded by Bonhoeffer (1930), the symptoms simulated a brain tumour, headache, vomiting, vertigo and failing vision being the chief symptoms. Ophthalmological examination, however, revealed the presence of retrobulbar neuritis.

Enlargement of the blind spot is believed by McDonald (1938) to be a significant early ocular sign in chronic carbon disulphide intoxication. He found it in 22·5 per cent of all the workers he examined. Diminution of the corneal reflex was present in about half the workers, but this was partly due to other factors. Disappearance of the corneal reflex and abnormalities of the pupil reaction to light and accommodation have also been noted as pathognomonic signs of poisoning (Bashore and Staley, 1938), but the former was found in only 3 cases out of 40 in a later series of 80 described by Warnecke (1941).

Gastro-intestinal symptoms

Slight gastro-intestinal symptoms, nausea, vomiting, gastric pain and constipation, are common early symptoms of chronic carbon disulphide poisoning, but, as pointed out by Weise (1933), they are often overshadowed by the more prominent nervous disturbances. Weise himself lays stress on the early appearance of these gastric disorders; he found a number of workers in the artificial silk and rubber industry complaining of quite severe symptoms, in some cases suggestive of gastric or duodenal ulcer. It should be noted that some of these workers had been exposed also to sulphuretted hydrogen.

Audo-Gianotti (1934), however, found "chronic gastritis" in several of his 77 chronic cases, and in several of the mild cases reported to the Home Office abdominal pain and discomfort, nausea, loss of appetite and vomiting were prominent symptoms, while 62 per cent of the workers examined by Constensoux and Heim (1910) suffered from loss of appetite, dyspepsia and sometimes nausea and vomiting. Audo-Gianotti related the frequency of gastric and duodenal ulcer and hyperacidity to a lability of the parasympathetic system, in itself an effect of carbon disulphide poisoning. In a detailed investigation of gastro-intestinal disturbance among carbon disulphide workers made by Zeglio in 1942, he found that while there was no definite type of gastric secretory change characteristic of carbon disulphide poisoning, hypochlorhydria was most often present. There was evidence of gastritis of less severity than in those workers complaining of acid regurgitations, epigastric pain, irregular bowel action, etc., even in a group of workers who were symptomless, and he suggests that functional disturbance is probably present in many symptomless cases. He found no relation between the time of exposure and the severity of the disturbance, but improvement or cure always followed cessation of work. Warnecke (1941) found gastro-intestinal symptoms to be the most prominent in 34 cases out of 100, loss of appetite, nausea, colic and constipation.

Anaemia

While carbon disulphide, unlike benzene, does not cause severe injury to the haemopoietic organs, it may, according to Ranelletti (1931), cause a severe anaemia. This he found to be the chief symptom in 17 out of his 100 cases. Tomassia (1882) and Bignami (1925) have also recorded blood changes in chronic poisoning—diminished erythrocytes and haemoglobin, anisocytosis and poikilocytosis. In the cases quoted by Jump and Cruice (1904) and by Francine (1905), both subjects had haemoglobin values of only 40 per cent, though the red cells were at a normal level. Warnecke (1941), on the contrary, found the principal changes in the blood picture to be those of hyperglobulia (R.B.C.s up to 7·45 millions), lymphocytosis and monocytosis. Changes in the leucocytes, according to Ranelletti, are not constant, but he observed a monocytosis from 12 to 20 per cent in some cases, as did Creskoff (1938), while Bignami (1925) found an eosinophilia up to 13 per cent. A reduced peroxidase reaction of the leucocytes was observed by Velicogna and Viziano (1932) in 8 out of 15 cases of chronic poisoning; this they correlate with a lesion of the mesencephalon.

Injury to the adrenals

The possibility that carbon disulphide may cause chronic insufficiency of the adrenals is put forward by Devoto (1934) on the evidence of one rather unusual case only. The patient was a doctor engaged in research, who was severely exposed to carbon disulphide vapour over a period of 6 hours. He had severe headache and giddiness the same evening, and loss of appetite, vomiting and slight jaundice during the next 2 days. A month later he developed typical symptoms of Addison's disease—pigmentation, general weakness and low blood pressure. He died nearly 3 years later and the diagnosis of adrenal disease due to carbon disulphide poisoning was accepted by the insurance company. Devoto suggests that the original severe exposure caused acute injury of the cells and blood vessels of the adrenals, followed by sclerosis and atrophy producing chronic insufficiency.

Disturbances of sexual function

Loss of sexual function, going on in some cases to impotence, has been recorded by some authors. Constensoux and Heim (1910) found diminution of genital function in 54 per cent of the workers exposed to concentrations of 1·5 mg. per l. and in 21 per cent of those exposed to concentrations of 0·133 to 0·218 mg. per l. Abe (1933) states that sexual activity may be increased in the first stages, but that later complete sterility may follow; total impotence is also stated to occur by Cazeneuve, Morel and de Leeuw (1932). Delpech (1856) observed arrested development of the testes in a boy aged twelve exposed to carbon disulphide, and stated that in women chronic poisoning might lead to metrorrhagia, and to abortion if pregnancy occurred.

Post-mortem Findings

In a few of the cases reported by earlier investigators where a post mortem had been performed, the chief findings were congestion of the cerebral cortex, multiple haemorrhages in the liver and spleen, and evidences of neuritis in some of the peripheral nerves (Ranelletti, 1931).

A more recent case, where death took place after many years of exposure and after typical symptoms of the Parkinsonian type of nervous manifestation had been present for about 6 years, has been investigated in great detail by Abe (1933). The chief pathological appearances in this case were: (1) regressive changes and fatty degeneration of the internal organs, especially liver, kidneys and heart; (2) degenerative changes in the central nervous system; these were greatest in the grey matter of the cerebrum and in the pons, where the pyramidal tracts showed severe degeneration, a finding which Abe correlates with the paresis and ataxia which were among the chief symptoms. The degenerative lesions were found in the nerve cells and nerve bundles with simultaneous proliferation of neuroglial tissue, but with no appearance of inflammatory reaction. Some of these changes were present in the corpus striatum, but there was little change in the globus pallidus.

Individual Susceptibility

The following factors appear to increase susceptibility to carbon disulphide poisoning:

(1) A previous attack (Mattei and Sédan, 1924). It will be noted that in the case of the man showing symptoms of Parkinsonian type reported to the Home Office in 1932 (p. 388), the second attack was much more severe than the first.

(2) The type of individual (psychic or hormonal). Rodenacker (1931) believes that the "pyknic" (broadly built in body and of the "cyclothymic" type of mentality) is specially sensitive to the nervous manifestations. In one of Monbrun and Facquet's (1932) most severe cases, the wife of the patient, who worked with him, was quite unaffected, while the 4 cases reported by Nectoux and Gallois (1931) were the only ones to occur among 20 persons exposed.

(3) According to Warnecke (1941) women and young persons are especially susceptible.

2. Acetic Acid

CH$_3$·COOH

Produced by the oxidation and fermentation of dilute alcohol, or synthetically from acetylene (via acetaldehyde), or by oxidation of ethyl alcohol in the presence of a catalyst.

PROPERTIES

A colourless liquid with a pungent odour. B.P. 118·5° C. Sp.Gr. 1·049. Fl.P. 45° C.

USES

(1) In the manufacture of cellulose acetate, artificial leather, pharmaceuticals, phenol condensation products.
(2) In the dyeing industry.
(3) In the printing of textiles.
(4) As a coagulant (dilute) for latex.

TOXICITY

Acetic acid is a strong irritant of the mucous membranes.

Maximum Permissible Concentration

Sterner (1949) states that 10 p.p.m. is the maximum permissible concentration and would be given by 0·69 ml. acetic acid volatilized and distributed equally at 25° C. in 1,000 c. ft. of air.

TOXIC EFFECTS IN ANIMALS

Inhalation of 12,000 to 14,000 p.p.m. (30 to 36 mg. per l.) causes irritation of the nasal mucous membrane; that of 19,000 to 35,000 p.p.m. causes hyperaemia of the tracheal mucous membrane, but no severe symptoms (Flury and Zernik, 1931).

TOXIC EFFECTS IN MAN

No injury from its industrial use has been reported, but severe irritation of mucous membranes, which becomes intolerable after a few minutes, occurs with exposure to 800 to 1,200 p.p.m. (2 to 3 mg. per l.) (Flury and Zernik, 1931).

3. Acetic Anhydride

C$_4$H$_6$O$_3$, i.e. (CH$_3$·CO)$_2$O

Produced by the catalytic dehydration of acetic acid; also by the action of sulphur dichloride and chlorine on sodium acetate.

PROPERTIES

A colourless liquid with a pungent odour. B.P. 139·4°–139·5° C. Fl.P. 60° C. (127° F.). Soluble in alcohol or ether; decomposed by water to acetic acid.

Inflammable in saturated air (concentrations of 2·67–10·13 per cent by volume) at temperatures between 117° and 166° F. (47° and 74° C.) (Jones, Scott and Scott, 1943).

USES

(1) As a reagent in organic synthesis and analysis.
(2) For acetylation.
(3) In the examination of fats.
(4) In the manufacture of artificial acetate silk and film.

TOXICITY

Acetic anhydride is an irritant to the skin and mucous membranes. It is uncertain whether the irritating properties are caused by the anhydride itself, or by the acetic acid into which it has been transformed by contact with moisture in the air and tissues.

TOXIC EFFECTS IN ANIMALS

Lethal Dose

By oral administration. For rats, 3·31 g. per kg. body weight; for mice, 4·96 g. per kg. body weight (Woodard, Lange, Nelson and Calvery, 1941).

TOXIC EFFECTS IN MAN

According to Oettel (1936) acetic anhydride has very little effect on the unbroken skin, but causes severe injury to mucous membranes. In the *Quarterly Safety Summary* (1938) issued by the Association of British Chemical Manufacturers, there is a report of severe injury to the eyes following an explosion during an acetylation experiment. Acetic anhydride had been projected into the eyes of a chemist, causing destruction of the surface layer of the conjunctiva and cornea. Treatment with dilute sodium bicarbonate resulted in satisfactory recovery.

4. Cresols

(*Cresylic Acid*)

$CH_3 \cdot C_6H_4 \cdot OH$

PROPERTIES

Cresylic acid, distilling between 195° and 205° C., is a mixture of the three isomers of cresol, *ortho-*, *meta-* and *para-*cresol:—

(1) *ortho*-Cresol (*o*-methylphenol), B.P. 190·5° C.; forms crystals of Sp.Gr. 1·0487.
(2) *meta*-Cresol (*m*-methylphenol), B.P. 202·8° C.; liquid, Sp.Gr. 1·034.
(3) *para*-Cresol (*p*-methylphenol), B.P. 202° C.; forms prisms of Sp.Gr. 1·035.
Solubilities in water at 25° C. lie between 2·0 and 2·5 g. per 100 ml.

USES

(1) As a constituent of antiseptics, such as lysol, creolin, etc. (Waldhecker, 1941).
(2) As an intermediate in the chemical industry.
(3) In the manufacture of plastics (Corcos, 1940).

TOXICITY

The cresols have a local corrosive action causing severe inflammation of the skin and mucous membranes; when taken by mouth there is a systemic action on the nervous system with severe depression and collapse. Less severe effects may be caused by inhalation of cresol vapour.

Comparison of Toxicity with that of Phenol

Inhalation by mice of air saturated with vapours of cresylic acid, if repeated, caused death, but single brief exposures to the cresols appear to be harmless to human beings (Deichmann, 1949). It appears that the toxicity of the cresols is not materially different from that of phenol (Klinger and Norton, 1945) but, according to Deichmann and Witherup (1944), the depressant effects are more severe with the cresols, although the convulsant action is less pronounced.

TOXIC EFFECTS IN ANIMALS

The toxicity in animals varies with the different species and with the method of administration. With oral administration dogs are more sensitive than cats; with subcutaneous injection cats are more sensitive than mice, rabbits or guinea-pigs; rats are the most resistant (von Oettingen, 1949).

Lethal Dose

By oral administration. 10 per cent emulsion in water; for rabbits, *ortho*-cresol, 0·949 mg. per kg.; *meta*-cresol, 1·40 mg. per kg.; *para*-cresol, 0·62 mg. per kg.

By subcutaneous injection. 10 per cent solution in olive oil; for cats, *ortho*-cresol, 0·055 g. per kg.; *meta*-cresol, 0·18 g. per kg.; *para*-cresol, 0·08 g. per kg.

By intravenous injection. 0·5 per cent solution in water; *ortho*-cresol, 0·78 g. per kg.; *meta*-cresol, 0·28 g. per kg.; *para*-cresol, 0·18 g. per kg. (Deichmann and Witherup, 1944).

Symptoms

Local irritation, like that produced by phenol, is manifested by white discoloration, followed by reddening, inflammation, and finally necrosis and sloughing (Goodman, 1933).

Systemic Effects

Convulsive movements, more marked depression of the central nervous system than with phenol, and muscular tremor follow injections of 0·025 g. per kg.

TOXIC EFFECTS IN MAN

Most of the observations of cresol poisoning have referred to the absorption of lysol and other antiseptic cresol preparations through accidental or suicidal ingestion.

Local Effects

Prolonged contact of cresols with the skin causes reddening, itching, vesiculation, and later, depending upon the concentration and the duration of contact, eczema and ulceration. Goodman (1933) states that as little as 0·063 per cent of cresol in cutting oil may cause a dermatitis. Contact with mucous membranes produced white burns.

Systemic Effects

Oral administration is followed by severe burning pain in the upper digestive tract, difficulty in swallowing, vomiting and diarrhoea. In very severe cases there may be immediate collapse and unconsciousness with cold, clammy skin and small, rapid pulse. Later complications may include pneumonia, pulmonary oedema, abscess of the lung (Koster, 1943), injury to the liver with jaundice and injury to the kidneys with albuminuria and the presence of casts and red blood cells.

Slight poisoning by inhalation has been reported by Duvoir, Hazemann, Deruelle and Fallot (1938) and by Corcos (1940). The chief symptoms were headache, nausea and vomiting; three cases showed definite tremors.

Absorption

Cresol may be absorbed through the skin, the gastro-intestinal tract, the lungs, and other cavities, such as the bladder and uterus. The rate of absorption through the intact skin depends more on the area covered than on the concentration (Deichmann and Witherup, 1944). Absorption through the gastro-intestinal tract takes place rapidly at first, then more slowly, and has been amply illustrated by many cases of accidental or suicidal poisoning. Absorption through the lungs by inhalation of the vapour has been found to be followed by a moderate degree of gastro-intestinal disturbance (Duvoir *et al.*, 1938; Corcos, 1940).

Metabolism and Excretion

Excretion is largely through the kidneys, though some may be excreted through the bile and some traces in expired air. On account of the methyl group present in the cresol ring, cresol is more readily oxidized than phenol. Excretion takes place chiefly with *m*-cresol unchanged in the form of ethereal sulphate. To a large extent this applies to *o*-cresol, which is also partially oxidized to toluhydroquinone, while *p*-cresol is oxidized to *p*-hydroxybenzoic acid.

5. Dimethyl Sulphate

$(CH_3)_2 SO_4$

PROPERTIES

An oily, colourless liquid, with a characteristic odour. B.P. 187·5° C. Sp.Gr. 1·25. At 50° C. gives off a greyish vapour. On prolonged contact with water it decomposes to give methyl hydrogen sulphate and methyl alcohol.

USES

(1) In the chemical industry as a methylating agent.
(2) In the analysis of benzene.
(3) In the manufacture of pharmaceuticals and perfumes.
(4) As a catalyst in the preparation of cellulose esters (Brina, 1946).

TOXICITY

It has a strong caustic effect on the skin and mucous membranes, producing blisters and ulcers which show difficulty in healing (von Nida, 1947). In addition to this caustic effect on the mucous membranes, the vapour has a general systemic effect, producing, after a latent period, injury to the kidneys and liver, and inflammatory lesions of the respiratory system, leading in fatal cases to paralysis of the respiratory centre. According to Lehmann and Flury (1943), the special danger of inhalation of methyl sulphate lies in its comparative lack of odour, absence of warning signs, and the latent period for the development of symptoms.

TOXIC EFFECTS IN ANIMALS

Lethal Dose

By stomach tube. For rabbits, 0·25 g. per kg. body weight after 1 hour; 0·05 g. per kg. body weight after 12 to 15 hours (Weber, 1902).

By inhalation. For cats after 11 minutes' inhalation, 0·9 mg. per l. (175 p.p.m.) causes death after 1½ weeks; for monkeys, 0·132 mg. per l. (26 p.p.m.) after 3 days (Lehmann and Flury, 1943).

Symptoms

Cramp and unconsciousness were observed; with stomach tube administration, diarrhoea and erosion of stomach also occur.

TOXIC EFFECTS IN MAN

When taken by mouth, dimethyl sulphate acts as a caustic, producing inflammation of the whole gastro-intestinal tract. Borner (cited by Lehmann and Flury, 1943) describes a case in which death was due to pneumonia, but there was also injury to the heart, liver and kidneys.

Fatal Cases

Among the early fatal cases quoted by Lehmann and Flury (1943), the majority were caused by inhaling the vapour from liquid which had been spilt, and death was primarily due to pneumonia. Von Nida (1947) describes a case similar in origin, but showing a latent period of some days before death which was due to oedema of the glottis.

Non-fatal Cases

All such cases are characterized by the latent period between exposure and the onset of symptoms. Balázs (1934) describes nine cases in which exposure occurred during the wiping-up of dimethyl sulphate overflowing from containers. Antonioli (1942) reports one case of inhalation from a broken container, and Brina (1946) two cases from wiping-up of spilt liquid.

Symptoms

Mucous membrane inflammation. Photophobia, lachrymation, swelling of the eyes, lips and tongue, dyspnoea and cough with yellow, frothy, blood-streaked sputum were reported by Antonioli (1942).

Renal involvement. Albuminuria and haematuria occurred in two cases (Balázs, 1934).

Metabolic and haematological injury. Hypoglycaemia, hyperproteinaemia and polycythaemia were recorded by Rastelli (1941); hypoglycaemia was found by Brina (1946) in only two cases; none of these signs was observed by Antonioli (1942).

Mode of Action

There has been some controversy as to whether the systemic symptoms are due to the action of the undecomposed molecule or to the products of its decomposition, i.e. methyl alcohol and sulphuric acid; Antonioli and Brina hold the latter view. All authorities agree that the toxic effect is a general one, not secondary to cutaneous lesions but due to the action of the vapour on the mucous membranes. According to von Nida (1947) the vapour attacks only the mucous membrane and not the skin, and therefore inhalation is probably more dangerous than actual contact of the liquid with the skin.

6. Silicones and Silane Intermediates
A. SILICONES

Silicones, or organopolysiloxanes, are organic derivatives of silica, SiO_2. They are obtained in the form of fluids, and resins. Only the fluid silicones will be considered here and these are almost certainly end-stopped, siloxane polymers of the type $R_3 Si \cdot O \cdot (SiR_2 \cdot O)_n \cdot SiR_3$.

PROPERTIES

A large number of liquid silicones can be made by varying the hydrocarbon groups attached to the silicon atoms and by varying the length of the chain. The most important commercially are the methyl or partly methyl and partly phenyl compounds, such as hexamethyldisiloxane and methylphenylsilicone.

Insoluble in water or alcohol, and neutral in reaction.

USES

(1) In the plastic industry as release agents in moulding organic plastics.
(2) In glass and ceramic manufacture as water repellents.
(3) As special high- and low-temperature lubricating oils.
(4) As hydraulic fluids.
(5) As liquid dielectrics.

TOXICITY

Most of the silicone fluids, being essentially non-volatile, present no significant toxic hazard. Hexamethyldisiloxane, though a good solvent and volatile (B.P. 99·5° C. at 760 mm.), is of a low order of toxicity, its chief hazard being that of irritation of the conjunctiva.

TOXIC EFFECTS IN ANIMALS

Oral Administration

Hexamethyldisiloxane in single doses of from 3 to 50 ml. per kg. produced no laxative action in rats, rabbits or guinea-pigs (Rowe, Spencer and Bass, 1948). In the higher dosages, mild irritation and subsequent nervous depression were noted some hours afterwards. Dodecamethylpentasiloxane caused some laxative action, but no central nervous depression. Other silicone fluids had about the same effect as that of ordinary mineral oil.

Intraperitoneal Administration

With doses of 1, 3, 4 and 10 ml. per kg. of hexamethyldisiloxane, animals became moribund within the first week; extensive adhesions and widespread local irritation were found at autopsy.

Intradermal and Subcutaneous Administration

Hexamethyldisiloxane caused some irritation with induration at the site of the injection.

Skin Application

Even repeated application of hexamethyldisiloxane caused no significant irritation.

Eye Instillation

A transitory conjunctival irritation followed instillation of several silicone fluids tested, but there was no evidence of corneal damage as tested by fluorescein staining.

Inhalation

Hexamethyldisiloxane when given in single exposures of 25,000 p.p.m. for 30 minutes or with repeated exposures of 4,400 p.p.m. produced no toxic effect. Saturated atmospheres (39,000 to 40,000 p.p.m.) in 15 to 20 minutes caused death, apparently due to respiratory failure (Rowe, Spencer and Bass, 1948).

TOXIC EFFECTS IN MAN

Rowe, Spencer and Bass (1948) consider the only probable hazard to human beings from the volatile liquid silicone, hexamethyldisiloxane, is that of a transitory conjunctival irritation, characterized by erythema of the conjunctival membranes, and frequently accompanied by oedema of the eyelids with a sensation similar to that of windburn.

With regard to inhalation of the vapour, concentrations of the magnitude required to kill rats in 30 minutes are completely non-respirable for human beings.

No cases of injury from the industrial use of silicones have been reported.

B. SILANE INTERMEDIATES

The hydrolysable silanes encountered in the production of silicones include:

(a) The chlorosilanes, e.g. silicon tetrachloride and the alkyl chlorosilanes (dichlorodimethylsilane, methyltrichlorosilane, dichlorodiethylsilane and ethyltrichlorosilane). These compounds are highly corrosive; silicon tetrachloride has been used in warfare as an irritant gas and in the preparation of smoke screens. All were found highly irritating to the skin of animals, causing denaturation in about 10 minutes. Inhalation of the vapour caused attempts to hold the breath, shallowness and difficulty in breathing.

(b) The ethoxysilanes include tetraethoxysilane, methyltriethoxysilane, diethoxydimethylsilane, and ethoxytrimethylsilane. According to Rowe, Spencer and Bass (1948) tetraethoxysilane is by far the most toxic, especially by inhalation. Concentrations of 100 p.p.m. caused kidney damage in the form of tubular degeneration and necrosis. These results confirmed those of Kasper, McCord and Frederick (1937) and Smyth and Seaton (1940).

TOXIC EFFECTS IN MAN

Rowe, Spencer and Bass (1948) point out that though the methylated ethoxysilanes do not present any serious problem in industry, the handling of chlorosilanes and of tetraethoxysilanes does, especially with regard to contact with the skin and to inhalation of the vapour. Particular care should be taken to avoid even the smallest splashing of the eye; inhalation of tetraethoxysilane must be carefully watched because hazardous concentrations are not sufficiently disagreeable to give warning of danger, as is the case with the less toxic diethoxydimethylsilane.

BIBLIOGRAPHY

ABE, M. (1933). Beitrag zur pathologischen Anatomie der chronischer Schwefelkohlenstoffvergiftung. *Jap. J. med. Sci., VIII, (Internal Medicine)*, **3**, 1.
ALPERS, B. J. and LEWEY, F. H. (1940). Changes in nervous system following CS_2 poisoning in animals and man. *Arch. Neurol. Psychiat., Chicago*, **44**, 725.
ANDRÉ, M. J. (1947). Quelques aspects neurologiques du sulfocarbonisme. *Brux. méd.*, **44**, 2398.
ANTONIOLI, E. (1942). Un caso di intossicazione da dimetisolfato. *Med. d. Lavoro*, **33**, 138.
AUDO-GIANOTTI, G. B. (1932). Le Parkinsonisme sulphocarboné professionnel. *Pr. méd.*, **40**, 1289.
AUDO-GIANOTTI, G. B. (1932). Richerche anatomo-patologiche sul' intossicazione sperimentali da solfuro di carbonio. *Rass. Med. Lav. industr.*, **3**, 434.
AUDO-GIANOTTI, G. B. (1934). Sulla patogenesi di gastriche e duodenali nell' intossicazione solfocarbonica professionali. *Rass. Med. Lav. industr.*, **5**, 446.
AUFFRET, J. (1946). L'industrie des fibres artificielles et des dangers. *Arch. Mal. prof.*, **7**, 181.
BAADER, E. W. (1932). An Hirntumor erinnernde Vergiftungserscheinungen durch Schwefelkohlenstoff. *Med. Klinik*, **28**, 1740.
BALÁZS, J. (1934). Dimethylsulfat-Vergiftung. *Samml. Vergiftungsf.*, **5**, 47, A414.
BARTHELEMY, H. L. (1939). Ten years' experience with industrial hygiene in connection with the manufacture of viscose rayon. *J. industr. Hyg.*, **21**, 141.
BASHORE and STALEY (1938). Survey of CS_2 and H_2S hazards in viscose rayon industry. Occupational Disease Prevention Bulletin, No. 46. Dept. Labor and Industry, Pennsylvania.
BIGNAMI, G. (1925). Modificazioni del sangue nell' avvelenamento da solfuro di carbonio. *Boll. Soc. med. chir. Pavia*, **37**, 745.
BINET, L. and BOURLIÈRE, F. (1944). Sur les modifications du sang au cours du sulfo-carbonisme chronique. *Arch. Mal. prof.*, **6**, 12.
BONHOEFFER, K. (1930). Über die neurologischen und psychischen Folgeerscheinungen der Schwefelkohlenstoffvergiftung. *Mschr. Psychiat. Neurol.*, **75**, 195.
BÖRNER. Cited by Lehmann and Flury, 1943, in Toxicology and hygiene of industrial solvents. Williams and Wilkins, Baltimore.
BOURRET, J. and KOHLER, C. (1946). Deux cas de polynévrites par sulfocarbonisme professionnel. *Arch. Mal. prof.*, **7**, 294.
BRIEGER, G. (1941). The effects of carbon disulfide on the blood corpuscles. *J. industr. Hyg.*, **23**, 388.
BRINA, A. (1946). Duo casi di intossicazione da solfatodimetilico. *Med. d. Lavoro*, **37**, 225.
CAZENEUVE, P., MOREL, A. and DE LEEUW, H. (1932). L'hygiène et l'industrie de soie artificielle. *Chim. et Industr.*, **28**, 473.
CONSTENSOUX, M. G. and HEIM, M. F. (1910). Fréquence relative des stigmates nerveux dans le sulfo-carbonisme chronique. Question VI. *2me Congr. int. Mal. prof.*, Brussels.
CORCOS, A. (1940). Contribution to the study of occupation poisoning by creosols. *Rass. Med. Lav. industr.*, **11**, 55 (abstr. in *J. industr. Hyg.*, **22**, 124).
CRESKOFF, A. F. (1938). Survey of carbon disulphide and hydrogen sulphide in the viscose rayon industry. Pennsylvania, Dept. Labor and Industry. Occupation disease prevention division. *Bull.*, **46**.
DAVIS, P. A. (1929). Toxic substances in the rubber industry. *Rubb. Age, N.Y.*, **26**, 83.
DEICHMANN, W. B. (1949). Phenol and phenolic compounds: in Patty, F. A., Industrial hygiene and toxicology. Part II, p. 1023. Interscience Publishers, New York.
DEICHMANN, W. B. and WITHERUP, S. (1944). Phenol studies VI. The acute and comparative toxicity of phenol and o-, m- and p-cresols for experimental animals. *J. Pharmacol.*, **80**, 233.
DELPECH (1856). Accidents produits par l'inhalation du sulfure de carbone en vapeur: expériences sur les animaux. *Gaz. hebd. Méd. Chir.*, **3**, No. 1, 384.
DEPARTMENTAL COMMITTEE ON CERTAIN MISCELLANEOUS DANGEROUS TRADES. Final Report. H.M. Stationery Office, 1899.
DEPARTMENT OF SCIENTIFIC AND INDUSTRIAL RESEARCH (1939). Leaflet No. 6: Carbon disulphide vapour. H.M. Stationery Office.
DEVOTO, L. (1934). Schwefelkohlenstoff und Nebenniere (Addisons Krankheit?). *Arch. Gewerbepath. Gewerbehyg.*, **5**, 429.
DUVOIR, M., HAZEMANN, R., DERUELLE, H. and FALLOT, P. (1938). Sur l'intoxication professionnel du phénol. *Bull. Soc. méd. Hôp., Paris*, **54**, 106.
FLORET, F. (cited by Krause, F., 1931. Beitrag zur Frage der Schwefelkohlenstoffvergiftung. *Z. ges. Neurol. Psychiat.*, **134**, 139).
FLURY, F. and ZERNIK, F. (1931). Schädliche Gase, Dämpfe, Nebel, Rauch- und Staubarten. Springer, Berlin.
FOREMAN, W. (1886). Notes of a fatal case of poisoning by bisulphide of carbon; with postmortem appearances and remarks. *Lancet*, **ii**, 118.
FRANCINE, A. P. (1905). Acute carbon bisulfide poisoning. *Amer. Med.*, **9**, 871.
GOODMAN, H. (1933). Silk handlers disease of the skin, *Dermatitis venenata*, due to isomers of cresol. *Med. J. Rec.*, **138**, 349.

GORDY, S. T. and TRUMPER, M. (1938). Carbon disulphide poisoning, with a report of six cases. *J. Amer. med. Ass.*, **110**, 1543.
GORDY, S. T. and TRUMPER, M. (1940). Carbon disulphide poisoning. Report of 21 cases. *Industr. Med.*, **9**, 231.
HAAS, G. and HEIM, M. F. (1910). Manifestations oculaires du sulfocarbonisme professionnel. *2me Congr. int. Mal. prof.*, Brussels.
HARMSEN, E. (1905). Die Schwefelkohlenstoff Vergiftung im Fabrikbetriebe und ihre Verhütung. *Vjschr. gerichtl. Med.*, **30**, 149.
HEUPER, W. C. (1936). Etiologic studies on the formation of skin blisters in viscose workers. *J. industr. Hyg.*, **18**, 432.
HOME OFFICE (GREAT BRITAIN) FACTORY DEPARTMENT (1935). Memorandum on precautions against dangers of poisoning, fire and explosion in connection with use of carbon bisulphide in artificial silk, India rubber, and other works. H.M. Stationery Office.
JONES, G. W., SCOTT, F. E. and SCOTT, G. S. (1943). Limits of inflammability and ignition temperatures of acetic anhydride. *U.S. Bur. Mines, Rep. Invest.*, No. **3741**.
JUMP, H. D. and CRUICE, J. M. (1904). Chronic poisoning from carbon bisulphide. *Univ. Pa. med. Bull.*, **17**, 193.
KASPAR, J. A., MCCORD, C. P. and FREDERICK, W. G. (1937). Toxicity of organic silicon compounds. 1. Tetraethyl-ortho-silicate. *Industr. Med.*, **6**, 660.
KLINGER, M. E. and NORTON, J. F. (1945). Toxicity of cresylic acid-containing solvent. *U.S. Nav. med. Bull.*, **44**, 438.
KOSTER, E. F. (1943). Abscess of lung and of brain as complications of "lysol" poisoning. *Ohio St. med. J.*, **39**, 840.
KRANENBERG, W. R. H. and KESSENER, H. (1925). Schwefelwasserstoff- und Schwefelkohlenstoff-vergiftungen. *Zbl. GewHyg.*, **12**, 348.
KRAUSE, F. (1931). Beitrag zur Frage der Schwefelkohlenstoffvergiftung. *Z. ges. Neurol. Psychiat.*, **134**, 139.
LAUDENHEIMER, R. (1899). Die Schwefelkohlenstoffvergiftung der Gummi-arbeiter unter besonderer Berücksichtigung der psychischen und nervösen Störungen und der Gewerbehygiene. Veit, Leipzig.
LEHMANN, K. B. (1894). Experimentelle Studien über den Einfluss technisch und hygienisch wichtiger Gase und Dämpfe auf den Organismus. *Arch. Hyg., Berl.*, **20**, 26.
LEHMANN, K. B. (1908). Untersuchungen über die Absorption von Schwefelkohlenstoff. *Arch. Hyg., Berl.*, **67**, 93.
LEHMANN, K. B. and FLURY, F. (1943). Toxicology and hygiene of industrial solvents. (Tr. by E. King and H. F. Smyth, jr.) Williams and Wilkins, Baltimore.
LEWEY, F. H. (1939). Vitamin B deficiency and nervous diseases. *J. nerv. ment. Dis.*, **89**, 1, 174.
LEWEY, F. H. (1941). Neurological, medical and biochemical signs and symptoms indicating chronic industrial carbon disulphide absorption. *Ann. intern. Med.*, **15**, 869.
LEWEY, F. H., with the co-operation of ALPERS, B. J., BELLET, S., CRESKOFF, A. C., DRABKIN, D. L., EHRICH, W. E., FRANK, J. H., JONAS, L., MCDONALD, R., MONTGOMERY, E. and REINHOLD, J. G. (1941). Experimental chronic carbon disulfide poisoning in dogs; a clinical, biochemical and pathological study. *J. industr. Hyg.*, **23**, 415.
LUIG, B. (1913). Beiträge zur Schwefelkohlenstoff- und Benzolvergiftung in akuten und chronischen Versuchen. Dissertation, Würzburg.
MCDONALD, R. (1938). Carbon disulfide poisoning. *Arch. Ophthal., Chicago*, N.S., **20**, 839.
MACGREGOR, R. D. (1892). Supposed poisoning by the daily use of CS_2. *Aust. med. J.*, **14**, 622.
MATTEI, C. and SÉDAN, J. (1924). Contribution à l'étude de l'intoxication par le sulfure de carbone. *Ann. Hyg. publ., Paris, N.S.*, **2**, 385.
MONBRUN, A. and FACQUET, J. (1932). Névrite optique rétro-bulbaire par le sulfure de carbone. *J. Méd. Chir. prat.*, **103**, 657.
MONBRUN, A., RICHET, C. and FACQUET, J. (1932). La névrite optique rétro-bulbaire par sulfure de carbone. *Arch. Ophtal., Paris*, **49**, 697.
MUTSCHLECHNER, A. (1924). Seltenere Vergiftungen. 1. Schwefelkohlenstofftabes. *Dtschr. med. Wschr.*, **50**, 210.
NECTOUX, R. and GALLOIS, R. A. (1931). Quatre cas de névrite rétro-bulbaire par le sulfure de carbone. *Bull. Soc. Ophtal., Paris*, p.750.
NEGRO, F. (1930). Les syndromes Parkinsoniens par l'intoxication sulfo-carbonic. *Rev. neurol.*, **37**, 518.
VON NIDA, S. (1947). Tödliches Glottisödem nach Dimethylsulfatverätzung der oberen Verdauungswege. *Klin. Wschr.*, **24**, 633.
OETTEL, H. (1936). Einwirkung organische Flüssigkeiten auf die Haut. *Arch. exp. Path. Pharmak.*, **183**, 641.
VON OETTINGEN, W. F. (1949). Phenol and its derivatives. *Nat. Inst. Hlth. Bull.*, **190**.
OFFRET, A. (1906). Essai sur l'amblyopie par le sulfure de carbone. Thèse, Paris.
PALUCH, E. A. (1948). Two outbreaks of carbon disulfide poisoning in rayon staple fiber plants in Poland. *J. industr. Hyg.*, **30**, 37.
PETERSON, F. (1892). Three cases of acute mania from inhaling carbon bisulphide. *Boston med. surg. J.*, **127**, 325.
PIERRE-MARIE, M. (1888). Sulfure de carbone et hystérie. *Bull. Soc. méd. Hôp., Paris*, **5**, 445.

QUARELLI, G. (1930). Intossicazione da solfuro di carbonio nella lavorazione della seta artificiale. *Med. d. Lavoro*, **21**, 247.
QUARTERLY SAFETY SUMMARY (1938). **9**, 57. Fire and Explosion. Acetic anhydride. Laboratory accident during acetylisation. Published by Association of British Chemical Manufacturers.
QUENSEL, F. (1904). Neue Erfahrungen uber Geistesstörungen nach Schwefelkohlenstoffvergiftung. *Mschr. Psychiat. Neurol.*, **16**, 48.
RANELLETTI, A. (1931). Die berufliche Schwefelkohlenstoffvergiftung in Italien. Klinik und Experimente. *Arch. Gewerbepath. Gewerbehyg.*, **2**, 664.
RASTELLI (1941). (Cited by Antonioli, E., 1942, in: *Gazz. Osp. Clin.*, No. **48**.)
REDAELLI, P. (1925). Sull' anatomia patologica dell' avvelenamento cronico da solfuro di carbonio. *Boll. Soc. med.- chir.*, *Pavia*, **38**, 133.
RICHTER, R. (1945). Degeneration of basal ganglia from chronic carbon disulphide poisoning in monkeys. *J. Neuropath. exp. Neurol.*, **4**, 324.
RÖDENACKER (1931). Die Bedeutung der Konstitution für die Schwefelkohlenstofferkrankung. *Zbl. GewHyg.*, **18**, 17.
ROWE, V. K., SPENCER, H. C. and BASS, S. L. (1948). Toxicological studies on certain commercial silicons and hydrolyzable silane intermediates. *J. industr. Hyg.*, **30**, 332.
SCHRAMM, H. (1940). Eine seltene Schwefelkohlenstoff-Vergiftung. *Samml. Vergiftungsf.*, **10**, 213, A826.
SMYTH, H. F., jr. and SEATON, J. (1940). Acute response of guinea pigs and rats to inhalation of the vapours of tetraethyl ortho silicate (ethyl silicate). *J. industr. Hyg.*, **22**, 288.
STERNER, J. H. (1949). Aliphatic nitro, diazo and amino compounds in Patty, F. A., Industrial hygiene and toxicology. Part II, p. 967. Interscience Publishers, New York.
TERRIEN, F. (1920). Deux cas d'amblyopie de sulfure de carbone. *Paris méd.*, **35**, 317.
TOMASSIA, A. (1882). De l'intoxication suraiguë par le sulfure de carbone. *Ann. Hyg. publ.*, *Paris*, *3me sér.*, **7**, 292.
VELICOGNA, A. and VIZIANO, A. (1932). La reazione della perossidasi nel solfo-carbonismo. *Policlinico*, **39**, 297.
VIGLIANI, E. C. (1946). L'intossicazione cronica da solfuro da carbonio. *Med. d. Lavoro*, **37**, 165.
WALDHECKER, H. (1941). Ueber die chemischen Desinfektionsmittel. *Münch. med. Wschr.*, **88**, 949.
WALSHE, F. M. R. (1929). Carbon bisulphide intoxication. *Proc. R. Soc. Med.*, **23**, 89.
WARNECKE, F. (1941). Die gewerbliche Schwefelkohlenstoff-vergiftung. *Arch. Gewerbepath. Gewerbehyg.*, **11**, 198.
WARNECKE, F. (1941). Gesunderhaltung in der Gummiindustrie. *Zbl. GewHyg.*, **28**, 1.
WEBER, S. (1902). Über die Giftigkeit des Schwefelsäure-dimethylesters; Dimethylsulfates und einiger verwandter Ester der Fettreihe. *Arch. exp. Path. Pharmak.*, **47**, 113.
WEISE, W. (1933). Magen-Darm Erkrankungen durch chronische Schwefelkohlenstoff- und chronische Schwefelwasserstoff-Inhalation. *Arch. Gewerbepath. Gewerbehyg.*, **4**, 219.
WIENER, J. (1908). Quoted by Lehmann, K. B. (1908). *Arch. Hyg., Berl.*, **67**, 93.
WILEY, F. H., HUEPER, W. C. and VON OETTINGEN, W. F. (1936). Toxic effects of low concentrations of carbon disulfide. *J. industr. Hyg.*, **18**, 733.
WOODARD, G., LANGE, S. W., NELSON, K. W. and CALVERY, H. O. (1941). Acute oral toxicity of acetic, choracetic, dichloracetic and trichloracetic acids. *J. industr. Hyg.*, **23**, 78.
ZANGGER, H. (1930). Über die modernen organischen Lösungsmittel. *Arch. Gewerbepath. Gewerbehyg.*, **1**, 77.
ZANGGER, H. (1930). Weitere Mitteilungen über Vergiftungen durch flüchtige Gifte und deren Beziehung zu gewerblichen Vergiftungen. *Schweiz. med. Wschr.*, **60**, 1193.
ZEGLIO, P. (1942). Su di una complessa sindrome nervosa de intossicazione sulfocarbonica. *Med. d. Lavoro*, **33**, 121.
ZEGLIO, P. (1942). Le alterazioni della funzionalità gastrica nel solfocarbonismo cronico. *Med. d. Lavoro*, **33**, 217.
ZEGLIO, P. (1946). Sulla prognosi delle polineurite solfocarbonio. *Med. d. Lavoro*, **37**, 288.

INDEX

Acetal, 274
  properties of, 274
  toxic effects of, 275
  toxicity of, 275
  uses of, 275
Acetic acid, 392
Acetic anhydride, 392
  properties of, 392
  toxicity of, 393
  uses of, 393
Acetic ester (see ethyl acetate), 289
Acetic ether (see ethyl acetate), 289
Acetone, 320
  absorption of, by inhalation, 321, 325
  —, through skin, 323, 326
  accumulation of, in tissues, 325
  concentration of, maximum permissible, 321
  —, in blood, 326
  elimination of, 325
  estimation of, in air, 320
  explosive risks of, 321
  lethal dose of, 321
  narcotic dose of, 322
  production of, 320
  properties of, 320
  toxic effects of, in animals, 321
  —, —, acute, 322
  —, —, —, symptoms of, 322
  —, —, chronic, 323
  —, —, —, on internal organs, 324
  —, in man, 324
  —, —, acute, 325
  —, —, —, on internal organs, 325
  —, —, —, from skin absorption, 326
  —, —, —, symptoms of, 325
  —, —, chronic, 326
  —, —, —, blood changes from, 327
  —, —, —, eye irritation from, 326
  —, —, —, symptoms of, 326
  toxicity of, basic, 321
  —, comparison with other solvents, 321
  uses of, 320
Acetone oils, 328
Acetylene dichloride (see dichloroethylene), 165
Acetylene tetrachloride (see tetrachloroethane), 154
Actylol (see ethyl lactate), 313
Adronol (see *cyclo*hexanol), 236
Adronol acetate (see *cyclo*hexyl acetate), 306
"Alanol" (see tetrachloroethane), 154
Allyl alcohol, 241
  absorption of, 241
  properties of, 241
  toxic effects of, 241, 242
  toxicity of, 241
  uses of, 241
Amyl acetate, 299
  eosinophilia from, 306
  estimation of, in air, 300
  lethal dose of, 301
  maximum permissible concentration of, 301
  narcotic dose of, 301
  post-mortem changes by, 303
  properties of, 299
  toxic effects of, in animals, 301
  —, —, symptoms of, 302
  —, on blood, 305

—, in man, 303
  —, —, symptoms of, 304, 305
  —, —, on mucous membranes, 302, 303, 305
  toxicity of, 300
  urobilinuria from, 305
  uses of, 299
Amyl alcohol, 232
  absorption of, through skin, 235
  concentration of, lethal, 234
  —, maximum permissible, 234
  estimation of, in air, 234
  metabolism of, 234
  properties of, 232
  toxic effects of, 234
  toxicity of, 233
  —, relative to other alcohols, 233
  uses of, 233
  varieties of, 232, 233
Amyl chloride, 186
  properties of, 186
  uses of, 187
Amyl formate, 285
Amyl lactate, 314
Amyl propionate, 311
Amylene dichloride, 187
Anol (see *cyclo*hexanol), 236
Anon (see *cyclo*hexanone), 333
"Avantine" (see *iso*propyl alcohol), 224

Benzene, 3
  absorption of, by animals, 10
  —, by man, 26
  —, —, blood phenol in, 28
  concentration of, in factories, 5
  —, toxic, 6
  —, maximum allowable, 6
  distinction from benzine, 3
  distribution in body tissues of, 11
  effect of inhalation of, in animals, 13
  effect of injection of, in animals, 13
  estimation of, in air, 6
  —, in blood and tissues, 8
  excretion of, by animals, 12
  —, by man, 26
  —, urinary laevorotation in, 27
  —, urinary sulphur in, 12, 27
  explosive risk of, 5
  lethal and narcotic concentrations of, for animals, 14
  lethal and toxic concentrations of, for man, 28
  manufacture of, 4
  mixtures of, with other solvents, 15, 26
  poisoning, acute, after-effects of, 30
  —, —, in animals, 13
  —, —, —, effect of, on blood, 15, 19–24
  —, —, —, —, on body temperature, 15
  —, —, —, —, on metabolism, 25
  —, —, —, —, on resistance to infection, 24
  —, —, —, local irritation by, 15
  —, —, —, post-mortem findings in, 15
  —, —, —, symptoms of, 14
  —, —, in man, 28
  —, —, —, post-mortem findings in, 30
  —, —, —, symptoms of, 29
  —, —, predisposing factors in, 28
  —, —, treatment of, 31
  —, in animals, susceptibility to, 13, 20
  —, chronic, in animals, 15
  —, —, anaemia in, 22

402

INDEX

Benzene,—*continued*
  Poisoning, chronic, in animals, effect of, on alkali reserve, 25
  —, —, —, —, on blood circulation 19
  —, —, —, —, on haemoglobin level, 23
  —, —, —, —, on red blood cells, 16, 20, 22
  —, —, —, —, on white blood cells, 21
  —, —, —, —, on immunity reactions, 24
  —, —, —, —, on bone-marrow, 17
  —, —, —, —, on internal organs, 18
  —, —, —, leucocytosis in, 16
  —, —, —, leucopenia in, 16
  —, —, —, phosphatase activity in, 25
  —, —, —, symptoms of, 17
  —, —, factors related to, 32
  —, —, in man, 31
  —, —, —, anaemia in, 37
  —, —, —, aplastic anaemia in, 39
  —, —, —, blood groups in, 34
  —, —, —, duration of symptoms of, 37
  —, —, —, effect of, on blood, 37–44
  —, —, —, —, on blood coagulation, 42
  —, —, —, —, on blood platelets, 36, 41
  —, —, —, —, on differential count, 39
  —, —, —, —, on haemoglobin level, 41
  —, —, —, —, on internal organs, 44
  —, —, —, —, on red blood cells, 37, 40
  —, —, —, —, on skin, 37
  —, —, —, —, on white blood cells, 38
  —, —, —, lag phenomenon, 43
  —, —, —, leucocytosis in, 39
  —, —, —, leucopenia in, 38
  —, —, —, leukaemia in, 39
  —, —, —, persistence of effect of, on blood, 42
  —, —, —, purpuric manifestations in, 36
  —, —, —, regeneration of bone-marrow in, 37, 38
  —, —, —, symptoms of, 31
  —, —, —, terminal infection in, 36
  —, —, —, treatment of, 45
  —, differential diagnosis of, 45
  —, in man, susceptibility to, 29, 33
  —, and vitamin C deficiency, 25, 34
  properties of, 3
  toxic effects of, in animals, 10–26
  —, in man, 26–46
  toxicity of, 8
  uses of, 4
  varieties of, 3
Benzine, 79
  absorption of, 82
  addiction to, 91
  comparison of, with benzene, 80
  composition of, 79
  lethal dose of, 81
  mixtures of, with other solvents, 86, 87
  narcotic dose of, 82
  poisoning (see toxic effects)
  properties of, 79
  tolerance to, 85
  toxic concentrations of, 93
  toxic effects of, in animals, 81–88
  —, —, on red blood cells, 84, 86, 87
  —, —, symptoms of, 83
  —, —, on white blood cells, 84, 86, 87
  —, on blood, 83, 84, 86, 87, 91
  —, on bone-marrow, 88
  —, on fat metabolism, 88, 92
  —, on internal organs, 88
  —, leucocytosis, 84, 92
  —, leucopenia, 84, 87, 91
  —, lymphocytosis, 91, 92
  —, lymphopenia, 84, 87, 91
  —, in man, 88–92
  —, —, on red blood cells, 91, 92
  —, —, symptoms of, 89, 90, 91
  —, —, on white blood cells, 91, 92
  —, on mucous membranes, 91
  —, on nerves, 90, 91
  —, on permeability of spinal membrane, 83
  —, on resistance to infection, 85
  —, sequelae of, 90
  —, tendency to thrombosis from, 87
  toxicity of, 80
  —, influence of temperature on, 81
  —, relative to benzene, 80
  uses of, 80
  varieties of, 79, 82, 83, 84, 86, 87
Benzol (see benzene), 3
Benzoline (see petroleum spirit), 74
Benzyl acetate, 309
  lethal dose of, 310
  narcotic dose of, 310
  properties of, 309
  toxic effects of, in animals, 310
  —, in man, 311
  toxicity of, 310
  uses of, 309
Benzyl alcohol, 242
  properties of, 242
  toxic effects of, 243
  toxicity of, 242
  uses of, 242
Benzyl formate, 286
Butanol (see *n*-butyl alcohol), 227
Butanol-2 (see *secondary* butyl alcohol), 230
Butanone (see methyl ethyl ketone), 328
"Butol" (see *n*-butyl butyrate), 312
Butoxyl, 308
*iso*Butyl acetate, 298
n-Butyl acetate, 295
  concentration of, maximum permissible, 296
  —, toxic, 296
  lethal dose of, 296
  narcotic dose of, 296
  properties of, 295
  toxic effects of, in animals, 296
  —, —, symptoms of, 297
  —, in man, 297
  —, —, symptoms of, 297
secondary Butyl acetate, 298
*iso*Butyl alcohol, 231
n-Butyl alcohol, 227
  concentration of, maximum permissible, 228
  toxic effects of, 228
  —, symptoms of, 229
  toxicity of, 228
  uses of, 227
secondary Butyl alcohol, 230
tertiary Butyl alcohol, 231
  toxic effects of, 232
n-Butyl butyrate, 312
Butyl carbitol (see diethylene glycol mono-*n*-butyl ether), 360
*iso*Butyl carbinol (see amyl alcohol), 232
secondary Butyl carbinol (see amyl alcohol), 232

Butyl cellosolve (see ethylene glycol mono-*n*-butyl ether), 353
n-Butyl formate, 284
  concentration of, toxic, 285
  properties of, 284
  toxic effects of, in animals, 284
  —, in man, 285
  uses of, 284
Butyl lactate, 314
n-Butyl propionate, 311

Carbitol (see diethylene glycol monoethyl ether), 358
Carbon disulphide, 380
  absorption of, 384
  concentration of, maximum permissible, 381
  —, toxic, 380, 381, 382, 383
  estimation of, in air, 382
  —, in blood, 385
  excretion of, 384
  explosive risk of, 380
  lethal dose of, 382
  poisoning (see toxic effects)
  —, preclinical, 386
  —, susceptibility to, 391
  —, vitamin B_1 deficiency and, 384
  properties of, 380
  toxic effects of, 382
  —, on adrenals, 390
  —, anaemia from, 386, 390
  —, in animals, 382
  —, on blood, 383, 390
  —, on eyes, 383, 389
  —, hysterical manifestations, 387
  —, in man, 384
  —, on nervous system, 383, 386, 387
  —, Parkinsonian syndrome, 388
  —, polyneuritic syndrome, 388
  —, post-mortem findings, 382, 391
  —, psychoses, 386, 387
  —, —, mechanism of action of, 386
  —, on sexual function, 391
  —, on sulphur metabolism, 384
  —, symptoms of, 382, 385, 387
  toxicity of, 380
  uses of, 380
Carbon tetrachloride, 128
  dangerous concentrations of, 130
  estimation of, in blood, 132
  impurities in, 129, 130
  lethal dose of, 134
  narcosis from, 141
  narcotic dose of, 135
  phosgene formation from, 129, 132, 141
  poisoning (see toxic effects)
  —, methionine therapy in, 138, 140, 143, 148
  —, persistent effect of, 143
  —, post-mortem findings in, 140, 147
  —, predisposing factors to, 131, 133
  —, susceptibility to, 133, 149
  —, tolerance to, 149
  —, treatment of, 148
  preparation of, 128
  properties of, 128
  toxic effects of, 133
  —, acidosis from, 134
  —, amblyopia from, 147
  —, in animals, 133
  —, —, on blood, 137
  —, —, symptoms of, 135, 137
  —, on blood calcium, 147
  —, on blood urea, 141, 142, 145, 146
  —, effect of diet on, 138
  —, on internal organs, 133, 135, 137
  —, on kidneys, 135, 137, 139, 140, 142
  —, —, regeneration of, 138
  —, on liver, 133, 135, 136, 137, 145
  —, —, regeneration of, 138
  —, —, yellow atrophy of, 134, 135, 140
  —, on lungs, 141
  —, in man, 139
  —, —, on blood, 147
  —, —, symptoms of, 139, 140, 141, 144, 148
  —, on metabolism, 134, 142, 146
  —, on nerve tissue, 138
  —, on nervous system, 136, 145
  —, on skin, 147
  —, on urine, 142, 146, 148
  toxicity of, 130
  —, relative to chloroform, 136
  uses of, 130
  —, in fire extinguishers, 132, 140, 141, 142
Cellosolve (see ethylene glycol monoethyl ether), 348
Cellosolve acetate (see ethylene glycol monoethyl ether monoacetate), 351
Chlorex (see $\beta\beta'$-dichloroethyl ether), 266
Chlorobenzene (see monochlorobenzene), 187
Chlorobenzol (see monochlorobenzene), 187
2-Chloroethyl alcohol (see ethylene chlorohydrin), 246
Chloroform, 124
  absorption of, 127
  addiction to, 128
  as anaesthetic, 124
  concentration of, maximum permissible, 125
  estimation of, in brain, 127
  excretion of, 127
  lethal dose of, 125, 127
  narcotic dose of, 125, 127
  poisoning (see toxic effects)
  properties of, 124
  tolerance to, acquired, 126
  toxic effects of, in animals, 125
  —, —, symptoms of, 126
  —, hypoglycaemia, 126
  —, on liver, 126
  —, local irritation from, 126, 128
  —, in man, 127
  —, —, symptoms of, 128
  toxicity of, 124
  uses of, 124
Chloropropylene glycol (see monochlorohydrin), 250
Chlorosilanes, 398
Chlorylene (see trichloroethylene), 169
Coal tar solvent naphtha, 70
  distinction of, from petroleum naphtha, 70
  properties of, 70
  toxic effects of, 71
  toxicity of, 71
  uses of, 71
Columbia spirit (see methyl alcohol), 202
Cresols, 393
  absorption of, 395
  excretion of, 395

INDEX

Cresols,—*continued*
 lethal dose of, 394
 metabolism of, 395
 properties of, 393
 toxic effects of, 394
 —, local, 394
 —, systemic, 394, 395
 toxicity of, 394
 —, relative to phenol, 394
 uses of, 393
Cresylic acid (see cresols), 393
Cumene, 64
 concentration of, "hazard index", 65
 metabolism of, 65
 poisoning (see toxic effects)
 properties of, 64
 toxic effects of, in animals, 64
 toxicity of, 64
 uses of, 64

Decahydronaphthalene, 68
 concentration of, lethal, 69
 properties of, 68
 toxic effects of, in animals, 69
 —, in man, 69
 toxicity of, 69
 uses of, 69
Decalin (see decahydronaphthalene), 68
Diacetone alcohol, 244
 concentrations of, lethal and narcotic, 244
 properties of, 244
 toxic effects of, in animals, 244
 —, in man, 245
 toxicity of, 244
 uses of, 244
Dialkyl carbonates, 316
Diatol (see diethyl carbonate), 315
"Dichloren" (see under dichloroethylene), 168
ortho-Dichlorobenzene, 189
 concentration of, toxic, 191
 estimation of, in air, 190
 properties of, 189
 toxic effects of, in animals, acute, 190
 —, —, chronic, 191
 —, on internal organs, 191
 —, leucopenia, 191
 —, in man, 191
 toxicity of, 190
 —, relative to other solvents, 190
 uses of, 189
sym.-Dichloroethane, 149
 concentration of, maximum permissible, 150
 distinction of, from α-dichloroethane, 149
 —, from dichloroethylene, 149
 lethal dose of, 150
 narcosis due to, 150
 narcotic dose of, 151
 poisoning (see toxic effects)
 properties of, 149
 toxic effects of, in animals, 150
 —, —, symptoms of, 151
 —, on cornea, 152
 —, dermatitis, 153
 —, effect of diet on, 152
 —, on kidneys, 152, 153
 —, leucocytosis, 153
 —, on liver, 152, 153
 —, in man, 152

—, —, symptoms of, 153
—, post-mortem appearances in, 152, 153
—, on urine, 153
 toxicity of, 150
—, relative to other solvents, 150
 uses of, 149
ββ'-Dichloroethyl ether, 266
 concentration of, maximum permissible, 266
 properties of, 266
 toxic effects of, 266
—, symptoms of, 267
 uses of, 266
Dichloroethylene, 165
 concentrations of, lethal and narcotic, 166
 properties of, 165
 toxic effects of, 166, 168
 toxicity of, 166
 uses of, 166
Dichlorohydrin, 251
 properties of, 251
 toxic effects of, 252
 toxicity of, 252
 uses of, 251
Dichloromethane (see methylene dichloride), 122
1:2-Dichloropropane (see propylene dichloride), 185
1:3-Dichloropropan-2-ol (see dichlorohydrin), 251
Dichloro*iso*propyl alcohol (see dichlorohydrin), 251
Di*cyclo*hexylamine, 365
Di*cyclo*pentadiene, 107
Dielene (see dichloroethylene), 165
Dioform (see dichloroethylene), 165
Diethanolamine, 366
Diethyl acetal (see acetal), 274
Diethyl carbonate, 315
Diethyl cellosolve (see ethylene glycol diethyl ether), 352
Diethyl ether (see ethyl ether), 261
Diethyl oxalate, 316
 concentration of, toxic, 317
 properties of, 316
 toxic effects of, 317
—, on blood, 317
 toxicity of, 317
 uses of, 317
Diethylene dioxide (see dioxan), 268
Diethylene glycol, 356
 absorption of, by skin, 357
 properties of, 356
 toxic effects of, in animals, 357
—, on bladder, 357
—, on kidneys, 357
—, on liver, 357
—, in man, 357
 toxicity of, 357
—, relative to ethylene glycol, 357
 uses of, 356
—, in cigarettes, 357
Diethylene glycol monoacetate, 360
Diethylene glycol mono-*n*-butyl ether, 360
Diethylene glycol monoethyl ether, 358
 lethal dose of, 358
 properties of, 358
 toxic effects of, in animals, 358
—, from cosmetics, 358, 360
—, from ingestion, 359

Diethylene glycol monoethyl ether—*continued*
  toxic effects of, on kidneys, 359
  —, in man, 360
  —, on nervous system, 359
  toxicity of, 358
  uses of, 358
Dihydroxydiethyl ether (see diethylene glycol), 356
Dimethoxymethane (see methylal), 274
Dimethylacetonylcarbinol (see diacetone alcohol), 244
Dimethylbenzene (see xylene), 55
Dimethyl carbinol (see *iso*propyl alcohol), 224
Dimethyl ketone (see acetone), 320
Dimethyl sulphate, 395
  mode of action of, 396
  properties of, 395
  toxic effects of, 396
  toxicity of, 395
  uses of, 395
Dioform (see dichloroethylene), 165
Dioxan, 268
  absorption of, by skin, 272
  concentration of, maximum permissible, 269
  lethal dose of, 269, 271
  narcotic dose of, 269, 271
  properties of, 268
  tolerance to, 271
  toxic effects of, in animals, 269
  —, on blood, 273
  —, on internal organs, 270, 271, 273
  —, on kidneys, 270, 271, 272, 273
  —, in man, 272
  —, post-mortem findings in, 271, 272, 273
  —, symptoms of, 272
  toxicity of, 269
  uses of, 269
Dipentene, 106
Dipropylene glycol, 361
  toxic effects of, 361, 362
Dissolvan C.A. and D.N. (see acetal), 274
"Dukeron" (see under trichloroethylene), 171

"Emaillet" (see tetrachloroethane), 154
"Enodrin" (see under dichlorohydrin), 252
Estisol (see ethyl lactate), 313
Ethanol (see ethyl alcohol), 217
Ethanolamines, 366
  properties of, 366
  toxic effects of, 367
  toxicity of, 366
  uses of, 366
2-Ethoxyethanol (see ethylene glycol monoethyl ether), 348
Ethoxysilanes, 398
Ethyl acetate, 289
  concentration of, toxic, 292
  lethal dose of, 291
  narcotic dose of, 291
  properties of, 289
  toxic effects of, in animals, 291
  —, —, symptoms of, 291
  —, on blood, 292
  —, in man, 292
  —, —, symptoms of, 292
  toxicity of, 290
  uses of, 290
Ethyl alcohol, 217
  absorption of, by inhalation, 219

  —, by skin, 220
  concentration of, in internal organs, 220
  —, toxic, 219, 220
  estimation of, in air, 218
  lethal dose of, 218
  narcotic dose of, 218
  oxidation of, 217
  poisoning (see toxic effects)
  preparation of, 217
  properties of, 217
  toxic effects of, in animals, 218
  —, —, symptoms of, 218
  —, on internal organs, 219
  —, in man, 220
  —, —, symptoms of, 221
  toxicity of, 217
  —, relative to methyl alcohol, 217
  uses of, 217
Ethylbenzene, 63
  concentration of, dangerous, 64
  metabolism of, 64
  toxic effects of, in animals, 63
  —, in man, 64
  toxicity of, 63
Ethyl benzoate, 313
Ethyl ether, 261
  absorption of, 262
  concentration of, maximum permissible, 262
  decomposition of, products of, 261
  "ether habit", 265
  excretion of, 262
  explosive risk of, 261
  impurities of, 261
  lethal dose of, 263
  narcotic dose of, 263
  properties of, 261
  toxic effects of, in animals, 262
  —, on blood, 265
  —, in man, 264
  —, on metabolism, 263
  —, symptoms of, 263, 264, 265
  toxicity of, 262
  uses of, 262
Ethyl formate, 281
  concentration of, toxic, 283
  lethal dose of, 282
  narcotic dose of, 283
  properties of, 281
  toxic effects of, in animals, 282
  —, —, symptoms of, 283
  —, in man, 284
  toxicity of, 282
  uses of, 282
Ethyl glycol acetate (see ethylene glycol monoethyl ether monoacetate), 351
Ethyl hydroxy*iso*butyrate, 315
Ethyl lactate, 313
Ethyl oxybutyrate (see ethyl hydroxy*iso*butyrate), 315
Ethylene chlorohydrin, 246
  absorption of, by skin, 247, 250
  concentration of, maximum permissible, 246
  poisoning by, conditions conducive to, 250
  —, post-mortem findings in, 249
  properties of, 246
  toxic effects of, in animals, 246
  —, —, symptoms of, 247
  —, on internal organs, 248

Ethylene chlorohydrin—*continued*
  toxic effects of, in man, 248
  —, —, symptoms of, 249
  toxicity of, 246
  uses of, 246
Ethylene dichloride (see *sym.*-dichloroethane), 149
Ethylene glycol, 340
  concentration of, toxic, 344
  lethal dose of, 342
  properties of, 340
  toxic effects of, in animals, 342
  —, on internal organs, 344
  —, on kidneys, 343, 344
  —, local irritation from, 343
  —, in man, 344
  —, post-mortem findings in, 344
  —, symptoms of, 342, 344, 345
  —, on urine, 344, 345
  toxicity of, 340
  —, relative, 341
  uses of, 340
Ethylene glycol diacetate, 356
Ethylene glycol diethyl ether, 352
  lethal and narcotic doses of, 352
Ethylene glycol monoacetate, 355
Ethylene glycol mono-*n*-butyl ether, 353
  concentration of, toxic, 354
  "hazard index" of, 354
  lethal dose of, 353
  properties of, 353
  toxic effects of, in animals, 353
  —, on blood, 355
  —, on internal organs, 354
  —, in man, 355
  —, symptoms of, 354
  —, on urine, 354
  toxicity of, 353
  —, relative, 354
  uses of, 353
Ethylene glycol monoethyl ether, 348
  concentration of, toxic, 349
  lethal dose of, 350
  properties of, 348
  toxic effects of, in animals, 349
  —, on blood, 350, 351
  —, on internal organs, 350
  —, on kidneys, 348, 350
  —, in man, 350
  —, on urine, 351
  toxicity of, 348
  —, relative, 349
  uses of, 348
Ethylene glycol monoethyl ether monoacetate, 351
  properties of, 351
  toxic effects of, in animals, 351
  —, in man, 352
Ethylene glycol monomethyl ether, 345
  concentration of, in atmosphere, 345
  estimation of, in air, 346
  lethal dose of, 346
  properties of, 345
  toxic effects of, in animals, 346
  —, on blood, 347, 348
  —, on internal organs, 346
  —, on kidneys, 346, 347
  —, in man, 347
  —, on nervous system, 347
  —, on urine, 347

—, "toxic encephalopathy", 347
  toxicity of, 345
  uses of, 345
Ethylidene diethyl ether (see acetal), 274
Eusolvan (see ethyl lactate), 313
Formal (see methylal), 274

Gasoline (see petroleum spirit), 74
Glycol chlorohydrin (see ethylene chlorohydrin), 246

Heptanaphthene (see methyl*cyclo*hexane), 98
Hexahydrobenzene (see *cyclo*hexane), 94
Hexahydrocresol (see methyl*cyclo*hexanol), 239
Hexahydrophenol (see *cyclo*hexanol), 236
Hexalin (see *cyclo*hexanol), 236
Hexahydrotoluene (see methyl*cyclo*hexane), 98
Hexalin acetate (see *cyclo*hexyl acetate), 306
Hexamethyldisiloxane, 397, 398
Hexamethylene (see *cyclo*hexane), 94
*cyclo*Hexane, 94
  concentrations of, maximum permissible, 95
  —, lethal and narcotic, 95
  manufacture of, 94
  properties of, 95
  toxic effects of, in animals, acute, 95
  —, —, on blood, 97, 98
  —, —, chronic, 97
  —, —, on internal organs, 97
  —, —, on metabolism, 97, 98
  —, —, symptoms of, 96, 97
  toxicity of, 95
  uses of, 95
*cyclo*Hexanol, 236
  concentration of, maximum permissible, 237
  lethal dose of, 237
  narcotic dose of, 238
  properties of, 236
  toxic effects of, in animals, 237
  —, in man, 239
  —, on metabolism, 239
  —, symptoms of, 238, 239
  toxicity of, 237
  uses of, 237
Hexanon (see *cyclo*hexanone), 333
*cyclo*Hexanone, 333
  concentration of, maximum permissible, 333
  properties of, 333
  toxic effects of, 334
  —, on metabolism, 334
  toxicity of, 333
  uses of, 333
Hexone (see methyl *iso*butyl ketone), 331
*cyclo*Hexyl acetate, 306
  concentration of, toxic, 307, 308
  properties of, 306
  toxic effects of, in animals, 307
  —, in man, 308
  toxicity of, 307
  uses of, 306
secondary Hexyl acetate, 306
*cyclo*Hexylamine, 365
Hydrolin (see methyl*cyclo*hexanol), 239
4-Hydroxy-4-methylpentanone-2 (see diacetone alcohol), 244

INDEX

Industrial spirit (see ethyl alcohol), 217
"Inertol" (see under xylene), 62
Isophorone, 335
   concentration of, maximum permissible, 336
   properties of, 335
   toxic effects of, 336
   toxicity of, 336
   uses of, 336

"Kalosche" (see under benzine), 81
Ketols (see acetone oils), 328
Ketone oils (see acetone oils), 328

Limonene (see dipentene), 106
Lösungsmittel C (see acetal), 274

Mesityl oxide, 331
   concentration of, irritative, 333
   —, toxic, 332
   properties of, 331
   toxic effects of, in animals, 332
   —, on internal organs, 332
   —, in man, 333
   toxicity of, 332
   uses of, 332
"Methone" (see under methyl acetone), 327
Methanol (see methyl alcohol), 202
Methoxybutyl acetate (see butoxyl), 308
2-Methoxyethanol (see ethylene glycol monomethyl ether, 345
Methyl acetate, 286
   concentration of, toxic, 288, 289
   lethal dose of, 287
   narcotic dose of, 288
   properties of, 286
   toxic effects of, in animals, 287
   —, —, symptoms of, 288
   —, on blood, 289
   —, in man, 289
   —, —, symptoms of, 289
   —, post-mortem findings in, 288
   toxicity of, 287
   uses of, 286
"Methyl acetone", 327
Methylal, 274
Methyl-adronol (see methyl cyclohexanol), 239
Methyl alcohol, 202
   absorption of, by skin, 210, 213
   acidosis from, 206, 210
   amblyopia from, 212, 214
   concentration of, maximum permissible, 205
   —, toxic, 205, 210
   decomposition products of, 204
   estimation of, in air, 205
   excretion of, 206
   formaldehyde formation from, 206, 215, 216
   lethal dose of, 207
   metabolism of, 205, 215
   narcotic dose of, 207
   poisoning by (see toxic effects)
   —, industrial, 212
   properties of, 202
   susceptibility to, 211
   toxic effects of, 205–216
   —, in animals, 205
   —, —, symptoms of, 208

—, on blood, 206, 210
—, on eyes, 208, 212, 214
—, on internal organs, 209
—, local irritation, 214
—, in man, 211
—, —, symptoms of, 213
—, on nervous system, 210, 215
—, on retina, 215
—, on urine, 206
toxicity of, 203
—, relative, 203, 206
uses of, 202
Methyl amyl alcohol (see methyl isobutyl carbinol, 236
Methyl anon (see methyl cyclohexanone), 334
Methylated naphthalenes, 69
   toxicity of, 70
Methylated spirit (see ethyl alcohol), 217
Methylbenzene (see toluene), 46
Methyl benzoate, 312
Methylisobutylcarbinol, 236
Methyl isobutyl ketone, 331
Methyl cellosolve (see ethylene glycol monomethyl ether, 345
Methylcyclohexane, 98
   concentration of, maximum permissible, 99
   lethal dose of, 99
   narcotic dose of, 99
   properties of, 98
   toxic effects of, in animals, 99
   —, —, symptoms of, 99
   —, on internal organs, 99
   —, on metabolism, 99
   toxicity of, 98
   uses of, 98
Methylcyclohexanol, 239
   concentration of, maximum permissible, 240
   properties of, 239
   toxic effects of, 240
   toxicity of, 239
   uses of, 239
Methylcyclohexanone, 334
   concentration of, maximum permissible, 335
Methylcyclohexyl acetate, 308
Methyl ethyl carbinol (see secondary butyl alcohol), 230
Methyl ethyl ketone, 328
   absorption of, by skin, 330
   concentration of, maximum permissible, 329
   —, toxic, 330
   estimation of, in air, 329
   explosive risk of, 328
   toxic effects of, in animals, 329
   —, on cornea, 330
   —, dermatitis, 330
   —, symptoms of, 330
   toxicity of, 329
   uses of, 329
Methyl formate, 281
Methyl-hexalin (see methylcyclohexanol), 239
Methyl-hexalin acetate (see methylcyclohexyl acetate), 308
4-Methyl-2-pentanone (see methyl isobutyl ketone), 331
2-Methyl-2-pentenone-4 (see mesityl oxide), 331
m-Methyl phenol (see cresols), 393

INDEX

o-Methyl phenol (see cresols), 393
p-Methyl phenol (see cresols), 393
Methylphenylsilicone, 397
2-Methyl-2-propanol (see *tertiary* butyl alcohol), 231
Methylpyridine (see picoline), 371
Methylene chloride (see methylene dichloride), 122
Methylene dichloride, 122
  anaemia from, 124
  as anaesthetic, 122
  concentration of, maximum permissible, 122
  lethal and narcotic doses of, 123
  properties of, 122
  toxic effects of, in animals, 122
  —, on liver, 123
  —, in man, 123
  toxicity of, 122
  uses of, 122
Methylene glycol dimethyl ether (see methylal), 274
Monochlorobenzene, 187
  concentration of, lethal and narcotic, 188
  —, maximum permissible, 188
  toxic effects of, 188
  —, symptoms of, 189
  toxicity of, 187
  uses of, 187
Monochlorohydrin, 250
  properties of, 250
  toxic effects of, 251
Monoethanolamine, 366

Naphthene (see *cyclo*hexane), 94
Nitrobenzene, 376
  absorption of, 377
  properties of, 376
  toxic effects of, 377
  —, on blood, 378
  —, on nervous system, 377, 378
  —, methaemoglobinaemia, 376, 377, 378
  —, symptoms of, 378
  —, treatment of, 379
  —, on urine, 378
  toxicity of, 376
  uses of, 376
Nitrobutanes, 375
1-Nitrobutane, 375
2-Nitrobutane, 375, 376
Nitromethane, 373
Nitropropanes, 373
1-Nitropropane, 374
2-Nitropropane, 375
"Novania" (see tetrachloroethane), 154

Oil of Mirbane (see nitrobenzene), 376
Oil of Niobé (see methyl benzoate), 312

Paraldehyde, 276
  properties of, 276
  toxic effects of, in animals, 276
  —, in man, 277
  —, post-mortem findings in, 277
  —, symptoms of, 277, 278
  toxicity of, 276
  uses of, 276
Pentachloroethane, 163
  concentration of, toxic, 164
  estimation of, in animal tissues, 165
  lethal dose of, 163
  narcotic dose of, 164
  properties of, 163
  toxic effects of, 163
  —, acute, 163
  —, chronic, 165
  —, on internal organs, 164
  —, on mucous membranes, 164, 165
  toxicity of, 163
  —, relative, 163, 164, 165
  uses of, 163
*cyclo*Pentadiene, 106
  narcotic dose of, 107
  properties of, 106
  toxic effects of, in animals, 107
  —, on internal organs, 107
  —, in man, 107
  —, symptoms of, 107
  toxicity of, 107
Pentaline (see pentachloroethane), 163
$\Delta^{1:3}$-Pentamethylene (see *cyclo*pentadiene), 106
"Pentasol" (see under amyl alcohol), 233
Pentol (see *cyclo*pentadiene), 106
*cyclo*Pentylene (see *cyclo*pentadiene), 106
"Perawin" (see under perchloroethylene), 182
Perchloroethylene, 182
  concentration of, maximum permissible, 183
  lethal dose of, 183
  narcotic dose of, 183
  properties of, 182
  toxic effects of, in animals, acute, 184
  —, —, chronic, 184
  —, in man, 185
  —, symptoms of, 184, 185
  toxicity of, 182
  —, relative, 182, 183, 184, 185
  uses of, 182
"Perspirit" (see *iso*propyl alcohol), 224
"Petrohol" (see *iso*propyl alcohol), 224
Petrol (see petroleum spirit), 74
Petroleum spirit, 74
  concentration of, toxic, 75
  distinction of, from benzine, 74, 76
  properties of, 74
  toxic effects of, in animals, 75
  —, —, symptoms of, 76
  —, in man, 76
  —, —, on blood, 78
  —, —, symptoms of, 77, 78
  toxicity of, 75
  uses of, 74
Phenylcarbinol (see benzyl alcohol), 242
Phenyl ethane (see ethylbenzene), 63
Phenylmethanol (see benzyl alcohol), 242
Picoline, 371
Propanol (see *n*-propyl alcohol), 222
*iso*Propanol (see *iso*propyl alcohol), 224
Propan-2-one (see acetone), 320
Propenol (see allyl alcohol), 241
Propenyl alcohol (see allyl alcohol), 241
*iso*Propyl acetate, 294
n-Propyl acetate, 293
  concentration of, toxic, 293, 294
  properties of, 293
  toxic effects of, 293
*iso*Propyl alcohol, 224
  absorption of, 226

INDEX

*iso*Propyl alcohol—*continued*
  properties of, 224
  toxic effects of, 225
  —, symptoms of, 225, 226, 227
  toxicity of, 224
  —, relative to other alcohols, 225
  uses of, 224
n-Propyl alcohol, 222
  properties of, 222
  toxic effects of, 222, 223
  toxicity of, 222
  uses of, 222
secondary Propyl alcohol (see *iso*propyl alcohol), 224
*iso*Propylbenzene (see cumene), 64
*iso*Propyl ether, 267
  concentration of, maximum permissible, 268
  properties of, 267
  uses of, 268
Propylene dichloride, 185
  concentration of, maximum permissible, 185
  dermatitis from, 185, 186
  properties of, 185
  toxic effects of, in animals, 186
  —, —, on internal organs, 186
  —, —, symptoms of, 186
  toxicity of, 185
  uses of, 185
Propylene oxide, 268
Pyranton-A (see diacetone alcohol), 244
Pyridine, 367
  impurities in, 368, 369, 370
  lethal dose of, 368
  metabolism of, 369
  properties of, 367
  toxic effects of, in animals, 368
  —, on kidneys and liver, 369
  —, in man, 370
  —, —, symptoms of, 368, 370
  toxicity of, 368
  uses of, 368

"Quittnerlack" (see tetrachloroethane), 154

Sextate (see methyl*cyclo*hexyl acetate), 308
Sextol (see *cyclo*hexanol, methyl*cyclo*hexanol), 236, 239
Sextone (see *cyclo*hexanone), 333
Sextone B (see methyl*cyclo*hexanone), 334
Silane intermediates, 398
Silicones, 397
Solactol (see ethyl lactate), 313
"Solaesthin" (see under methylene dichloride), 122
Solvent G.C. (see ethylene glycol monoacetate), 355
Solvulose (see ethylene glycol monoethyl ether), 348
Spirits of wine (see ethyl alcohol), 217
Sulphuric ether (see ethyl ether), 261

Tetrachloroethane, 154
  absorption of, 159
  concentration of, maximum permissible, 157
  estimation of, 157
  lethal dose of, 157
  narcotic dose of, 157
  poisoning (see toxic effects)
  —, prophylactic measures against, 162
  —, treatment of, 162
  properties of, 154
  tolerance to, 162
  toxic effects of, in animals, 157
  —, —, on blood, 158
  —, —, on liver, 158
  —, —, symptoms of, 157
  —, ascites, 160
  —, on internal organs, 158, 160
  —, in man, 159
  —, —, on blood, 161
  —, —, on liver, 160
  —, —, symptoms of, 159, 162
  —, —, toxaemic jaundice, 160, 161
  —, on nervous system, 162
  —, post-mortem findings in, 159, 160
  —, on skin, 161
  —, on urine, 158, 160
  toxicity of, in industrial processes, 155
  —, relative, 156
  uses of, 154
Tetrachloroethylene (see perchloroethylene), 182
Tetrachloromethane (see carbon tetrachloride), 128
Tetrahydronaphthalene, 65
  excretion of, 68
  properties of, 65
  toxic effects of, in animals, 66
  —, —, symptoms of, 67
  —, on kidneys, 67, 68
  —, in man, 68
  —, on metabolism, 67
  —, on urine, 68
  toxicity of, 66
  uses of, 65
Tetralin (see tetrahydronaphthalene), 65
"Tetralix" (see under perchloroethylene), 182
Toluene, 46
  absorption of, 48
  excretion of, 48
  lethal dose of, 49
  manufacture of, 47
  narcotic dose of, 49
  poisoning (see toxic effects)
  properties of, 46
  toxic effects of, in animals, 48
  —, —, on blood, 51, 52
  —, —, symptoms of, 50
  —, on bone-marrow, 50
  —, on immunity reactions, 52
  —, on internal organs, 50, 54
  —, local irritation, 54
  —, in man, 52
  —, —, on blood, 54
  —, —, on liver, 54
  —, —, symptoms of, 53
  toxicity of, 47
  —, in industrial processes, 48
  —, relative, 47, 49
  uses of, 47
Toluol (see toluene), 46
Trichloroethylene, 169
  absorption of, 182
  —, by skin, 174
  addiction to, 177
  decomposition of, 179, 181
  estimation of, in air, 172

Trichloroethylene—*continued*
   estimation of, in blood, 172
   intolerance to, 181
   lethal dose of, 172
   metabolism of, 182
   narcosis from, 170, 175
   narcotic dose of, 173
   poisoning (see toxic effects)
   properties of, 169
   toxic effects of, in animals, 172
   —, —, symptoms of, 173
   —, —, on blood, 174, 180
   —, on internal organs, 173, 174, 177
   —, in man, 175
   —, —, symptoms of, 176
   —, on mucous membranes, 178
   —, nervous disorders, 176, 180
   —, optic atrophy, 176
   —, post-mortem findings in, 175
   —, on skin, 178, 181
   toxicity of, in industrial processes, 171
   —, relative, 170
   trichloroacetic acid from, 182
   uses of, 169
Triethanolamine, 366
Trilene (see trichloroethylene), 169
Trimethylcarbinol (see *tertiary* butyl alcohol), 231
1:1:3-Trimethyl*cyclo*hex-3-en-5-one (see isophorone), 335
Turpentine, 100
   absorption of, 101
   anaemia from, 104
   concentration of, maximum permissible, 101
   estimation of, 101
   excretion of, 101, 104
   lethal dose of, 101
   poisoning (see toxic effects)
   properties of, 100
   toxic effects of, in animals, 101
   —, —, symptoms of, 101
   —, on blood, 102, 105
   —, haematuria, 103, 104
   —, on internal organs, 102
   —, on kidneys, 102, 103

—, leucocytosis, 105
—, in man, 102
—, —, symptoms of, 103
—, on mucous membranes, 103
—, nephritis, 103, 104, 105
—, post-mortem findings in, 102, 105
—, on skin, 103, 105
—, on urine, 103, 104
toxicity of, 101
uses of, 100
varieties of, 100

"Westron" (see tetrachloroethane), 154
White spirit, 93
   properties of, 93
   toxic effects of, 94
   uses of, 93
Wood alcohol (see "wood spirit"), 216
Wood naphtha (see "wood spirit"), 216
"Wood Spirit", 216

Xylene, 55
   anaemia from, 60, 61
   concentration of, in industrial processes, 56
   —, maximum permissible, 56
   estimation of, in air, 56
   excretion of, 57
   lethal dose of, 57
   mixtures of, with other solvents, 61
   narcotic dose of, 57
   properties of, 55
   toxic effects of, in animals, 57
   —, —, symptoms of, 57
   —, on blood, 58, 60, 61
   —, on bone-marrow, 58
   —, on internal organs, 58
   —, on kidneys, 58, 63
   —, local irritation, 59
   —, in man, 58
   —, —, symptoms of, 59, 62, 63
   toxicity of, 56
   —, comparative, 56, 57, 58, 60
   uses of, 55
Xylol (see xylene), 55